Pharmaceutical Dosage Forms and Drug Delivery

Pharmaceutical Dosage Forms and Drug Delivery

Third Edition: Revised and Expanded

by

Ram I. Mahato, PhD

Department of Pharmaceutical Sciences,
University of Nebraska Medical Center
Omaha, NE

Ajit S. Narang, PhD

Small Molecule Pharmaceutical Sciences, Genentech, Inc.
One DNA Drive, South San Francisco, CA

CRC Press
Taylor & Francis Group
Boca Raton London New York

CRC Press is an imprint of the
Taylor & Francis Group, an **informa** business

CRC Press
Taylor & Francis Group
6000 Broken Sound Parkway NW, Suite 300
Boca Raton, FL 33487-2742

© 2018 by Taylor & Francis Group, LLC
CRC Press is an imprint of Taylor & Francis Group, an Informa business

International Standard Book Number-13: 978-1-4822-5362-7 (Hardback)

Library of Congress Cataloging-in-Publication Data

Names: Mahato, Ram I., author. | Narang, Ajit S., author.
Title: Pharmaceutical dosage forms and drug delivery : third edition, revised and expanded / Ram I. Mahato and Ajit S. Narang.
Description: Third edition. | Boca Raton : CRC Press, [2017] | Includes bibliographical references and index.
Identifiers: LCCN 2017017702| ISBN 9781482253627 (hardback : alk. paper) | ISBN 9781315156941 (ebook)
Subjects: LCSH: Drugs--Dosage forms. | Drug delivery systems.
Classification: LCC RS200 .M34 2017 | DDC 615.1/9--dc23
LC record available at https://lccn.loc.gov/2017017702

Visit the Taylor & Francis Web site at
http://www.taylorandfrancis.com

and the CRC Press Web site at
http://www.crcpress.com

Printed and bound in the United States of America by
Edwards Brothers Malloy on sustainably sourced paper

I dedicate this book to my wife Subhashini, my children Kalika and Vivek for their love and support; my late mother Sarswati for believing in me; and to my students and mentors who have always helped me in my quest for learning and in achieving higher goals.

Ram I. Mahato

To Tirath Singh and Gurdip Kaur, my parents, who taught me simplicity, sincerity, and hard work.

Ajit S. Narang

Contents

2 Drug development   25

10 Surfactants and micelles   241

PART III
Dosage forms   385

16 Suspensions   387

17 Emulsions   407

20 Tablets   479

Foreword

Education and practice of the pharmaceutical sciences are rapidly evolving in the modern times, not only with the new ways of doing things enabled by technological advancements of the twenty-first century—but also by the advent of new medicines and modalities of drug therapy. As much as the basic sciences and the fundamental principles of pharmaceutical sciences remain sound, there is an increasing need to integrate the education of dosage forms and drug delivery with the current research and contemporary practices in the current biopharmaceutical industry. It is here that this book—coauthored by leading scientists in both academia and industry—makes a strong impact by presenting the basic principles in a succinct manner that affords a deeper understanding of their reduction to practice.

The revised and expanded third edition of *Pharmaceutical Dosage Forms and Drug Delivery* provides greater emphasis to the multidisciplinary nature of drug discovery and development. The third edition has a new chapter on "Drug Discovery" and the authors have significantly updated and modified the chapter on "Drug Development and Regulatory Processes." These revised chapters provide not only an industry perspective of current practices in pharmaceutical sciences, they also share interdisciplinary nature of *what it takes to discover and bring a new drug to the market* by equally emphasizing the role of regulatory authorities and decision-making criteria in a stage-gate process of drug development.

Modern drug discovery is facing increasing challenges of identifying new drug targets and accessing the difficult-to-reach targets. Although early years of drug discovery focused predominantly on extracellular and cell-surface targets, recent efforts have refocused on new mechanisms of disease progression and targets that are intracellular and often also intraorganelle. These new frontiers are exemplified by the treatment of hepatitis C infection and the emergence of immuno-oncology and antibody–drug conjugates. Pharmaceutical scientists are increasingly using targeted drug-delivery strategies to address challenges inherent with evolving targets and increasingly complex nature of the therapeutic moieties. This edition includes a new chapter on "Targeted Drug Delivery" that provides a concise and in-depth overview of different strategies as applied in current practice.

A new chapter has been added on "Radiopharmaceuticals" to familiarize the students with the basic principles; their utilization is currently embedded in the current practice of diagnostic and therapeutic medicine, and the emerging paradigms of their use in new drug development. This revised edition has many more changes to improve and update several other chapters of this book.

The recent years have seen an explosive increase in the information and progress in various subdisciplines of pharmaceutical sciences as well as the basic sciences, which forms the foundation of this applied science. This up-to-date book incorporates both the basics of the underlying sciences and the changes currently underway in a balanced fashion to enable a holistic development of the next generation of students.

I am confident that this book will enable a deeper understanding of fundamental principles and a wider perspective of their practice to our next generation of students.

Happy reading!

Courtney Fletcher
University of Nebraska Medical Center

Preface

This book is designed as a textbook for teaching basic principles of pharmaceutics, dosage form design, and drug delivery to the Doctor of Pharmacy (Pharm. D.) students in the United States and Bachelor of Pharmacy (B. Pharm.) students in other countries. Although there are numerous books on the science of pharmaceutics and dosage form design, including *Martin's Physical Pharmacy* and *Pharmaceutical Sciences* by Sinko, *Physicochemical Principles of Pharmacy* by Florence and Attwood, *Pharmaceutics: The Science of Dosage form Design* by Aulton, *Theory and Practice of Contemporary Pharmaceutics* by Ghosh and Jasti, and *Pharmaceutical Sciences* by Remington, these books cover different areas of the discipline in varying depths and provide limited insight into contemporary practices and practical applications. Each of these textbooks, by themselves, does not provide *an integrated approach* to the students. This leads to the students as well as the teachers to refer many textbooks to develop an overall understanding of the basic physicochemical principles and their applications to the design and development of different pharmaceutical dosage forms. In an attempt to overcome these challenges, this book provides a unified perspective of the overall field to the students as well as instructors.

The students need to know the basic physicochemical principles, application of these principles to the design of dosage forms, and the relevance of these principles to the biopharmaceutical aspects of drugs. Another important aspect of teaching that is urgently needed in our Pharm. D. curricula is to expose students to the *latest developments* in the application of biomaterials as well as protein and nucleic acid-based pharmaceutical dosage forms and therapeutics, and various biotechnology-based developments. Various books that are currently taught to students miss these latest developments in the field of pharmaceutics and drug delivery. Exposure of students to these latest developments is critical to the successful training of future pharmacists, because these therapeutic modalities and options are likely to be clinically significant in the future. All these principles and applications need to be integrated in a single textbook, so that the student develops a better and overall understanding of the principles involved in dosage form design and drug delivery. This book covers an in-depth discussion on what physiochemical parameters can be used for the design, development, and evaluation of biotechnological dosage forms for delivery of proteins, peptides, oligonucleotides, and genes.

What's new in the third edition of Pharmaceutical Dosage Forms and Drug Delivery? This edition is significantly revised and expanded in the content matter of each chapter—to make sure the most recent progress in the field and pharmaceutical research is captured adequately—and expanded to include additional chapters that provide contemporary practices. These new chapters include "Drug Discovery" and "Drug Development" as expanded separate chapters, "Organ Specific Drug Delivery," and "Radiopharmaceuticals."

You have an updated, contemporary, new book that can serve as a textbook for Pharm. D. students and a valuable resource for novice in the pharmaceutics and drug delivery fields.

Acknowledgments

Writing the third edition of this book has taken several years of reading and teaching—and has benefited from the real-life application of these pharmaceutical principles to the contemporary industrial discovery and development of new drug products. Many people have contributed to our thinking and have been the invisible guide to this book. In particular, we acknowledge with grateful appreciation the major contributions and foresight of Professors Vincent H. L. Lee, Mitsuru Hashida, and Sung Wan Kim. Their leadership in this field over the years has significantly contributed to our current understanding of pharmaceutical dosage forms and drug delivery. This is a subject of paramount significance in clinical drug therapy as well as new drug discovery and development, and must be taught to our Pharm. D. students.

We extend our gratitude to our students and colleagues who have shared their thoughts with us on the third edition of this book. We hope that we have been successful in responding to their suggestions. We gratefully acknowledge several graduate students and postdoctoral fellows of our department, who contributed in many ways to the development of this book through their critiques, review, and suggestions on individual chapters.

We especially thank the staff members at the CRC Press who have contributed so expertly to the planning, preparation, and production of this book. Specifically, we would like to recognize Hilary Lafoe and Natasha Hallard for their help in the assembly of the draft. Last but not the least, we thank our wonderful wives, Subhashini (RM) and Swayam (AN), and our fabulous children, Kalika (RM), Vivek (RM), Manvir (AN), and Arjun (AN), for their support and patience during the long hours away from them to work on this edition.

Ram I. Mahato and Ajit S. Narang

About the authors

Ram I. Mahato is Professor and Chairman of the Department of Pharmaceutical Sciences, University of Nebraska Medical Center, Omaha, Nebraska. He was a Professor at the University of Tennessee Health Science Center, Memphis, Tennessee, Research Assistant Professor at the University of Utah, Salt Lake City, Utah, Senior Scientist at GeneMedicine, Inc., The Woodlands, Texas, and as a postdoctoral fellow at the University of Southern California in Los Angeles, Washington University in St. Louis, and Kyoto University, Japan. He earned his PhD in drug delivery from the University of Strathclyde, Glasgow, UK and BS from China Pharmaceutical University, Nanjing, China. He has published 132 papers, 17 book chapters, holds 2 U.S. patents, and has edited/written 8 books and 10 journal issues (total Google citations = 8358 and h-Index = 54). He was the feature editor of the *Pharmaceutical Research* (2006–2013) and editorial board member of 8 journals. He is a Controlled Release Society (CRS) and American Association of Pharmaceutical Scientists (AAPS) fellow, permanent member of Bioengineering, Technological and Surgical Sciences (BTSS)/National Institutes of Health (NIH) Study section, and American Society of Gene and Cell Therapy (ASGCT) scientific adviser. He is applying sound principles in pharmaceutical sciences in the context of the latest advances in life sciences and material sciences to solve challenging drug delivery problems in therapeutics.

Ajit S. Narang works for the Department of Small Molecule Pharmaceutical Sciences, Genentech, Inc., in South San Francisco, California, responsible for the pharmaceutics and biopharmaceutics of small molecules in preclinical and early clinical development of various dosage forms of small-molecule drugs. He also serves as an adjunct faculty at the Universities of Tennessee and Phoenix; an industrial advisory board member of Western Michigan University; a panel member of the Biopharmaceutics Technical Committee (BTC) of the Pharmaceutical Quality Research Institute (PQRI) in Arlington, Virginia; a panel member of the International Pharmaceutics Excipient Council (IPEC) committees; vice-chair of the Formulation Design and Delivery (FDD) section of the American Association of Pharmaceutical

Scientists (AAPS), and committee member of a Master's Degree Program at Campbell University, North Carolina. He holds more than 15 years of pharmaceutical industry experience in the development and commercialization of oral and parenteral dosage forms and drug delivery platforms. In addition to Genentech, he has worked for Bristol–Myers Squibb, Co., New Brunswick, New Jersey; Ranbaxy Research Labs (currently a subsidiary of Daiichi Sankyo, Japan), Gurgaon, India; and Morton Grove Pharmaceuticals (currently, Wockhardt U.S.), Gurnee, Illinois. He holds undergraduate pharmacy degree from the University of Delhi, India and graduate degree in Pharmaceutics from the Banaras Hindu University, India and the University of Tennessee Health Science Center (UTHSC), Memphis, Tennessee. He currently serves the vice-chair of the Formulation and Drug Delivery (FDD) section of the American Association of Pharmaceutical Scientists (AAPS). Ajit has contributed to several preclinical, clinical, and commercialized drug products including New Drug Applications (NDAs), Abbreviated New Drug Applications (ANDAs), and 505B2s. He is credited with more than 40 peer-reviewed articles, 2 books, 7 patent applications, 20 invited talks, and 60 presentations at various scientific meetings. His current research interests are translation from preclinical to clinical and commercial drug product design; incorporation of Quality by Design (QbD) elements in drug product development; and mechanistic understanding of the role of material properties on product performance.

Part I

Introductory chapters

Chapter 1

Drug discovery

LEARNING OBJECTIVES

On the completion of this chapter, the students should be able to

1. Differentiate between drug discovery and development processes.
2. Outline key elements of the contemporary process of drug discovery.
3. Identify different functional areas involved in drug discovery and their roles.
4. Differentiate drug discovery and development paradigms for small- and large-molecule drugs.

1.1 INTRODUCTION

Throughout history, new drug discovery has enabled human well-being and advancement, not only by ensuring survival in the wake of infections and debilitating diseases but also by steadily advancing the quality of life. For example, the discovery of penicillin by Alexander Fleming in 1928 saved countless lives and paved the way for the antibiotic medicines. More recent examples of life-saving medicines include the discovery and commercialization of statins for the management of hypercholesterolemia and humanized monoclonal antibodies targeting the immune checkpoints for the immunotherapy of cancer. Prominent examples among the quality-of-life improving therapeutics are the drugs for the management of pain, hypertension, gastrointestinal disorders, and countless others. New drug discovery remains a continuing priority and exhibits relentless effort of governments, private corporations, and academia alike. This chapter will outline contemporary practices in new drug discovery and development, with an emphasis on the interdisciplinary process, stage-gate paradigm, and the key features that ensure safety and efficacy of every new drug through the earliest stages of drug discovery by commercialization.

1.1.1 Elements of drug discovery

Early discovery starts with the identification of disease area and molecular targets that may be useful in the clinical setting to modulate a particular disease condition. Several drug candidates that may be able to produce the desired outcome at the chosen drug targets are produced. The activity of these candidates is assessed in *in vitro* assays, and candidates presenting high efficacy are shortlisted. Structure–activity relationships (SARs) are developed to help narrow down the drug candidates to the ones with the greatest potential for efficacy at the lowest dose.

Early discovery is involved not only in the generation of drug candidates but also in the assessment of their activity and toxicity in cell culture–based, *in vitro*, and *in vivo* systems (animal species). These often require development of disease models on which compounds can be tested. Assessment of relative activity and potential of different compounds must be made under a multitude of criteria, thus requiring an effort to *optimize* the relative performance of a compound in several attributes that are important for success in the overall discovery and development process.

Typical drug discovery and development process (Figure 1.1) involves a series of iterative work streams that develop and mature in parallel. Target selection and validation often precede first discovery efforts and continue to

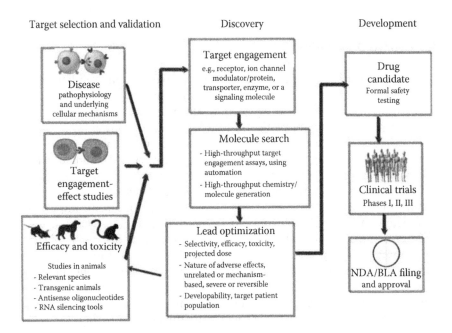

Figure 1.1 Key activities involved in drug discovery and development research.

develop concurrently with the discovery drug candidate screening. Target screening and validation involve the following steps:

- Understanding disease *pathophysiology* and underlying cellular or subcellular mechanisms.
- *Target engagement–effect studies* in cell culture or *in vitro* systems, with a view to develop a high-throughput assay that can be used for the screening of new drug candidates.
- *Efficacy, toxicity*, and preliminary pharmacokinetic studies in animals relevant to disease pathophysiology and candidate drugs. These studies may involve the use of transgenic animals that exhibit the disease or modulation of particular biochemical processes and the use of antisense oligonucelotides or RNA-silencing approaches to modulate subcellular genetic pathways.

Discovery of new drug candidates relies on target screening and validation efforts to not only identify the right target but also provide a high-throughput screening assay to shortlist compounds from vast institutional libraries and synthetic capabilities. Drug discovery efforts focus on the following:

- *Target engagement and evaluation* of efficacy with a particular target identity and quantitative evaluate drug binding to target proteins in cells and tissues. Quantitation of target engagement can be in terms of percentage. The ability of a quantitated extent of interaction to produce the desired efficacy in animal models is assessed.
- *Molecule search* is an interdisciplinary involvement of quantitative SAR (QSAR) assessment; *in silico* molecular modeling studies that try to assess molecular mechanism and location of engagement on the target, which can help guide QSAR efforts; and synthesis/generation of new molecules that can be tested based on these assessments.
- Once one or more series of molecules have been identified with key structural features needed for target engagement and efficacy, *lead optimization* efforts focus on developability assessment (identification of any liabilities that can hinder development of the compound during later stages in the development) and appropriate modification of molecular design or delivery strategy design to enable the progression of a molecule through later stages of drug development. These assessments also include dose projections, identification of appropriate route of drug administration, and an assessment of the observed or predictable liabilities of adverse effects.

The drug *development* studies follow the identification of a single lead molecule or drug candidate. These studies are aimed at formal toxicity assessment that can enable initiation of first-in-human (FIH) studies and at enabling clinical trials through various phases of drug development into generating the needed data package for the regulatory approval and commercialization. The details of the drug development studies will be a subject of the Chapter 2.

1.1.2 Sources of drugs

Identification of new molecules with the potential to produce a desired therapeutic effect involves a combination of (1) molecular physiology and pathophysiology; that is, research on the molecular mechanisms of biological process and disease progression; (2) review of known therapeutic agents; and/or (3) conceptualization and synthesis/procurement of potential new molecules that may also involve random selection and broad biological screening.

The sources of new drugs are varied (Figure 1.2). NMEs can be of synthetic or natural origin, the latter involving inorganic compounds or compounds purified from plants or animals.

1.1.2.1 Plant sources

Natural compounds extracted from plants have often provided novel structures for therapeutic applications. For example, vincristine is derived from the periwinkle plant *Vinca rosea*, etoposide is from the mandrake plant *Podophyllum peltatum*, taxol is from the pacific yew *Taxus brevifolia*, doxorubicin is a fermentation product of the bacteria *Streptomyces*, L-asparaginase is from *Escherichia coli* or *E. cartovora*, rhizoxin is from the fungus *Rhizopus chinensis*, cytarabine is from the marine sponge *Cryptotethya crypta*, and bryostatin is from the sea moss *Bugula neritina*. Another example is paclitaxel (Taxol®), prepared from the extract of the pacific yew, used in the treatment of *ovarian cancer*. Digoxin is one of

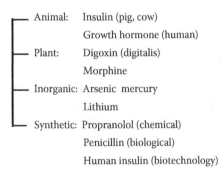

Figure 1.2 Different sources of drug molecules.

the most widely used drugs in the management of congestive heart failure, weakened heart, and irregular heart beat (arrhythmia). The common garden plant, the foxglove or *Digitalis purpurea*, is the source of digoxin.

1.1.2.2 Organic synthesis

Chemical synthesis could involve (a) synthesis of analogs of natural compounds in an effort to improve affinity, specificity, or potency to improve the safety and efficacy profile of the original natural compound; (b) synthesis of a natural molecule from a more abundantly available intermediate to reduce cost and/or improve purity (e.g., taxotere was developed to overcome the supply problems with taxol); or (c) synthesis of a new, unique chemical structure.

Synthesis of analogs of natural compounds is exemplified by the following: carboplatin—an analog of cisplatin with reduced renal toxicity, doxorubicin—an analog of daunomycin with lower cardiotoxicity, and topotecan—an analog of camptothecin with lower toxicity. Synthesis of analogs of known drugs is sometimes aimed at improving the targeting and the pharmacokinetics of a drug. The tauromustine couples a nitrosourea anticancer agent to a brain-targeting peptide. Synthesis of new molecular entities (NMEs) that are analogs of known compounds or completely novel structures involves computer modeling of drug–receptor interactions, followed by synthesis and evaluation by using tools such as solid-state and combinatorial chemistry. For example, methotrexate and 5-fluorouracil were developed as analogs of natural compounds that demonstrated anticancer activity.

1.1.2.3 Use of animals

The use of animals in the production of various biologic products, including serum, antibiotics, and vaccines, has life-saving significance. Hormonal substances, such as thyroid extract, insulin, and pituitary hormones obtained from the endocrine glands of cattle, sheep and swine, are life-saving drugs used daily as replacement therapy.

1.1.2.4 Genetic engineering

In addition to the use of whole animals, cultures of cells and tissues from animal and human origin are routinely used for the discovery and development of new drugs—both small molecules and biologicals, such as vaccines. Drugs that were traditionally produced in animals are increasingly being synthesized by using cell and tissue cultures. The two basic technologies that drive the genetic field of drug development are recombinant DNA technology and monoclonal antibody production. Recombinant DNA technology involves the manipulation of cellular DNA to produce desired proteins, which may then be extracted from cell cultures for therapeutic use. Recombinant DNA technology has the potential to produce a wide variety of proteins. For example, human insulin, human growth hormone, hepatitis B vaccine, and interferon are produced by recombinant DNA technology.

A growing class of biologics is *monoclonal antibodies* against cellular targets aimed for destruction, such as molecular markers on tumors. Monoclonal antibodies target a single epitope, an antigen surface recognized by the antibody, as against natural polyclonal antibodies, which bind to different epitopes on one or more antigen molecules. This confers a high degree of specificity to monoclonal antibodies. While recombinant DNA techniques usually involve protein production within cells of lower animals, monoclonal antibodies are produced in cells of higher animals, sometimes in the patient, to ensure the lack of patient immune reaction against these macromolecules on administration. Monoclonal antibodies are used as anticancer therapeutics, in home pregnancy testing products, and for drug targeting to specific sites within the body. In home pregnancy testing products, the monoclonal antibody used is highly sensitive to binding at one site of the human chorionic gonadotropin (HCG) molecule, a specific marker to pregnancy because HCG is synthesized exclusively by the placenta.

1.1.2.5 Gene therapy

Gene therapy is the process of correction or replacement of defective genes and has the potential to be used to prevent, treat, cure, diagnose, or mitigate human disease caused by genetic disorders. Oligonucleotides and small interfering RNA (siRNA) are used to inhibit aberrant protein production, whereas gene therapy aims at expressing therapeutic proteins inside the body.

1.2 COLLABORATING DISCIPLINES

New drug discovery and development is a long, complex process that involves multidisciplinary scientists working together in diverse, interconnected, and interdependent teams that coordinate their activities and where decision by each team potentially affects every other team working on a given project. A consistent feature of new drug discovery research is the cross-disciplinary progress and collaboration. This interdisciplinary collaboration is partly accomplished by scientists working together in different departments within a pharmaceutical or a biopharmaceutical company. Each of these disciplines is responsible for one or more aspects of the drug discovery and development process as it moves along the pipeline. These functions are usually known by different names in different organizations but have common underlying themes, such as discovery chemistry, discovery biology, preclinical development, toxicology, pharmaceutical development, clinical development, analytical and bioanalytical sciences, regulatory sciences, manufacturing operations, quality operations, and commercialization functions.

This chapter will briefly highlight the roles and responsibilities of each of these functions, the key methodologies that are adopted in fulfilling those responsibilities, and the underlying scientific disciplines of study that contribute to drug discovery.

Various scientific disciplines that work within this paradigm are briefly discussed in the following subsections.

1.2.1 Biology

The scientists in this function are responsible for the identification of drug targets for chosen disease indications and assessment of efficacy of compounds against those targets. The efficacy may be tested in assays that may be conducted *in vitro*, *ex vivo*, or *in vivo*. These assays, for example, could be target-binding assays for isolated receptors, cellular response assays in cell culture models, or animal studies. Dose–response curves are generated in cell culture models (Figure 1.3) that help rank different compounds

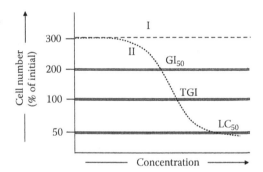

Figure 1.3 An illustration of the dose–response curves generated during cell culture prescreening. This example illustrates the dose-dependent cytotoxic effect of drugs on cells cultured *in vitro* in cell culture dishes. Cell cultures that are not exposed to the drug (I) grow to a hypothetical 3-fold, or 300%, of their initial numbers on culturing in a growth-promoting media for a fixed amount of time. Thus, these cells show 200% growth, or 200% increase in viable cell count. However, the cells exposed to the drug (II) have less number of viable cells on culture under similar conditions for the same amount of time. The number of viable cells in the drug-exposed culture dish depends on the drug concentration in a manner illustrated by curve II. The drug concentration, at which the viable cell count after culture remains the same as the initial, that is, at 100%, is defined as the total growth inhibitory (TGI) concentration. Drug concentration that halves the growth of cells in culture, that is, increase in cell number to half of the levels seen without drug (which was 200%), is defined as the GI_{50} (growth inhibition to 50% level). Similarly, the concentration of drug that halves the viable cell count from its initial level (which is 100%) is defined as LC_{50} (lethal concentration to 50% level). (With kind permission from Springer Science+Business Media: *Pharmaceutical Perspectives of Cancer Therapeutics,* Anticancer drug development, 2009, 49–92, Narang A.S. and Desai, D.S.)

in the development pipeline through various metrics of their effectiveness. New drug discovery research should meet multiple criteria, such as clinical novelty, commercial opportunities, meeting unmet clinical need, and to build intellectual property.

This group also gets involved in identifying biomarkers and indicators of efficacy and toxicity in preclinical species, for potential utilization as surrogates for efficacy and toxicity assessment in the clinical studies. The discovery biology group works closely with bioanalytical scientists, who get involved in developing assays for compounds and physiological markers. The data generated by these functions are critically analyzed by pharmacokineticists and toxicologists to differentiate compounds for prioritization for advancement to the next stages of the drug discovery and development process.

1.2.2 Chemistry

Working closely with the discovery biology function, the discovery chemists provide an array of purified compounds for investigation. Discovery chemists, typically synthetic organic chemists who are involved in chemical synthesis, purification, and characterization of new chemical entities (NCEs), form the bulk of this function. This function is involved in identifying chemical compounds that may have drug-like properties and would respond in desirable ways (such as agonist or antagonist) on chosen enzymes, receptors, or other targets. The drug design roles of these scientists frequently involve molecular modeling of the target receptor and *in silico* assessments of receptor binding of potential chemical structures and structural modifications.

This group gets closely involved in the identification of potential measures of efficacy, as well as in the pharmaceutical developability assessment of new drug candidates to identify lead compounds that potentially maximize efficacy and minimize potential developability risks and toxicity. Developing structure–activity and structure–toxicity relationships is a key function of this group of scientists. High-throughput synthesis and purification technologies are commonplace in modern-day discovery chemistry laboratories.

The discovery chemistry function works closely with analytical scientists to assess the purity and identify of their compounds and with the discovery biology function to assess the efficacy of animal, cell culture, or *in vitro* models. Working coherently, discovery chemistry and discovery biology functions shortlist a few candidates that enter preclinical assessment and optimization.

1.2.3 Pharmaceutics

Preclinical pharmaceutical development involves optimization of potential drug candidates for drug-like properties. Development of pharmaceuticals comprises diverse group of scientists that assess both pharmaceutical dosage form developability through a series of assessments of solubility as a function of pH and in various biorelevant fluids, chemical stability of the compound to various stresses such as temperature and humidity, and polymorph stability. This function also undertakes pharmacokinetic and toxicokinetic assessment, in close collaboration with discovery biology colleagues, in one or more species, in an effort to identify potential starting dose and efficacious dose range for the FIH administration and dose escalation clinical studies. Pharmaceutical development scientists work closely with all functions involved in drug characterization and administration to address three key aspects of any new drug: stability, bioavailability, and manufacturability.

Preclinical optimization, discovery chemistry, and discovery biology functions work together as a team and may identify several molecules that go through developability assessments and are compared against each other for an array of desirable physicochemical and biological properties. The outcome of this exercise is the identification of one lead candidate that provides optimum balance of desirable attributes, while avoiding the undesirable ones. Often, one or more backup candidates are identified in case any significant undesirable observation (such as toxicity) is observed with the lead candidate.

1.2.4 Animal toxicology

As drug candidates advance toward FIH, formal toxicological evaluation in animal species is initiated. These studies are guided by the compound characteristics, target therapeutic area and biological receptor, as well as regulations that govern toxicity assessment in the animal species. Typically, toxicity is studied in two species, one rodent and one nonrodent, with an intent to identify target organs and organ systems that may exhibit toxicities at higher doses. These studies involve increasing drug dosing and exposures in the animal species until toxicity is observed and carefully documented. These studies frequently combine plasma concentration assessment and biomarker studies, if a biomarker has already been identified.

Toxicologists, working with diverse teams of professionals, are involved in the design and conduct of animal studies, as well as in the interpretation of observations. The goal of toxicological assessments is to rule out any significant toxicities, identify a starting dose for the clinical studies, and outline a monitoring strategy for potential toxicities during clinical studies. Toxicology studies seek to identify a maximum tolerated dose (MTD) that

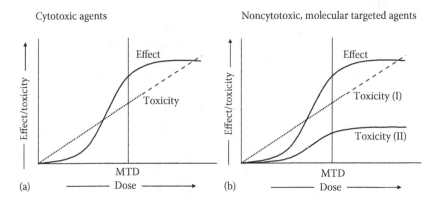

Figure 1.4 Hypothetical dose–effect and dose–toxicity curves for cytotoxic (a) and noncytotoxic, molecularly targeted anticancer (b) agents. The cytotoxic agents are known for their dose-dependent toxicity, which closely follows the dose–effect curve. Noncytotoxic agents, on the other hand, could have a linear dose–toxicity relationship similar to the cytotoxic agents (I) or a nonlinear profile with dose–toxicity curve lower than the dose–effect curve (II). *MTD* represents the maximum tolerated dose for the cytotoxic agent. (With kind permission from Springer Science+Business Media: *Pharmaceutical Perspectives of Cancer Therapeutics*, Anticancer drug development, 2009, 49–92, Narang A.S. and Desai, D.S.)

presents an acceptable balance of desired therapeutic effect and toxicity. Dose–effect and dose–toxicity relationships are delineated to identify the balance of effect and toxicity as a function of dose. As exemplified in Figure 1.4, these can be different for different compounds, based on their mechanism of action and target specificity.

1.2.5 Clinical pharmacology

Although no clinical studies are carried out with test candidates during drug discovery stages, clinical pharmacologists get involved in understanding the emerging profile of new drug candidates being screened. Aspects of drug discovery that can impact later stages of drug development, such as the relevance of animal models to human disease state and projected clinical doses or administration profile (e.g., route or frequency of administration), are critically assessed.

1.2.6 Analytical and bioanalytical sciences

Analytical and bioanalytical sciences form the core indispensable component of all functions involved in drug discovery and development. Analyses of drug content, purity, and any changes during storage are an essential part of identification of new chemical candidates through all stages of drug development, including commercialization. All drugs are required by federal regulations to have specification controls to ensure their identity, purity, and quality. The analytical methodologies utilized for ensuring these attributes could be spectroscopic or wet analytical techniques such as titrations and chromatography. For example, an oral solid dosage form must be tested for drug content, purity, water content, and drug release. A parenteral biologic drug product must be tested for drug content, purity by different methods (charge or size-based separation techniques), receptor binding, bioactivity, pH, osmolality, and structure/isoforms. In addition, all the starting (raw) materials, intermediates for synthesis, and excipients used in formulations must be rigorously tested to ensure these attributes and consistent quality across different batches. The assurance of maintenance of all quality attributes of the drug substance and drug product over the duration of storage (stability testing) utilizes analytical testing at various time points under different storage conditions (such as temperature, humidity, and light exposure). These functions are carried out by analytical scientists.

Bioanalytical sciences focus on analyses of drug content, metabolites, and any drug-related substances (such as both the parent compound and prodrug in the case of prodrug administration) in both animal and human studies. Bioanalytical sciences present significant and unique challenges due to complex, multicomponent nature of the biological fluids in which specific compounds must be tested—often without rigorous separation—and the very low concentrations of the target compounds (often in micromolar quantities). These analyses are typically carried out using highly sensitive analytical methods such as high-performance or ultra-high performance liquid chromatography (HPLC/UPLC), followed by tandem mass spectrometry (MS/MS).

All analytical methods are required to be qualified and validated for a host of criteria that ensure their robust and reproducible performance across potentially multiple testing sites, laboratories, and personnel.

1.2.7 Regulatory sciences

All drug products developed in modern biopharmaceutical settings are designed for global patient populations. Government regulatory agencies that monitor and control (regulate) the commercialization and utilization of drug products vary by each country and as do their requirements for the testing and commercialization of new drug products. As much harmonization of international regulations is being advocated and implemented (e.g., by the International Council on Harmonisation [ICH]), each country thus maintains its sovereign control over access to its markets and the requirements, which are often embedded in historical idiosyncrasies and scientific elements that may be unique to each region or subpopulation. The government regulatory agencies include, for example, the federal Food and Drug

Administration (FDA) in the United States and the European Medicines Agency (EMA) for several countries in the European Union; each of these countries do have their own drug regulatory agencies.

To ensure access of new drug products to patient populations globally, regulatory scientists work in the biopharmaceutical industry to proactively understand the regulations of the targeted markets. The regulatory sciences have evolved into a complex field that addresses not only the diversity of regulatory requirements but also the variations in scientific perspectives and understanding of different regions and countries. With ever-evolving analytical methodologies, drug development paradigms, and accelerating scientific growth in multiple disciplines involved in drug discovery and development, the regulatory scientists also form the interface of biopharmaceutical companies with the regulatory agencies and seek to educate and influence regulatory policy to evolve with the times.

1.3 SMALL- VERSUS LARGE-MOLECULE PARADIGMS

Drugs are typically divided into small and large in the contemporary vernacular based on their molecular size. The small-molecule drugs are typically the low molecular weight compounds, such as acetaminophen and ibuprofen, with molecular weight typically being less than or equal to 400 Da. The large-molecule drugs are typically the high molecular weight compounds, such as antibodies and antibody–drug conjugates, with molecular weight typically being greater than about 4,000. This classification is based on the significantly different molecular, biopharmaceutical, and pharmacokinetic characteristics of compounds that fall in either of the two categories. In addition, the access to biological targets usually differs significantly between these two broad drug classes, with large hydrophilic antibody drugs predominantly confined to extracellular or cell surface receptors, while the small hydrophobic drugs being able to permeate through cell membranes and access intracellular targets. The drugs that fall between these two broad classes in terms of molecular size and properties are typically defined and developed on a case-by-case basis, since their properties can be a unique mix, with dominance of one or the other characteristic.

Although the overall requirements and goals for new drug development are similar between the small- and large-molecule drug candidates, there are significant differences in the technical depth and detail as well as the pathway for these drug candidates. For the purpose of a general comparison, a typical small-molecule drug is exemplified by a hydrophobic crystalline compound intended for oral administration, whereas a large-molecule drug is exemplified by a monoclonal human antibody intended for parenteral administration. These differences are summarized in Table 1.1 and are briefly outlined below.

Table 1.1 Typical differences in the drug discovery process between small- and large-molecule drugs

Paradigm	Property	Small-molecule drugs	Large-molecule drugs
Clinical or therapeutic	Drug target	They are both intracellular and extracellular.	Typically, they are extracellular or cell surface.
	Target engagement	They typically have very fast receptor on and off rates. This results in close temporal correlation of a drug's pharmacodynamics with its pharmacokinetics. Thus, any desire for reduction in the dosing frequency of a drug generally requires the use of sustained- or controlled-release drug delivery technologies.	Antibodies can have very slow off-rates on target receptors, enabling longer duration activity after single dose administration. This leads to a potential disconnect between the pharmacokinetics and pharmacodynamics, since receptor engagement can be much longer than the presence of drug in the blood. Thus, the large-molecule drugs may have once-a-week dosing, without the need for sustained- or controlled-release drug delivery technologies.
Early development	Source of compounds	Most small molecules are synthetic in origin.	Usually, recombinant DNA technology is utilized in cell culture–based systems to produce the large molecules.
	Animal toxicology studies	They are usually carried out in two species, one rodent and one nonrodent.	They usually include a primate species, such as monkeys, since human sequences for 'arge-molecule drugs generally elicit an immune response in nonprimate species.
	Target engagement	They have fast on/fast off rates, leading to drug pharmacokinetics being closely linked to drug pharmacodynamics.	They have fast on/slow off rates, leading to longer duration of target engagement and action than the dosing frequency.
	Interspecies scaling and prediction of human dose	They are usually based on dose per unit body weight.	They are usually based on dose per unit body surface area.

(Continued)

Table 1.1 (Continued) Typical differences in the drug discovery process between small- and large-molecule drugs

Paradigm	Property	Small-molecule drugs	Large-molecule drugs
	Molecular identity	Small molecules have very precise identity, down to the atomic level.	Large molecules are difficult to characterize down to the atomic level. Normal levels of variation may exist within a pool of antibodies, for example, without affecting stability or activity. Multiple clones are sometimes used during early development, of which one clone may later be selected for commercial development.
	Key toxicological risks	Off-target toxicity is usually the most common observation with small-molecule new chemical entities (NCEs).	Off-target toxicities are less likely with antibody drugs. Immunogenicity is a significant concern throughout drug product development.
Pharmaceutical development	Physicochemical properties	Most contemporary small-molecular NCEs are hydrophobic, crystalline compounds.	Most large-molecule drugs, such as antibodies, are hydrophilic drugs that exist in solution or amorphous matrices with other excipients.
	Characterization techniques	Robust and precise identification of related substances, metabolites, and impurities with mechanism of their formation guides drug product development.	Complex analytical methods that usually provide bulk characterization rather than exact identity at the atomic level. Comparability characterization between batches used in clinical studies is often used as a criterion of uniformity across a product's clinical development and life cycle.
	Product presentation	A solid dosage form for oral administration is the most common or targeted product presentation for most small molecules.	A ready-to-use parenteral solution presentation is the most common or targeted product presentation for most large-molecule drugs.

1.3.1 Preclinical discovery and toxicology assessment

Large-molecule drugs are typically produced in cell culture systems by utilizing the recombinant DNA technology, while most small-molecule drugs are synthetic compounds produced by using one of the laboratory high-throughput manufacturing technologies. The synthetic chemistry utilized for generation of small-molecule drugs are based on the principles of organic chemistry and reaction kinetics. The utilization of cell culture for biologic manufacturing is based on increasing scale of cell culture in bioreactors in what is termed the upstream manufacturing operations, while the downstream manufacturing operations focus on purifying the target protein of interest (harvesting) from the cells (Figure 1.5).

The preclinical toxicology assessment of antibody therapeutics is typically carried out in primates to ensure relative tolerance to human sequences and closeness of animal physiology with the human physiology that is being targeted. In addition, more than one clone of the antibody may be tested in animal species in certain cases in early stages of development. In the case of small molecules, very precise crystalline form of a highly pure compound is usually tested in the animal species to allow scaling of the observations to human administration.

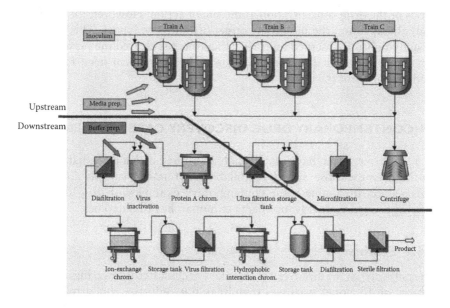

Figure 1.5 A typical process train for a biologic-manufacturing operation showing cell culture in progressively increasing tank sizes and cell culture volumes (upstream), and purification of the target protein from the cultured cells (downstream) through a series of operations yielding the target protein of interest. (From http://www.mdpi.com/2306-5354/1/4/188/htm.)

1.3.2 Pharmaceutical development

Small- and large-molecule drugs are inherently different in that the typical small molecules are hydrophobic, crystalline, well-characterized compounds, whereas the typical large molecules are hydrophilic species that exist in solution. A particular batch of an antibody may not be exactly uniform in all the molecules that exist in solution. Antibody product presentations are usually a solution, whereas preferred product presentations for small molecules are typically in some solid-state dosage form intended for oral administration. The ready-to-use solution presentations of antibodies are typically intended for either direct parenteral administration or dilution in intravenous fluids before administration. In certain cases, for ensuring maximum product stability, antibodies may be lyophilized in the presence of excipients to present a solid dosage form in a sterile vial for reconstitution before administration.

1.3.3 Clinical development

Target engagement, on and off rates, and frequency of administration vary significantly between antibodies and small-molecule drugs. Small-molecule drugs typically have fast on/fast off rates of target receptor occupation. This results in their target activity being closely linked to the pharmacokinetics or the drug concentration at the site of action. However, for the antibody therapeutics, the molecular target off rates are longer. This result in bioactivity being observed long after the drug has cleared the plasma. This can result in longer duration of action and less frequent need for repeat administration of the drug.

1.4 CONTEMPORARY DRUG DISCOVERY CASE STUDIES

Drug discovery often happens through an unpredictable path that depends on the deep expertise and close collaboration among various disciplines. This section highlights some contemporary examples of drug discovery that led to significant advancements in human health outcomes.

1.4.1 Statins

The pathway[1] to the discovery of this class of cholesterol-lowering drugs—a subject of 13 nobel prizes—exemplifies not only the concurrent cross-disciplinary progress but also the magnitude of effort required for such a breakthrough. Lovastatin, the first compound in this series, was commercialized by Merck in 1987 and was followed by the commercialization of semisynthetic (simvastatin and pravastatin) and synthetic (fluvastatin, atorvastatin, rosuvastatin, and pitavastatin) statins.

The intertwined pathway to the discovery of these drugs starts with the discovery of cholesterol in 1784 from gallstones by the French chemist Francois Poulletier. The proof of structure of cholesterol was established in the early twentieth century. Cholesterol was linked to atherosclerosis in 1910 by an investigation that showed significantly higher (~20-fold) cholesterol levels in human atherosclerotic plaques. This was followed by an experimental production of atherosclerosis in rabbits by the consumption of cholesterol-rich diet. A Norwegian clinician first proposed the link between cholesterol and heart attacks in several large families in 1939. A genetic understanding of this phenomenon, now known as familial hypercholesterolemia, was established in 1960s. Further, epidemiological studies carried out in 1950s and 1960s established a link between cholesterol and coronary artery disease.

Meanwhile, the complex 30-enzyme biochemistry of cholesterol's synthesis and regulation of metabolism was elucidated in 1950s. With the cholesterol synthetic pathway known and the connection to human disease fairly plausible, several companies invested in research on molecules that would block one or more of these steps through synthesis of analogs of natural substrates. Some of these drugs, such as triparanol, commercialized in 1959 and withdrawn from the market in 1960s, were effective in reducing cholesterol level but had serious side effects due to the accumulation of other sterols in the synthetic pathway. Cholesterol-lowering drugs such as nicotinic acid, clofibrate, and cholestyramine started becoming available in late 1950s to early 1960s. The first molecule that inhibited HMG CoA reductase enzyme, citrinin, a key enzyme in the cholesterol synthetic and metabolic regulatory pathway, was isolated from mold in 1971, and the second compound, compactin, was isolated in 1972. The preclinical and clinical results with compactin in the 1970s inspired compactin was also isolated from mold Penicillium citrinum chemists from Merck to discover lovastatin in 1978–1979, whose development was a bit delayed due to the initial suspicion of carcinogenicity of this class of compounds based on observations in dogs at high doses. Nevertheless, promising clinical results with low side effect profile resulted in full development of lovastatin, with new drug application (NDA) filing in 1986 and commercialization in 1987.

1.4.2 Immuno-oncology

Immuno-oncology drugs started transforming clinical treatment of cancer for a wide spectrum of patients with the regulatory approval and commercialization of the following humanized recombinant monoclonal antibodies (mAbs): ipilimumab (Yervoy®) by Medarex, Inc., and Bristol-Myers Squibb, Co. (BMS), targeting the CD28-cytotoxic T-lymphocyte-associated antigen 4 (CTLA4) receptor, in 2011; and pembrolizumab (Keytruda®) by Merck and nivolumab (Opdivo®) by BMS, both targeting the programmed cell death 1 (PD-1) receptor, in 2014–2015. The discovery of these revolutionary medicines exemplifies not only cross-functional development, but also the necessity of continuous evolution of drug discovery and development paradigms. These shifting paradigms are exemplified by the following[2]:

- Immunotherapy agents are unique in that they do not directly attack the tumor but rather mobilize the body's innate and adaptive immunity. The contemporary clinical new cancer drug development paradigm was based on the historical experience with chemotherapy drugs, whereby reduction in tumor size was a measured endpoint, and clinical success was measured in terms of progression-free survival (PFS). Increase in tumor size during clinical trial over a relatively short period of time, about 2 months, would typically lead to cessation of dosing with the experimental drug. In the clinic, the immunotherapy drugs take longer to reach time-to-event endpoint, with delayed separation of survival curves. They also produced unique immune-related adverse effects that were reversible and clinically manageable. Clinical success of ipilimumab was a result of modification of phase III clinical trial endpoint during the study to overall survival (OS) and prolonged dosing, with clinical management of immune-specific side effects. A previous clinical trial of another immunotherapy agent, the anti-CTLA4 mAb tremelimumab, which operated with the chemotherapy paradigm of clinical development, failed to demarcate itself in the clinic to gain regulatory approval.
- Host immunity being consequential to tumor treatment had been observed and experimented with for a long time. German pathologist Rudolf Virchow observed immune infiltration of human tumors in the nineteenth century. American surgeon William Coley observed clearance of cancer on intratumoral injection of bacterial broth in some cases. The discovery of immune checkpoints to tumor progression in the twentieth century, cloning of the *CTLA4* gene in 1987, and demonstration of the value of immune checkpoint inhibitors in cancer treatment in mouse models in 1990s provided impetus to the discovery of immune-checkpoint modifying drugs.
- Antibodies for human use long suffered from immune rejection, until progressive development from polyclonality to monoclonality and humanization through recombinant DNA technologies made possible the development of therapeutic antibodies that would not elicit an immune response.
- Development of antibodies as drugs went through a long journey, with the difficulty of exact characterization of physicochemical properties and structural features. The number, type, and quantity of size and charge variants of these large-molecule drugs are often used for purity assessment instead of exact chemical structures of compounds. The development paradigms deviate from the conventional small-molecule drugs in utilizing comparability among clinical batches, instead of absolute chemical purity, as a benchmark during drug development.

The development of immuno-oncology drugs exemplifies the value of concurrent significant advancements in different fields and their interplay in facilitating the discovery and commercialization of an entirely new class of drugs.

1.5 SUMMARY AND OUTLOOK TO THE FUTURE

Drug discovery is inherently a complex, multidisciplinary endeavor that requires close collaboration of highly skilled professionals, integration of each disciplines output, and relatively long timelines with different stages of development that have stage gates to check the progression of the compound to later stages of development and commercialization. This process is also continuously evolving as the drug targets change, pressures on the biopharmaceutical industry increase, and our understanding of the fundamental mechanisms improve. Teamwork and collaboration form the hallmarks of cultural norms required in this high-paced environment. In addition, idiosyncratic cross-functional application plays a unique role in breakthrough drug discovery.

The influences of evolving technologies on drug discovery are evident at all stages. For example, big data and trend analysis is helping in decision making at all stages of development. Greater ability for clinical monitoring and preclinical toxicology assessment is enabling the development of safer drugs and early assessment of potential toxicological liabilities. Increasingly, drugs are being developed in patient-centric manner to serve the needs of given subpopulations of patients.

Recent trends in drug development also highlight the increasing use of the outsourcing model to allow investment of scarce internal resources on the highest-value items. Typical modern-day biopharmaceutical companies collaborate with several contract research organizations (CROs), contract manufacturing organizations (CMOs), and academia alike to execute just about every step of the drug discovery and development process. Organizations increasingly focus on generating intellectual property and maximizing the speed to market while maintaining and improving product quality outcomes. These internal emphases with external forces make the industry a continuously evolving endeavor.

REVIEW QUESTIONS

1.1 What is the one significant difference between antibodies from hydrophobic small-molecule drugs?
   A. Antibodies are hydrophilic.
   B. Antibodies are large molecular weight compounds.
   C. Antibodies typically have slow off rates on receptors that they occupy.
   D. All of the above.

1.2 What is the sequence of activities in the new drug development?
  A. Preclinical development > clinical development > compound characterization > commercialization
  B. Compound characterization > preclinical development > clinical development > commercialization
  C. Clinical development > preclinical development > compound characterization > commercialization

1.3 All drugs intended for human administration must be manufactured and released by utilizing practices that conform to which of the following? Check any two that apply.
  A. Good manufacturing practices
  B. Good clinical practices
  C. Good laboratory practices
  D. The International Council on Harmonisation

1.4 Please categorize the number of subjects tested in clinical trials in the increasing order of the different phases of clinical studies.
  A. Phase 1 > phase 2 > phase 3
  B. Phase 1 > phase 3 > phase 2
  C. Phase 2 > phase 1 > phase 3
  D. Phase 2 > phase 3 > phase 1
  E. Phase 3 > phase 2 > phase 1
  F. Phase 3 > phase 1 > phase 2

1.5 Typical toxicology studies must be carried out in how many different species of animals before first human administration of a new drug candidate?
  A. 1
  B. 2
  C. 3
  D. 4
  E. 5

1.6 Which of the following is not an analytical method used to test the quality of a drug?
  A. Purity
  B. Potency
  C. Quantity
  D. Water content

REFERENCES

1. Hoos A (2016) Development of immuno-oncology drugs–from CTLA4 to PD1 to the next generations. *Nat Rev Drug Discov* 15(4): 235–247.
2. Endo A (2010) A historical perspective on the discovery of statins. *Proc Jpn Acad Ser B Phys Biol Sci* 86(5): 484–493.

FURTHER READING

Drews J (2000) Drug discovery: A historical perspective. *Science* **287**: 1960–1964.

Endo A (2010) A historical perspective on the discovery of statins. *Proc Jpn Acad Ser B Phys Biol Sci* **86**(5): 484–493.

Hughes JP, Rees S, Kalindjian SB, and Philpott KL (2011) Principles of early drug discovery. *Br J Pharmacol* **162**: 1239–1249.

Narang AS and Desai DS (2009) Anticancer drug development. In Lu Y and Mahato RI (Eds.) *Pharmaceutical Perspectives of Cancer Therapeutics*, New York: Springer, pp. 49–92.

FURTHER READING

Chapter 2

Drug development

2.1 INTRODUCTION

The process of discovery and development of safe and effective new medicines is long, difficult, and expensive. A new molecular entity (NME), sometimes also called a new chemical entity (NCE), is characterized for its potential therapeutic applications and toxicological profile in nonprimate species, which is followed by extensive animal and human testing. On average, it costs a company more than $1 billion and 10–15 years to get one drug from the laboratory to patients. Only five in ~5000 compounds that enter preclinical testing make it to human testing. Only one of those five drugs entering human clinical trials is approved for commercialization.

New drugs include prescription drugs, over-the-counter (OTC) medications, generic drugs, biotechnology products, veterinary products, and/or medical devices. Over-the-counter drugs do not require a physician's prescription.

Drug development also focuses on new dosage forms, routes of administration, and delivery devices for existing drugs. A typical drug discovery process entails target identification, such as a protein or an enzyme whose inhibition may help in a disease state. The structural features necessary in a potential drug candidate are identified using *in silico* molecular modeling. Several drugs may be synthesized using combinatorial chemistry and screened for *in vitro* activity in high-throughput assays. The lead candidates are then synthesized in larger quantities, screened for biological activity, and further optimized to maximize the affinity, specificity, and potency. A highly specific compound that only binds the target site is likely to have minimal nontarget effects, which often lead to adverse effects and toxicity related to the mechanism of drug action. High affinity for the target site, often resulting in high potency (low dose for the desired pharmacological effect), minimizes the required dose of a compound, which can reduce adverse effects and toxicities not associated with the drug's mechanism of action. Such a drug candidate is then identified as an NME (Figure 2.1), which enters the development pipeline.

Drug development studies include preclinical studies, whereby a compound is thoroughly characterized for physicochemical characteristics and is tested in animal models for toxicity and activity. This is followed by first-in-human (FIH) phase I studies that test the safety of a compound through sequential dose-escalating studies, which could be single-dose (called single ascending dose, SAD) and/or multiple-dose (called multiple ascending dose, MAD) studies. In these studies, a subject is administered a SAD or MAD at predefined intervals, with each successive cohort receiving doses higher than the previous cohort (hence the term ascending dose studies). Increasingly extensive and thorough toxicological, pharmacological, and pharmacokinetic characterization is carried out in humans during phase II and phase III clinical trials (Figure 2.2).

Stages of drug development that precede human testing are termed *preclinical development*, while human testing stage of a drug is termed *clinical development*. The transition from preclinical development to clinical development requires regulatory approval through an investigational new drug (IND) application. In the United States, the sponsor of an NME files an IND with the Food and Drug Administration (FDA) with preclinical data and proposed protocol for clinical testing. Unless the FDA has a question or objection to the proposal within 30 days of filing, the drug compound enters FIH (phase I) clinical studies.

Clinical evidence of safety and efficacy of drug products forms the cornerstone of regulatory approval of any new drug product. The clinical evidence is typically gathered in a phased manner, wherein the toxicity or adverse events

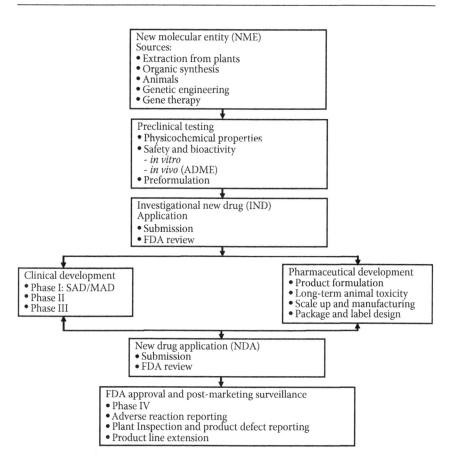

Figure 2.1 Stages of drug development. Drug discovery efforts lead to a new molecular entity (NME) that is identified for development. The NME enters preclinical testing and development, which include efficacy and safety testing in animal species. This is followed by the submission of an investigational new drug (IND) application to the regulatory agency, such as the U.S. Food and Drug Administration (FDA). The drug then enters clinical trials: phase I, phase II, and phase III. Pharmaceutical development proceeds concurrent with clinical development and with the objectives of supporting the ongoing clinical studies (providing information, documentation, and the drug product for administration to the subjects) and preparation for commercialization of the product. Following successful clinical and pharmaceutical development, a data package is submitted to the FDA for approval; this is called the new drug application (NDA). Approval of the NDA is required for the commercialization of a new drug product. Several activities on the drug product continue after commercialization, such as adverse event monitoring, development of line extension products, and additional clinical trials to support label claims or expand target patient populations.

Preclinical testing	Clinical trials research and development	FDA	Postmarketing surveillance

Synthesis
-Identify a lead compound
Characterization
-Physicochemical properties
Toxicity and bioactivity
  -*In vitro* (cell culture)
  -*In vivo* (short term)
  -ADME/Tox
	Phase I
	-Healthy volunteers (20–80)
	-Safety profiles
	-Drug tolerance
	Phase II
	-Patients (100–300)
	-Controlled, randomized trials
	-Double-blinded
	Short-term side effects
	-Decision on final dosage form
	Phase III
	Patients (1,000–3,000)
	-Expanded and uncontrolled trials
	-Monitor adverse reactions
	-Confirm effectiveness
	-Decision on physician labeling
		Review and approval	Phase IV
			-Postmarketing testing
			-Report adverse effects
			-Report product defects

Average 3.5 years	1.5 + 2 + 4 = 7.5 years	6–10 months	
Evaluation of thousands of compounds	<1% enter trials	1 approved	

↑ IND submission   ↑ NDA filing   ↑ NDA approval

Figure 2.2 Timeline of different phases of drug development. Discovery and preclinical testing to identify a lead compound and its detailed characterization for toxicity and bioactivity *in vitro* and *in vivo* can take a few years, such as about 3–4 years. A significant amount of time is taken by different phases of clinical trials, about 7–8 years, with increasing duration of time and number of patients required for higher stages of clinical trials. Agency review of the submitted new drug application (NDA) can take several more months, depending on prioritization and workload considerations. A typical medium- to large-sized biopharmaceutical company has several pipeline candidates that are at various stages of development.

of a drug are assessed in increasing number of subjects as a molecule progresses through the development timeline (Figure 2.3).

Pharmaceutical development follows a parallel track with preclinical and clinical development. Pharmaceutical development is responsible for chemistry, manufacturing, and control of both the drug substance and the drug product throughout the life cycle of a compound. This function provides a robust dosage form that meets three key requirements of a drug product: (a) stability, (b) bioavailability, and (c) manufacturability. The key roles of pharmaceutical development are to provide a suitable drug product in a stage-appropriate manner while also ensuring path to future development and commercialization and to bridge the drug product used during different stages of development.

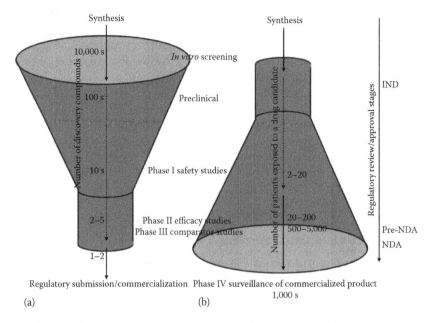

Figure 2.3 Number of compounds proceeding through various stage of drug development—an upright funnel (a), and the number of patients exposed to a given drug through its development—an inverse funnel (b). This figure also illustrates the steps where regulatory review and/or approval are required in the United States, such as the investigational new drug (IND) application submission before initiating phase I studies, a pre-NDA meeting with the FDA after the phase II studies, and new drug application (NDA) submission for drug approval for marketing after the completion of phase III studies. (With kind permission from Springer Science+Business Media: *Pharmaceutical Perspectives of Cancer Therapeutics*, Anticancer drug development, 2009, 49–92, Narang A.S. and Desai, D.S.)

2.2 STAGE-GATE PROCESS OF NEW DRUG DISCOVERY AND DEVELOPMENT

New drug discovery and development process can be divided into distinct sequential phases that evaluate and develop the drug-like characteristics of a potential new compound being considered a drug candidate (Figure 2.1). The progression of new drug candidates through various stages of this sequential process depends on successful demonstration of drug-like characteristics in each of these phases. Scientists working in a wide array of disciplines are responsible for both characterization and enablement of drug-like properties in new drug candidates throughout these stages of drug development. These stages include early discovery, preclinical development, FIH, registrational clinical studies, commercialization, and life cycle management. There is a progressive reduction in the number of molecules that progress through this stage-gate process, concurrent with increasing knowledge, patient exposure, and understanding of the molecule (Figure 2.3).

These stages are bounded by distinct boundaries, where the governance leadership of an organization must make a decision whether to continue to invest in a molecule or not. These are called *decision-points* or *stage-gate checkpoints*. At these points, interdisciplinary discussions that include both technical (e.g., developability risk) and nontechnical (e.g., commercial potential and competition) aspects identify risk–reward balance and the path to commercialization of an asset, which feed into governance decision for a given asset or a set of assets.

2.2.1 Preclinical development

Preclinical development involves in-depth characterization of a few select potential drug candidates, the NCEs or NMEs, before first human administration. The preclinical stage includes detailed physicochemical characterization of the compound and animal studies. This stage has the objectives of identifying developability and clinical risks to the compound and of identifying a viable development path to a commercial drug product. In addition, a critical decision of the starting human dose is made based on the animal studies carried out during the preclinical phase.

2.2.1.1 Physicochemical characterization

New chemical entities are characterized for their pH–solubility profile, pH–stability profile, solid form and form stability, excipient compatibility, dissolution rate, photostability, supersaturation on pH transfer, and other physicochemical characteristics. Prototype formulations are designed for use in animal efficacy studies, toxicological assessment, and phase I clinical studies. The ability to produce a commercially viable dosage form of the compound is assessed at this stage.

2.2.1.2 Efficacy studies

A key goal of preclinical testing is to determine whether a compound exhibits a pharmacological activity and is reasonably safe for initial testing in humans. Following identification of some lead compounds, the pharmacological and toxicological effects of these compounds are determined. These tests involve the use of laboratory animals, cell culture, enzymes, and receptors, as well as computer models. Animal testing may be carried out, for example, in transgenic mice or other animals, that exhibit the pharmacology of a particular disease state and/or drug target that defines the target patient population.

2.2.1.3 Toxicology studies

Animal toxicology assessment is carried out in at least two different animal species. These toxicology studies are intended to assess the organs or organ systems in which a particular compound tends to exhibit toxicity as well as to identify the doses at which the toxicity appears. These animal studies are used to understand how the drug is absorbed, distributed, and metabolized in the body; ascertain its metabolites; and determine how quickly the drug is excreted from the body. Pharmacokinetic and pharmacodynamic studies are conducted to analyze the *absorption, distribution, metabolism, excretion,* and *toxicological* effects of the drug, commonly known as *ADME/Tox* prediction.

Animal doses are translated into the FIH dose with normalization based on body weight or surface area. Interspecies dose scaling for small-molecule compounds is generally carried out using body weight as a metric, for example, mg dose per kg body weight is kept constant across species. The metric for biologic compounds, for example, therapeutic proteins and antibodies, is generally dose per unit body surface area (e.g., mg dose per m^2). The FIH doses are typically a fraction of the lowest toxic dose observed in any animal species.

2.2.2 Clinical development

Clinical development can be divided into four phases: phase I, II, III, and IV, with progressively increasing number of subjects exposed to the drug. Such division, through distinct clinical protocols, is intended to utilize the results of the previous phase or subphase of the study to inform the design of the next phase of the study.

Phase I clinical studies are aimed at identifying dose-limiting toxicities, toxicological profile, dose–exposure relationship, and drug pharmacokinetics in a small group of healthy or patient volunteers. Phase II trials are designed to confirm toxicological findings and generate information on pharmacologic activity and pharmacokinetics. Phase II studies also aim to define the appropriate dose for larger phase III trials. Phase III trials are the clinical studies in a large group of patients. These studies are often also called registrational clinical trials, since the results of these studies are submitted (registered) to the FDA as a new drug application (NDA) or a biologics license application (BLA) to gain regulatory approval for commercialization of a new drug. Phase IV studies are postapproval clinical studies that might include specific targeted studies that verify or expand the current spectrum of claims on the label of the drug.

The FIH clinical studies, also called phase I studies, are typically carried out in a very small cohort of human subjects, often normal healthy volunteers. These studies are intended to assess the safety of the compound. The drug is administered at very low doses, based on the observations in animal studies.

The doses are increased in single- or multiple-dose studies in a systematic fashion under close clinical observations to identify potential adverse events, monitor drug pharmacokinetics and biochemical markers for those adverse events, assess target engagement, and identify doses that may be utilized in subsequent larger-scale phase II and phase III studies.

The large late-stage clinical studies that form the basis of a drug's approval are considered *registrational*. Phase I studies are typically carried out in healthy volunteers to assess toxicity and define doses for human administration. Phase II studies may involve patients and are utilized to identify adverse events in a broader population set and to more closely define doses for the registrational studies. Registrational phase III studies often involve comparison against a placebo or current standard-of-care treatment. This may vary depending on the therapeutic category and indication of the drug. For example, studies with cytotoxic anticancer drugs are not carried out in healthy volunteers, and placebo may not be used as a comparator for patients with serious diseases. Increasingly, the demarcation between different phases of clinical development are getting blurred with the key defining criterion of clinical studies being restricted to nonregistrational dose-escalation studies and *registrational* studies.

In addition, several smaller, focused clinical studies are carried out at different points in time during drug development to support a drug product's labeling and interchangeability. For example, bioequivalence studies may be carried out to *bridge* a phase I formulation with a phase III formulation or to support a formulation change during clinical development. Examples of other focused studies include food effect evaluation, drug–drug interaction studies, and studies in special populations such as in renally compromised patients.

The larger-scale studies are carried out in an increasing number of subjects (Figure 2.3). As clinical trials progress, the dose and dosing regimen are optimized to the effective levels that present an acceptable toxicologic profile. In addition to the characterization of the clinical profile of the drug candidate (such as the dose and frequency of administration), these studies seek to differentiate the new drug candidate from the existing therapeutic options for a given set of patients. Thus, studies may be carried out to compare the efficacy and toxicity of the NCE with the current standard of care.

Phase IV studies involve postcommercialization monitoring of drug effects and are designed to monitor a drug's long-term safety and effectiveness, as well as to uncover any rare but serious side effects that may not have been evident in earlier, relatively smaller pool of healthy and patient volunteer studies. In addition, studies may be carried out in special patient populations—such as pediatrics, geriatrics, and renally compromised patients—to more closely define the dosage and risk–benefit profile in those patient populations.

2.2.2.1 Phase I clinical trials

Phase I studies are the FIH studies, which involve the first introduction of an IND into humans. These studies are closely monitored and may be conducted in patients (when ethically required, e.g., anticancer drugs), but are usually conducted in healthy subjects. Phase I clinical trials are relatively short (several months) and involve relatively less number of human volunteers (6–20). These studies are designed to identify a drug's safety profile, including the safe dosage range. The purpose of a phase I clinical trial is to establish the tolerance of the drug in healthy human subjects at different doses and define its pharmacologic effects. Thus, these studies often involve *dose escalation* within a clinical trial.

These studies also involve measurement of plasma drug concentration to determine how a drug is absorbed, distributed, metabolized, and excreted (ADME), as well as the duration of its action. The purpose of a phase I clinical trial is to establish a safe dosage range by determining the tolerance of the drug in healthy human subjects at different doses and define its pharmacological effects. Information about the pharmacokinetic profiles of the drug in humans is used to design appropriate dosing regimens for the next phase of clinical trials.

2.2.2.2 Phase II clinical trials

Once an experimental drug has proven to be safe and well tolerated in phase I healthy volunteer studies, it is tested in patients in phase II studies. Phase II clinical trials involve controlled studies on 100–300 volunteer patients to assess the effectiveness of the drug for a particular indication(s) and reconfirm toxicological profile. These studies are usually longer and larger than phase I studies (1+ years). This phase of testing also helps determine the common short-term side effects and risks associated with the drug.

Two key aspects of late-stage clinical development are (1) randomization, and (2) blinding.

- *Randomization*: Most phase II studies are randomized trials against a control group; that is, one group of patients receives the experimental drug, while a second *control* group receives the treatment that represents a current standard of care, or placebo. Placement of the subject into the drug treatment or control group is done by random assignment. The randomization of subject assignment is an important statistical control to obviate any bias in study design.
- *Blinding*: An important methodology to avoid any bias during clinical testing in terms of patient perception and response is for the patient to not know whether he or she is receiving the drug or the control option. Such a clinical study in which the patient does not know the therapy but the healthcare providers, including the physician, may know what is being administered is called a single-blind study.

However, often, the clinical studies are *double-blinded*; that is, neither the patient nor the physician knows which patients are getting the experimental drug. A double-blinding approach overcomes bias in physician's influence on patient perception and response as well as physician's reporting of patient experience. The drug product manufacturer often carries out double blinding by providing two look-alike medicinal products that are only coded differently. In certain cases, the blinding may be carried out by the on-site healthcare professional, such as the pharmacist, who prepares drug products for administration and prepares the blinded labels.

Phase II studies are designed to determine the correct dose and, thus, are often referred to as dose-ranging studies. During phase II clinical trial, the final dosage form is selected and developed for phase III trials. Phase II studies are sometimes divided into phase IIA and phase IIB studies. When defined distinctly, phase IIA studies are considered exploratory evaluation of clinical efficacy, with pharmacodynamics or biological activity as the primary endpoint. Phase IIB studies are then defined as the definite dose-range-finding study that identifies doses for the large-scale, registrational phase III studies.

2.2.2.3 Phase III clinical trials

A phase III clinical trial is an expanded controlled clinical trial of a drug's safety and efficacy in large and diverse patient populations. This phase usually lasts several years and involves approximately 500–3,000 patients in clinics and hospitals. Physicians monitor patients closely to determine efficacy and identify adverse reactions. Phase III studies gather precise information on the drug's effectiveness for specific indications, determine whether the drug produces a broad range of adverse effects than those exhibited in the small study populations of phase I and II studies, and identify the best way of administrating and using the drug for the purpose intended. Phase III studies also provide an adequate basis for extrapolating the results to the general population and transmitting that information in the physician labeling.

2.2.2.4 Evolving clinical development paradigms

In cases where phase I studies involve patients and are adequately extensive and detailed, phase II studies may be bypassed. Similarly, in cases where phase II studies are extensive, they may be used as registrational studies, thus bypassing the need for phase III studies. Such decisions are made on a case-by-case basis, depending on the exact disease area and the therapeutic profile of the investigational drug candidate.

As mentioned earlier, the division of clinical studies into distinct phases is based on the generation of distinct clinical protocols, with an intention to utilize the results of the previous phase or subphase of the study to inform the design of the next phase of the study. In cases where the decision tree for how the previous phase of the study would influence the design of the succeeding phase can be constructed before actually carrying out the first phase of the study, clinical study designs for different phases can be made *seamless*. Such clinical studies continue the treatment of a selected cohort of patients enrolled in a phase II study into the extended phase III study, while enrolling additional patients to meet the requirements of the number of subjects needed for the phase III study.

In addition, clinical study designs can be *adaptive*. Adaptive clinical trial designs allow changes in the design or analyses while the study is in progress. These changes are guided by the accumulated data at an interim point in the trial. Such designs make the clinical study more efficient in their ability to reduce the duration or the number of subjects required for the study.

2.3 PHARMACEUTICAL DEVELOPMENT

Pharmaceutical development provides the drug product needed for preclinical and clinical studies to identify the biological mechanism of a new drug and its clinical utility. In designing the drug product, functions of pharmaceutics seek to fulfill three key requirements: (a) manufacturability to robust and reproducible quality; (b) stability to the worst-case shipping, storage, and use conditions; and (c) adequate bioavailability with a desired, reproducible pharmacokinetic profile. Pharmaceutics work is carried out through all stages of drug discovery and development to provide stage-appropriate drug product for preclinical and clinical testing, to bridge the studies carried out at different stages of development, and to enable commercialization of a product and process that ensures reproducible manufacture of a high-quality drug product.

Pharmaceutical scientists work on developing suitable dosage forms for drug administration at different stages of drug development. These might include, for example, a parenteral solution formulation during efficacy and toxicology studies *in vitro* and in animal models. During phase I studies, the formulation could be a suspension, drug-in-capsule (DIC), or drug-in-bottle (DIB) formulation. During phase II studies, a more representative tablet or capsule formulation might be developed, which is further refined for phase III dosing and commercialization.

In designing the drug product, pharmaceutical development considerations include the target population (children or adults), the amount of drug to be given in each dose, storage stability of the drug product, the characteristics of the drug and disease state, preferred route of administration, drug stability, and robustness of the manufacturing process. An early assessment of the properties of the desired dosage form can contribute greatly to the speed of the drug development process.

2.3.1 Preformulation and formulation studies

Preformulation studies are initiated to define the physical and chemical properties of the agent, followed by *formulation* studies to develop the initial features of the proposed pharmaceutical product or dosage form (e.g., liquid, tablet, capsule, topical ointment, intravenous [IV] solution, and transdermal patch). The final formulation includes substances called *excipients* in addition to the active pharmaceutical ingredient (API). Preformulation and formulation studies take approximately 3 years and occur concurrently with preclinical (animal) and clinical studies. Depending on the design of the clinical protocol and desired final product, pharmaceutical scientists are called upon to develop specific dosage forms of one or more dosage strengths for administration of the drug. The initial formulations prepared for phases I and II of the clinical trials should be of high pharmaceutical quality, meet all product specifications, and be stable for the period of use.

Three key goals of pharmaceutical development are to ensure the delivery of stage-appropriate drug product with acceptable (a) stability, (b) bioavailability, and (c) manufacturability.

2.3.2 Stage-appropriate drug product design

The drug product used for testing at earlier stages of development, such as preclinical or phase I, is generally different than the one used at later stages. Stage-appropriate drug product design takes into consideration the objectives and requirements for each stage. For example, during animal toxicology studies, the objective is primarily to maximize exposure and allow the administration of large doses. At this stage of development, storage stability requirements are minimal and the drug product can be handled in carefully controlled manner in the laboratory setting. Therefore, a high-concentration solution dosage form may be preferred at this stage of development. The objective in the later stages of development, such as phases II and III studies, is to be as similar as possible to the final commercial formulation and process. Thus, a market-image formulation is generally developed for those later stages of development.

2.3.3 Stability

A drug product is expected to maintain the chemical purity (i.e., chemically unchanged API) and physical integrity (e.g., the same polymorphic form) of the drug, physical integrity of the dosage form (e.g., no breakage of tablets), and reproducible drug release from the dosage form throughout the projected storage period under recommended storage conditions.

The stability requirements for drug product are different at each stage of drug development and depend primarily on the anticipated duration of storage and the storage conditions (e.g., refrigerated or room temperature) for the animal and human studies. For commercialization, the stability requirements are based on the target shelf life at the desired FDA-approved label storage conditions. Generally, no less than 18 months of shelf life is considered commercially viable.

Harmonization of stability requirements across the companies involved in new drug development for product commercialization across different regions of the world is carried out through the guidelines provided by the International Council on Harmonisation (ICH). These guidelines define the storage conditions that can be considered representative of year-round weather in different regions of the world. For example, for the United States and Western Europe, normal room temperature storage conditions have been identified as 25°C, with a relative humidity of 60%.

2.3.4 Bioavailability

A vital aspect of any dosage form is to be consistent (dose-to-dose and batch-to-batch) in delivering the total amount of drug into the systemic circulation and the rate at which it is delivered (bioavailability) from the drug product. Optimization and control of drug product properties that ensure robust manufacturing, physicochemical stability, and reproducible drug release help ensure consistent bioavailability. Drug substance and drug product attributes that impact drug release and bioavailability are identified, and the mechanistic basis of their impact are studied. *In vitro* assays are developed to measure drug release, and their results are correlated with *in vivo* performance. Such a correlation between *in vitro* and *in vivo* performance is termed *in vitro–in vivo* correlation (IVIVC).

In silico modeling of drug absorption is commonly used to understand and predict a drug's behavior after administration. These models are a complex array of equations that are solved simultaneously using a computing software, such as the commercially available GastroPlus® or Simcyp®, to identify pharmacokinetic properties (e.g., ADME rates) as an outcome of drug (e.g., solubility), dosage form (e.g., dissolution rate), mode of administration, and species characteristics.

In addition to achieving reproducible bioavailability of a given dosage form, pharmaceutical scientists pay attention to changes in bioavailability through different phases of drug development due to change in animal species (e.g., bioavailability differs among rats, dogs, monkeys, and humans, even with the same dosage form, due to differences in physiology), translation of animal data into humans (e.g., solution administration to animals by IV route vs. oral solid dosage form for human administration), changes in dosage form (e.g., capsules in phase I vs. tablets in phase II), changes in human patient populations (e.g., normal healthy volunteers vs. patients suffering from a chronic disease state such as renal impairment), or other factors of human drug administration (e.g., bioavailability in the fasting state can be different than that in the fed state and in special patient populations, such as pediatric and geriatric patients). Extensive dosage form characterization and *bridging* studies (e.g., relative bioavailability of two different dosage forms) are carried out whenever any significant change is made to the dosage form.

2.3.5 Manufacturability

The ability to reproducibly manufacture drug substance and drug product with predefined acceptable quality attributes in a robust manner at a stage-appropriate scale of manufacture is critical to ensuring that consistent dosage form is used throughout development. For drug products in late-stage development, such as phase III clinical trials, and in preparation for commercialization, in-depth investigations are carried out to understand and carefully control the incoming raw materials, manufacturing process, and the quality of the output drug substance or drug product through well-designed mechanistic and statistically controlled design-of-experiment (DoE) studies.

Critical quality attributes (CQAs) of the drug product are identified. These are the quality attributes that can impact the patient, such as delivered dose uniformity or the content of impurities. Critical material attributes (CMAs) of incoming raw materials, such as excipients, are delineated. These are the physicochemical properties of the raw materials that impact the CQAs of the drug product. In addition, critical process parameters (CPPs) of the manufacturing process, the parameters that impact the CQAs, are identified. A control strategy is then put into place. It identifies how the CPPs and CMAs would be controlled, so that the CQAs would be predictably within the specifications from batch to batch. These represent the quality-by-design (QbD) development paradigm and are communicated to the regulatory agencies in an NDA or a BLA filing.

2.4 REGULATORY APPROVAL

Regulation and control of new drugs in the United States are the responsibilities of the federal FDA (http://www.fda.gov). The FDA regulates the following:

- Initiation of new clinical trials and clinical trials with new drug candidates by requiring the filing and approval of an IND.
- Marketing of a new drug product or an existing drug product for a new application by requiring the filing and approval of an NDA or a BLA.

The agency, a frequently used synonym for the FDA, has various constituent centers, including the Center for Biologics Evaluation and Research

(CBER), the Center for Drug Evaluation and Research (CDER), the Center for Devices and Radiological Health (CDRH), and the Center for Food Safety and Applied Nutrition (CFSAN). The CDER evaluates prescription, generic, and OTC drug products for safety and efficacy before they can be marketed. It also monitors all human drugs and biopharmaceuticals once they are in the market. The CBER regulates biologics not reviewed by the CDER, such as vaccines, blood and blood products, gene therapy products, and cellular and tissue transplants. Many biopharmaceuticals fall under the responsibilities of both the CBER and the CDER. The Office of Regulatory Affairs (ORA) is responsible for monitoring sites and facilities in which pharmaceuticals are manufactured. The FDA has the authority to enforce withdrawal or recall of those drug products from the market that do not meet quality, safety, and efficacy requirements.

A typical process for the discovery and commercialization of new drug products in the United States generally follows the following pathway:

- Preclinical laboratory tests and *in vivo* preclinical studies in animals.
- Submission of an IND application to the FDA for clinical testing.
- Clinical trials for establishing product safety and efficacy.
- Submission of an NDA to the FDA for a BLA.
- Approval of the NDA or BLA by the FDA before any commercial sale.

2.4.1 Investigational new drug application

After completing preclinical testing, the sponsor of a potential NDA/BLA makes the decision about whether or not the drug has enough potential to proceed to *in vivo* studies in humans. To initiate the clinical study, the sponsor needs to file an IND application with the agency. An IND application has several components, including chemistry, manufacturing, and control (CMC) of the test article or drug product, clinical study plan, and investigator's brochure (IB). The IND application presents results of previous experiments; how, where, and by whom the new studies will be conducted; the chemical structure of the compound; its mechanism of action in the body; any toxic effects found in animal studies; and how the compound is manufactured.

The IB is the document that the sponsor of the clinical study provides to the physician and healthcare professionals to successfully execute the clinical study. In addition to gaining the FDA's approval for the IND, the IB must also be reviewed and approved by the institutional review board (IRB) of each clinical site (e.g., a hospital and medical center) where the proposed clinical trials will be conducted. Once an IND application is filed, the FDA has 30 days to respond to this initial admission. The IND application is considered approved if the FDA does not get back to the sponsor within that time.

2.4.2 New drug application

After successful completion of phase I through phase III clinical development, a drug's sponsor submits the result of all the studies to the FDA in an NDA to obtain approval for marketing of the new drug. The NDA is a formal request to the FDA to approve a new drug product for sale and marketing in the United States. Technically, the FDA regulates inter-state transport of medicinal products, which is what it approves. Each state has its own regulatory body for new drug products that can be marketed within its territory. Usually, the regulations of each state reflect that of the U.S. FDA. Therefore, approval of the FDA is considered a benchmark for the ability to market a new drug in all states.

2.4.2.1 Basis of approval

The NDAs are usually comprehensive documents that detail all studies carried out and can run over 100,000 pages in print; however, recent electronic submissions have eliminated the need for printing. The average NDA review time for NMEs approved in 2016 was 10.1 months. The NDA must contain all of the scientific information that the company has gathered. The data gathered during the animal studies and human clinical trials of an IND become the part of the NDA. The goals of the NDA are to provide enough information to permit the FDA reviewers to reach the following key conclusions:

- Whether the drug is safe and effective for its proposed use(s), and whether the benefits of the drug outweigh the risks.
- Whether the drug's proposed labeling is appropriate, and what it should contain.
- Whether the methods used in manufacturing the drug and the controls used to ensure the drug's quality are adequate to preserve the drug's identity, strength, quality, and purity.

2.4.2.2 Role of advisory committees

The FDA often constitutes advisory committees consisting of experts in respective areas in several disciplines, such as clinical practitioners in a specific disease area, to assist in the review of an NDA. Committees are typically asked to comment on whether the approval, clearance, or licensing of a medical product for marketing is supported by adequate data. The primary role of an advisory committee is to provide independent advice that will contribute to the quality of the agency's regulatory decision-making and lend credibility to the drug product review process. In this way, the FDA can make sound decisions about new medical products and other public health issues.

Although advisory committees have a prominent role in the approval of new drug products, they may also be called in earlier in the product development cycle or asked to consider issues relating to products that are already in the market. Committees are typically asked to comment on whether the submitted data adequately support approval, clearance, or licensing of a medical product for marketing. Advisory committees may also recommend that the FDA request additional studies or suggest changes to a product's labeling. Their recommendations are nonbinding advice to the agency. While committee discussions and final votes are very important to the FDA, the final regulatory decision rests with the agency.

2.4.3 Biologics license application

A BLA is an application for marketing authorization of a biologic drug product, such as proteins, antibodies, antibody–drug conjugates, vaccines, and gene and cell therapy products. These products are historically unique not only in their origin but also in the physicochemical characteristics and the extent of characterization contemporarily possible. For example, while the small-molecule drugs are well characterized to an atomic level, with crystal structures of crystalline drugs elucidated, the large-molecule compounds are generally not crystalline and are difficult to isolate in solid state as pure compounds. In their solution state, they are quite big (moluecular weight >10 kDa) and exact characterization of each atom and bond is currently not possible. Thus, while the impurities of small-molecule compounds are known and characterized to exact molecular structure, the structural variants of large-molecule compounds are generally characterized only as size or charge variants.

Accordingly, the criteria for comparability of different drug substance and drug product batches, product scale-up and manufacturing control, scaling of dose across species, and analytical characterization of drug substances and drug product differ significantly between the small- and large-molecule drug products. Although the process of drug development remains the same for both small- and large drug molecules, different experts within the FDA review BLA and NDA.

2.4.4 Abbreviated new drug application

An abbreviated NDA (ANDA) is used to gain approval for a generic equivalent of a drug product that is already approved and is being marketed by the pioneer or original sponsor of the drug. Generic drugs are defined as products containing the same active ingredient as the branded drug, in the same dosage form, and intended for administration by the same route. Generic drug products may have different inactive ingredients and/or product-manufacturing process and controls.

Generic products offer low-cost alternatives to branded medicines once the patent life of the molecule expires. The underlying precept in the approval of generic drug products is that drug products with similar drug pharmacokinetics will have similar efficacy and toxicity profile. These pharmacokinetic parameters include area under the curve (AUC) and maximum plasma concentration (Cmax). Therefore, clinical safety and efficacy testing for generic drug products are waived on the basis of their bioequivalence to the branded or innovator drug product. The CMC requirements for the generic drug products do not change.

2.4.5 Biosimilars

Biosimilars are the generic equivalents of biologic drug products. Often called follow-on biologics, the equivalence requirements for generic biologics are still evolving. The clinical proof of equivalency for biosimilars currently involves abbreviated safety and efficacy studies. In addition, similarity criteria for biosimilar drug products to branded drugs are considered. These may include, for example, a combination of relatively proportion of size and charge variants in the molecule.

2.4.6 Approval and postmarketing surveillance

Once the FDA approves an NDA or a BLA, the drug's sponsor can sell the new medicine to the public in the United States. Approval of the FDA for marketing of a new drug product does not end a sponsor's responsibility toward clinical investigation of the drug. Continued clinical investigation, often called phase IV studies, may contribute to the understanding of the drug's mechanism or scope of action, indicate possible new therapeutic uses, and/or investigate the need for additional dosage strengths, dosage forms, or routes of administration. Phase IV monitoring in commercial use may also reveal additional side effects, especially the rare events that may not be detected even in large-scale clinical trials.

The sponsor is required to submit periodic reports to the FDA, including any adverse event reports, internal quality investigations, and/or changes to manufacturing and controls since the NDA's approval. If an adverse effect is identified with a marketed drug, the Office of Drug Safety (ODS) can take one or more of the following actions: labeling changes, boxed warnings, product withdrawals, and medical and safety alerts.

2.4.7 Accelerated development/review

Accelerated development/review is a specialized mechanism for speeding up the development of drugs that promise significant benefit over the existing therapy for serious or life-threatening diseases for which no therapy exists. This process incorporates several elements, such as abbreviated clinical studies, aimed at accelerating drug development. Safeguards to protect the patients and the integrity of the regulatory process balance regulatory review. The fundamental element of this process is that the manufacturers must continue to test after approval to demonstrate that the drug indeed provides therapeutic benefit to the patient.

2.5 POSTCOMMERCIALIZATION ACTIVITIES

Commercialization of a drug product involves many different functions in addition to the research and development (R&D). Key commercialization functions that intercept with R&D are marketing, finance, and manufacturing. The marketing function identifies the regions and countries where the drug product should be marketed, as well as the country of first approval, the competition in the same therapeutic category, and the desired product characteristics. Finance function provides input regarding cost-of-goods analyses and feasibility of a particular product and technology. The manufacturing function is responsible for day-to-day operations in reproducibly producing the drug product to meet market demand.

The postcommercialization involvement of R&D with a particular drug molecule focuses on three areas: manufacturing support, intellectual property, and life cycle management.

2.5.1 Manufacturing support

As the manufacturing operations produce multiple batches, they may face unforeseen problems that require R&D input and troubleshooting. These problems often come from changes in the input raw material properties (CMAs) or some drift in the process parameters (CPPs) and reflect as an impact on the quality attributes of the drug product (CQAs). Although the manufacturing personnel themselves usually address common problems, the ones that require additional research are usually brought to the interface with R&D. Additional development activities often result in the identification or refinement of current CMAs or CPPs.

The regulatory bodies restrict postapproval changes to the drug substances or drug products, or their manufacturing process. Thus, a sponsor of an NDA or a BLA is required to manufacture the drug product within the confines of what was filed at the time of seeking the approval. If a significant change is required post approval, depending on the nature and extent of change and its impact on the drug product, the sponsor must seek the FDA's approval before implementing this change in commercial operations.

2.5.2 Intellectual property

Intellectual property, in the form of patent rights, forms the cornerstone of ensuring profitability on commercialization of new medicines. This mechanism serves a significant social need of providing sufficient incentive to encourage individuals and organizations to invest in developing new medicines for unmet medical needs.

A new compound, when entering the development pipeline, is patented by the innovator company. A typical patent life consists of discovery and development of the compound through various stages of clinical trials. On commercialization, the patent provides exclusive marketing rights to the sponsor and allows the sponsor to recoup the cost of developing the drug. The commercial return on investment increases with time as the drug gains increasing acceptance. On expiration of the patent, the profits of the innovator company decline significantly due to generic competition and price reduction. Thus, an innovator company is under constant pressure to continue the discovery and development of new medicines to replace the ones that would predictably expire at the end of their patent life.

In addition to pursuing new compounds, innovator companies also seek to extend patent life by filing additional patents that would provide significant value addition to the molecule. These fall in the category of life cycle management.

2.5.3 Life cycle management

Typical life cycle of an NCE/NME is several decades, with different stages that can generally be characterized as follows:

- An NCE/NME is first conceptualized and synthesized during discovery. A patent that assigns the commercial rights of this compound to the sponsor is filed.
- The compound is nurtured through different stages of clinical trials and commercialization, during which process it evolves into a drug product.
- The drug product is commercialized, it gains widespread acceptance and use, and the sponsor is able to recoup the investment in the compound through exclusive commercialization for the duration of the patent life.
- On expiration of the patent, several generic competitors enter the market, which significantly reduces the profitability of the sponsor in this compound. Market viability and lifetime of the compound then depend on market conditions such as the disease state, other molecules or therapies available, and patients' need.

The R&D efforts postcommercialization of an NCE or NME are targeted on improving the value proposition from the compound, with a goal of increasing patent life. These could include, for example:

- Improving some aspect of the drug product that increases the value proposition, such as higher bioavailability or reduction in variability in drug absorption. For example, an improvement in the drug product, such as micronization of the drug, that overcomes the effect of food or gastric pH on oral absorption.

- Different route of administration. For example, conversion of a previously IV route of administration to subcutaneous route can lead to significant patient benefit, in terms of convenience, and the ability to self-administer the drug in an outpatient setting.
- Combination of drug product with another compound that leads to synergism in therapy. For example, amoxicillin is often coadministered with clavulanic acid, saxagliptin is coadministered with metformin, and several antihypertensive drugs are coadministered in a single-dosage unit.
- Introduction of drug–device combination products, such as insulin pens or pumps.
- Increasing label claim to include coverage of additional disease states. For example, an oncology drug approved for a given indication can increase its market by seeking approval for additional indications if clinical studies are done to prove its efficacy.

REVIEW QUESTIONS

2.1 Which of the following is true for the drug development and regulatory process?
   A. A drug's sponsor must submit an IND application before an FIH trial of a drug
   B. An IND application must precede an NDA submission to the FDA
   C. An NDA approval must precede a corresponding ANDA submission
   D. All of the above
   E. None of the above

2.2 Indicate which of the following statements is TRUE and which is FALSE.
   A. The FDA can approve new formulations without phase III clinical trials.
   B. In phase III clinical trials, only a small number of patients are enrolled.
   C. New drug substances are extracted from plants or animals or synthesized in laboratories.
   D. The CDER is responsible for the approval of vaccines.
   E. The ANDA requires full clinical and nonclinical testing.
   F. The ANDA can be filed for biological products.
   G. The BLA is approved by the CBER, whereas the NDA is approved by the CDER.

2.3 A. Define the following terminologies: FDA, IND, NDA, CDER, FIH, CBER, and BLA.
   B. List the different steps involved in the drug development and approval process.
   C. In which phase of drug development are healthy subjects evaluated?

2.4 A. What are the specific responsibilities of the CDER and the CBER?
   B. What information does the FDA require in an IND application?

C. What are the goals of phase I, II, and III trials?

D. Why is the postmarketing surveillance necessary?

2.5 What are the three key components of pharmaceutical development?

A. Bioavailability: To ensure that the drug has reproducible and clinically desired bioavailability from a dosage form.

B. Stability: To ensure that the drug product is stable at the labeled storage conditions for the duration of its assigned shelf life.

C. Manufacturability: To ensure that the drug product can be manufactured reproducibly and robustly at a commercial scale.

D. Tolerability: To ensure that the drug is tolerated by the subjects and there are no significant adverse effects.

E. All of the above

2.6 Match the stage of drug development in the left column with the key deliverables of that stage in the right column. Write the letter of row in the left column in front of the corresponding row in the right column.

	Stage of drug development		Key deliverables of this stage
1	Preclinical	A	Prove safety and efficacy of the new drug product in a wide patient population (number of subjects usually in 1000s)
2	Clinical: phase I	B	Conversion of a drug substance to a drug product with the assurance of stability, bioavailability, and manufacturability at scales needed for the given stage of product development
3	Clinical: phase II	C	Identification of potential ways to enhance the value of an existing commercial drug product, such as by coming up with new dosage form, route of administration, combination drug product, indication, or intellectual property
4	Clinical: phase III	D	First-in-human safety studies in a limited number of subjects
5	Pharmaceutical development	E	Efficacy and safety evaluation in a larger patient population (number of subjects usually in 100s)
6	Life cycle management	F	Physicochemical characterization along with safety and efficacy studies in animal species in preparation for the first-in-human dosing of a drug candidate

2.7 Interspecies dose scaling for small-molecule compound is generally carried out using which metric?

A. Body weight

B. Body surface area

C. Body fluid volume

D. Muscle weight

E. Fat tissue weight

2.8 Clinical studies carried out in which phase are also called first-in-human studies?
   A. Phase I
   B. Phase II
   C. Phase III
   D. Phase IV
2.9 Which of the following may not be a typical postcommercialization activity?
   A. Investigation of a new drug–drug combination product
   B. Investigation of a new drug–device combination product
   C. First phase III clinical trial to support commercialization of an NCE
   D. Investigation of an approved drug's ability to treat a new disease indication

FURTHER READING

Allen LV, Popovich NG, and Ansel HC *Ansel's Pharmaceutical Dosage Forms and Drug Delivery Systems*, 8th ed., New York: Lippincott Williams & Wilkins, 2005.

Bashaw ED (2004) Drug and dosage form development: Regulatory perspectives. In Ghosh TK and Jasti BR (Eds.) *Theory and Practice of Contemporary Pharmaceutics*, Boca Raton, FL: CRC Press, pp. 257–275.

http://www.fda.gov/ (last accessed August 15, 2016).

Narang AS and Desai DD, Anticancer drug development: Unique aspects of pharmaceutical development. In Mahato, RI and Lu Y (Eds.) *Pharmaceutical Perspectives of Cancer Therapeutics*, New York: AAPS-Springer Publishing Program, 2009.

Pandit NK (2007) *Introduction to the Pharmaceutical Sciences*, Philladelphia, PA: Lippincott Williams & Wilkins.

Welling PG, Lasagna L, and Banakar UV (Eds.) *The Drug Development Process*, New York: Marcel Dekker, 1996.

Chapter 3

Pharmaceutical considerations

LEARNING OBJECTIVES

On completion of this chapter, the students should be able to

1. Describe the pH-partition theory as it applies to drug absorption.
2. Describe factors influencing pharmaceutical dosage forms.
3. Describe the Henderson–Hasselbalch equation.
4. Describe the pH-partition theory and its limitation.

3.1 INTRODUCTION

Drug substances are seldom administered alone; rather, they are given as part of a formulation, in combination with one or more nonmedical agents (known as *pharmaceutical ingredients* or *excipients*) that serve varied and specialized pharmaceutical functions. Commonly used pharmaceutical ingredients are listed in Table 3.1. Pharmaceutical ingredients solubilize, suspend, thicken, dilute, emulsify, stabilize, preserve, color, flavor, and fashion medicinal agents into efficacious and appealing dosage forms.

Drug absorption depends on its lipid solubility, formulation, and the route of administration. The proper design and formulation of a dosage form require a thorough understanding of the physical, chemical, and biologic characteristics of the drug substances as well as that of the pharmaceutical ingredients to be used in fabricating the product. The drug and pharmaceutical ingredients must be compatible with one another to produce a drug product that is stable, efficacious, attractive, easy to administer, and safe.

Table 3.1 List of pharmaceutical ingredients

Ingredients	Definition	Examples
Antifungal preservatives	Used in liquid and semisolid formulations to prevent growth of fungi	Benzoic acid, butylparaben, ethylparaben, sodium benzoate, and sodium propionate
Antimicrobial preservatives	Used in liquid and semisolid formulations to prevent growth of microorganisms	Benzalkonium chloride, benzyl alcohol, cetylpyridinium chloride, phenylethyl alcohol
Antioxidant	Used to prevent oxidation	Ascorbic acid, ascorbyl palmitate, sodium ascorbate, sodium bisulfite, sodium metabisulfite
Emulsifying agent	Used to promote and maintain dispersion of finely divided particles of a liquid in a vehicle in which it is immiscible	Acacia, cetyl alcohol, glyceryl monostearate, sorbitan monostearate
Surfactant	Used to reduce surface or interfacial tension	Polysorbate 80, sodium lauryl sulfate, sorbitan monopalmitate
Plasticizer	Used to enhance coat spread over tablets, beads, and granules	Glycerin, diethyl palmitate
Suspending agent	Used to reduce sedimentation rate of drug particles dispersed throughout a vehicle in which they are not soluble	Carbopol, hydroxymethyl cellulose, hydroxypropyl cellulose, methylcellulose, tragacanth

3.2 ADVANTAGES OF PHARMACEUTICAL DOSAGE FORMS

A pharmaceutical dosage form is the entity that is administered to patients so that they receive an effective dose of a drug. Some common examples are tablets, capsules, suppositories, injections, suspensions, and transdermal patches. Besides providing the mechanism for the safe and convenient delivery of accurate dosage, pharmaceutical dosage forms are needed for the following additional reasons:

- To protect the drug substance from the destructive influence of atmospheric oxygen or humidity (*coated tablets*).
- To protect the drug substance from the destructive influence of gastric acid after oral administration (*enteric-coated tablets*).
- To conceal the bitter, salty, or offensive taste or odor of a drug substance (*capsules, coated tablets,* and *flavored syrup*).
- To provide liquid preparations of substances that are either insoluble or unstable in the desired vehicle (*suspensions*).

- To provide rate-controlled drug action (*various controlled-release tables, capsules,* and *suspensions*).
- To provide site-specific and local drug delivery (e.g., rectal and vaginal suppositories).
- To target the drug at the desired site of action (e.g., nanoparticulate systems, liposomes, etc.).

3.3 INFLUENTIAL FACTORS IN DOSAGE FORM DESIGN

Each drug substance has intrinsic chemical and physical characteristics that must be considered before the development of its pharmaceutical formulation. Among these characteristics are the particle size, surface area, the drug's solubility, pH, partition coefficient, dissolution rate, physical form, and stability. All these factors are discussed below, except the particle size and dissolution rate, which will be discussed in the next chapter.

3.3.1 Molecular size and volume

Molecular size and volume have important implications for drug absorption. Tight junctions can block the passage of even relatively small molecules, whereas gap junctions are looser. Molecules up to 1,200 Da can pass freely between cells; however, larger molecules cannot pass through gap junctions. Drug diffusion in simple liquid is expressed by Stokes–Einstein equation:

$$D = \frac{RT}{6\pi\eta r}$$

where:
D is the diffusion of drugs
R is the gas constant = 8.313 $JK^{-1}mol^{-1}$
T is the temperature (Kelvin)
η is the solvent viscosity
r is the solvated radius of diffusing solute

Since volume $(V) = (4/3)\ \pi r^3$, the above equation suggests that drug diffusivity is inversely proportional to the molecular volume. Molecular volume is dependent on molecular weight, conformation, and heteroatom content. Molecules with a compact conformation will have a lower molecular volume and thus a higher diffusivity. As shown in Figure 3.1, the diffusion and permeability of the endothelial monolayer to molecules decreased with increasing molecular weight.

A drug must diffuse through a variety of biological membranes after administration into the body. In addition, drugs in many controlled-release

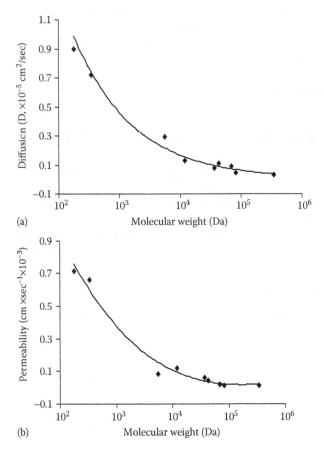

Figure 3.1 Diffusion (a) and permeability (b) of compounds with different molecular weight across an endothelial monolayer at 37°C.

systems must diffuse through a rate-controlling membrane or matrix. The ability of a drug to diffuse through membranes is a function of its molecular size and volume. For drugs with a molecular weight greater than 500, diffusion in many polymeric matrices is very small. Lipinski devised the so-called *Rule of 5*, which refers to drug-like properties of molecules. It states that poor oral absorption is more likely when the drug molecule has:

- More than five hydrogen-bond donors (–OH groups or –NH groups).
- A molecular weight greater than 800.
- A log $P > 5$.
- More than 10 H-bond acceptors.

However, this rule is not applicable to the compounds that are substrates for transporters.

3.3.2 Drug solubility and pH

Pharmacological activity is dependent on solubilization of a drug substance in physiological fluids. Therefore, a drug substance must possess some aqueous solubility for systemic absorption and therapeutic response. Enhanced aqueous solubility may be achieved by forming salts or esters, by chemical complexation, by reducing the drug's particle size (i.e., micronization), or by creating an amorphous solid. One of the most important factors in the formulation process is pH, as it affects solubility and stability of weakly acidic or basic compounds. Changes in pH can lead to ionization or salt formation. Adjustment in pH is often used to increase the solubility of ionizable drugs, because the ionized molecular species have higher water solubility than their neutral species. According to Equations 3.1 and 3.2, the total solubility, S_T, is the function of intrinsic solubility, S_0, and the difference between the molecule's pK_a and the solution pH. The intrinsic solubility is the solubility of the neutral species. Weak acids can be solubilized at pHs below their acidic pK_a, whereas weak bases can be solubilized at pHs above their basic pK_a. For every pH unit away from the pK_a, the weak acid–base solubility increases 10-fold. Thus, solubility can be achieved as long as the formulation pH is at least 3 units away from the pK_a. Adjusting solution pH is the simplest and most common method to increase water solubility in injectable products.

$$\text{For a weak acid } S_T = S_0(1 + 10^{pH-pK_a}) \tag{3.1}$$

$$\text{For a weak base } S_T = S_0(1 + 10^{pK_a-pH}) \tag{3.2}$$

Unlike a weak acid or base, the solubility of a strong acid or base is less affected by pH. The drugs without ionizable groups are often solubilized by the combination of an aqueous solution and water-soluble organic solvent/surfactant. Frequently, a solute is more soluble in a mixture of solvents than in one solvent alone. This phenomenon is known as *cosolvency*, and the solvents that in combination increase the solubility of the solute are called *cosolvents*. The addition of a cosolvent can increase the solubility of hydrophobic molecules by reducing the *dielectric constant*, which is a measure of the influence by a medium on the energy needed to separate two oppositely charged bodies. Some of the cosolvents commonly being used in pharmaceutical formulations include ethyl alcohol, glycerin, sorbitol, propylene glycol, and polyethylene glycols (PEGs). Polyethylene glycol 300 or 400, propylene glycol, glycerin, dimethylacetamide (DMA), N-methyl 2-pyrrolidone (NMP), dimethyl sulfoxide (DMSO), Cremophor, and polysorbate 80 are often used for solubilization of drugs that have no ionizable groups. As shown in Figure 3.2, the solubility of phenobarbital is, for example, significantly increased in a mixture of water, alcohol, and glycerin compared with

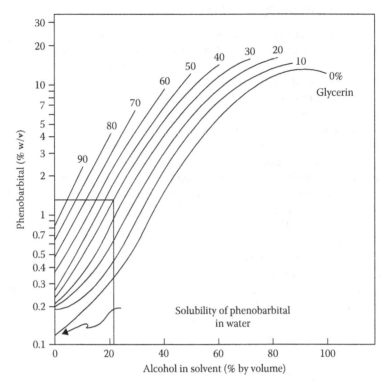

Figure 3.2 Effect of cosolvents on the solubility of phenobarbital in a mixture of water, alcohol, and glycerin at 25°C. (Reproduced from Krause, G.M. and Gross, J.M., *J. Am. Pharm. Assoc. Ed.*, 40, 137, 1951. With permission.)

one of these solvents alone. However, the use of cosolvents often leads to the precipitation of the drug on dilution during the administration of the drug solution into the body, resulting in pain or tissue damage.

Excipients that solubilize a molecule via specific interactions, such as complexation with a drug molecule in a noncovalent manner, lower the chemical potential of the molecules in solution. These noncovalent solubility-enhancing interactions are the basis of the phenomenon that *like dissolves like* and include van der Waals forces, hydrogen bonding, dipole–dipole, ion–dipole interactions, and, in certain cases, favorable electromagnetic interactions. Solutes dissolve better in solvents of similar polarity. Therefore, to dissolve a highly polar or ionic compound, one should use a solvent that is highly polar or has a high dielectric constant. On the contrary, to dissolve a drug that is nonpolar, one should use a solvent that is relatively nonpolar or has a low dielectric constant.

Table 3.2 Water solubility of different substituent groups

Hydrophobic substituent groups

$-CH_3$

$-CH_2^-$

$-Cl, -Br$

$-N(CH_3)_2$

$-SCH_3$

$-OCH_2CH_3$

Hydrophilic substituent groups

$-OCH_3$

$-NO_2$

$-CHO$

$-COOH$

$-COO-$

$-NH_2$

$-NH_3^+$

$-OH$

Drug solubility can also be enhanced by altering its structure; this is one basis for the use of *prodrugs*. A prodrug is a drug that is therapeutically inactive when administered but becomes activated in the body by chemical or enzymatic processing. The addition of polar groups, such as carboxylic acids, ketones, and amines, can increase aqueous solubility by increasing the hydrogen bonding and the dipole–dipole interaction between the drug molecule and the water molecules. Table 3.2 lists different substituents that will have significant influence on the water solubility of drugs. Substituents can be classified as either hydrophobic or hydrophilic, depending on their polarity. The position of the substituents on the molecule can also influence its effect.

3.3.3 Lipophilicity and partition coefficient

Partitioning is the ability of a compound to distribute in two immiscible liquids. When a weak acid or base drug is added to two immiscible liquids, some drug goes to the nonpolar phase and some drug goes to the aqueous layer. Because *like dissolves like*, the nonpolar species migrates (partitions) to the nonpolar layer and the polar species migrate to the polar aqueous layer.

To produce a pharmacologic response, a drug molecule must first cross a biologic membrane, which acts as a lipophilic barrier to many drugs. Since passive diffusion is the predominant mechanism by which many drugs are transported, the lipophilic nature of the molecules is important. A drug's partition coefficient is a measure of its distribution in a lipophilic–hydrophilic phase system, and it indicates the drug's ability to penetrate biologic multiphase systems. The octanol–water partition coefficient is commonly used in formulation development and is defined as:

$$P = \frac{\text{Concentration of drug in octanol or nonpolar phase}}{\text{Concentration of drug in water or polar phase}}$$

For an ionizable drug, the following equation is applicable:

$$P = \frac{\text{Concentration of drug in octanol or nonpolar phase}}{(1-\alpha)(\text{Concentration of drug in water of polar phase})}$$

In this equation, α is equal to the degree of ionization. The concentration in aqueous phase is estimated by an analytical assay, and concentration in octanol or other organic phases is deduced by subtracting the aqueous amount from the total amount placed in the solvents. Partition coefficient can be used for drug extraction from plants or biologic fluids, drug absorption from dosage forms, and recovery of antibiotics from fermentation broth.

The logarithm of partition coefficient (P) is known as *log P*. Log P is a measure of lipophilicity and is used widely, since many pharmaceutical and biological events depend on lipophilic characteristics. Often, the log P of a compound is quoted. Table 3.3 lists the log P values of some representative compounds. For a given drug:

If log $P = 0$, there is equal drug distribution in both phases.
If log $P > 0$, the drug is lipid soluble.
If log $P < 0$, the drug is water soluble.

In general, the higher the log P, the higher the affinity for lipid membranes and thus the more rapidly the drug passes through the membrane via passive diffusion. However, there is a parabolic relationship between log P and drug activity when percentage of drug absorption is plotted against log P values (Figure 3.3). The parabolic nature of bioactivity and log P values is due to the fact that drugs with high log P values, protein binding, low solubility, and binding to extraneous sites cause them to have a

Table 3.3 Log *P* values of representative drugs

Drug	Log P
Acetylsalicylic acid	1.19
Amiodarone	6.7
Benzocaine	1.89
Bromocriptine	6.6
Bupivacaine	3.4
Caffeine	0.01
Chlorpromazine	5.3
Cortisone	1.47
Desipramine	4.0
Glutethimide	1.9
Haloperidol	1.53
Hydrocortisone	4.3
Indomethacin	3.1
Lidocaine	2.26
Methadone	3.9
Misoprostol	2.9
Ondansetron	3.2
Pergolide	3.8
Phenytoin	2.5
Physostigmine	2.2
Prednisone	1.46
Sulfadimethoxine	1.56
Sulfadiazine	0.12
Sulfathiazole	0.35
Tetracaine	3.56
Thiopentone	2.8
Xamoterol	0.5
Zimeldine	2.7

lower bioactivity. Decrease in activity is due to the limitation in solubility beyond a certain log *P* value. If a drug is too lipophilic, it will remain in the lipid membrane and not partition out again into the underlying aqueous environment. On the other hand, very polar compounds with very high log *P* values are not sufficiently lipophilic to be able to pass through lipid membrane barriers.

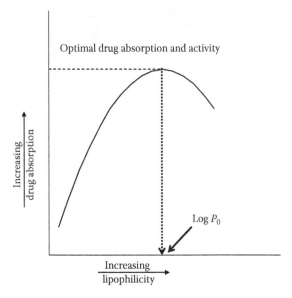

Figure 3.3 Relationship between drug absorption and log *P*. Decrease in the drug absorption beyond a certain log *P* value is probably due to its binding to plasma proteins, reduction in free drug levels, or its binding to extraneous sites.

3.3.4 Polymorphism

The capacity of a substance to exist in more than one solid state forms is known as *polymorphism*, and the different crystalline forms are called *polymorphs*. If the change from one polymorph to another is reversible, the process is *enantiotropic*. However, if the transition from a metastable to a stable polymorph is unidirectional, the system is *monotropic*. Polymorphic forms may exhibit detectable differences in some or all of the following properties: melting point, dissolution rate, solubility, and stability. Drug substances can be amorphous (i.e., without regular molecular lattice arrangements), crystalline (which are more oriented or aligned), polymorphic, anhydrous, or solvated. An important form in the formulation is the crystal or amorphous form of the drug substance. Many drug substances can exist in more than one crystalline form, with different lattice arrangements. This property is termed polymorphism. Drugs may undergo a change from one metastable polymorphic form to a more stable polymorphic form. Various drugs are known to exist in different polymorphic forms (e.g., cortisone and prednisolone). Polymorphic forms usually exhibit different physicochemical properties, including melting point and solubility, which can affect the dissolution rate and thus the extent of their absorption. The amorphous form of a compound is always more soluble than the corresponding crystal form. Changes in crystal characteristics can influence bioavailability and stability and thus can have important implications for dosage form design. For example, insulin exhibits a differing degree of activity, depending on its state. The amorphous form of insulin is rapidly absorbed and has short duration of action, whereas the large crystalline product is more slowly absorbed and has a longer duration of action.

3.3.5 Stability

The chemical and physical stability of a drug substance alone and when combined with formulation components is critical to preparing a successful pharmaceutical product. Drugs containing one of the following functional groups are liable to undergo hydrolytic degradation: ester, amide, lactose, lactam, imide, or carbamate. Drugs that contain ester linkages include acetylsalicylic acid, physostigmine, methyldopa, tetracaine, and procaine. For example, the hydrolysis of acetylsalicylic acid (commercially known as aspirin) is represented in Figure 3.4. Aspirin is hydrolyzed to salicylic acid and acetic acid.

Nitrazepam, chlordiazepoxide, penicillins, and cephalosporins are also susceptible to hydrolysis. Several methods are available to stabilize drug solutions that are susceptible to hydrolysis. For example, protection against moisture in formulation, processing, and packaging may prevent decomposition. Suspending drugs in nonaqueous solvents such as alcohol, glycerin, or propylene glycol may also reduce hydrolysis.

After hydrolysis, *oxidation* is the next most common pathway for drug degradation. Drugs that undergo oxidative degradation include morphine, dopamine, adrenaline, steroids, antibiotics, and vitamins. Oxidation can be minimized by storage under anaerobic conditions. Since it is very difficult to remove all of the oxygen from a container, antioxidants are often added to formulations to prevent oxidation.

Excipients used to prepare a solid dosage can also affect the drug's stability, possibly by increasing the moisture content of the preparation. Excipients, such as starch and povidone, have very high water contents.

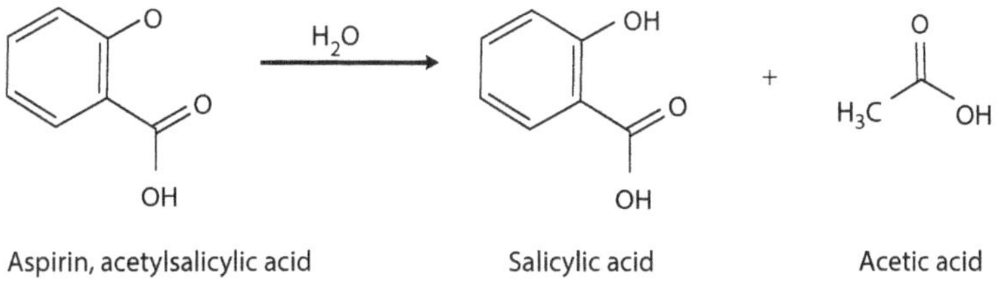

Aspirin, acetylsalicylic acid

Salicylic acid

Acetic acid

Figure 3.4 Hydrolysis of aspirin. Acetylsalicylic acid (aspirin) is hydrolyzed to salicylic acid and acetic acid.

Povidone contains about 28% equilibrium moisture at 75% relative humidity. However, the effect of this high moisture content on the stability of a drug will depend on how strongly it is bound and whether the moisture can come in contact with the drug. Effects of tablet excipients on drug decompositions are widely reported in the literature. For an example, magnesium trisilicate is known to cause increased hydrolysis of aspirin in the tablet because of its high moisture content.

3.3.6 pK_a/Dissociation constants

Many drug substances are either weak acids or weak bases and thus undergo a phenomenon known as dissociation when dissolved in liquid medium. If this dissociation involves a separation of charges, then there is a change in the electrical charge distribution on the species and a separation into two or more charged particles, or ionization. The extent of ionization of a drug has an important effect on the formulation and pharmacokinetic profiles of the drug. The extent of dissociation or ionization is dependent on the pH of the medium containing the drug. Table 3.4 lists the normal pHs of some organs and body fluids, which are used in the prediction of the percentage ionization of drugs *in vivo*. In a formulation, often, the vehicle is adjusted to a certain pH to obtain a certain level of ionization of the drug for solubility and stability. The extent of ionization of a drug has a strong effect on its extent of absorption, distribution, and elimination.

Acids tend to donate protons to a system at pH greater than 7, and bases tend to accept protons when added to acidic system (i.e., at pH < 7). Many drugs are weak acids or bases and therefore exist in both unionized and

Table 3.4 Nominal pH values of some body fluids and sites

Sites	Nominal pH
Aqueous humor	7.21
Blood	7.40
Cerebrospinal fluid	7.35
Duodenum	7.35
Ileum	8.00
Colon	5.5–7
Lacrimal fluid (tears)	7.4
Saliva	6.4
Semen	7.2
Stomach	1–3
Urine	5.7–5.8
Vaginal secretions, pre-menopause	4.5
Vaginal secretions, postmenopause	7.0

ionized forms; the ratio of these two forms vary with pH. The fraction of a drug that is ionized in solution is given by the dissociation constant (K_a) of the drug. Such dissociation constants are conveniently expressed in terms of pK_a values for both acidic and basic drugs. For a weak acidic drug, HA (e.g., aspirin and phenylbutazone), the equilibrium is presented by:

$$HA \leftrightarrow H^+ + A^-$$

The symbol ($\leftrightarrow$) indicates that equilibrium exists between the free acid and its conjugate base. According to the *Bronsted–Lowry theory* of acids and bases, an acid is a substance that will donate a proton and a base is a substance that will accept a proton. Based on this theory, the conjugate base A^- may accept a proton and revert to the free acid. Therefore, the dissociation constant for this reaction is:

$$K_1[HA] \leftrightarrow K_2[H^+][A^-]$$

$$Or\, K_a = \frac{K_1}{K_2} = \frac{[H^+][A^-]}{[HA]}$$

Taking logarithms of both sides:

$$\log K_a = \log[H^+] + \log[A^-] - \log[HA]$$

The signs in this equation may be reversed to give the following equation:

$$-\log K_a = -\log[H^+] - \log[A^-] + \log [HA]$$

$$\therefore pK_a = \frac{pH + \log[HA]}{[A^-]}$$

This is a general equation applicable for any weakly acidic drugs.

Similarly, for a weak basic drug (e.g., chlorpromazine) $\rightarrow B + H^+ = BH^+$
$\therefore pK_a = pH + \log[BH^+]/[B]$ for a weakly basic drug.

These equations are known as *Henderson–Hasselbalch equations*. This equation describes the derivation of pH as a measure of acidity (using pK_a) in biologic and chemical systems. The equation is also useful for estimating the pH of a buffer solution and finding the equilibrium pH in acid–base reactions. Bracketed quantities such as [*Base*] and [*Acid*] denote the molar concentration of the quantity enclosed. Based on these equations, it is apparent that the pK_a is equal to the pH when the concentration of the ionized and nonionized species is equal (i.e., log1 = 0). It is therefore important to realize that a compound is only 50% ionized when the pK_a is equal to the pH. Ionization constants are usually expressed in terms of pK_a values for both acidic and basic drugs. The strength of acid is inversely related to the magnitude of its pK_a. The lower is the pK_a, the stronger is the acid. Conversely, the strength of a base is directly related to the magnitude of its pK_a. The pK_a of a strong base is high. The pK_a values of a series of drugs are listed in Table 3.5. Acidic drugs are completely unionized at pHs up to 2 units below their pK_a and are completely ionized at pHs greater than 2 units above their pK_a. Conversely, basic drugs are completely ionized at pH up to 2 units below their pK_a and are completely unionized when the pH is greater than 2 units above their pK_a. Both types of drugs are exactly 50% ionized at their pK_a values. Some drugs can donate or accept more than one proton, and so, they may have several pK_a values.

For either weak acid or base, the ionized species, BH^+ and A^-, have very low solubility and are virtually unable to permeate membrane, except where specific transport mechanisms exist. The lipid solubility of the uncharged drugs will depend on the physicochemical properties of the drug.

Proteins and peptides contain both acidic (–COOH) and basic (–NH$_2$) groups. The pK_a values of ionizable groups in proteins and peptides can be significantly different from those of the corresponding groups when they

Table 3.5 pK_a values of typical acidic and basic drugs

Drugs	pK_a
Acidic drugs	
Acetylsalicylic acid	3.5
Barbital	7.9
Phenobarbital	7.4
Penicillin G	2.8
Phenytoin	8.3
Theophylline	8.6
Tolbutamide	5.3
Basic drugs	
Amphetamine	9.8
Atropine	9.7
Quinine	4.2, 8.8
Codeine	7.9
Morphine	7.9
Procaine	9.0
Verapamil	8.8

Source: Martindale, W. and Reynolds, J.E.F., *Martindale: The Extra Pharmacopeia*, 30th ed., The Pharmaceutical Press, London, UK, 1993.

are isolated in solution. Therefore, these compounds are often referred to as amphoteric in nature. The pH of a solution determines the net charge on the molecule and ultimately the solubility. Since water is a polar solvent and ionic species are more water soluble than the nonionic ones, a conjugate acid (BH+) and a conjugate base (A-) are generally more water soluble than the corresponding free base (B) or free acid (HA).

Example 3.1

The pK_a value of aspirin, which is a weak acid, is about 3.5. What are the ratios of unionized and ionized forms of this drug in the stomach (pH 2) and in the plasma (pH 7.4)? Why does aspirin often cause gastric bleeding?

Answer:

According to the Henderson–Hasselbalch equation,

$$pK_a = pH + \log \frac{[HA]}{[A^-]}$$

$$\log \frac{C_u}{C_i} = pK_a - pH = 3.5 - 2.0 = 1.5$$

where C_u is the concentration of unionized drug and C_i is the concentration of ionized drug.

$$\therefore \frac{C_u}{C_i} = \text{antilog} \, 1.5 = 31.62{:}1$$

In the plasma,

$$\log \frac{C_u}{C_i} = pK_a - pH = 3.5 - 7.4 = -3.9$$

$$\therefore \frac{C_u}{C_i} = \text{antilog}(-3.9) = 1.259 \times 10^{-4} : 1$$

Therefore, most of the administered aspirin remains unionized in the stomach, and thus, it is rapidly taken up by the stomach, leading to gastric bleeding.

3.3.7 Degree of ionization and pH-partition theory

For a drug to cross a membrane barrier, it must normally be lipid soluble to get into the biological membranes. The ionized forms of acidic and basic drugs have low lipid:water partition coefficients compared with the coefficients of the corresponding unionized molecules. Lipid membranes are preferentially permeable to the latter species. Thus, an increase in the fraction of a drug that is unionized will increase the rate of drug transport across the lipid membrane. This phenomenon can be explained by the *pH-partition theory*, which states that drugs are absorbed from biological membranes by passive diffusion, depending on the fraction of the unionized form of the drug at the pH of that biological membrane. Based on the *Henderson–Hasselbalch equation*, the degree of ionization of a drug will depend on both its pK_a value and the solution's pH.

The gastrointestinal (GI) tract acts as a lipophilic barrier, and thus, ionized drugs, which will be more hydrophilic, will have minimal membrane transport compared with the unionized form of the drug. The solution pH will affect the overall partition coefficient of an ionizable substance. The *pI* of the molecule is the pH at which there is a 50:50 mixture of conjugate acid–base forms. The conjugate acid form will predominate at a pH lower than the pK_a, and the conjugate base form will be present at a pH higher than the pK_a.

3.3.7.1 Limitations of pH-partition theory

Although the pH-partition theory is useful, it often does not hold true. For example, most weak acids are well absorbed from the small intestine, which is contrary to the prediction of the pH-partition hypothesis. Similarly, quaternary ammonium compounds are ionized at all pHs but are readily absorbed from the GI tract. These discrepancies arise because *pH-partition theory does not take into account the following*:

- The small intestine has a large epithelial surface area for drug absorption to take place. This large epithelial area results from mucosa, villi, and microvilli (Figure 3.5). The large mucosal surface area compensates for ionization effects.
- Drugs have a relatively long residence time in the small intestine, which also compensates for ionization effects.

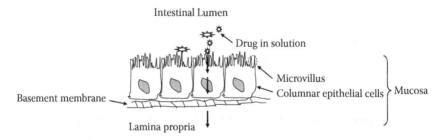

Figure 3.5 Drug absorption across small intestine. The small intestine has a large epithelial surface area due to mucosa, villi, and microvilli. This large surface area compensates the effect of drug ionization on its absorption across the small intestine and invalidates pH-partition theory of drug absorption.

- Charged drugs, such as quaternary ammonium compounds and tetracyclines, may interact with oppositely charged organic ions, resulting in a neutral species, which is absorbable.
- Some drugs are absorbed via active pathways.
- Many more.

REVIEW QUESTIONS

3.1 Which of the following statements is FALSE?
   A. The partition coefficient is the ratio of drug solubility in n-octanol to that in water.
   B. Absorption of a weak electrolyte drug does not depend on the extent to which the drug exists in its unionized form at the absorption site.
   C. Amorphous forms of drug have faster dissolution rates than crystalline forms.
   D. All of the above.
3.2 The pH of a buffer system can be calculated with:
   A. Henderson–Hasselbalch equation
   B. Noyes–Whitney equation
   C. Michaelis–Menten equation
   D. Yang's equation
   E. All of the above
3.3 Indicate which of the following statement are TRUE and which are FALSE:
   A. Henderson–Hasselbalch equation describes the effect of physical parameters on the stability of pharmaceutical suspensions.
   B. The passive diffusion rate of hydrophobic drugs across biological membranes is higher than that of hydrophilic compounds.
   C. Factors influencing dosage form design do not include drug solubility and pH but include partition coefficient and pK_a values.
   D. Drug solubility can be enhanced by salt formation, use of cosolvent, complex formation, and micronization.
3.4 A. What is the difference between drug adsorption and drug absorption?
   B. Describe the pH-partition theory and its limitation in relation to drug absorption across the GI tract.
   C. Compare any two compounds differing in the following characteristics and suggest which one would be absorbed more efficiently and why:
      i. A water-insoluble compound versus a highly soluble compound.
      ii. A low molecular weight compound versus a high molecular weight compound.

3.5 A. Why do we need to formulate a drug into a pharmaceutical dosage form?

B. Define partition coefficient and log P.

C. Define electrolytes and nonelectrolytes.

3.6 A. Enlist eight intrinsic characteristics of a drug substance that must be considered before the development of its pharmaceutical formulation.

B. Enlist two limitations of pH-partition theory.

3.7 Define pH-partition theory. The pK_a value of aspirin, which is a weak acid, is about 3.5. What are the ratios of ionized and unionized forms of the drug in the stomach (pH 2) and in the plasma (pH 7.4)? Why does aspirin often cause gastric bleeding?

3.8 Enlist six physicochemical properties of a drug that influence absorption. How can the physicochemical properties be improved to increase drug absorption?

3.9 The pK_a of pilocarpine is 7.15 at 25°C. Compute the mole percentage of free base present at 25°C and a pH of 7.4.

3.10 Calculate the percentage of cocaine existing as the free base in a solution of cocaine hydrochloride at pH 4.5 and pH 8.0. The pK_a of cocaine is 5.6.

3.11 For a weak acid with a pK_a of 6.0, calculate the ratio of acid to salt at pH 5.

FURTHER READING

Allen LV, Popovich NG, and Ansel HC (Eds.) (2005) *Ansel's Pharmaceutical Dosage Forms and Drug Delivery Systems*, 8th ed., New York: Lippincott Williams & Wilkins.

Aulton ME (Ed.) (1988) *Pharmaceutics: The Science of Dosage Form Design*, New York: Churchill Livingstone.

Banker GS and Rhodes CT (Eds.) (2002) *Modern Pharmaceutics*, 4th ed., New York: Marcel Dekker.

Block LH Collins CC (2001) Biopharmaceutics and drug delivery systems. In Shargel L, Mutnick AH, Souney PH and Swanson LN (Eds.) *Comprehensive Pharmacy Review*, New York: Lippincott Williams & Wilkins, pp. 78–91.

Block LH and Yu ABC (2001) Pharmaceutical principles and drug dosage forms. In Shargel L, Mutnick AH, Souney PH and Swanson LN (Eds.) *Comprehensive Pharmacy Review*, New York: Lippincott Williams & Wilkins, pp. 28–77.

Carter SJ (1986) *Cooper and Gunn's Tutorial Pharmacy*. New Delhi, India: CBS Publishers & Distributors.

Florence AT and Attwood D (2006) *Physicochemical Principles of Pharmacy*, 4th ed., London: Pharmaceutical Press.

Gennaro A (Ed.) (2000) *Reminton's The Science and Practice of Pharmacy*, 20th ed., Easton, PA: Lippincott, Williams and Wilkins.

Hillery AM. (2001) Advanced drug delivery and targeting: An introduction. In Hillery AM, Lloyd AW and Swarbrick J (Eds.), *Drug Delivery and Targeting: For Pharmacists and Pharmaceutical Scientists*, New York: Taylor & Francis, pp. 63–82.

Hogben CAM, Tocco DJ, Brodie BB and Schanker LS (1959) On the mechanism of intestinal absorption of drugs. *J Pharmacol Exp Ther* **125**: 275–282.

Mahato RI (2005) Dosage forms and drug delivery systems, In Gourley DR (Ed.) *APhA's Complete Review for Pharmacy*, 3rd ed., New York: Castle Connolly Graduate Medical Publishing, pp. 37–63.

Shore PA, Brodie BB and Hogben CAM (1957) The gastric secretion of drugs. A pH partition hypothesis. *J Pharmacol Exp Ther* **119**: 361–369.

Sinko PJ (Ed.) (2006) *Martin's Physical Pharmacy and Pharmaceutical Sciences*, 5th ed., New York: Lippincott Williams & Wilkins.

Strickley RG (2004) Solubilizing excipients in oral and injectable formulations. *Pharm Res* **21**: 201–230.

Biopharmaceutical considerations

4.1 INTRODUCTION

To achieve an optimal drug response from a dosage form, a drug should be delivered to its site of action at a concentration that both minimizes its side effects and maximizes its therapeutic effects. The drug concentration at its target site depends on the dose of the drug administered and the rate and extent of its absorption, distribution, metabolism, and elimination (ADME/pharmacokinetics). For a drug molecule to exert its biological effect, it must be absorbed unaltered in a significant quantity (absorption) and transported by the body fluids. Moreover, it must escape widespread distribution to unwanted tissues, traverse the biological membrane barriers, penetrate in adequate concentration to the sites of action (distribution), escape metabolism and excretion (metabolism and elimination), and interact with its target in a specific fashion to cause the desired alteration of cellular function (pharmacodynamics).

The biological effect of a drug depends on three components:

1. Biopharmaceutical factors
   a. Dose and dosing frequency
   b. Route of administration
   c. Drug release from the delivery system
2. Pharmacokinetic factors (what the body does to the drug)
   a. Absorptin
   b. Distribution
   c. Metabolism
   d. Elimination
3. Pharmacodynamic factors (what the drug does to the body)
   a. Concentration–effect relationship at the site of action
   b. Specificity of drug action

All these factors must be taken into consideration when making dosage form decisions. For example, a high-dose drug may not be a suitable candidate for an oral sustained-release dosage form due to tablet size restrictions. Similarly, a drug with a long plasma elimination half-life and low frequency of dosing, such as once in a day, would not be considered a good candidate for a sustained-release oral dosage form.

Physicochemical properties of the drug, such as solubility, partition coefficient ($K_{o/w}$), pK_a value, diffusion rate, and intrinsic dissolution rate, primarily determine the *biopharmaceutical aspects of dosage form design*. In this chapter, we will discuss these aspects, with an emphasis on oral solid dosage forms, such as tablets, since these are the most commonly used drug delivery system. We will discuss drug release from the dosage form (diffusion and dissolution) and absorption across the biological membranes.

4.2 DIFFUSION

4.2.1 Drug transport across a polymeric barrier

Drug transport through a polymeric or biological barrier may occur by simple molecular permeation known as *molecular diffusion* or by movement through pores and channels known as *pore diffusion* (Figure 4.1).

4.2.1.1 Molecular diffusion

The transport of a drug molecule through a polymeric membrane that involves dissolution of the drug in the matrix of the membrane, followed by its diffusive transport to the surrounding bulk liquid, is an example of simple molecular diffusion (Figure 4.1a). The release rate of drug by diffusive transport through the polymeric matrix depends on the size and

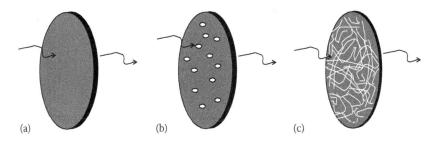

Figure 4.1 An illustration of passive diffusion processes. (a) Diffusion through the homogeneous film, (b) diffusion through solvent (usually water)-filled pores, and (c) diffusion through and/or between the fibrous membrane strands.

shape of the diffusing molecules, drug solubility in the polymeric matrix, partition coefficient of the drug between the polymeric matrix and the bulk liquid, and the degree of stirring of the bulk liquid at the interface.

4.2.1.2 Pore diffusion

Pore diffusion involves passage of drug through the solvent-filled pores in the polymeric membrane (Figure 4.1b). In pore diffusion, the release rate of dissolved drug is affected by porosity of the membrane, pore structure, surface functional groups (e.g., hydrophobic or hydrophilic), tortuosity, and length of pores.

The molecules may also pass through the tortuous gaps between the overlapping strands of the polymer (Figure 4.1c). In the cases of both molecular diffusion and pore diffusion, the drug must be available in a dissolved state. This would be the case if the drug product is formulated as a drug solution in the polymer. If a formulation consists of a suspension of drug particles in the polymer, another kinetic step of dissolution of the drug into the polymer or the solvent is involved. The rate of *dissolution* of a drug would depend on the degree of crystallinity, crystal size, and surface area of the drug; intrinsic dissolution rate of the drug in the polymer and/or the solvent; degree of swelling of the polymer with the solvent; and the extent of mechanical agitation in the system. Drug dissolution from its particles will be discussed in the next section.

4.2.1.3 Matrix erosion

In addition to molecular diffusion and pore diffusion, erosion of the polymeric matrix may often be involved in the case of biodegradable polymers. The kinetic contribution of matrix erosion to the drug release rate would depend on the relative rates of drug dissolution, polymer erosion, drug dissolution in the polymer, drug dissolution in the bulk solvent, molecular diffusion, and pore diffusion.

4.2.2 Principles of diffusion

Passive diffusion leads to change in concentration in a region over a period of time and space. Fick's laws of diffusion quantitate the amount of solute diffusing per unit time and area as a function of concentration gradient of solute in the direction of diffusion. These laws also relate the changes in solute concentration in a given region over time to the change in concentration gradient of the solute in that region.

4.2.2.1 Fick's first law

Fick's law of diffusion postulates that the diffusing molecules go from regions of high concentration to regions of low concentration. The rate of diffusion, the amount of material (M) flowing through a unit cross-section (S) of a barrier in unit time (t) is defined as the flux (J). Flux is related to the concentration gradient ($dC = C_1 - C_2$) between the donor region at a higher concentration (C_1) and the receiving region at a lower concentration (C_2) per unit distance (x) by the following expression:

$$J = \frac{dM}{dt} \times \frac{1}{S} \tag{4.1}$$

where:
  J is the flux, in g/(cm^2 s)
  S is the cross-section of the barrier, in cm^2
  dM/dt = rate of diffusion, in g/s (M = mass in g; t = time in s)

The flux is proportional to the concentration gradient, dC/dx:

$$J = -D \times \frac{dC}{dx} \tag{4.2}$$

where:
  D is the diffusion coefficient of a penetrant, in cm^2/s
  C is the concentration, in g/cm^3 or g/mL
  x is the distance perpendicular to the surface of the barrier, in cm

Thus,

$$\frac{dM}{dt} = -D \times S \times \frac{dC}{dx}$$

The negative sign in this equation signifies that diffusion occurs in a direction of decreasing concentration. Thus, the flux is always a positive quantity. Although the diffusion coefficient, D, or diffusivity, as it is often called, appears to be a proportionality constant, it does not remain constant. It is affected by changes in concentration, temperature, pressure, solvent properties, molecular weight, and chemical nature of the diffusant. For example, the larger the molecular weight, the lower the diffusion coefficient.

4.2.2.2 Fick's second law

Fick's second law predicts changes in solute concentration over time caused by diffusion. It states that the change in concentration with time in a particular region is proportional to the change in the concentration gradient at that region in the system. Concentration of solute or diffusant, C, in the volume of the region, x, changes as a result of the net flow of diffusing molecules in or out of the region. This change in concentration with time, t (i.e., dC/dt), is proportional to the change in the flux of diffusing molecules, J, per unit distance, x (i.e., dJ/dx).

$$\frac{dC}{dt} = -\frac{dJ}{dx} \tag{4.3}$$

Differentiating the equation of flux, J, as per Fick's first law of diffusion ($J = -D \times dC/dx$), with respect to x, we obtain:

$$-\frac{dJ}{dx} = D\frac{d^2C}{dx^2} \tag{4.4}$$

Therefore, concentration and flux are often written as $C(x, t)$ and $J(x, t)$, respectively, to emphasize that these parameters are functions of both distance, x, and time, t. Substituting dC/dt for $-dJ/dx$, Ficks's second law of diffusion can be expressed as:

$$\frac{dC}{dt} = D\frac{d^2C}{dx^2} \tag{4.5}$$

This equation represents diffusion only in one direction. To express concentration changes of diffusant in three dimensions, Fick's second law of diffusion is written as:

$$\frac{dC}{dt} = D\left[\frac{d^2C}{dx^2} + \frac{d^2C}{dy^2} + \frac{d^2C}{dz^2}\right] \tag{4.6}$$

4.2.3 Diffusion rate

Fick's first law of diffusion describes the diffusion process under steady state when the concentration gradient (dC/dx) does not change with time. The second law refers to a change in the concentration of diffusant with time at any distance (i.e., a nonsteady state). Diffusive transport from a dosage form is usually slow, leading to most of the drug transport happening under steady-state conditions. Therefore, it is important to understand the diffusive conditions under a steady state.

4.2.3.1 Diffusion cell

Figure 4.2 shows the schematic of a *diffusion cell*, with a diaphragm of thickness h and cross-sectional area S separating the two compartments. A concentrated solution of drug is loaded in the donor compartment and allowed to diffuse into the solvent in the receptor compartment. The solvent in both the compartments is continuously mixed and sampled frequently to quantitate drug transport across the membrane.

Equating both equations for flux, Fick's first law of diffusion may be written as:

$$J = \frac{dM}{dt} \times \frac{1}{S} = D \times \frac{(C_1 - C_2)}{h} \tag{4.7}$$

in which $(C_1 - C_2)/h$ approximates dC/dx. Concentrations C_1 and C_2 within the membrane (Figure 4.2) are determined by the partition coefficient of the solute ($K_{membrane/solvent}$) multiplied by the concentration in the donor compartment (C_{donor}) or in the receptor compartment ($C_{receptor}$). Thus, we have:

$$C_1 = C_{donor} \times K_{membrane/solvent} \tag{4.8}$$

and

$$C_2 = C_{receptor} \times K_{membrane/solvent} \tag{4.9}$$

Therefore, the partition coefficient is given by:

$$K_{membrane/solvent} = \frac{C_1}{C_{donor}} = \frac{C_2}{C_{receptor}} \tag{4.10}$$

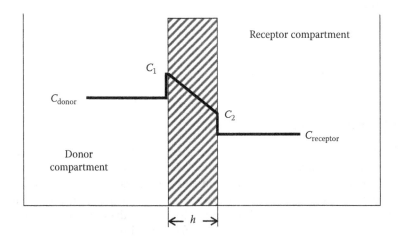

Figure 4.2 Drug concentrations in a diffusion cell.

Hence,

$$\frac{\mathrm{d}M}{\mathrm{d}t} = \frac{D \times S \times K_{\mathrm{membrane/solvent}} \times (C_{\mathrm{donor}} - C_{\mathrm{receptor}})}{h} \tag{4.11}$$

Under sink conditions, the drug concentration in the receptor compartment is maintained much lower than the drug concentration in the donor compartment, such that $C_{\mathrm{receptor}} \to 0$. Therefore, the above equation can be simplified as:

$$\frac{\mathrm{d}M}{\mathrm{d}t} = \frac{DSK_{\mathrm{membrane/solvent}}C_{\mathrm{donor}}}{h} \tag{4.12}$$

This equation can also be expressed in terms of the permeability coefficient, P, in cm/s, which is defined as:

$$P = \frac{D \times K_{\mathrm{membrane/solvent}}}{h} \tag{4.13}$$

as,

$$\frac{\mathrm{d}M}{\mathrm{d}t} = P \times S \times C_{\mathrm{donor}} \tag{4.14}$$

4.2.3.2 Spherical membrane-controlled drug delivery system

Following the same principles as outlined above for a diffusion cell, diffusive drug release from a spherical rate-limiting membrane enclosing a drug solution can be defined in terms of the surface area of the membrane at the center point of its thickness and the linear distance of drug diffusion across the membrane, x. Given the inner-boundary radius of the membrane as r_{inner} and the outer-boundary radius of the membrane as r_{outer}, the surface area of the sphere at its mean radius is given by:

$$S = 4 \times \pi \times \left(\frac{r_{\mathrm{outer}} + r_{\mathrm{inner}}}{2} \right)^2 \tag{4.15}$$

The surface area of the sphere may be approximated by:

$$S = 4 \times \pi \times r_{\mathrm{outer}} \times r_{\mathrm{inner}} \tag{4.16}$$

And the linear distance for solute diffusion across the membrane is given by:

$$x = r_{\mathrm{outer}} - r_{\mathrm{inner}} \tag{4.17}$$

Thus, the expression for the drug-release rate from a sphere is:

$$\frac{\mathrm{d}M}{\mathrm{d}t} = 4 \times \pi \times r_{\mathrm{outer}} \times r_{\mathrm{inner}} \times \frac{D \times K_{\mathrm{membrane/solvent}} \times \Delta C}{\left(r_{\mathrm{outer}} - r_{\mathrm{inner}} \right)} \tag{4.18}$$

where ΔC is the concentration gradient between the inside and the outside of the membrane.

Thus, permeability depends on both the properties of the diffusing solute (partition coefficient and diffusion coefficient) and the properties of the membrane (thickness and surface area).

4.2.3.3 Pore diffusion

In microporous reservoir systems, drug molecules are released by diffusion through the solvent-filled micropores. Drug transport across such porous membranes is termed pore diffusion. In this system, the pathway of drug transport is no longer straight but tortuous. The rate of drug transport is directly proportional to the porosity, ε, of the membrane and inversely proportional to the tortuosity, τ, of the pores. In addition, the partition coefficient of the drug between the membrane and the solvent ($K_{membrane/solvent}$) is no longer a factor, since drug dissolution in the membrane is not required. Therefore,

$$\frac{dM}{dt} = \frac{DSK_{membrane/solvent}C_{donor}}{h} \tag{4.12}$$

which is modified as:

$$\frac{dM}{dt} = \frac{D_sSC_{donor}\varepsilon}{h\tau} \tag{4.19}$$

In this modified equation, D_s is the drug diffusion coefficient in the solvent.

4.2.3.4 Determining permeability coefficient

Using the permeability coefficient:

$$\frac{dM}{dt} = \frac{D_sSC_{donor}\varepsilon}{h\tau} \tag{4.19}$$

where, surface area is given by:

$$S = 4 \times \pi \times r_{outer} \times r_{inner} \tag{4.16}$$

and concentration gradient is given by $\Delta C = C_{donor} - C_{receptor}$, with the assumption that $C_{receptor} \to 0$ under sink conditions; thus, if $\Delta C \cong C_{donor}$, we obtain:

$$\frac{dM}{dt} = PSC_{donor} \tag{4.20}$$

or

$$dM = PSC_{donor}dt \tag{4.21}$$

Thus, the value of permeability coefficient, P, can be obtained from the slope of a linear plot of M versus t, provided that C_{donor} remains relatively constant.

4.2.3.5 Lag time in non-steady state diffusion

A sustained-release dosage form may not exhibit a steady-state phenomenon from the initial time of drug release. For example, the rate of drug diffusion across a membrane slowly increases to steady-state kinetics (Figure 4.3). As shown in this figure, the curve is convex to the time axis in the early stage and then becomes linear. This early stage is the nonsteady-state condition. Later, the rate of diffusion is constant, the curve is essentially linear, and the system is at a steady state. When the steady state portion of the line is extrapolated to the time axis, the point of intersection represents the time of zero diffusion concentration if the system had been at the steady state all along. This time period between the actual nonsteady state and the projected steady-state time at zero diffusion concentration is known as the lag time. This is the time required for a penetrant to establish a uniform concentration gradient within the membrane that separates the donor from the receptor compartments.

4.2.3.6 Matrix (monolithic)-type non-degradable system

In a matrix-type polymeric delivery system, the drug is distributed throughout a polymeric matrix. The drug may be dissolved or suspended in the polymer. Regardless of a drug's physical state in the polymeric matrix, the

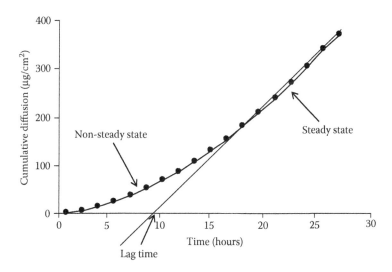

Figure 4.3 Drug diffusion rate across a polymeric membrane.

release of the drug decreases over time. In these systems, drug molecules can elute out of the matrix only by dissolution in the surrounding polymer (if drug is suspended) and by diffusion through the polymer structure. Initially, drug molecules closest to the surface are released from the device. As drug release continues, molecules must travel a greater distance to reach the exterior of the device. This increases the diffusion time required for drug-release. This increase in diffusion time results in a decrease in the drug-release rate from the device with time.

In an insoluble matrix-type system, the drug-release rate decreases over time as a function of the square root of time (Higuchi, 1963).

$$\frac{dM}{dt} = k_{device}\sqrt{t} \tag{4.22}$$

where k_{device} is a proportionality constant dependent on the properties of the device.

This release kinetics is observed for the release of the first 50%–60% of the total drug content. Thereafter, the release rate usually declines exponentially. Thus, the reservoir system can provide constant release with time (zero-order release kinetics), whereas a matrix system provides decreasing release with time (square root of time-release kinetics).

4.2.3.7 Calculation examples

4.2.3.7.1 Drug-release rate

Calculation of diffusion coefficient across and partition coefficient into a membrane barrier of a drug delivery system is often undertaken to simulate drug-release rate kinetics under different formulation conditions, such as drug loading. The diffusion rate may be calculated by using experimental data in one set of experiments. For example, if it were known that the diffusion coefficient of tetracycline in a hydroxyethyl methacrylate–methyl methacrylate copolymer film is $D = 8.0(\pm4.7) \times 10^{-9}$ cm²/s and the partition coefficient, K, for tetracycline between the membrane and the reservoir fluid of the drug delivery device is $6.8(\pm5.9) \times 10^{-3}$, drug-release rate can be calculated if the design parameters of the device are known. If the membrane thickness, h, of the device is 1.4×10^{-2} cm and the concentration of tetracycline in the core, C_{core}, is 0.02 g/cm³, tetracycline-release rate, dM/dt, may be calculated as follows.

$$\frac{dM}{dt} = \frac{DSK_{membrane/solvent}C_{donor}}{h} \tag{4.12}$$

$$\frac{dM}{dt} = \frac{8.0\times10^{-9} \text{ cm}^2/\text{s}\times1 \text{ cm}\times6.8\times10^{-3}\times0.02 \text{ g/cm}^3}{1.4\times10^{-2}\text{cm}} = 3.1\times10^{-10}\text{g/cm}\cdot\text{s}$$

To obtain the results in micrograms per day,

$$\frac{dM}{dt} = 3.1 \times 10^{-10} \text{ g/cm} \cdot \text{s} \times 10^{6} \text{ µg/g} \times 60 \times 60 \times 24 \text{ s/day} = 26.85 \text{ µg/day} \cdot \text{cm}$$

4.2.3.7.2 Partition coefficient

Knowing the permeability coefficient and the diffusion coefficient of a drug in a membrane, its partition coefficient can be calculated. For example, if a new glaucoma drug diffuses across a barrier of 0.02 cm with a permeability coefficient of 0.5 cm/s and the diffusion coefficient is 4 cm²/s, its permeability coefficient may be calculated as follows:

$$P = \frac{D \times K_{\text{membrane/solvent}}}{h} \tag{4.13}$$

$$0.5 \text{ cm/s} = \frac{4 \text{ cm}^2/\text{s} \times K_{\text{membrane/solvent}}}{0.02 \text{ cm}}$$

or

$$K_{\text{membrane/solvent}} = 0.5 \text{ cm/s} \times \frac{0.02 \text{ cm}}{4 \text{ cm}^2/\text{s}} = 0.0025$$

4.3 DISSOLUTION

For most drugs, the rate at which the solid drug dissolves in a solvent (dissolution) is often the rate-limiting step in the drug's bioavailability.

4.3.1 Noyes–Whitney equation

Noyes–Whitney equation correlates the dissolution rate of a drug with the particle surface area (S), the thickness of the unstirred solvent layer on the particle surface (h), diffusion coefficient of the drug (D), and the concentration gradient, that is, the difference in the concentration of drug solution at the particle surface (C_s) and the bulk solution (C).

$$\frac{dM}{dt} = \frac{DS}{h}(C_s - C) = kS(C_s - C) \tag{4.23}$$

or

$$\frac{dC}{dt} = \frac{DS}{Vh}(C_s - C) \tag{4.24}$$

where:

dM/dt is the mass rate of dissolution (mass of the drug dissolved per unit time, e.g., in mg/min)

D is the diffusion coefficient of solute in solution, in cm²/s

S is the surface area of the exposed solid, in cm²

k is the dissolution rate constant ($k = D/h$, in cm/s)

h is the thickness of the unstirred layer at the solid surface, in cm

C_s is the drug solubility at the particle surface, in g/mL

C is the drug concentration in bulk solution at time t, in g/mL

The quantity, dC/dt, represents the change in drug concentration in the bulk solution per unit time, or the dissolution rate, and V is the volume of solution (mL). Thus, $C = M/V$.

Under sink conditions, $C \ll C_s$. Therefore, the Noyes–Whitney equation can be simplified as:

$$\frac{dM}{dt} = \frac{DSC_s}{h}$$

or

$$\frac{dC}{dt} = \frac{DSC_s}{Vh}$$

4.3.2 Calculation example

Knowledge of dissolution rate constant, k, allows simulation of the rate of drug dissolution by using different quantities of drug substance, changes in the particle size and surface area of the drug, and dissolution conditions, such as volume. A simulation of these results can assist in dissolution method development by minimizing the number of experiments needed under different conditions. Dissolution rate constant can be calculated using dissolution data collected from a well-defined system.

For example, a preparation of drug particles weighing 550 mg and having a total surface area of 0.28×10^4 cm² was allowed to dissolve in 500 mL of water at 37°C. Assuming that analysis of bulk dissolution sample showed that 262 g had dissolved after 10 min, if the saturation solubility of the drug in water is 1.5 mg/mL at 37°C, k can be calculated as follows:

According to the Noyes–Whitney equation,

$$\frac{dM}{dt} = \frac{DS}{h}(C_s - C) = kS(C_s - C) \tag{4.23}$$

or

$$\frac{550 \text{ mg}}{10 \text{ min}} = k \times 0.28 \times 10^4 \text{ cm}^2 \times \left(1.5 \text{ mg/mL} - \frac{262 \text{ mg}}{500 \text{ mL}}\right)$$

$$k = \frac{550\,\text{mg}}{10\,\text{min}} \times \frac{1}{0.28 \times 10^4\,\text{cm}^2} \times \frac{1}{\left(1.5\,\text{mg/cm}^3 - \dfrac{262\,\text{mg}}{500\,\text{cm}^3}\right)} = 0.0201\,\text{cm/min}$$

The dissolution rate constant is related to the diffusion constant of the drug through the solvent (D) and the diffusion layer thickness (h):

$$k - \frac{D}{h} \tag{4.25}$$

Therefore, if the diffusion layer's thickness could be estimated, the diffusion coefficient of the drug can be calculated. Thus, if the diffusion layer's thickness were 5×10^{-3} cm, the diffusion coefficient (D) would be given by:

$$0.0201\,\text{cm/min} = \frac{D}{5 \times 10^{-3}\,\text{cm}}$$

or

$$D = 0.0201\,\text{cm/min} \times 5 \times 10^{-3}\,\text{cm} = 1.01 \times 10^{-4}\,\text{cm}^2/\text{min}$$

4.3.3 Factors influencing dissolution rate

The main biopharmaceutical and physiological factors that influence the dissolution rate of a drug can be summarized as follows:

1. *Drug solubility*: The greater the drug solubility, the greater the drug's dissolution rate. This is evident in the Noyes–Whitney equation. The solubility and dissolution rates of acidic drugs are low in acidic gastric fluids, whereas the solubility and dissolution rates of basic drugs are high. Similarly, the solubility and dissolution rates of basic drugs are low in basic intestinal fluids, whereas those of acidic drugs is high.
2. *Viscosity* (of the dissolving medium): The greater the viscosity of the dissolving liquid, the lower the diffusion coefficient of the drug and hence the lower the dissolution rate. Viscosity of the dissolving bulk medium and/or the unstirred layer on the surface of the dissolving formulation can be affected by the presence of hydrophilic polymers in the formulation, which dissolve to form a viscous solution. *In vivo*, the viscosity may be affected by the food intake.
3. *Diffusion layer's thickness*: The greater the diffusion layer's thickness, the slower the dissolution rate. The thickness of the diffusion layer is influenced by the degree of agitation of the dissolving medium, both *in vitro* and *in vivo*. Hence, an increase in gastric and/or intestinal motility may increase the dissolution rate of poorly soluble drugs. For example, food and certain drugs can influence gastrointestinal (GI) motility.

4. *Sink conditions*: Removal rate of dissolved drugs by absorption through the GI mucosa and the GI fluid volume affect drug concentration in the GI tract.

5. *pH* (of the dissolving medium): The drug dissolution rate is determined by the drug solubility in the diffusion layer surrounding each dissolving drug particle. The pH of the diffusion layer has a significant effect on the solubility of a weak electrolyte drug and its subsequent dissolution rate. The dissolution rate of a weakly acidic drug in GI fluid (pH 1–3) is relatively low because of its low solubility in the diffusion layer. If the pH in the diffusion layer could be increased, the solubility exhibited by the weak acidic drug in this layer (and hence the dissolution rate of the drug in GI fluids) could be increased. The potassium and sodium salt forms of the weakly acidic drug have a relatively high solubility at the elevated microenvironmental pH in the diffusion layer due to the strong counterion bases, KOH and NaOH, respectively. Thus, the dissolution of the drug particles takes place at a faster rate.

6. *Particle size and surface area*: An increase in the specific surface area (surface area per unit mass) of a drug in contact with GI fluids would increase its dissolution rate. Generally, the smaller a drug's particle size, the greater its specific surface area and the higher the dissolution rate. However, particle size reduction may not always be helpful in increasing the dissolution rate of a drug and hence its oral bioavailability. For example:

   • *Porosity* of drug particles plays a significant role. Thus, smaller particles with lower porosity may have lower surface area compared with larger particles with greater porosity. The dissolution rate depends on the *effective* surface area, which includes the influence of particle porosity.

   • In some cases, particle size reduction may cause particle *aggregation*, thus reducing the effective surface area. To prevent the formation of aggregates, small drug particles are often dispersed in polyethylene glycol (PEG), polyvinylpyrrolidone (PVP), dextrose, or surfactants such as polysobrates. For example, micronized griseofulvin is dispersed in PEG 4000.

   • In addition, certain drugs such as penicillin G and erythromycin are *unstable* in gastric fluids and do not readily dissolve in them. For such drugs, particle size reduction may increase not only the rate of drug dissolution in gastric fluids but also the extent of drug degradation.

7. *Crystalline structure*: Amorphous (noncrystalline) forms of a drug may have faster dissolution rate compared with the crystalline forms. Some drugs exist in a number of crystal forms or polymorphs. These different forms may have significantly different drug solubility and dissolution rates.

a. Dissolution rate of a drug from a crystal form is a balance between the energy required to break the intermolecular bonds in the crystal and the energy released on the formation of the drug–solvent intermolecular bonds. Thus, stronger crystals may have lower intrinsic dissolution rate.

b. *Intrinsic dissolution rate* reflects the dissolution rate of a drug crystal or powder normalized for its surface area. It is expressed in terms of mass per unit time per unit surface area. Drug forms that have higher intrinsic dissolution rate are expected to have higher dissolution rates.

c. The greater strength of a crystalline polymorph, sometimes evident by its high *melting point* and sometimes by the rank order, correlates with its lower intrinsic dissolution rate.

d. Similarly, amorphous solids, which lack a long-range order that defines crystalline structure, tend to have higher intrinsic dissolution rates.

8. *Temperature*: An increase in temperature leads to greater solubility of a solid, with *positive heat of the solution*. Heat of solution indicates release of heat on dissolving. Positive heat of solution is indicative of a greater strength of solute–solvent bonds formed (which release energy) compared with the solute–solute bonds broken (which take energy). The solid will therefore dissolve at a more rapid rate if the system is heated. Therefore, *in vitro* dissolution studies are carried out at 37°C to simulate body temperature and *in vivo* dissolution condition.

9. *Surfactants*: Surface-active agents increase the dissolution rate by (a) lowering the *interfacial tension*, which lowers the *contact angle* of the solvent on the solid surface and increases wetting of the drug particle and penetration of the solvent inside the dosage form, and (b) increasing the saturation solubility of the drug in the dissolution medium. Surfactants such as sodium lauryl sulfate (SLS) and Triton X-100 are frequently used to achieve sink conditions and rapid dissolution during *in vitro* dissolution method development.

4.4 ABSORPTION

Bioavailability is the fraction of an ingested dose of a drug that is absorbed into the systemic circulation, compared with the same dose of the compound injected intravenously, which is directly injected into the systemic circulation. Bioavailability of a drug is determined during new product development.

Bioequivalence, on the other hand, is a comparison of relative bioavailability of two dosage forms in terms of the rate and extent of the drug levels achieved in the systemic circulation and the maximum drug concentration reached. Generic drugs are required to satisfy statistical criteria of bioequivalence to the branded version before they can be considered equivalent.

In the case of oral dosage forms, drug bioavailability depends on the rate and extent of drug absorption from the GI tract. *Drug absorption from the gut depends on* many factors, such as the drug's solubility and intrinsic dissolution rate in the GI fluids, which are influenced by GI pH and motility, and the drug's particle size and surface area. Thus, an interplay of physicochemical properties of the drug and physiological properties of the GI tract determines the outcome of factors that determine drug absorption.

Drug absorption is affected not only by the properties of drug and its dosage forms but also by the nature of the biological membranes. Drugs pass through living membranes by the following processes (Figure 4.4):

1. Passive diffusion
   a. Simple diffusion
   b. Facilitated diffusion
      i. Channel-mediated transport
      ii. Carrier-mediated transport
2. Active transport

Passive diffusion can also be classified as paracellular or transcellular, depending on the route of drug absorption across the epithelial cell barrier. The surface lining of the GI tract consists of epithelial cells attached to each other by tight junctions formed through their membranes. Drug transport across the tight junctions between cells is known as *paracellular transport.* It involves both diffusion and the convective flow of water accompanying water-soluble drug molecules. Drug transport by absorption into the

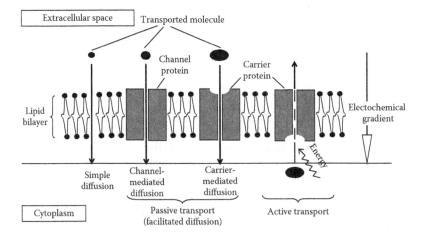

Figure 4.4 An illustration of the main transport processes across cellular membranes.

epithelial cell from the gut's lumen side, followed by release of the drug molecule from the epithelial membrane on the other side of the epithelial cell into the systemic circulation, is known as *transcellular transport.*

4.4.1 Passive transport

Passive transport can be divided into simple diffusion, carrier-mediated diffusion, and channel-mediated diffusion (Figure 4.4).

4.4.1.1 Simple diffusion

Biological membranes are lipoidal in nature; that is, they are made of lipid bilayers, with hydrophobic tails in the center and hydrophilic heads facing the aqueous environment on either side. Therefore, hydrophobic lipid-soluble drugs of low molecular weight can pass through membranes by simple diffusion. Passive transport by simple diffusion is driven by differences in drug concentration on the two sides of the membrane. In intestinal absorption, for example, the drug travels by passive transport from a region of high concentration in the GI tract to a region of low concentration in the systemic circulation. Given the instantaneous dilution of the absorbed drug once it reaches the bloodstream, sink conditions are essentially maintained at all times.

4.4.1.2 Carrier-mediated transport

Carrier-mediated transport is a passive diffusion process that involves facilitation or increase of diffusion rate by the involvement of a carrier protein embedded in the biological membrane. It differs from active transport in that the drug moves along a concentration gradient (i.e., from a region of high concentration to the one of low concentration) and that this system does not require energy input; that is, the carrier does not use energy, such as adenosine triphosphate (ATP), to transport the drug. Carrier-mediated transport is saturable, structurally selective for the drug, and shows competition kinetics for drugs of similar structures. Carrier-mediated transport does not require the substrate to be lipophilic: both hydrophilic and lipophilic solutes can be transported in this manner.

Amino acid transporters, oligopeptide transporters, glucose transporters, lactic acid transporters, phosphate transporters, bile acid transporters, and other transporters facilitate drug transport across the GI tract, especially the small intestine. *Transporters* are specific proteins in the biological membranes that transport the molecules (e.g., glucose) across the membrane. Transporters bind to the molecule, transport the molecule across the membrane, and then release it on the other side. The transporter remains unchanged after the process.

4.4.1.3 Channel-mediated transport

A fraction of the cell membrane is composed of aqueous-filled pores or channels, which are continuous across the membrane. These pores offer a pathway parallel to the diffusion pathway through the lipid bilayer. Channel-mediated transport (also known as port or convective transport) plays an important role in the transport of ions and charged drugs, especially in the case of renal excretion and hepatic uptake of drugs. Certain transport proteins may form an open channel across the lipid membrane of the cell. Small molecules, including drugs, move more rapidly through the channel by diffusion than by simple diffusion across the membrane due to facilitation by the solvent and if their diffusion rate in the solvent is higher than in the lipoidal membrane.

4.4.1.4 Fick's laws of diffusion in drug absorption

Transport of a drug by diffusion across a membrane such as the GI mucosa is represented by Fick's law equation:

$$\frac{dM}{dt} = \frac{D_m S_{membrane} K_{membrane/intestinal\cdot fluid}}{h_{membrane}} (C_{gut} - C_{plasma})dx \tag{4.26}$$

where:

M is the amount of drug in the gut compartment at time t

D_m is the diffusion coefficient or diffusivity of the drug in intestinal membrane

$S_{membrane}$ is the surface area of the membrane

$K_{membrane/intestinal\cdot fluid}$ is the partition coefficient of the drug between the membrane and the aqueous intestinal fluid

$h_{membrane}$ is the membrane's thickness

C_{gut} is the drug concentration in the gut or intestinal compartment

C_{plasma} is the drug concentration in plasma compartment

Since the absorbed drug is instantaneously diluted and rapidly removed from the absorption site by the systemic circulation, $C_{plasma} \rightarrow 0$. Therefore, Equation 4.26 becomes:

$$\frac{dM}{dt} = \frac{D_m S_{membrane} K_{membrane/intestinal\cdot fluid}}{h_{membrane}} C_{gut} \tag{4.27}$$

The left-hand side of Equation 4.27 can be converted into concentration units, since:

$$C_{gut} = \frac{M}{V}$$

On the right-hand side of the equation, the diffusion coefficient, membrane area, partition coefficient, and membrane thickness can be combined to yield a permeability coefficient.

$$P_{gut} = \frac{D_m S_{membrane} K_{membrane/intestinal \cdot fluid}}{h_{membrane}} \qquad (4.28)$$

Therefore,

$$-V_{gut} \frac{dC_{gut}}{dt} = P_{gut} C_{gut} \qquad (4.29)$$

or

$$-V_{plasma} \frac{dC_{plasma}}{dt} = P_{plasma} C_{plasma} \qquad (4.30)$$

where C_{gut} and P_{gut} are the concentration and permeability coefficient, respectively, for drug passage from intestine to plasma. Similarly, C_{plasma} and P_{plasma} are the concentration and permeability coefficient, respectively, for the reverse passage of drug from plasma to intestine. These equations demonstrate that the ratio of absorption rates in the intestine-to-plasma and the plasma-to-intestine directions depends on the ratio of permeability coefficients, drug concentrations, and volumes of drug distribution.

4.4.2 Active transport

Active transport involves the use of transmembrane proteins that require the use of cellular energy (usually ATP) to actively pump substances into or out of the cell. In active transport, the molecules usually move from regions of low concentration to those of high concentration. The most well-known active transport system is the sodium–potassium–ATPase pump (Na⁺/K⁺ ATPase), which maintains an imbalance of sodium and potassium ions inside and outside the membrane, respectively, for neuronal signal transmission. The Na⁺/K⁺ pump is an antiport; it transports K⁺ into the cell and Na⁺ out of the cell at the same time, with no expenditure of ATP. Other active transport systems include the sodium–hydrogen ion pump of the GI tract, which maintains gastric acidity while absorbing sodium ions, and the calcium ion pump, which helps maintain a low concentration of calcium in the cytosol.

REVIEW QUESTIONS

4.1 The characteristics of an active transport process include all the following, except:
    A. Active transport moves drug molecules against a concentration gradient.
    B. It follows Fick's law of diffusion.

    C. It is a carrier-mediated transport system.

    D. It requires energy.

    E. Active transport of drug molecules may be saturated at high drug concentrations.

4.2 The passage of drug molecules from a region of high drug concentration to the region of low drug concentration is known as:

    A. Active transport

    B. Simple diffusion or passive transport

    C. Pinocytosis

    D. Bioavailability

    E. Biopharmaceutics

4.3 Which of the following is true about Fick's first law of diffusion?

    A. It refers to a nonsteady-state flow.

    B. The amount of material flowing through a unit cross-section of a barrier in unit time is known as the concentration gradient.

    C. Flux of material is proportional to the concentration gradient.

    D. Diffusion occurs in the direction of increasing concentration.

    E. All of the above.

4.4 Which equation describes the rate of drug dissolution from a tablet?

    A. Fick's law

    B. Henderson–Hasselbalch equation

    C. Michaelis–Menten equation

    D. Noyes–Whitney equation

    E. All of the above

4.5 The diffusion coefficient of a permeant depends on:

    A. Diffusion medium

    B. Diffusion length

    C. Temperature

    D. All of the above

4.6 The rate of drug dissolution from a tablet dosage form will increase with:

    A. The particle size of the drug

    B. The surface area of drug particles

    C. The disintegration time

    D. The amount of excipients to dilute the drug

4.7 The permeability coefficient of a weak electrolyte through a biological membrane will increase if:

    A. The particle size of the drug increases.

    B. The surface area of drug particles increases.

    C. The partition coefficient increases.

    D. The drug dissolution rate increases.

4.8 Indicate which statement is true and which is false.

    A. Fick's first law of diffusion states that the amount of material flowing through a unit cross-section of a barrier in unit time is proportional to the concentration gradient.

B. The diffusion rate of molecules with a larger particle size is less than that of those with a smaller particle size.

C. Under the sink condition, the drug concentration in the receptor compartment is lower than that in the donor compartment.

4.9 Define Fick's first law of diffusion. Describe how Fick's first law is expressed in the Noyes–Whitney equation for dissolution. Calculate the diffusion coefficient of the new diet drug Lipidease® across a diffusion cell, given the following information: mass rate of diffusion = 5×10^{-4} g/s, cross-section of barrier = 1.0 cm², and concentration gradient = -175 g/cm⁴.

4.10 Calculate the rate of dissolution (dM/dt) of drug particles with a surface area of 2.5×10^3 cm³ and a saturated solubility of 0.35 mg/mL at room temperature. The diffusion coefficient is 1.75×10^{-7} cm²/s, and the thickness of the diffusion layer is 1.25 μm. The drug concentration in the bulk solution is 2.1×10^{-4} mg/mL.

4.11 The diffusion coefficient of tetracycline in a hydroxyethyl methacrylate—methyl methacrylate copolymer film is $D = 8.0\ (\pm 4.7) \times 10^{-9}$ cm²/s and the partition coefficient, k, for tetracycline between the membrane and the reservoir is $6.8\ (\pm 5.9) \times 10^{-3}$. The membrane thickness, h, of the trilaminar device is 1.40×10^{-2} cm, and the concentration of tetracycline in the concentration, C_0, is 0.02 g/cm³ of the core material. Calculate the release rate, Q/t, in units of mg/cm² of tetracycline per day.

4.12 Drug A weighs 0.5 g and has a total surface area of 0.3 m². In an experiment, it was found that 0.15 g of A (C) dissolved in 1000 mL of water in the first 2 min. Sink conditions were present. The saturation solubility was found to be 1.2×10^{-3} g/cm³. Calculate the dissolution rate constant in cm/min. Assume that saturation solubility, $C_{sat,}$ is much greater than the value C.

FURTHER READING

Allen LV, Popovich NG, and Ansel HC (2005), *Ansel's Pharmaceutical Dosage Forms and Drug Delivery Systems*, 8th ed., New York: Lippincott Williams & Wilkins.

Aulton ME (Ed.) (1988), *Pharmaceutics: The Science of Dosage Form Design*, New York: Churchill Livingstone.

Banker GS and Rhodes CT (Eds.) (1995), *Modern Pharmaceutics*, 3rd ed., New York: Marcel Dekker.

Block LH and Collins CC (2001), Biopharmaceutics and drug delivery systems. In Shargel L, Mutnick AH, Souney PH, and Swanson, LN (Eds.), *Comprehensive Pharmacy Review*, New York: Lippincott Williams & Wilkins, pp. 78–91.

Block LH and Yu ABC (2001), Pharmaceutical principles and drug dosage forms. In Shargel L, Mutnick AH, Souney PH, and Swanson LN (Eds.) *Comprehensive Pharmacy Review*, New York: Lippincott Williams & Wilkins, pp. 28–77.

Carter SJ (1986), *Cooper and Gunn's Tutorial Pharmacy*, New Delhi, India: CBS Publishers and Distributors.

Higuchi T (1963), Mechanism of sustained-action medication. Theoretical Analysis of rate of release of solid drugs dispersed in solid matrices. *J Pharm Sci.* 52: 1145–1149.

Hillery AM (2001), Advanced drug delivery and targeting: An introduction. In Hillery, AM, Lloyd AW, and Swarbrick J (Eds.), *Drug Delivery and Targeting: For Pharmacists and Pharmaceutical Scientists*, New York: Taylor and Francis, pp. 63–82.

Mahato RI (2005), Dosage forms and drug delivery systems. In Gourley, DR (Ed.), *APhA's Complete Review for Pharmacy*, 3rd ed., New York: Castle Connolly Graduate Medical Publishing, pp. 37–63.

Mayersohn M and Milo Gibaldi M (1971), Drug transport III: Influence of various sugars on passive transfer of several drugs across the everted rat intestine. *J Pharm Sci.* 60: 225–230.

Sinko PJ (Ed.) (2006), *Martin's Physical Pharmacy and Pharmaceutical Sciences*, 5th ed., New York: Lippincott Williams & Wilkins.

Chapter 5

Pharmacy math and statistics

LEARNING OBJECTIVES

On completion of this chapter, the students should be able to

1. Identify common systems of measure and differentiate between them.
2. Interconvert common systems of measure.
3. Differentiate between precision and accuracy.
4. Use ratio and proportion in different calculations.
5. Interconvert various units of concentration.
6. Calculate dilution requirements using alligation method.
7. Calculate salt requirement for preparing isotonic solutions.
8. Calculate clinical dose based on body weight.
9. Describe various types of sample distributions.
10. Identify and use appropriate tests of significance, depending on the situation.
11. Define variance of a sample distribution and be able to conduct analysis of variance (ANOVA).

5.1 INTRODUCTION

Mathematical calculations are an essential part of the practice of pharmacy. Calculations are required not only for the accurate preparation and dispensing of medications but also for clinical dose calculations and adjustments for individual patient needs. In this chapter, the common calculations encountered in the practice of pharmacy and their basic principles are summarized. This chapter assumes the background knowledge of mathematics such as mathematical functions with fractions, interconversions of fractions and decimals, natural and log exponential functions, and basic algebraic principles.[1]

5.2 SYSTEMS OF MEASURE

The metric system of measurements is based on the principle of multiples of 10 to define different ranges of quantities. Prefixes in the metric system indicate that the mentioned numeric value be multiplied by nth power of 10. For example, the represented multipliers for the common prefixes are as follows: nano (prefix: μ) is 10^{-9}, micro (prefix: μ) is 10^{-6}, milli (prefix: m) is 10^{-3}, centi (prefix: c) is 10^{-2}, deci (prefix: d) is 10^{-1}, deca (prefix: dk) is 10^{1}, hecto (prefix: h) is 10^{2}, and kilo (prefix: k) is 10^{3}. Therefore, 1 kg = 1,000 g = 1,000,000 mg = 1,000,000,000 μg = 1,000,000,000,000 ng.

In addition, the avoirdupois system (e.g., ounces and pounds) is the commonly used system in everyday life, and the apothecary system (meaning pharmacy) (e.g., quarts and pints) is commonly used in the practice of pharmacy.

- In the avoirdupois system, weight is expressed in grain (gr), ounce (oz), and pound (lb). The interconversions between these units and their relationship to the metric system are as follows:
  1 kg = 2.2 lb
  1 lb = 16 oz
  1 oz = 437.5 gr = 28.4 g
  1 gr = 65 mg
- In the apothecary system, volume is expressed in the units of fluid dram (dr), fluid ounce (oz), pint (pt), quart (qt), and gallon (gal). The interconversions between these units and their relationship to the metric system are as follows:
  1 gal = 4 qt = 3,785 mL
  1 qt = 2 pt = 946 mL
  1 pt = 16 oz = 473 mL
  1 oz = 30 mL (more accurately, 29.57 mL)
  3 (fluid dram) = 5 mL

Therefore, the sign "3" appearing in the sign of the prescription as "3i" indicates 1 teaspoonful (tsp) or 5 mL. Note that 1 tablespoon is 15 mL and is symbolized as "3ss" in the prescription. Similarly, the dispensing instruction "3V" means to dispense 5 f3 or 150 mL.

The laws of ratios and proportions can be used to interconvert units during calculations. For example, to convert 2.5 mg/5 mL into g/gal:

$$\frac{2.5\,mg}{5\,mL} = \frac{2.5\,mg}{5\,mL} \times \frac{3785\,mL}{1\,gal} \times \frac{1\,g}{1000\,mg} = 1.8925\,g/gal$$

Ratio and proportion can also be used for the reduction and enlargement of formulas for dispensing the required quantity of a prescription. In addition, a *conversion factor* can be derived, which then becomes the multiplier for every ingredient in the formulation to dispense a given quantity.

$$\text{Conversion factor} = \frac{\text{Volume to be dispensed}}{\text{Volume in the (unit) formula}}$$

For example, to dispense 200 mL of a prescription with a unit formula for 5 mL quantity, the conversion factor would be 200/5 = 40. Therefore, the quantity of every ingredient would be multiplied by 40 to make a 200-mL dispensed quantity.

5.2.1 Volume and weight interconversions

The interconversions of weight and volume are useful in pharmacy dispensing to aid accuracy of measurement. Interconversions of weight for volume of liquids can be done using their density, which is weight per unit volume. Therefore, 1 mL of glycerol, of density 1.26 g/mL at room temperature, is equivalent to 1.26 g of glycerol. Alternatively, 1 g of glycerol is $^{1}/_{1.26} = 0.79$ mL of glycerol. Note that 1 cubic centimeter (cc or cm^3) volume = 1 mL.

Sometimes, the information on specific gravity of a substance is available, which can be used to perform similar calculations. Specific gravity is the ratio of weight of a substance to the weight of an equal volume of distilled water at 25°C. Therefore, it does not have any units. Since 1 mL of water = 1 g of water at 25°C, specific gravity represents the number of grams of a substance per unit volume of that substance in mL at 25°C. Density, on the other hand, is usually determined at ambient temperature or at the temperature at which measurements are to be made.

5.2.2 Temperature interconversions

Interconversions of temperature between the Celsius (sometimes also known as centigrade), Fahrenheit, and Kelvin scales can be carried out using the following equations:

$$\frac{C}{5} = \frac{F - 32}{9} \tag{5.1}$$

$$K = C + 273.15 \tag{5.2}$$

Interestingly, $-40°C = -40°F$.
At temperatures less than $-40°C$, °F < °C.
At temperatures greater than $-40°C$, °F > °C.

Although the Celsius and Fahrenheit scales are more commonly encountered in routine use, the Kelvin scale is used more commonly in the derivation and use of scientific equations.

5.2.3 Accuracy, precision, and significant figures

Accuracy represents the degree of closeness of a measurement to the desired, target, or actual quantity. Thus, if the target quantity to be weighed is 125 mg and the actual weighed quantities are 121 and 123 mg in two different trials, the latter would be considered more accurate than the former. Accuracy is a measure of distance from the target.

Precision, on the other hand, represents the reproducibility or repeatability of a measurement. It represents the relative closeness of individual measurements to the average of these measurements when the measurements are carried out more than once. Precision is an indication of variability of a measurement or, said differently, of the confidence in the exactness of a measurement. For example, a balance that can weigh ±0.01 g of a target weight would lead to a more precise measurement than a balance that weighs ±0.1 g of the target weight. In pharmacy practice, both accuracy and precision are needed.

Significant figures, or the number of digits in the decimal places, represent the precision of a measurement by indicating the least amount that could be measured. For example, a weight of 1.0 g represents ±0.1 g precision of the balance on which the weight was taken. Thus, the actual weight of the substance that is labeled as 1.0 g could be anything in the range of 0.9–1.1 g (a range of 0.2 g). Similarly, a recorded weight of 1.000 g represents ±0.001 g precision of the balance on which the weight was taken. Thus, the actual weight of the substance that is labeled as 1.000 g could be anything in the range of 0.999–1.001 g (a range of 0.002 g).

The concept of significant figures is utilized in rounding off considerations. For example, numerical calculations of quantities can introduce additional digits at the tailing end of the calculated number. These numbers should then be rounded off to the significant digits of original measurement when communication of precision is important. For example, splitting a tablet labeled 125 mg has the precision of dose measurement of ±1 mg. When this tablet is split in half, each half can be considered to contain 125/2 = 62.5 mg of the drug. However, the number 62.5 mg indicates a precision of ±0.1 mg, which does not accurately represent the precision of dose measurement.

5.3 RATIO AND PROPORTION

Ratio represents a quantitative relationship between two quantities. It can be expressed as a fraction (e.g., ½, ¼, and so on) or as a ratio (e.g., 1:2, 1:4, and so on).

A proportion represents the equality of two ratios. Thus,

$$\frac{1}{2} = \frac{2}{4}$$

The equality of two ratios can be checked by cross multiplying the numerator of the first with the denominator of the second. For example,

$$\frac{1}{2} = \frac{2}{4} \text{ but } \frac{1}{2} \neq \frac{2}{6}$$

because

$$1 \times 4 = 2 \times 2 \text{ but } 1 \times 6 \neq 2 \times 2$$

Proportions are commonly utilized to find an unknown quantity or variable when three other related quantities or variables are known. For example,

$$\text{If } \frac{x}{4} = \frac{2}{9}, \quad \text{then } x = \frac{2 \times 4}{9} = \frac{8}{9}$$

In these calculations, caution must be exercised to ensure that the numerators and denominators have the same units on both sides of the proportion. For example, if the pharmacist wishes to substitute 100-mg strength tablets with 200-mg strength tablets for a patient who was prescribed four tablets of 100-mg strength, then the number of tablets of 200-mg strength can be calculated as:

$$\frac{4 \text{ tablets}}{100 \text{ mg/tablet}} = \frac{x \text{ tablets}}{200 \text{ mg/tablet}}, \quad \text{then } x = \frac{4 \times 200}{100} = 8 \text{ tablets}$$

A good practice in carrying out these calculations is to always label the units in the proportions.

5.4 CONCENTRATION CALCULATIONS

A formulation is essentially a multicomponent mixture. The relative amount of a substance in a multicomponent system represents its concentration. It could be the concentration of a dissolved drug in a solution, a suspended drug in a suspension, or a drug powder in a triturate of solid powders. The expression of concentration, its relation to the total amounts, and calculations involving changes to the concentration or total amount are an essential part of pharmacy practice. This section discusses the common ways of expressing concentrations, their basic principles, and the calculations involving drug amounts in such preparations.

5.4.1 Percentage solutions

Concentrations of ingredients in a formula are often represented as a percentage (%). Percentage represents parts of 100 (cent). In liquid preparations, percentage values can represent % weight/weight (% w/w, e.g., 2 g of solid in a 100 g of liquid = 2% w/w), % weight/volume (% w/v, e.g., 2 g of solid in 100 mL of liquid = 2% w/v), or % volume/volume (% v/v, e.g., 2 mL of liquid A in 100 mL of liquid B = 2% v/v of liquid A).

Calculations for the exact amount of an ingredient to be used in a formulation when the percentage composition of the formula is known can be made using ratio and proportion. Thus, to dispense 240 mL of a 10% w/v solution of a drug substance, the amount of drug substance needed can be calculated as:

$$\frac{10\,g}{100\,mL} = \frac{x\,g}{240\,mL} \quad \text{Therefore, } x = \frac{10\,g}{100\,mL} \times 240\,mL = 24\,g$$

5.4.2 Concentrations based on moles and equivalents

Molecular weights or moles of a compound are more useful for calculations when two or more chemical compounds are to be compared for a given attribute. Thus, during drug discovery, relative potencies of different compounds are compared on a molar basis.

The concepts of solution concentrations of compounds are based on their molecular or equivalent weights, which can be defined as follows:

- The *molecular weight* of a compound represents the weight of one mole (abbreviation: mol) of a compound, in grams. Thus, 1 mole of glucose is 180.16 g of glucose. The molar mass of glucose is thus represented as 180.16 g/mol.
- An *equivalent weight* of a compound represents its molecular weight divided by the number of valence or ionic charges in solution. It takes into account the chemical activity of an electrolyte. One equivalent (abbreviation: Eq), in grams, of a compound represents 1 mole of compound in grams divided by its valence. Thus, molecular weight of Mg^{2+} ions is 24.3 g, indicating that 24.3 g of Mg^{2+} ions represents 1 mole of Mg^{2+}. On the other hand, the equivalent weight of Mg^{2+} ions is 24.3/2 = 12.15 g, since there are two charges on Mg^{2+} ions. Thus, when used for charge-neutralization calculations, 1 mole or molecular weight of Mg^{2+} ions represents two equivalents.

Solutions of electrolytes are often prepared in terms of molarity, molality, and normality.

- *Molarity* (abbreviation: M) is defined as the moles of solute per liter of solution. Therefore, 1 M of sulfuric acid solution represents 98 g (molecular weight) of H_2SO_4 dissolved in 1 L of solution.
- *Normality* (abbreviation: N) represents gram equivalent weight of solute per liter of solution. The difference between molarity and

normality is representative of the difference between moles and equivalents of a compound. Thus, 1 N of sulfuric acid solution represents 49 g (equivalent weight) of H_2SO_4 dissolved in 1 L of solution. The equivalent weight of H_2SO_4 is half its molecular weight, since H_2SO_4 is a diprotic acid (i.e., dissociates to release 2 H^+ ions in solution).

- *Molality* (abbreviation: m) is a less frequently used term that represents the number of moles of solute per kilogram of solvent.
- *Formality* (abbreviation: F) is another less frequently used term that represents the formula weight of a compound in 1 L of solution. It differs from molarity in indicating the amount of solute added to the solution, but it does not consider the nature of the chemical species that actually exist in solution. For example, when 1 mole of sodium carbonate (Na_2CO_3) or sodium bicarbonate ($NaHCO_3$) is dissolved in a total of 1 L of an acidic solution of hydrochloric acid (HCl), the concentration of Na_2CO_3 or $NaHCO_3$ may be represented as 1 F (indicates the amount added) but not as 1 M (indicates amount in solution)—even though quantitatively they would be the same—since the compound reacts with acid in the solution and does not remain as the same species that was added.
- The amount of a solute may also be represented as its *mole fraction.* The mole fraction of a solute is the number of moles of solute as a proportion of the total number of moles (of solute + solvent) in a solution. For example, a mole fraction of 0.2 indicates 2 moles of solute dissolved in 8 moles of solvent. Mole fraction is a dimensionless quantity. Mole fraction is frequently used to represent the relative amount of two different solutes in a system.

Concentrations and amounts can be represented in fractions by using the prefixes used in the metric system of measure. Thus, 1 mEq is one milliequivalent of a solute, thus representing 1/1,000th of an equivalent weight of the solute. Similarly, 1 μM would represent 1 micromolar, or 1/1,000,000th of a molar (1 mol/L) concentration of a solute.

5.4.3 Parts per unit concentrations

Parts per unit concentrations are commonly expressed for very low concentrations of solutes. The commonly used parts per unit concentrations are as follows:

- Parts per million (ppm) represents 1 part of a substance in 1 million (10^6) parts of the total mixture. Parts per million (ppm) is dimensionless, since the parts of both the substance and the total mixture are represented in the same units. In addition, this measure is applicable to both solutions and solids. Thus, 1 ppm of NaCl in a solid powder may represent 1 µg/g or 1 mg/kg of NaCl. Moreover, 1 ppm of NaCl solution can represent 1 µL/L of NaCl in water.

- Parts per billion (ppb) represents 1 part of a substance in 1 billion (10^9) parts of the total mixture. Similar to ppm, it is dimensionless and does not represent a state (solid or liquid) of the substance.
- Other less commonly used parts per unit measures are parts per thousand, parts per trillion (ppt, 1 in 10^{12}), and parts per quadrillion (ppq, 1 in 10^{15}).

5.4.4 Dilution of stock solutions

A stock solution is a concentrated solution of a substance that can be diluted to a lower, desired concentration by adding the solvent immediately before use or dispensing. Stock solutions are frequently used in pharmacy dispensing to increase the efficiency, ease, and accuracy of dispensing, as well as the space and cost advantages with the transportation and storage of lower-volume concentrated solutions.

A common calculation required for the dilution of a concentrated stock solution to a desired concentration is the amount of solvent needed to achieve the desired concentration. This can be derived from the formula for concentration (c) based on the volume (v) of solution and the weight (w) of the substance.

$$c \, (\text{in g/mL}) = \frac{w \, (\text{in g})}{v \, (\text{in mL})}$$

Therefore, if the stock solution were designated by the subscript "1" and the final solution to be prepared by the subscript "2,"

$$\frac{c_1}{c_2} = \frac{w_1/v_1}{w_2/v_2} = \frac{w_1}{v_1} \times \frac{v_2}{w_2}$$

Since it is the same weight of the solute that would be transferred from the stock solution into the final, diluted solution,

$$w_1 = w_2$$

Therefore,

$$c_1 \times v_1 = c_2 \times v_2 \tag{5.3}$$

This formula can be used to calculate the volume of solvent required to make a diluted solution. For example, to dilute a 50% w/v stock solution to make 200 mL of a 5% w/v solution, $c_1 = 50$, $c_2 = 5$, and $v_2 = 200$.

$$v_1 = \frac{c_2 \times v_2}{c_1} = \frac{5 \times 200}{50} = 20 \, \text{mL}$$

Hence, the amount of stock solution needed = 20 mL and the amount of solvent needed = 200 − 20 = 180 mL to make a total of 200 mL of the diluted solution.

The measurements can also be carried out in weight rather than in volume for the stock and the diluted solutions. Thus,

$$c_1 \times w_1 = c_2 \times w_2 \tag{5.4}$$

5.4.5 Mixing solutions of different concentrations

Often, mixing of two products made of the same solute but having different concentrations is required. A convenient approach to solve these problems is the alligation method. Two kinds of alligation methods are commonly used: alligation medial and alligation alternate.

5.4.5.1 Alligation medial

5.4.5.1.1 For two ingredients

This method is based on finding the proportion for a formula. The strength (e.g., % w/w) of an ingredient is multiplied by its amount (e.g., quantity in grams) to obtain the product of each ingredient. The products of all ingredients and their quantities in the original formula are added together separately. Dividing the sum of products by the sum of quantities in the original formula gives a quotient, which represents the strength (e.g., % w/w) of the final mixture.

For example, to calculate the strength of the final mixture when 12 g of a 10% w/v sucrose solution is mixed with 24 g of a 40% w/v sucrose solution, one would write the alligation medial method as indicated in Table 5.1. Working through this table, the final solution would be of 30.0% w/w strength.

5.4.5.1.2 For more than two ingredients

This method is also applicable for more than two ingredients. For example, to calculate the strength of the final mixture when 12 g of a 10% w/v sucrose

Table 5.1 Alligation medial method for two ingredients

Ingredient	Strength (% w/w)	Quantity (g)	Product of strength and quantity (% w/w * g)
A	10	12	120
B	40	24	960
Sum		36	1080
Quotient = 1080/36 = 30			

Table 5.2 Alligation medial method for more than two ingredients

Ingredient	Strength (% w/w)	Quantity (g)	Product of strength and quantity (% w/w * g)
A	10	12	120
B	40	24	960
C	5	36	180
Sum		72	1260

Quotient = 1260/72 = 17.5

solution is mixed with 24 g of a 40% w/v and 36 g of a 5% w/v sucrose solution, one would write the alligation medial method as indicated in Table 5.2. Working through this table, the final solution would be of 17.5% w/w strength.

5.4.5.2 Alligation alternate

5.4.5.2.1 For two ingredients

This method can be used to calculate the amount of a diluent, solute, or different-concentration product that would need to be added to a given concentration product to make a new concentration preparation. The number of parts required for the lower- and higher-concentration preparations to make the target-concentration preparation is obtained by constructing a matrix and doing the calculation as shown in Table 5.3.

Thus, subtracting the target concentration from the lower concentration gives the target amount of the higher-concentration preparation, and subtracting

Table 5.3 Alligation alternate method for two ingredients

Cell #A1	Cell #A2	Cell #A3
Write numerical concentration value of the higher-concentration preparation		Subtract cell #C1 from cell #B2 and write the numerical value here
Cell #B1	Cell #B2	Cell #B3
	Write numerical concentration value of the target-concentration preparation	
Cell #C1	Cell #C2	Cell #C3
Write numerical concentration value of the lower-concentration preparation		Subtract cell #B2 from cell #A1 and write the numerical value here

the higher concentration from the target concentration gives the target amount of the lower-concentration preparation. Thus, the total amount of the target-concentration preparation that would be prepared can be obtained by adding together the target amounts of higher- and lower-concentration preparations needed. If the required amount of the target-concentration preparation is different than the amount obtained by the formula, the principles of proportion, discussed earlier, can be used to calculate the quantities needed for the required total amount of the target-concentration preparation.

For example, to prepare 200 mL of a 12% w/v sucrose solution using a 40% w/v and another 5% w/v sucrose solution, one would write the alligation matrix as shown in Table 5.4.

Thus, combining 7 mL of 40% w/v solution with 28 mL of 5% w/v solution would give 7 + 28 = 35 mL of 12% w/v solution. To make 200 mL of 12% w/v solution, one would use the principles of proportion as follows:

For the quantity of 40% w/v solution,

$$\frac{7 \text{ mL}}{35 \text{ mL}} = \frac{x \text{ mL}}{200 \text{ mL}} \quad \text{Hence, } x = \frac{7}{35} \times 200 = 40 \text{ mL}$$

For the quantity of 5% w/v solution,

$$\frac{28 \text{ mL}}{35 \text{ mL}} = \frac{x \text{ mL}}{200 \text{ mL}} \quad \text{Hence, } x = \frac{28}{35} \times 200 = 160 \text{ mL}$$

Alternatively, a conversion factor could be derived for the calculation:

$$\text{Conversion factor} = \frac{200 \text{ mL}}{35 \text{ mL}} = 5.714$$

The required quantities of low- and high-concentration solutions can then simply be obtained by multiplying their quantities obtained by the alligation formula by this factor. Thus, the quantity of 40% w/v solution required = 7 × 5.714 = 39.998 = 40 mL. Therefore, the quantity of the 5% w/v solution required = 200 − 40 = 160 mL or 28 × 5.714 = 159.992 = 160 mL.

Table 5.4 An example of alligation alternate method for two ingredients

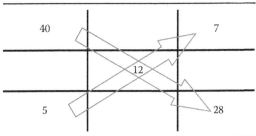

Table 5.5 An example of alligation alternate method for more than two ingredients

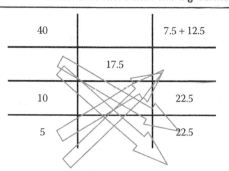

5.4.5.2.2 For more than two ingredients

The alligation alternate method can be used for more than two ingredients by *pairing off* the values of one higher (than the desired) strength ingredient with two lower (than the desired) strength ingredients, or vice versa. This is illustrated by the following example:

To prepare a 17.5% w/w solution using a 10% w/v, a 40% w/v, and a 5% w/v sucrose solution, one would write the alligation alternate method, as shown in Table 5.5.

Thus, combining 20 mL of 40% w/v solution with 22.5 mL of 10% w/v solution and 22.5 mL of a 5% w/w solution would give 20 + 22.5 + 22.5 = 65 mL of 17.5% w/v solution. The laws of proportion can be used, as described earlier, to calculate specific quantities of starting solutions that would be needed to prepare a desired quantity of the final solution.

The alligation alternate method for more than two ingredients can use any pairing of higher (than the desired) strength ingredient(s) with lower (than the desired) strength ingredient(s). The pairings can be any number depending on the number of ingredients.

The alligation methods are applicable to all forms of preparations, including powders. In addition, the alligation method can also be used for calculating the required quantities for dilution of a preparation with the solvent or diluent alone by making the concentration of the lower-concentration preparation zero.

5.4.6 Tonicity, osmolarity, and preparation of isotonic solutions

Of the two compartments of solution separated by a semipermeable membrane, the solvent tends to flow from the solution with the lower solute concentration to the solution with the higher solute concentration. If uninterrupted flow of solvent is allowed, it would result in the equalization of concentration across the membrane. This phenomenon is called *osmosis*. The pressure of solvent involved in this phenomenon is termed *osmotic pressure.* A solution containing a nonpermeable solute creates a pressure for the inward flow of solvent across the semipermeable membrane. Thus, osmotic pressure can also be defined as the pressure that must be applied to a solution to prevent the inward flow of solvent across a semipermeable membrane.

Tonicity is the osmotic pressure of two solutions separated by a semipermeable membrane. Tonicities of solutions are often represented with reference to that of normal body fluids. Thus, solutions that exert lower osmotic pressure than the body fluids are termed *hypotonic,* while solutions that exert higher osmotic pressure than the body fluids are termed *hypertonic.* Hypotonic solutions have lower and hypertonic solutions have higher impermeable solute concentration than the body fluids. Two solutions that have the same osmotic pressure are termed *isosmotic,* while a solution that has the same osmotic pressure as a reference body fluid is termed *isotonic.*

To define the osmotic amount and concentration of a solute without referring to another solution, the concepts of osmole, osmolarity, and osmolality are introduced. An *osmole* is the amount of a substance that represents the number of moles of particles that it forms in a solution. For a nondissociating substance, that is, a nonelectrolyte such as dextrose, 1 osmole = 1 mole. Thus, 1 osmole of dextrose = 186 g (molecular weight) of dextrose.

Similar to the concept of molarity, *osmolarity* (abbreviation: Osmol) is defined as the osmoles of solute per liter of solution. Therefore, 1 Osmol of glucose solution represents 186 g (molecular weight) of glucose dissolved in 1 L of solution. Similar to the concept of molality, *osmolality* is defined as the osmoles of solute per kg of solvent. These quantities can be used with prefixes in the metric system, such as milli and micro. Thus, a commonly used term is milliosmole (abbreviation: mOsmol), which represents 1/1,000[th] of an Osmol. Moreover, while osmole represents the quantity of solute in grams, Osmol represents the concentration of solute in a solution.

For a dissociating solute, such as an electrolyte, 1 mole ≠ 1 Osmol and 1 M solution ≠ 1 Osmol solution. The osmoles and osmolarity of such a solute are calculated by multiplying with the number of particles formed on dissociation and the fractional degree of dissociation of a substance in solution. Thus, assuming complete dissociation, NaCl, $CaCl_2$, and $FeCl_3$ form 2, 3, and 4 particles in solution. Thus, 1 mM solution of NaCl, $CaCl_2$, or $FeCl_3$ represents their 2, 3, or 4 mOsmol solution, respectively. Assuming, 80% degree of dissociation for dilute solutions, 2 M of NaCl, $CaCl_2$, and $FeCl_3$ solutions represent:

$$2 \times \left(1 + \frac{80}{100}\right) = 3.6 \text{ Osmol of NaCl solution}$$

$$2 \times \left(1 + \frac{80}{100} + \frac{80}{100}\right) = 5.2 \text{ Osmol of } CaCl_2 \text{ solution}$$

$$2 \times \left(1 + \frac{80}{100} + \frac{80}{100} + \frac{80}{100}\right) = 6.8 \text{ Osmol of FeCl}_3 \text{ solution}$$

The normal serum osmolality is in the range of 275–300 mOsmol/kg. osmolality of solutions can be measured in the laboratory by using an osmometer.

Tonicity is an important concept in the administration of ophthalmic and parenteral solutions. Hypertonic solutions tend to draw fluids out of body tissues, leading to irritation and dehydration. Hypotonic solutions, on the other hand, can provide excess fluid to the body tissues. However, since the volume of the administered solution is much lower than that of body fluids and fluid elimination is a regulated physiological phenomenon, hypotonic solutions are relatively inconsequential. Thus, administration of hypertonic solutions tends to be more tissue-damaging and painful than the administration of hypotonic solutions. Nonetheless, isotonic solutions are better tolerated by patients than either extremes of tonicity.

Preparation of isotonic solutions requires the use of one of the colligative properties of solutions. Colligative properties are the solution properties that depend on the number of molecules of solvent in a given volume of solution but are independent of the properties of the solute. These properties include lowering of vapor pressure, elevation of boiling point, osmotic pressure, and depression of freezing point of a solution with increasing solute concentration. Of these, the depression of freezing point is conveniently used to calculate the amount of solute required to prepare an isotonic solution.

For example, given that the freezing point of blood serum and ophthalmic lachrymal fluid is −0.52°C and that 1 M aqueous solution of a nonelectrolyte depresses the freezing point of water by 1.86°C, we can calculate the amount of glucose (molecular weight: 180 g/mol) required to prepare an isotonic solution by solving for the amount of glucose that would produce a freezing point depression of 0.52°C. Thus, to make 1 L of isotonic glucose solution, the amount of glucose required (x) can be calculated as:

$$\frac{1.86 \,°\text{C}}{0.52 \,°\text{C}} = \frac{180 \text{ g}}{x \text{ g}} \quad \text{Therefore, } x = 180 \times \frac{0.52}{1.86} = 50$$

This corresponds to 5% w/v glucose solution. The commonly available dextrose solution for intravenous (IV) administration has this concentration. Similar concentration for an electrolyte, such as sodium chloride, should take into consideration the dissociation constant of the solute and the number of species produced in the solution. Thus, assuming that NaCl in weak solutions is about 80% dissociated, the total number of solutes in solution would be 1.8 times the number of molecules added to the solution. This (1.8) *dissociation factor* (abbreviation: i) is used in the calculation of isotonic concentrations of electrolytes. Thus, to make a 1 L isotonic NaCl (molecular weight: 58 g/m) solution, the amount of NaCl required (x) can be calculated as:

$$\frac{1.86\ °C \times 1.8}{0.52\ °C} = \frac{58\ g}{x\ g} \quad \text{Therefore, } x = \frac{58 \times 0.52}{1.86 \times 1.8} = 9$$

This corresponds to 0.9% w/v NaCl solution, which is commonly available as an isotonic solution for experiments involving living cell and tissues. From these calculations, note that 50 g/L of glucose solution is *isotonic* to 9 g/L of NaCl solution. Therefore, in quantities of solutes, 50 g of glucose is *tonic equivalent* to 9 g of NaCl. The tonic equivalence of two substances represents their amounts that would produce the same osmotic pressure. Thus, the quantity of any substance divided by its dissociation factor, i, represents its tonic equivalent quantity to any other substance. This principle is used in the preparation of isotonic solutions by the addition of NaCl to hypotonic drug solutions to increase the tonicity to the physiological equivalent of 0.9% w/v NaCl. Using the above conversion of tonic equivalents, *NaCl equivalents* (E values) of various substances are known in the literature. The number of grams of all ingredients in a prescription is multiplied by their E values and added together to determine the osmotic equivalent of NaCl amount represented by the substances. In addition, the amount of NaCl that would be required to make a 0.9% w/v solution of the same volume as the prescription is determined. Subtracting the former from the latter gives the amount of NaCl needed to make the solution isotonic. Any substance other than NaCl, such as dextrose, can also be used to increase the tonicity of a solution by dividing the amount of NaCl needed by the NaCl equivalent of the other substance.

For example, to compound 10 mL of an ophthalmic preparation of 3% w/v pilocarpine nitrate, we first determine the amount of drug in 10 mL of solution.

$$\text{Drug amount} = \frac{3}{100} \times 10 = 0.3\ g$$

The NaCl equivalent (E value) of pilocarpine nitrate (molecular weight 271, dissociates into 2 ions, and i value = 1.8) can be read from the literature or calculated as:

$$\text{E value} = \frac{58.5 / 1.8}{271 / 1.8} = 0.216$$

Now, we multiply the E value with the drug amount in solution to get NaCl equivalents represented by the drug amount in the solution:

$$\text{NaCl equivalent in prescription} = 0.3 \times 0.216 = 0.0648\ g$$

This is the amount of particles in solution equivalent to NaCl, which must be subtracted from the amount of NaCl that would be needed to make an isotonic solution of the same volume as the prescription (i.e., 10 mL). This is calculated as:

$$\text{Total amount of NaCl needed for isotonicity} = \frac{0.9}{100} \times 10 = 0.09 \text{ g}$$

Hence, the amount of NaCl that must be added to the prescription to make an isotonic solution = 0.09−0.0648 = 0.0252 g.

If a prescription contains multiple components, NaCl equivalent for each component is calculated separately and added together to make the total NaCl equivalents in the prescription. This total amount is then subtracted from the total NaCl that would be needed for isotonicity of the volume of prescription to obtain the amount of NaCl that must be added to the prescription.

5.5 CLINICAL DOSE CALCULATIONS

The dose of a drug represents the amount of the drug substance that a patient must take at one time. This amount is designed with an expectation of producing the optimum therapeutic effect while minimizing the unwanted side effects. In the current pharmacokinetic paradigm, the designed therapeutic dose for a patient is usually based on the desired target concentration of the drug substance in the patient's central compartment body fluids, that is, blood, or the target site of action.

Any changes in the patient's profile or pathophysiological status that may affect the drug's pharmacokinetics can change the drug's concentration reached in the patient's body fluids for the same dose. These changes, for example, can include patient-to-patient differences in body weight, body surface area (BSA), age, and renal function. The usual adult dose mentioned for most medications reflects the amount of drug required for an average 180-lb adult with normal body functions. The drug's dose for an individual patient is often adjusted to reflect one more of these differences, so as to optimize the patient's exposure to the drug substance.

5.5.1 Dosage adjustment based on body weight or surface area

In many cases, the target dose is expressed in terms of BSA or body weight. For example, meperidine hydrochloride (Demerol®) has a dose of 6 mg/kg/day in divided doses to be taken 4–6 times daily, while isoniazid has a recommended daily dose of 450 mg/m² BSA/day to be administered in a single dose. Therefore, the daily dose is calculated based on the patient's weight or BSA and divided by the number of doses per day to determine an individual dose amount. A set of doses administered over a period of time as a part of a treatment plan is termed dosage regimen.

For example, for a patient of 180-lb body weight, the daily dose of meperidine hydrochloride would be $220/2.2 \times 6 = 600$ mg. For a patient recommended q.i.d. (Latin, quaque in die, four times a day) dosing for 3 days, the dose would be $600/4 = 150$ mg/dose. Therefore, the patient may take three tablets of 50 mg four times a day. The total number of tablets to be dispensed would be $= 3 \times 4 \times 3 = 36$ tablets.

Dosage calculation based on the BSA is often utilized for the IV administration of drugs and fluids. The BSA can be calculated by Mosteller's formula, using the body weight and height information as follows:

$$\text{BSA} = \sqrt{\frac{\text{weight (kg) height (cm)}}{3600}} \tag{5.5}$$

Another, more common, approach to the estimation of BSA is the use of a nomogram (graphical calculation device). Figure 5.1 illustrates a typical adult nomogram. To estimate the surface area, use a ruler to mark the patient's height and weight in his or her respective scales in a straight line. The point at which this straight line intersects the surface area line is the BSA of the patient.

5.5.2 Calculation of children's dose

In addition to the height and the weight, the BSA is also a function of the age and gender of an individual. For example, the average BSA of an adult men (~1.9 m²) is higher than that of an adult women (~1.6 m²) and children (~1.1 – 1.3 m² for 9- to 13-year-old children). An average adult's (150–154 lbs) BSA is assumed to be 1.73 m². This is often used in the calculation of children's doses. For example,

$$\text{Child's dose} = \text{Adult dose}\ \frac{\text{Child's BSA in m}^2}{1.73\,\text{m}^2} \tag{5.6}$$

Estimation of BSA for children uses a different nomogram, illustrated in Figure 5.2.

Less frequently, a child's dose is also calculated using the age of child in months (Fried's rule) or years (Young's rule) or using the weight of the child in pounds (Clark's rule). The formulas are illustrated as follows:

$$\text{Child's dose} = \text{Adult dose} \times \frac{\text{Child's age in months}}{150\ \text{months}}$$

$$\text{Child's dose} = \text{Adult dose} \times \frac{\text{Child's age in years}}{\text{Child's age in years} + 12\ \text{years}}$$

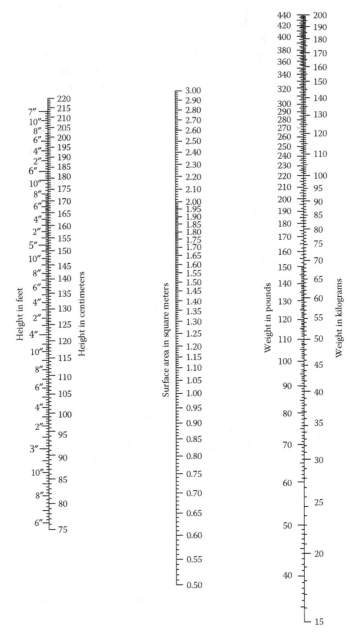

Figure 5.1 Example of a typical adult nomogram for the calculation of body surface area for patients weighing more than 65 lbs or 3-feet tall. (From http://www.smm.org.)

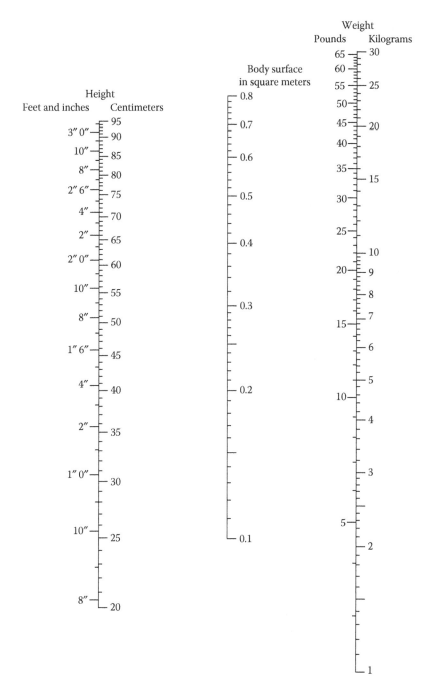

Figure 5.2 **Example of a typical child nomogram for the calculation of body surface area for patients weighing less than 65 lbs or 3 feet tall. (From http://www.smm.org.)**

$$\text{Child's dose} = \text{Adult dose} \ \frac{\text{Child's weight in pounds}}{150 \text{ pounds}}$$

The choice of a formula for dose calculation depends on the conventional practice of the pharmacy or hospital for a given drug. Attention should also be paid to the overall metabolic status of the patient and the therapeutic index of the drug. For drugs eliminated by the kidney, the renal function, measured by creatinine clearance (Cr Cl), plays an important role in dose adjustment of potent compounds. Creatinine clearance of greater than 80 mL/min is considered normal. For compromised Cr Cl, the formularies usually have a recommended table of doses, depending on the therapeutic index of the drug and the percentage of drug eliminated by the kidney.

5.5.3 Dose adjustment for toxic compounds

For the administration of highly toxic compounds with a narrow therapeutic window, such as the cytotoxic anticancer compounds, dosage calculation becomes very critical.[2] These compounds are dosed at very high levels, close to but lower than their maximum tolerated dose (MTD), to maximize their therapeutic benefit to the patient. Therefore, interpatient variability in drug exposure has serious implications on drug effectiveness and toxicity to the patients. The variation in drug exposure arises from differences in drug metabolism and elimination. For example, the total body clearance of carboplatin can range from 20 to 200 mL/min owing to interpatient differences in renal function, since most of the drug is eliminated by glomerular filtration.[3] Similarly, topotecan clearance correlates with renal function.[4]

Different dosage adjustment strategies are followed in these cases, depending on the drug being administered. For drugs with clinically established exposure–physiological parameter correlations, dosage adjustment for an individual patient is done *a priori*, based on the patient's physiological parameters, such as genotype and/or phenotype of the metabolizing enzymes, renal clearance, serum protein, or hepatic function. In addition, for drugs that are dosed repeatedly or continuously, dosage modification can be based on the measurement of blood levels of the drug and toxicities in the patient, for example, for etoposide and fluorouracil.[5] Another dose individualization strategy involves administration of a low test dose of the compound to determine the exact pharmacokinetic parameters for an individual patient, followed by modifying the dose to achieve a target drug exposure. In other cases, clinical oncologists frequently use the BSA for drug dose scaling between individuals. Other physiological scaling parameters, such as age, gender, weight, and body mass index, are also used in specific circumstances.[6]

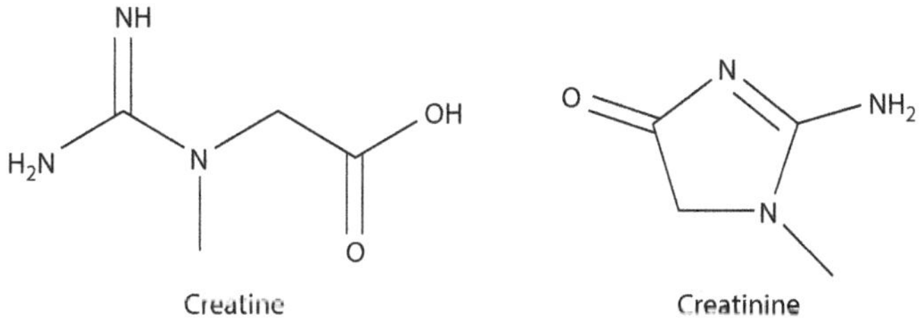

Figure 5.3 Structures of creatine and creatinine.

5.5.4 Dose adjustment based on creatinine clearance

Renal function is often determined in terms of a patient's Cr Cl. Creatinine is a cyclic derivative of the nitrogenous organic acid, creatine (Figure 5.3), found in the muscle. Creatinine is eliminated by filtration through the kidneys and is not reabsorbed. Therefore, the correlation of its blood and urine levels is an indication of the rate of filtration of blood plasma through the kidney (glomerular filtration rate [GFR]), which indicates renal function. Glomerular filtration rate can be calculated using the concentration of a chemical, such as inulin, that is freely filtered through the kidney but not secreted or reabsorbed.

$$GFR = \frac{\text{Urine concentration} \times \text{Urine flow}}{\text{Plasma concentration}}$$

The use of creatinine is preferred over inulin, since extraneous administration is not required for creatinine. However, a small amount of creatinine is also secreted by peritubular capillaries, which can contribute to some error (overestimation) in the calculation of Cr Cl. However, this error becomes significant only in the cases of severe renal dysfunction.

Creatinine clearance is estimated by determining blood creatinine concentration (which is relatively steady) and the amount of creatinine secreted in urine collected over a period of 24 h. For example, if 2 mg/mL of creatinine is detected in 1 L of urine collected over a period of 24 h and the blood creatinine concentration is 0.01 mg/mL, then:

$$Cr\ Cl = \frac{(2\ \text{mg/mL} \times 1000\ \text{mL}/24\ \text{h} \times 60\ \text{min/h})}{0.01\ \text{mg/mL}} = \frac{1.4\ \text{mg/min}}{0.01\ \text{mg/mL}} = 140\ \text{mL/min}$$

This is indicative of the rate of filtration of plasma volume through the kidneys per unit time. Creatinine clearance is often also corrected for the BSA to normalize dose calculation. Assuming 1.73 m² as the average-sized man's BSA, Cr Cl is expressed as:

$$Cr\ Cl\ (corrected) = Cr\ Cl \times \frac{1.73}{BSA}\ mL/min/1.73m^2$$

Creatinine clearance estimation requires and assumes complete urine collection over a 24 h period. To avoid this assumption for outpatients, creatinine clearance can be estimated on the basis of serum creatinine level alone. For example, Cockcroft–Gault formula estimates creatinine clearance as:

$$Cr\ Cl = \frac{(140 - age) \times weight(kg) \times [0.85, if\ female]}{72 \times serum\ creatinine\ level\ (mg/dL)}$$

The normal range of GFR is 100–130 mL/min/1.73 m². It varies with age, race, and kidney function. Glomerular filtration rate correlates with different stages of chronic kidney disease (CKD) as follows:

Stage 1 CKD—GFR greater than 90 mL/min/1.73 m²: normal
Stage 2 CKD—GFR 60–89 mL/min/1.73 m²: mild
Stage 3 CKD—GFR 30–59 mL/min/1.73 m²: moderate
Stage 4 CKD—GFR 15–29 mL/min/1.73 m²: severe
Stage 5 CKD—GFR less than 15 mL/min/1.73 m²: kidney failure

Dose adjustment based on Cr Cl are provided for most drugs by the manufacturers based on the results of clinical trials. These are mainly based on the percentage of drug eliminated by the kidneys. For highly toxic compounds, Cr Cl is utilized for the calculation of pharmacokinetic parameter, such as the elimination rate constant, which is then used with the drug's pharmacokinetic model for dose calculation.

The dosage regimen for a renal-compromised patient is usually adjusted by either reducing the dose or prolonging the dosing interval. Reduction in dose is recommended for cases where relatively constant blood level is desired, for example, β-lactam antibiotics. For drugs whose efficacy may be related to their peak level, for example, fluoroquinolone antibiotics, prolongation of the dosing interval is recommended.

5.6 STATISTICAL MEASURES

A pharmacist needs to be aware of how the scientific data are generated and interpreted in the modern *evidence-based medicine*. This is important not only for the adequate appreciation and interpretation of new research findings but also for an understanding of conventionally well-established practices in medicine. This section outlines the basic concepts utilized in the generation and interpretation of data. It assumes the background knowledge of experimental design and random sampling.

5.6.1 Measures of central tendency

When a collection of data is available, it can be arranged in an *array*. An array is a collection of data arranged in a systematic manner, such as listing a set of values in an ascending or descending order of their magnitude. The data can be analyzed in terms of their *frequency distribution*. The frequency distribution is constructed by identifying the number of times a value repeats itself (frequency of occurrence of such value). This information can be plotted in a two-dimensional x–y plot, with the x-axis representing the increasing order of values and the y-axis representing their frequency of occurrence. The frequency distribution can also be organized to represent a set of ranges of values, rather than individual values, with the frequency representing all data points that fall within the given ranges. An x–y plot of this range of values can produce a series of columns, called a *histogram*. These approaches both reduce and organize the data for easy interpretation.

Frequently, when the data are organized in a frequency distribution, a normal distribution is obtained (Figure 5.4).

A review of the normal distribution curve indicates that the data tend to be more frequent for a given set of values, which are usually toward the center of the numerical distribution of data values. This is called *central tendency*. The numeric location of the central tendency can be stated in one of three ways: mean, median, and mode.

- *Mean*: The arithmetic mean of a data is the sum of observations divided by the number of observations. The mean describes the central location of the data.

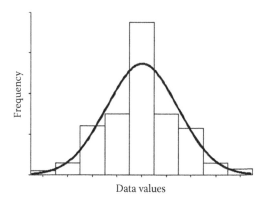

Figure 5.4 A normal distribution. Normal distribution of data can be represented by (a) a frequency distribution (histogram), (b) a curve passing through the medians of the frequency distribution, and (c) discrete data points.

- *Median*: The median is the numeric value of a data point that falls in the middle when counting the set of values after arranging them in an ascending or descending order.
- *Mode*: Mode is the value that occurs most frequently in a set of data.

Either of these values tends to indicate the numeric point in the spread of the data that all observations tend to lean toward, which can be interpreted as the expected value of a data set. The *expected value* of a distribution is the average, or the first moment, over the entire distribution. Each and every value in the data set is not the expected value due to random variation or errors in experimentation or data collection.

5.6.2 Measures of dispersion

In addition to knowing the central tendency of the data, one needs to appreciate the level of *distribution* or variation in the individual data values. This indicates how closely the data set represents a central tendency or value. For example, the four sets of data represented by the normal distribution curves in Figure 5.5 show increasing level of dispersion from the central tendency in the order a < b < c < d.

Distribution of a set of data can be quantified by one or more of the following numerical values:

- *Range*: It represents the difference between the highest and the lowest values in a data set.

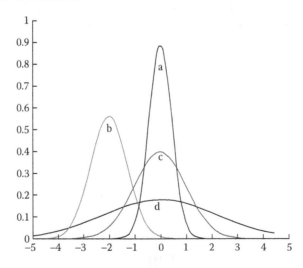

Figure 5.5 Illustration of variability in four different data sets following normal distribution. The level of dispersion from the central tendency is d > c > a, even though their means are the same. Data set b represents a difference of mean in addition to dispersion.

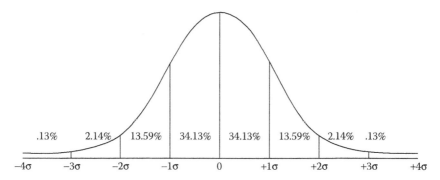

Figure 5.6 Illustration of spread of data (from the hypothetical mean of 0) in a normal distribution as a function of the standard deviation of the population (σ). The probability of finding data values at illustrated multiples of standard deviation is indicated in the figure as a percentage number.

- *Variance and standard deviation*: Variance represents the mean of square of deviation of all individual values in the data set from the mean of the set of data set. It is calculated by subtracting each individual value from the mean, squaring it, and dividing the sum of this squared difference by $n-1$, where n is the number of samples in the data set. Standard deviation is the square root of the variance.

Standard deviation is commonly used to interpret the spread of the data. As indicated in Figure 5.6, assuming a normal sample distribution, the standard deviation of a sample set (symbol: s) indicates the percentage of data set values that fall on either side of the mean value of this data set. As illustrated in the figure, 68.26% of values fall within ±1 s of the mean, 95.44% fall within ±2 s of the mean, and 99.72% fall within ±3 s of the mean. It would be noted that the greater the value of s compared with the mean, the more the spread of the data. This could indicate either lower precision of measurement and/or greater error in data collection.

5.6.3 Sample probability distributions

A probability distribution represents the probability of occurrence of each value of a discrete random variable or the probability of each value of a continuous random variable falling within a given interval. Hence, a probability distribution can be either:

- *Discrete probability distribution*: It reflects a finite and countable set of data whose probability is one.
- *Continuous probability distribution*: It reflects the probability of occurrence of a value in terms of its probability density function, which can be defined within an interval.

5.6.3.1 Normal distribution

The preceding examples assumed a normal frequency or probability distribution of the data set. Normal distribution, also known as the Gaussian distribution, reflects the tendency of the data to cluster around the mean from both directions. It is a continuous probability distribution and forms a typical bell-shaped curve. A data set following a normal distribution is indicative of the additive nature of underlying factors.

5.6.3.2 Log-normal distribution

A log-normal distribution refers to the probability distribution of a variable whose logarithm is normally distributed, such that for a variable y, log y is normally distributed. The base of the logarithmic function does not make a difference to the distribution pattern of the variable. A log-normal distribution typically represents a multiplicative effect of underlying factors.

5.6.3.3 Binomial distribution

Binomial distribution is a discrete probability distribution that reflects the number of a given outcome in a sequence of experiments with only two outcomes, each of which yields a given outcome with a defined probability. Such an experiment is frequently called a success/failure experiment or Bernoulli experiment, with n repetitions and p as the probability of each successful outcome.

5.6.3.4 Poisson distribution

Poisson distribution represents the probability of n occurrences of an event over a period of time or space, given the average number of occurrences of the event. For example, if the lyophilization process fails, on an average, in five batches per year, the Poisson distribution can be used to calculate the probability of 0, 1, 2, 3, 4, 5, ... failed lyophilization processes for a given year. Although both Poisson and binomial distributions are based on discrete random variables, the binomial distribution assumes a finite number of possible outcomes, while the Poisson distribution does not. The Poisson distribution is usually applied in cases where the mean is much smaller than the maximum data value possible, such as in radioactive decay.

5.6.3.5 Student's t-distribution

The Student's t-distribution is a continuous probability distribution that is used to estimate the mean of a normally distributed population when the sample size is small (population standard deviation is unknown). The t-distribution is based on the central limit theorem that the sampling distribution of a sample statistic, such as the sample mean (x), follows a normal distribution as n gets large. The t-distribution is a continuous probability distribution of the t-statistic or t-score, defined as:

$$t = \frac{x - \mu}{s/\sqrt{n}}$$

where:
    μ is the population mean
    s is the sample standard deviation
    n is the sample size

The shape of the t-distribution varies with the sample size or the number of degrees of freedom (df) of the sample. The degrees of freedom represent the number of values in the final calculation of a statistic that can freely vary and is calculated as $n-1$ for n number of samples. It is used as a measure of the amount of data that is used for the estimation of a given statistical parameter.

The t-distribution is characterized by having a mean of 0 and variance of always greater than 1. The variance approaches 1, and the t-distribution approaches the standard normal distribution at high sample sizes.

Knowing the sample mean, standard deviation, size, and the (assumed) population mean, a t-score or t-statistic can be calculated. Each t-score is associated with a unique cumulative probability of finding a sample mean less than or equal to the chosen sample mean for a random sample of the same size. The term t_α denotes a t-score that has a cumulative probability of (1α). For example, for a cumulative probability of occurrence of 95%, $\alpha = (1-95/100) = 0.05$. Hence, the t-score corresponding to this probability would be represented as $t_{0.05}$. The t-score for a given probability varies with the degrees of freedom (DF) of the sample. Thus, $t_{0.05}$ at DF of 2 is 2.92, whereas $t_{0.05}$ at DF of 20 is 1.725. In addition, since t-distribution is symmetric with a mean of zero, $t_{0.05} = -t_{0.95}$, or vice versa.

The t-statistic helps determine the probability of occurrence of a given sample mean when the (hypothetical or target) population mean is known. In other words, it can help determine the probability that the selected sample comes from the population with the given (hypothetical or target) mean. For example, during tablet compression for a target average tablet weight of 100 mg, a sample of 10 tablets is weighed. The average weight of 10 tablets was 90 mg, with a standard deviation of 35 mg. What is the probability that the tablet compression operation is proceeding at its target average tablet weight of 100 mg? To compute this probability, a t-score can be calculated as follows:

$$t = \frac{x - \mu}{s/\sqrt{n}} = \frac{90 - 100}{35/\sqrt{10}} = -0.9035$$

This t-score corresponds to 19% probability of occurrence (using standard probability distribution tables). Thus, if the tableting operation is performing at target, then there is a 19% chance that the sample mean would fall below 90, based on a sample of 10 tablets. Therefore, there is no evidence that the machine is off target. However, due to the large variability and small sample size, we cannot say that it is at target. A confidence interval would show that the target mean could be any value over a large range, which would include 100. Thus, it is likely that the tableting unit operation is performing at the target average tablet weight of 100 mg. On the other hand, if the sample of 10 tablets had a standard deviation of 15 mg, the t-score would be 2.1082, which corresponds to the probability of occurrence of 3%. These data would indicate that the tableting unit operation is probably not performing at its target average tablet weight of 100 mg.

This distribution forms the basis of the t-test of significance, which can help determine the following:

- Statistical significance of the difference between two sample means.
- Confidence intervals for the difference between two population means.

5.6.3.6 Chi-square distribution

Chi-square (χ^2) distribution represents the squared ratio of sample to population standard deviation as a function of the sample size used for computing the sample standard deviation. This distribution is used to estimate the probability ranges for the standard deviation values for a given sample size.

Mathematically, the chi-square distribution represents the distribution of the chi-square statistic, which represents the squared ratio of the standard deviation of a sample (s) to that of the population (σ), multiplied by the degrees of freedom of the sample.

$$\chi^2 = (n-1) \times \frac{s^2}{\sigma^2}$$

The shape of the chi-square distribution curve varies as a function of the sample size or the degrees of freedom. As the number of degrees of freedom increases, the chi-square curve approaches a normal distribution.

The chi-square distribution is constructed such that the total area under the curve is 1. This allows the estimation of cumulative probability of a given value of the chi-square parameter. Given this value, the probability of occurrence of the chi-square parameter above the obtained value can be obtained.

For example, for a population of N = 100 with the population standard deviation of 5, the probability of obtaining a sample standard deviation of 6 when testing n = 10 samples is given by the chi-square parameter,

$$\chi^2 = (n-1) \times \frac{s^2}{\sigma^2} = (10-1) \times \frac{6^2}{5^2} = 12.96$$

Using the chi-square distribution for the given degrees of freedom, the probability of occurrence of chi-square parameter less than 12.96 is 0.84. Hence, the probability of occurrence of $s > 6$ is $1 - 0.84 = 0.16$, or 16%.

5.7 TESTS OF STATISTICAL SIGNIFICANCE

The need for the statistical tests of significance is exemplified by questions posed in comparing two data sets. The tests of statistical significance are intended to compare two sets of data to address the question whether these data sets represent two different populations, that is, whether they are inherently different or not. A data set is a sample presumed to be taken from an infinite population of data that would represent infinite repetitions of the experiment. If two samples are taken from the same population, they would have a greater overlap with each other than if the samples belong to two different populations. As shown in Figure 5.7, samples 1 and 2 apparently come from two different populations in subfigure A but not in subfigure B. However, it is difficult to comment on whether the samples

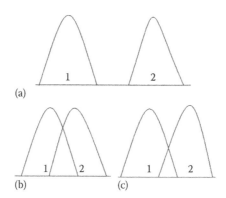

Figure 5.7 Three scenarios that may be encountered when comparing data sets from two samples, 1 and 2. In case A, the sample values of the two samples are significantly different by a large numeric value, indicating that the samples most likely represent two different populations. In case B, the sample values are so close to each other that it is very likely that both samples came from the same population and are not different from each other. In case C, the differences in sample values are intermediate. In the case of scenario C, it is difficult to make an assessment whether the two samples are really different from each other. In such cases, the tests of significance provide a statistical basis for decision making.

come from different populations in subfigure C. The tests of statistical significance are designed to answer questions such as these.

5.7.1 Parametric and nonparametric tests

A sample or a population can be described by the mean and variance of all observations, which represent statistical parameters, with an assumption of a known underlying population distribution. Alternatively, a nonparametric measure, such as median, can be used, which assume an underlying population distribution but not necessarily a known distribution.

Accordingly, statistical tests of significance can be parametric or nonparametric:

- Parametric tests of significance are based on parametric measures of distribution of data, viz., mean and variance of the data set. They assume a specific and known distribution of the underlying population.
- Nonparametric tests of significance are based on nonparametric descriptors of distribution of data, viz., median and ranks of the data values. They do not make the assumption about that the underlying distribution of the population is known.

Parametric tests are more powerful (less probability of type II error, described later) than the nonparametric tests, since they use more information about the samples. They are frequently used to provide information, such as interaction between two variables in a factorial design of experiments. However, they are also more sensitive to skewness in the distribution of data and the presence of outliers in the samples. Therefore, nonparametric tests may be preferred for skewed distributions.

Parametric tests are exemplified by t-test, chi square test, and analysis of variance (ANOVA). Nonparametric tests are exemplified by Wilcoxon, Kruskal– Wallis, and Mann– Whitney tests. The parametric tests will be described in more detail in the following sections.

5.7.2 Null and alternate hypothesis

Statistical tests of significance are designed to answer this and similar questions with a given level of confidence and power, expressed in numerical terms. Statistical tests of significance can be used, for example, to test the hypothesis that (a) a sampled data set comes from a single population or that (b) two sampled data sets come from a single population. A statistical hypothesis represents an assumption about a population parameter. This assumption may or may not be true and is sought to be tested using the statistical parameters obtained from a sample. For example, if the statistical tests of significance test the hypothesis that a given variation within or among data sets occurred purely by chance, it would be termed the *null hypothesis*. In this case, therefore, the null hypothesis is the hypothesis of no difference. If the null hypothesis cannot be proven at the selected levels of confidence and power of the test, the *alternate hypothesis* is assumed to hold true. The alternate hypothesis indicates that the sample observations are influenced by some nonrandom cause.

5.7.3 Steps of hypothesis testing

The process of testing a hypothesis involves the following general steps:

1. Ask the question (for a practical situation) that can be addressed using one of the statistical tests of significance.
2. Select the appropriate test of significance to be used and verify the validity of underlying assumptions.
3. State null and alternate hypothesis.
4. Define significance level (e.g., $\alpha = 0.01$, 0.05, or 0.1, which indicates 1%, 5%, or 10% probability of occurrence of given differences just by chance). Lower the significance level, greater the chance of not detecting the differences when they actually do exist.
5. Define sample size. Sample size affects the power of the significance test. Higher the sample size, higher the power, that is, greater the chance of detecting the differences when they actually do exist.
6. Compute the test statistic.
7. Identify the probability (p) of obtaining a test statistic as extreme as the calculated test statistic for the calculated degrees of freedom, using standard probability distribution tables.
8. Compare this probability with the level of significance desired. If $p_{\text{sample at }\alpha} < p$, null hypothesis is rejected. If $p_{\text{sample at }\alpha} \geq p$, null hypothesis cannot be rejected.

5.7.4 One-tailed and two-tailed hypothesis tests

The null and alternate hypotheses can be stated such that the null hypothesis is rejected when the test statistic is higher or lower than a given value, or both. The first two are called one-tailed hypothesis, while the latter is termed two-tailed hypothesis. For example, if μ_1 and μ_2 represent the means of two populations and H_0 represents the null hypothesis, (H_0: $\mu_1 - \mu_2 \geq d$) or (H_0: $\mu_1 - \mu_2 \leq d$) would be one-tailed hypothesis, since H_0 would be rejected when ($\mu_1 - \mu_2 < d$) and ($\mu_1 - \mu_2 > d$), respectively. However, (H_0: $\mu_1 - \mu_2 = d$) is a two-tailed hypothesis, since the null hypothesis would be rejected in both cases of ($\mu_1 - \mu_2 < d$) and ($\mu_1 - \mu_2 > d$).

The appropriate statement of null hypothesis depends on the practical situation being addressed. For example,

- If a sample of tablets were collected during a production run of tableting unit operation and tested for average tablet weight, the question could be asked whether the average tablet weight is the target tablet weight. In this case: (H_0: $Weight_{sample} - Weight_{target} = 0$) or ($H_0$: $Weight_{sample} = Weight_{target}$) would be a two-tailed hypothesis test, since the null hypothesis would be rejected when the sample weight is both higher than or lower than the target weight.
- If a sample of tablets were collected during a production run of the coating unit operation and tested for coating weight build-up on the tablets, the question could be asked whether the coating weight build-up has reached the target weight build-up of 3% w/w. In this case: (H_0: $Weight_{sample} - Weight_{target} \geq 0$) would be a one-tailed hypothesis test, since the null hypothesis would be rejected only if the sample weight is less than the target weight.

5.7.5 Regions of acceptance and rejection

The regions of acceptance and rejection of a hypothesis refer to regions in the probability distribution of the sample's test statistic. Assuming that the null hypothesis is true, a sample's test statistic is normally distributed, with the shape of the distribution defined by the degrees of freedom of the sample. Therefore, the probability of finding a given value of the test statistic can be defined by this distribution curve. For example, Figure 5.8a shows the normal distribution of a test statistic, with a vertical line to the right indicating the value of the test statistic associated with a probability of occurrence (α) of 0.05, or 5%, by random chance, or P_α. Decreasing the level of significance (α) increases the rigor of the test; that is, the differences must be really significant to be detected.

For a one-tailed hypothesis test (Figure 5.8a), the region of rejection lies on one (right) side of this distribution. If the test statistic value obtained for the sample in question is higher than P_α, the test statistic in the sample is assumed to lie in the region of rejection and the null hypothesis is rejected at the chosen level of significance (α). Region of acceptance in this case is defined as ($-\infty$ to P_α).

For a two-tailed hypothesis test (Figure 5.8b), the region of rejection lies on either side of the distribution. If the test statistic value obtained for the sample in question is higher than P_α or lower than $-P_\alpha$, the test statistic in the sample is assumed to lie in the region of rejection and the null hypothesis is rejected at the chosen level of significance (α). Region of acceptance in this case is defined as ($-P_\alpha$ to P_α).

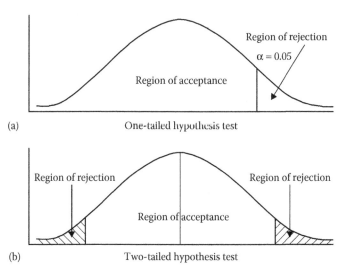

(a)   One-tailed hypothesis test

(b)   Two-tailed hypothesis test

Figure 5.8 An illustration of regions of acceptance and rejection in a normal probability distribution. Knowing the probability of occurrence of sample values at either extremes from the mean as a function of the standard deviation (Figure 5.6), a given level of significance (e.g., $\alpha = 0.05$) can quantify a *cut-off point*, indicated by a vertical line in the plot. This vertical line in Figure 5.8a represents 5% chance of occurrence of data values. Therefore, any value higher than the indicated α line has a lower than 5% chance of occurrence and is said to fail in the region of rejection. This is one-tailed hypothesis, since data values on only one side of the mean are being considered for hypothesis testing. This side could be the positive side, as indicated in Figure 5.8a, or the negative side, which would be indicated by the α line on the left of the mean. In a two-tailed hypothesis testing (Figure 5.8b), data values on both positive and negative sides of the mean are considered. Data values that are more extreme than the α line are said to fall in the region of rejection. All other data values are considered in the region of acceptance.

5.7.6 Probability value and power of a test

The level of significance of test results is indicated by the probability value (abbreviated as *p-value*). The *p*-value is the fractional probability of accepting the null hypothesis, assuming that the null hypothesis is true. In other words, lower the *p*-value of the test, expressed as fractional probability (e.g., 0.01, 0.05, or 0.1, representing 1%, 5%, or 10% probability, respectively), greater the chance of accepting the null hypothesis and not detecting differences between two samples. Lower *p*-value indicates greater difference between two samples. The commonly used probability level for accepting the null hypothesis is 5%, corresponding to the *p*-value of 0.05.

The power of a test of significance is the probability of rejecting the null hypothesis, assuming that the null hypothesis is not true. In other words, higher the power of the test, expressed in %, greater the chance that true differences between two different sample sets would be detected. Power of a test can be increased by increasing the sample size. The commonly accepted power of a test is 80%.

5.7.7 Types of error

Conducting a test of significance can result in two types of errors in assessing the difference in the chosen test statistic:

- Type I error is a false positive in finding the difference and inappropriately rejected null hypothesis. This is the error of rejecting a null hypothesis when it is actually true. In other words, type I error is the error of finding the difference between the two samples when they are actually not different. The probability of type I error is denoted by α.

  The probability of type I error is higher when the chosen level of significance, α, is higher. Therefore, using lower α tends to reduce the probability of a type I error.
- Type II error is a false negative in finding the difference and inappropriately failing to reject null hypothesis. This is the error of not rejecting a null hypothesis when it is actually not true. In other words, type II error is the error of not finding difference between the two samples when they are actually different. The probability of type II error is denoted by β.

  The probability of type II error is higher when the chosen power of the test, β, is lower. Therefore, using higher β tends to reduce the probability of a type II error.

5.7.8 Questions addressed by tests of significance

Tests of significance are designed to answer specific types of questions based on a selected test statistic and a probability distribution of the test statistic. For example, the differences between means are tested using t-test, the differences between proportions are tested using z-test, and the differences in the frequency of a categorical variable are tested using χ^2 test. Commonly used tests of significance, an example situation, underlying assumptions of tests, statement of null hypothesis, and calculations of the test statistic are summarized in Table 5.6.

It should be noted that these tests of significance invariably involve:

- The calculation of a test statistic, which represents the difference between the expected and the observed values, or the values of two samples. It also takes into account the variability in the sample through incorporation of standard error. The calculation of test statistic involves quantifying the extent of observed differences vis-à-vis the variability.

Table 5.6 Statistical tests of significance

Test question or situation and the test of significance to use	Example	Statement of hypothesis (for the example given)	Equations and abbreviations	Underlying assumptions
To test difference between two means, use two-sample t-test.	Two batches of tablets were manufactured, with an average tablet weight of 200 mg. A sample of 100 tablets each was tested from each of these batches. Do the two batches have different average tablet weight?	$H_0: \mu_1 = \mu_2$ $H_1: \mu_1 \neq \mu_2$ OR $H_0: \mu_1 - \mu_2 = d$ $H_1: \mu_1 - \mu_2 \neq d$ where, $d = 0$	$t = \dfrac{(mean_1 - mean_2) - d}{SE}$ where, $SE = \sqrt{\dfrac{sd_1^2}{n_1} + \dfrac{sd_2^2}{n_2}}$ and DF = $(n_1 - 1)$ or $(n_2 - 1)$, whichever is smaller. H_0, null hypothesis; H_1, alternate hypothesis; m, population mean; subscripts refer to different batches, populations, or samples; d, difference between the two means; sd, standard deviation; n, number of observations; subscripts refer to different batches, populations, or samples; mean, sample mean; SE, standard error of the sampling distribution; DF, degrees of freedom; and t, test statistic for the t-distribution.	• Random sampling • Independent samples • Population follows a normal or near-normal distribution • Population size is at least 10-fold higher than the sample size
To test difference between matched pairs, use matched-pairs t-test.	Tablet friability test was conducted on a batch on 10 different occasions. Total tablet weight was recorded before and after the friability test in each case. Is tablet friability >1%?	$H_0: \mu_d > D$ $H_1: \mu_d \leq D$ where, $D = 1$	$t = \dfrac{\bar{d} - D}{SE}$ where, $SE = \sqrt{\dfrac{\sum(d_i - \bar{d})^2/(n-1)}{n}}$ and DF = $n - 1$, null hypothesis; H_1, alternate hypothesis; D, difference between the two population means; D, hypothesized value of the mean difference between the matched pairs; sd(D), standard deviation of differences of matched pairs; d, difference for the matched pair i; $\bar{d}$, mean of difference between all matched pairs; n, number of pairs; SE, standard error; DF, degrees of freedom; and $\bar{d}$, mean difference between matched pairs.	• Random sampling • Data sets not independent • Population follows a normal or near-normal distribution

(Continued)

Table 5.6 (Continued) Statistical tests of significance

Test question or situation and the test of significance to use	Example	Statement of hypothesis (for the example given)	Equations and abbreviations	Underlying assumptions
To test difference between two proportions, use the two-proportion z-test.	Edge-chipping defects in tablets were counted in a sample of 400 coated and 350 uncoated tablets. Twenty-five uncoated and 22 coated tablets had this defect. Is edge chipping more likely for the coated or the uncoated tablets?	$H_0: P_1 = P_2$ $H_1: P_1 \neq P_2$ OR $H_0: P_1 - P_2 = d$ $H_1: P_1 - P_2 \neq d$ where, $d = 0$	$z = \dfrac{p_1 - p_2}{SE}$, where $SE = \sqrt{P_{pooled} \times (1 - P_{pooled}) \times \left(\dfrac{1}{n_1} + \dfrac{1}{n_2}\right)}$ $P_{pooled} = \dfrac{P_1 \times n_1 + P_2 \times n_2}{n_1 + n_2}$ H_0, null hypothesis; H_1, alternate hypothesis; P, proportion of observations in sample; subscripts refer to different batches, populations, or samples; P_{pooled}, pooled sample proportion; d, difference between the two proportions; and SE, standard error of the pooled sample proportion.	• Random sampling • Independent samples • Sample includes at least 10 events and 10 nonevents for calculating the proportion • Population size is at least 10-fold higher than sample size
To test whether a categorical variable follows a hypothesized frequency distribution, use the chi-square goodness-of-fit test.	A controlled-release capsule formulation uses drug microspheres encapsulated in hard gelatin capsules. Three types of microspheres are encapsulated: 30% w/w of immediate release, 35% w/w of delayed release by 2 h; and 35% w/w of delayed release by 4 h. In an analysis of 20 capsules, the proportions of these components were 23.2% w/w, 37.9% w/w, and 38.9% w/w. Does this sample represent the targeted amount for each capsule in the formulation?	$H_0: P_s = P_h$ $H_1: P_s \neq P_h$	$\chi^2 = \sum \dfrac{(O_i - E_i)^2}{E_i}$ where, $E_i = n \times P_i$ and DF $= k - 1$ H_0, null hypothesis; H_1, alternate hypothesis; P_s, proportion in the sample; P_h, hypothesized proportion in the population; O_i, observed proportion of the ith categorical variable in the sample; E_i, expected proportion of the ith categorical variable in the sample; n, sample size; P_i, hypothesized proportion of the ith categorical variable in the population; DF, degrees of freedom; and k, number of categorical variables in the sample (e.g., $k = 3$ for the example cited in column 2).	• Random sampling • Categorical variable • Population size is at least 10-fold higher than sample size • Expected value for each categorical variable is at least 5

(Continued)

Table 5.6 (Continued) Statistical tests of significance

Test question or situation and the test of significance to use	Example	Statement of hypothesis (for the example given)	Equations and abbreviations	Underlying assumptions
To test whether a categorical variable follows the same frequency distribution in two or more populations, use the chi-square test of homogeneity.	An antihypertensive drug was tested in 320 male and 290 female human volunteers. Three effects of this drug were tracked—reduction in blood pressure of at least 20 mm Hg, and skin rashes and nausea as adverse events. The proportion of populations showing these events were 220, 12, and 18 for males, and 236, 9, and 19 for females, respectively. Is it likely that the drug's effects are affected by gender?	$H_0: P_{i,r} = P_{i,r}$ $H_1: P_{i,r} \neq P_{i,r}$ for each categorical variable i in each population r.	$$\chi^2 = \sum \frac{(O_{i,r} - E_{i,r})^2}{E_{i,r}}$$ where, $E_{i,r} = \dfrac{n_i \times n_r}{n}$ and $DF = (i-1)(r-1)$ H_0, null hypothesis; H_1, alternate hypothesis; $P_{i,r}$, proportion of ith variable in rth population; $O_{i,r}$, observed proportion of the ith variable in rth population; $E_{i,r}$, expected proportion of the ith variable in rth population; n_i, total number of observations of the ith variable across all populations; n_r, total number of observations of the rth population; n, total number of observations of all populations; and DF, degrees of freedom.	• Random sampling • Categorical variable • Population size is at least 10-fold higher than sample size • Expected value for each categorical variable in each sample is at least 5

- Identifying the probability value associated with the test statistic at a given level of significance (P_α) for the given degrees of freedom. The degree of freedom is calculated based on the sample size and sometimes also the number of variables studied. The degrees of freedom affect the distribution plot of the test statistic and thus the P_α value for a given α.

Having calculated the P_α value and the test statistic, the given test of significance is carried out per the steps outlined earlier. For example, if the value of test statistic obtained for a given test of significance is 0.942 and the P_α value at the desired probability of error of 5% is 1.347, the test statistic falls in the region of acceptance. Hence, the null hypothesis cannot be rejected. On the other hand, if the test statistic value were higher than 1.347, the test statistic would fall in the region of rejection. Hence, the null hypothesis would be rejected.

5.7.9 Analysis of variance

The analysis of variance uses differences between means and variances to quantify statistical significance between means of different samples. Any number of samples or subgroups may be compared in an ANOVA experiment. ANOVA is based on the underlying explanation of variation of sample values from the population mean as being a linear combination of the variable effect and random error.

The number of variables (also termed treatments or factors) in an ANOVA experiment can be one (one-way ANOVA), two (two-way ANOVA), or more. Each variable or factor can be studied at different *levels*, indicating the intensity. For example, a clinical study that evaluates one dose of an experimental drug is a one-variable one-level experiment. A study that evaluates two doses of an experimental drug would be a one-variable two-level study. Another study that evaluates three doses of two experimental drugs would be a two-variable three-level study. The level may be a quantitative number, such as the dose in the above examples, or it may be a numerical designation of the presence or intensity of an effect, such as "0" and "1."

5.7.9.1 One-way ANOVA

5.7.9.1.1 Model equation

When sample sets are treated with a single variable at i different levels ($i = 1$, 2, 3, ..., k), the value of each data point is explained as:

$$y_{ij} = \mu + \tau_i + \varepsilon_{ij}$$

where:

y_{ij} represents the jth observation of the ith level of treatment of the variable

μ is the mean of all samples in the experiment
τ_i is the ith treatment effect
ε_{ij} represents random error

Hence, the value of each data point in an experiment is represented in terms of the mean of all samples and deviations arising from the effect of treatment or variable being studied (τ_i) and random variation (ε_{ij}). This equation represents a *one-way ANOVA* model.

5.7.9.1.2 Underlying assumptions

ANOVA is used to test hypotheses regarding means of two or more samples, assuming the following:

- The underlying populations are normally distributed.
- Variances of the underlying populations are approximately equal.
- The errors (ε_{ij}) are random and have a normal and independent distribution, with a mean of zero and a variance of σ_ε^2.

5.7.9.1.3 Fixed- and random-effects model

The one-way ANOVA model quantifies variation in each data point (y_{ij}) from the mean of all data points (μ) as a combination of random variation (ε_{ij}) and the effect of a known variable or treatment (τ). Different subgroups of the experimental data points can be subjected to different levels of the treatment, τ_i, where $i = 1, 2, 3, \dots k$. If the levels of the treatment are fixed, the model is termed *fixed-effects model*. On the other hand, if the levels of the treatment are randomly assigned from several possible levels, the model is termed *random-effects model*.

Whether the levels of a variable or treatment are fixed or random depends on the design of the experiment. A fixed effects model is exemplified by three subgroups of a group of 18 volunteers chosen for a pharmacokinetic study of a given drug at dose levels of 0, 50, and 100 mg. A random effects model would be exemplified by three subgroups of a group of 18 volunteers chosen for a pharmacokinetic study of three different drugs A, B, and C at unknown and variable dose levels (e.g., dose titration by the physician for individualization to the patient). The effects are assumed to be random in the latter case, since the level of the drug is not fixed.

The selection of a study design as a fixed- or random-effects model is critical to the accuracy of data interpretation. The calculation of variance between treatment groups is different between fixed- and random-effects model.

5.7.9.1.4 Null and alternate hypothesis

The null hypothesis (H_0) for a one-way ANOVA experiment would be no difference between the population means of samples treated with different levels of the selected factor. The alternate hypothesis (H_1) states that the means of underlying populations are not equal.

5.7.9.1.5 Calculations for fixed-effects model from first principles

ANOVA is based on the calculation of ratio of variance introduced by the factor and random variations. Although many software tools are currently available that reduce the requirement for tedious calculations, it is important to understand the calculations of statistical tests of significance from first principles.

1. Mean of all samples in the experiment (μ) is calculated by adding all observations and dividing by the total number of samples in the experiment.

$$\mu = \frac{\sum\limits_{i=1}^{k}\sum\limits_{j=1}^{n} y_{ij}}{N}$$

where, y_{ij} represents the jth observation of the ith level of treatment of the variable, there being a total of k treatments ($i = 1, 2, 3, \ldots k$) and n samples per treatment level ($j = 1, 2, 3, \ldots n$), and N being the total sample size, including all treatments and levels.

2. Total sum of squares (SS_T) of all observations is calculated by squaring all observations and subtracting from the mean of all samples in the experiment (μ).

$$SS_T = \sum\limits_{i=1}^{k}\sum\limits_{j=1}^{n} (y_{ij} - \mu)^2$$

3. Sum of squares for the factor studied is the sum of squares between the columns ($SS_{between}$) if each level of the factor is arranged in a column. It is calculated by subtracting the mean value for each column from the mean of all samples, squaring this value, and adding for all columns.

$$SS_{between} = \sum\limits_{j=1}^{n} j \times \left(\frac{\sum\limits_{i=1}^{k} y_j}{k} - \mu \right)^2$$

4. Sum of squares for the random error (SS_{error}) is the difference between the total sum of squares and the sum of squares between and within the columns.

$$SS_{error} = SS_{total} - SS_{between}$$

5. Degrees of freedom are calculated as follows:
Degrees of freedom between groups ($DF_{between}$):

$$DF_{between} = k - 1$$

Degrees of freedom for the error term (DF_{error}):

$$DF_{error} = N - k$$

6. Mean squares for the random error (MS_{error}) and the factor studied ($MS_{between}$) are calculated by dividing their respective sum of squares by their DF.

$$MS_{between} = \frac{SS_{between}}{DF_{between}}$$

$$MS_{error} = \frac{SS_{error}}{DF_{error}}$$

7. An F-ratio is computed as the ratio of mean squares of factor effect to the mean square of error effect.

$$F = \frac{MS_{between}}{MS_{error}}$$

8. Determine critical F-ratio at ($DF_{between}$ and DF_{error}) degrees of freedom for $\alpha = 0.05$.
9. Test the hypothesis. The F-ratio is compared to the P_α value for the F-test at designated degrees of freedom to determine the significance of observed results. Statistical significance of results would indicate that the contribution of the factor's or variable's effect on the observations is significantly greater than the variation that can be ascribed to random error.

5.7.9.1.6 Example of calculations for fixed-effects model

The computation of statistical significance by one-way ANOVA can be illustrated by a case of administration of two doses of a test antihyperlipidemic compound and a placebo to a set of six patients in each group. Hypothetical results of this study in terms of reduction of blood cholesterol level are

Table 5.7 A hypothetical example of a one-way ANOVA experiment

Subject #	Dose = 0	Dose = 50 mg	Dose = 100 mg
1	20	18	28
2	18	25	22
3	14	36	46
4	30	28	29
5	5	15	24
6	12	12	15

Table 5.8 Rephrasing the data in statistical terms for a hypothetical
example of a one-way ANOVA experiment

Subject #	Factor A, level 1 $i = 1$	Factor A, level 2 $i = 2$	Factor A, level 3 $i = 3\ (k = 3)$
$j = 1$	$y_{1,1} = 20$	$y_{2,1} = 18$	$y_{3,1} = 28$
$j = 2$	$y_{1,2} = 18$	$y_{2,2} = 25$	$y_{3,2} = 22$
$j = 3$	$y_{1,3} = 14$	$y_{2,3} = 36$	$y_{3,3} = 46$
$j = 4$	$y_{1,4} = 30$	$y_{2,4} = 28$	$y_{3,4} = 29$
$j = 5$	$y_{1,5} = 5$	$y_{2,5} = 15$	$y_{3,5} = 24$
$j = 6\ (n = 6)$	$y_{1,6} = 12$	$y_{2,6} = 12$	$y_{3,6} = 15$
a	16.5	22.3	27.3
b	30.9	0.1	27.9
c	185.2	0.5	167.1
d	352.8		

summarized in Table 5.7. These data can be rephrased in statistical terms, as presented in Table 5.8.

a. Mean of observations in each group $(i) = \dfrac{\sum\limits_{j=1}^{n} y_i}{n} = \dfrac{\sum\limits_{j=1}^{6} y_i}{6}$

b. $SS_{between,i}\ (for\ each\ i) = \left(\dfrac{\sum\limits_{j=1}^{n} y_i}{n} - \mu \right)^2 = \left(\dfrac{\sum\limits_{j=1}^{6} y_i}{6} - 22.1 \right)^2$

c. $SS_{between,i} \times no.\ of\ obsvns_i = j \times \left(\dfrac{\sum\limits_{j=1}^{n} y_i}{n} - \mu \right)^2 = 6 \times \left(\dfrac{\sum\limits_{j=1}^{6} y_i}{6} - 22.1 \right)^2$

$$\text{d. } SS_{\text{between}} = \sum_{i=1}^{3} j \times \left(\frac{\left(\sum_{j=1}^{n} y_i\right)^2}{n} - \mu \right) = (\text{sum of all } c \text{ values})$$

1. Mean of all samples in the experiment (μ):

$$\mu = \frac{\sum_{i=1}^{k} \sum_{j=1}^{n} y_{ij}}{N} = \frac{397}{6 \times 3} = 22.1$$

2. Total sum of squares of variation in all data points (SS_T):
   The calculations are illustrated in Table 5.9.
   Squaring ($y_{ij}-\mu$) values and adding them together,

$$SS_T = \sum_{i=1}^{k} \sum_{j=1}^{n} (y_{ij} - \mu)^2 = 1656.9$$

3. Sum of squares of variation coming from the factor studied (SS_{between}):
   As calculated in Table 5.8.
4. Sum of squares of variation coming from random error (SS_{error}):
   As calculated in Table 5.8.

$$SS_{\text{total}} = S_{\text{between}} + SS_{\text{error}}$$

$$SS_{\text{error}} = SS_{\text{total}} - SS_{\text{between}}$$

$$SS_{\text{error}} = 1656.9 - 352.8 = 1304.2$$

5. Degrees of freedom (DF):
   Degrees of freedom between groups (DF_{between}):

$$DF_{\text{between}} = k - 1 = 3 - 1 = 2$$

Table 5.9 Calculations for a hypothetical example of a one-way ANOVA experiment

Subject #	Factor A, level 1 $i = 1$	Factor A, level 2 $i = 2$	Factor A, level 3 $i = 3$
$j = 1$	$y_{1,1} - \mu = 20 - 22.1 = -2.1$	$y_{2,1} - \mu = 18 - 22.1 = -4.1$	$y_{3,1} - \mu = 28 - 22.1 = 5.9$
$j = 2$	$y_{1,2} - \mu = 18 - 22.1 = -4.1$	$y_{2,2} - \mu = 25 - 22.1 = 2.9$	$y_{3,2} - \mu = 22 - 22.1 = -0.1$
$j = 3$	$y_{1,3} - \mu = 14 - 22.1 = -8.1$	$y_{2,3} - \mu = 36 - 22.1 = 13.9$	$y_{3,3} - \mu = 46 - 22.1 = 23.9$
$j = 4$	$y_{1,4} - \mu = 30 - 22.1 = 7.9$	$y_{2,4} - \mu = 28 - 22.1 = 5.9$	$y_{3,4} - \mu = 29 - 22.1 = 6.9$
$j = 5$	$y_{1,5} - \mu = 5 - 22.1 = -17.1$	$y_{2,5} - \mu = 15 - 22.1 = -7.1$	$y_{3,5} - \mu = 24 - 22.1 = 1.9$
$j = 6$	$y_{1,6} - \mu = 12 - 22.1 = -10.1$	$y_{2,6} - \mu = 12 - 22.1 = -10.1$	$y_{3,6} - \mu = 15 - 22.1 = -7.1$

Degrees of freedom for the error term (DF_{error}):

$$DF_{error} = N - k = 18 - 3 = 15$$

6. Mean squares (MS) of variation:
   Mean square between groups ($MS_{between}$):

$$MS_{between} = \frac{SS_{between}}{DF_{between}} = \frac{352.8}{2} = 176.4$$

Mean square for the error term (MS_{error}):

$$MS_{error} = \frac{SS_{error}}{DF_{error}} = \frac{1304.2}{15} = 86.9$$

7. F-ratio:

$$F = \frac{MS_{between}}{MS_{error}} = \frac{176.4}{86.9} = 2.0$$

8. Determine the critical F-ratio at the chosen P_α value. Determine the critical F-ratio at (2, 15) degrees of freedom for $\alpha = 0.05$ is 3.7.
9. Test the hypothesis: Since the obtained F-value is lower than the critical F-value, the null hypothesis (no difference) cannot be rejected. In this example, although the data do look significantly different when reviewed without statistical analysis, the high random error in the observations leads to lack of statistical significance.

An alternate means to test the hypothesis is to use the standard tables to determine the p-value associated with the observed F-value. If the observed p-value is less than the chosen P_α value (e.g., 0.05), the null hypothesis is rejected. For example, in the above calculations, the p-value associated with the observed F-ratio is 0.17. Since this is higher than 0.05, the null hypothesis cannot be rejected.

5.7.9.1.7 Calculations using Microsoft Excel

An alternate to calculations from first principles is to use one of the available software tools for calculations. As an illustration, when Microsoft Excel's data analysis add-in function is utilized for single-factor ANOVA calculations, the software provides a tabular output of calculated values illustrated in Table 5.10.

This tabular output of results summarizes statistical parameters associated with the data, followed by a summary of calculated results in a tabular format. The critical F-value and the p-value associated with the calculated F-value are indicated to facilitate hypothesis testing.

Table 5.10 Statistical results for a hypothetical example of a one-way ANOVA experiment using Microsoft Excel

Summary

Groups	Count	Sum	Average	Variance
Dose = 0	6	99	16.5	71.1
Dose = 50	6	134	22.33333	81.06667
Dose = 100	6	164	27.33333	108.6667

Anova

Source of variation	SS	DF	MS	F	p-value	F-crit
Between groups	352.7778	2	176.3889	2.028754	0.166031	3.68232
Within groups	1304.167	15	86.94444			
Total	1656.944	17				

5.7.9.2 *Two-way ANOVA: Design of experiments*

Two-way ANOVA deals with investigation of effects of two variables in a set of experiments. ANOVA with two or more variables (also called treatments or factors) is most commonly utilized in the design of experiments.

5.7.9.2.1 *Factorial experiments*

When the effects of more than one factor are studied at one or more levels, the factorial experiment is defined as an L^F-factorial experiment. For example, three factors evaluated at two different levels would be a 2^3 factorial experiment and two factors evaluated at three different levels would be a 3^2 experiment. An example of such studies is the effect of temperature and pressure on the progress of a reaction. If an experiment is run at two temperature and pressure values, it is a 2^2 factorial experiment, with the total number of runs $= 2 \times 2 = 4$. If the experiment were run at three levels of temperature and pressure, it would be a 3^2 factorial experiment, with the total number of experimental runs $= 3 \times 3 = 9$. Conversely, if three factors (e.g., temperature, pressure, and reactant concentration) were studied at two levels each, it would be a 2^3 factorial experiment, with $2 \times 2\,2 = 8$ experimental runs. The experiments could be full-factorial or partial-factorial.

- A full-factorial experiment is one in which all combinations of all factors and levels are studied. For example, a full-factorial four-factor, two-level study would involve $2^4 = 2 \times 2 \times 2 \times 2 = 16$ experimental runs. Full-factorial experiments provide information on both the main effects of various factors and the effects of their interactions. Design and interpretation of a two-factor, two-level experiment are illustrated in the two-way ANOVA model.

- A partial-factorial experiment is one in which half the combinations of levels of all factors are studied. For example, a partial-factorial four-factor, two-level study would involve $2^{4-1} = 16/2 = 8$ experimental runs. Partial-factorial experiments provide information on the main effects of various factors but not on the interaction effects. Design and interpretation of partial-factorial experiments are beyond the scope of this chapter.

5.7.9.2.2 Model equation

If there are two variables or treatments being studied in the experiment, the value of each data point is explained as:

$$y_{ijk} = \mu + \tau_i + \beta_j + \gamma_{ij} + \varepsilon_{ijk}$$

where:

y_{ijk} represents the jth observation of the ith level of treatment of the first variable and kth treatment of the second variable
μ is the mean of all samples in the experiment
τ_i is the ith treatment effect of the first variable
β_j is the jth treatment effect of the second variable
ε_{ijk} represents random error

Hence, the value of each data point in an experiment is represented in terms of the mean of all samples and deviations arising from the effect of treatment or variable being studied (τ_i) and random variation (ε_{ij}). Hence, the value of each data point in an experiment is represented in terms of the mean of all samples and deviations arising from the effect of two treatments or variables being studied (individual or main effects, τ_i and β_j and effects arising from interaction of these variables, γ_{ij}) and random variation (ε_{ij}). The variables in this experiment are commonly termed *factors*, and the experiment is termed a *factorial experiment*. This equation represents a *two-way ANOVA* model.

5.7.9.2.3 Null and alternate hypotheses

The null hypotheses (H_0) for a two-way ANOVA experiment studying factors A and B could be the following:

- No difference between the population means of samples treated with different levels of factor A. The alternate hypothesis (H_1) would be that the means of underlying populations are not equal.
- No difference between the population means of samples treated with different levels of factor B. The alternate hypothesis (H_1) would be that the means of underlying populations are not equal.

- No difference between the population means of samples treated with different combinations of different levels of factors A and B (For example, if both factors A and B had two levels - high and low - the combinations could be high [A] with low [B] versus low [A] with high [B]. A study of this interaction reveals whether the effect of factor A is different when factor B is low versus high or not.). The alternate hypothesis (H_1) would be that there is an interaction between factors A and B.

5.7.9.2.4 Calculations

The calculations for a two-way ANOVA experiment are similar to the one-way ANOVA, with the inclusion of the case of a second variable B at levels 1 through b. The equations for the one-way ANOVA in the corresponding previous section are considered as the effect of variable A. The equations are modified as below for inclusion of the effect of variable B.

1. Mean of all samples in the experiment (μ) is calculated by adding all observations and dividing by the total number of samples in the experiment.

$$\mu = \frac{\sum_{i=1}^{k}\sum_{j=1}^{n}\sum_{B=1}^{b} y_{ijB}}{N}$$

where, y_{ijk} represents the jth observation of the ith level of treatment of the variable A and bth level of treatment of variable B, there being a total of k treatments ($i = 1, 2, 3, \ldots k$) and n samples per treatment level ($j = 1, 2, 3, \ldots n$) for variable A and b treatments ($B = 1, 2, 3, \ldots b$) and n samples per treatment level ($j = 1, 2, 3, \ldots n$) for variable B; N is the total sample size, including all treatments and levels.

2. Total sum of squares (SS_T) of all observations is calculated by squaring all observations and subtracting from the mean of all samples in the experiment (μ).

$$SS_T = \sum_{i=1}^{k}\sum_{j=1}^{n}\sum_{B=1}^{b} (y_{ijB} - \mu)^2$$

3. Sum of squares for the factor is the sum of squares between the columns ($SS_{between}$) if each level of the factor is arranged in a column. It is calculated by subtracting the mean value for each column from the mean of all samples, squaring this value, and adding for all columns.

$$SS_{\text{between},i} = n \times b \times \sum_{j=1}^{n} \sum_{B=1}^{b} \left(\frac{\sum_{i=1}^{k} y_j}{k} - \mu \right)^2$$

$$SS_{\text{between},B} = n \times k \times \sum_{j=1}^{n} \sum_{i=1}^{k} \left(\frac{\sum_{B=1}^{b} y_B}{b} - \mu \right)^2$$

Sum of squares for interaction between factors A and B is determined by:

$$SS_{\text{between},i,B} = n \times \sum_{i=1}^{k} \sum_{B=1}^{b} \left(\left(\frac{\sum_{B=1}^{b} y_{iB}}{n \times b} - \frac{\sum_{i=1}^{k} y_i}{n \times k} \right)^2 - \left(\frac{\sum_{B=1}^{b} y_B}{n \times b} - \frac{\sum_{i=1}^{k} \sum_{j=1}^{n} \sum_{B=1}^{b} y_{ijB}}{n \times k \times b} \right)^2 \right)$$

4. Sum of squares for the random error (SS_{error}) is the difference between the total sum of squares and the sum of squares between and within the columns.

$$SS_{\text{error}} = SS_{\text{total}} - SS_{\text{between},i} - SS_{\text{between},B}$$

5. Degrees of freedom are calculated as follows:
Degrees of freedom between groups (DF_{between}):

$$DF_{\text{between},i} = k - 1$$

$$DF_{\text{between},B} = b - 1$$

Degrees of freedom for the error term (DF_{error}):

$$DF_{\text{error}} = N - k \times b$$

Degrees of freedom for the interaction term ($DF_{\text{interaction}}$):

$$DF_{\text{interaction}} = (k - 1)(b - 1)$$

6. Mean squares for the random error (MS_{error}) and the factor studied ($MS_{between}$) are calculated by dividing their respective sum of squares by their degrees of freedom.

$$MS_{between,i} = \frac{SS_{between,i}}{DF_{between,i}}$$

$$MS_{between,B} = \frac{SS_{between,B}}{DF_{between,B}}$$

$$MS_{error} = \frac{SS_{error}}{DF_{error}}$$

7. An F-ratio is computed as the ratio of mean squares of factor effect to the mean square of error effect.

$$F_i = \frac{MS_{between,i}}{MS_{error,i}}$$

$$F_B = \frac{MS_{between,B}}{MS_{error,B}}$$

8. Determine critical F-ratio at ($DF_{between}$ and DF_{error}) degrees of freedom for $\alpha = 0.05$.
9. Test the hypothesis: The F-ratio is compared to the P_α value for the F-test at designated degrees of freedom to determine the significance of observed results. Statistical significance of results would indicate that the contribution of the factor's or variable's effect on the observations is significantly greater than the variation that can be ascribed to random error.

5.7.9.2.5 Calculations using Microsoft Excel

As an illustration of two-way ANOVA calculations using Microsoft Excel's data analysis add-in tool, the example summarized in Table 5.11 provides a tabular output of calculated values listed in Table 5.12.

This tabular output of results summarizes statistical parameters associated with the data, followed by a summary of calculated results in a tabular format. The critical F-value and the p-value associated with the calculated F-value are indicated to facilitate hypothesis testing. Two-way ANOVA results provide information about statistical significance of differences attributable to both factors. Thus, in this example, the contribution

Table 5.11 A hypothetical example of a two-way
ANOVA experiment. Yield of a chemical
synthesis reaction was studied as a
function of temperature and pressure in a
2^2 full-factorial study without replication.
The data, in terms of percentage yield,
are summarized in the table

Temperature (°C)	Pressure: I atm	Pressure: 2 atm
40	95.4	95.8
60	91.9	92.1

Table 5.12 Statistical results for a hypothetical example of a two-way ANOVA
experiment using Microsoft Excel

ANOVA: Two-factor without replication

Summary	Count	Sum	Average	Variance
Row I	2	191.2	95.6	0.08
Row 2	2	184	92	0.02
Column I	2	187.3	93.65	6.125
Column 2	2	187.9	93.95	6.845

ANOVA

Source of variation	SS	DF	MS	F	p-value	F crit
Rows	12.96	I	12.96	1296	0.017679	161.4476
Columns	0.09	I	0.09	9	0.204833	161.4476
Error	0.01	I	0.01			
Total	13.06	3				

of columns (pressure) to variation has a p-value of 0.20, while the contribution of rows (temperature) has a p-value of 0.02. Given the α value of 0.05, the contribution of temperature is significant, while that of pressure is not.

REVIEW QUESTIONS

5.1 Amoxicillin suspension.
   A. How much water would need to be added to a bottle containing 12.5 g of dry powder for reconstitution into a 250 mg/5 mL suspension?
   Hint: Use ratio and proportion, and remember to use the same units.

B. How many milliliters of amoxicillin suspension containing 250 mg/5 mL must be administered to a patient in need of a 400-mg dose of amoxicillin?

Hint: Use ratio and proportion.

C. If each 5 mL of a 250 mg/5 mL reconstituted amoxicillin suspension contains 0.15 mEq of sodium, how much sodium does it represent in mg?

Hint: Use the atomic weight of sodium.

D. Given your answers to (a) and (b) above, how much sodium would the patient be taking per day if the patient is dosed 400 mg t.i.d.?

Hint: Use ratio and proportion.

5.2 Cyclophosphamide tablets.

A. Cyclophosphamide is available as 50-mg tablets and has a recommended dose of 5 mg/kg o.d. What would be the daily dose for a 175-lb patient?

Hint: Use proportion after converting every quantity to same units.

B. How many tablets should be dispensed for a dosage regimen of 10 days?

Hint: Calculate the number of tablets per day first.

5.3 Dosage for Children.

For a drug with the adult dose of 100 mg/kg, what would be the dose for a 4-feet-tall 8-year-old child weighing 80 lbs? Calculate using the nomogram, Fried's rule, Young's rule, and Clark's rule.

5.4 Tonicity adjustment.

A. Calculate the NaCl equivalents (*E value*) for the following three drugs, given that NaCl has a molecular weight of 58.5 and dissociates into two ions, with a dissociation constant (i) of 1.8.

Drug A	Molecular weight = 220, ions = 3, i = 2.6
Drug B	Molecular weight = 180, ions = 1, i = 1
Drug C	Molecular weight = 140, ions = 2, i = 1.9

B. For the prescription noted below, calculate the NaCl equivalents present in the formulation.

Drug A	40 mg
Drug B	25 mg
Drug C	100 mg
Water q.s.	10 mL

C. Calculate the amount of NaCl equivalents that would need to be added to the above formulation to make it isotonic for ophthalmic administration.

D. If NaCl were incompatible with one or more of drugs, how much dextrose (molecular weight = 180) may be used instead.

5.5 Volume and weight interconversions

A. Glycerin is a highly viscous liquid that may be weighed instead of measured in volume. How much of weight of glycerin would be equivalent to 4.6 mL of its volume, given that the density of glycerin at room temperature is 1.26 g/cm³?

B. Ethanol is a low-viscosity liquid that is easier measured in volume than in weight. Given that its density is 0.78 g/cc, how much volume of ethanol is needed to prepare 25 mL of a 5% v/v solution?

C. Ethanol is a low-viscosity liquid that is easier measured in volume than in weight. Given that its density is 0.78 g/cc, how much volume of ethanol is needed to prepare 25 g of a 5% w/w solution?

5.6 Concentration calculations

A. What would be the equivalent weight of calcium chloride ($CaCl_2$) if its molecular weight is 111 g/mol?

B. What amount of $CaCl_2$ would be needed to make 50 mL of a 0.5 M solution?

C. What amount of $CaCl_2$ would be needed to make 50 mL of a 0.5 N solution?

D. A drug product was found to contain 40 ppm of an impurity during analysis. How many milligrams of this impurity might be ingested by an average 150-lb adult human being if the drug is to be administered in doses of 5 mg/kg/day in four divided doses?

E. What is the mole fraction of an isotonic NaCl solution? Molecular weight of NaCl is 58.5 and that of water is 18.

Hint: Isotonic NaCl solution has 0.9% w/v salt concentration.

F. How much purified water is needed to prepare 200 mL of 0.1 N HCl solution from its 5 N stock solution?

G. How much of the 0.1 N HCl solution would be needed to prepare 200 mL of a 2 N solution, using the 5 N stock solution of HCl?

5.7 Calculate the mean, median, variance, and standard deviation of following sets of values:

A. 2.6, 4.2, 3.7, 1.7, 3.2

B. 12.8, 9.6, 15.7, 14.8, 13.2

C. 3.2, 4.9, 12.4, 16.8, 9.3

D. By reviewing above results, which of the three data sets has the highest spread around the central tendency?

E. By reviewing the above results, which of the three data sets has the least spread around the central tendency?

F. By reviewing the above results, the differences between the means of which two data sets are most likely to be statistically significant?

G. By reviewing the above results, the differences between the means of which two data sets are least likely to be statistically significant?

REFERENCES

1. Dowdy S, Weardon S, and Chilko D (2004) *Statistics for Research*, Hoboken, NJ: Wiley-Interscience; Fulcher RM and Fulcher EM (2006) *Math Calculations for Pharmacy Technicians: A Worktext*, St. Louis, MO: Saunders; Hopkins WA (2005) *APhA's Complete Math Review for the Pharmacy Technician*, Washington DC: APhA Publications; Po ALW (1998) *Statistics for Pharmacists*, Oxford: Wiley-Blackwell.
2. Narang AS and Desai DS (2009) Anticancer Drug Development. In Mahato RI and Lu Y (Eds.) *Pharmaceutical Perspectives of Cancer Therapeutics*, New York: AAPS-Springer, p. 49.
3. Chatelut E, Canal P, Brunner V et al., (1995) *J Natl Cancer Inst* 87(8): 573.
4. Reilly SO, Rowinsky E, Slichenmyer W et al., (1996) *J Natl Cancer Inst* 88(12): 817.
5. Canal P, Chatelut E, and Guichard S, (1998) *Drugs* 56(6): 1019.
6. Hempel G and Boos J (2007) *Oncologist* 12(8): 924.

Part II

Physicochemical principles

Chapter 6

Complexation and protein binding

LEARNING OBJECTIVES

On completion of this chapter, the students should be able to

1. Define and exemplify coordination and molecular complexes.
2. Describe the types of molecular forces involved in the formation of coordination and molecular complexes.
3. Describe the influence of plasma–protein binding on the plasma concentration and biodistribution of drugs.
4. Describe the factors affecting complexation and protein binding of drugs.

6.1 INTRODUCTION

Complexation is a phenomenon that binds closely one or more molecules of two compounds—a ligand and a substrate by noncovalent attractive forces of interaction. The resulting structure, in which the ligand is bound to the substrate, is called a complex. In the case of an administered drug binding a physiological protein, the drug is the ligand and the protein is called the substrate. A ligand generally has the ability to complex different types of substrates with similar binding site—in terms of molecular size, geometry, and charge distribution. Similarly, a substrate can bind multiple different ligands of similar size, shape, and surface properties.

In the case of binding of two small-molecule compounds, either of the two compounds can be called a ligand or a substrate, depending on the molecular mechanism of interaction. Thus, a drug molecule can be either a ligand or a substrate. For example, complex formation of two molecules of theophylline (substrate) with one molecule of ethylenediamine (ligand) leads to the formation of the bronchodilator drug aminophylline (Figure 6.1), which has higher solubility than theophylline alone. On the other hand,

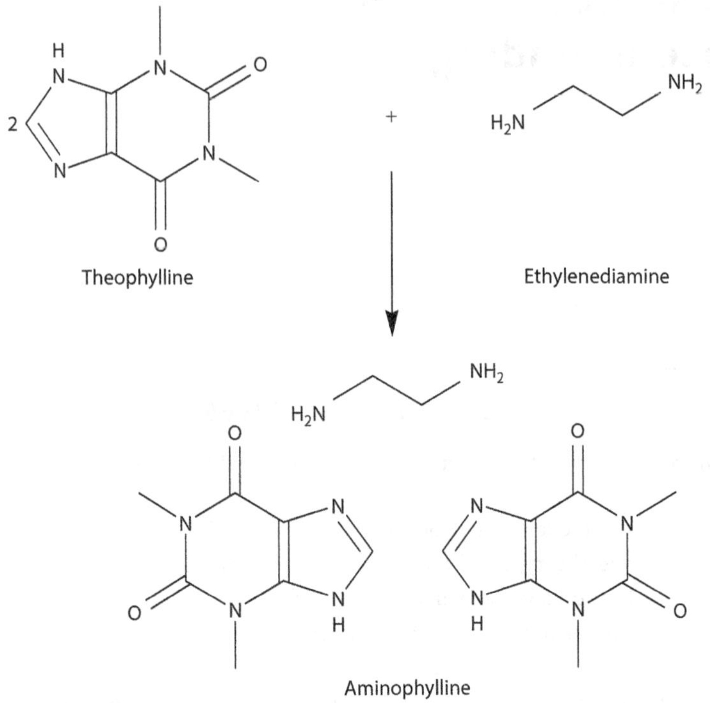

Figure 6.1 Example of complexation. Theophylline and ethylenediamine complex to yield the bronchodilator drug aminophylline.

aqueous solubility of oxytetracycline (a drug and a ligand) decreases when it complexes with calcium ions (substrate) (Figure 6.2), leading to low oral drug absorption of this antibiotic with dairy products. Thus, on complexation, properties of the drug such as solubility, stability, partitioning (hydrophilicity/lipophilicity), and absorption are altered.

Plasma–protein binding (PPP) is usually a reversible interaction of a drug with one or more of plasma proteins *in vivo*. The molecular forces and mechanisms involved in PPP are similar to those in the complexation phenomenon. Many drugs strongly bind to plasma proteins, such as albumin and alpha globulin. Since only the unbound drug is pharmacologically active and can diffuse out of the bloodstream into various tissue compartments, PPP can influence a drug's biodistribution (i.e., distribution inside the body compartments such as the proportion of the drug in the plasma or the central compartment compared with that in the tissue or the peripheral compartment), free drug concentration in plasma, and the duration of drug action. In addition, PPP can lead to drug–drug interactions when two or more coadministered drugs compete for the same binding site on the protein. For example, the anticoagulant drug warfarin (Figure 6.2) is ~97% bound to plasma protein and can be displaced by other highly protein-bound drugs,

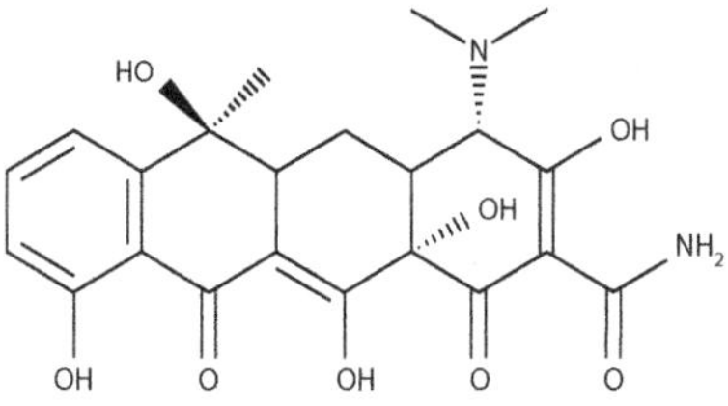

Oxytetracycline

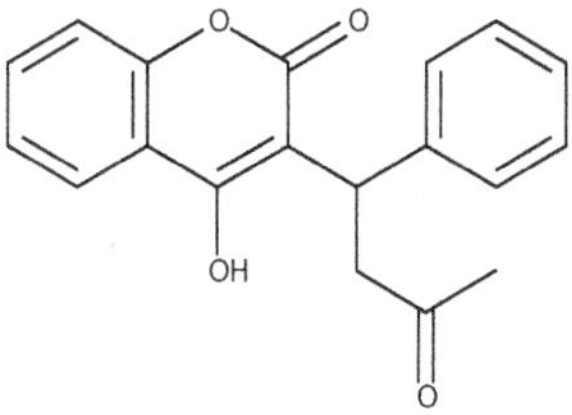

Warfarin

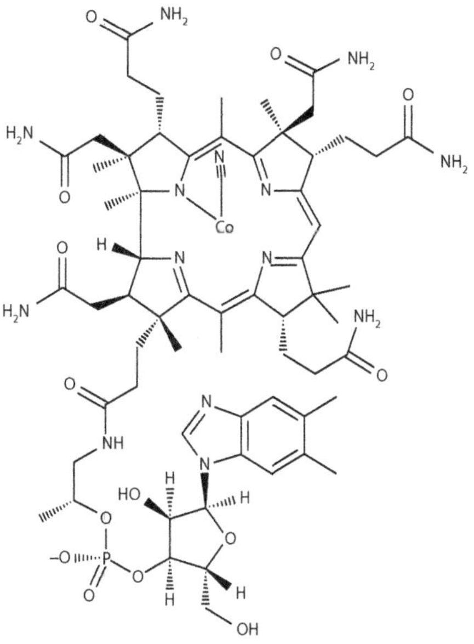

Cyanocobalamin (vitamin B$_{12}$)

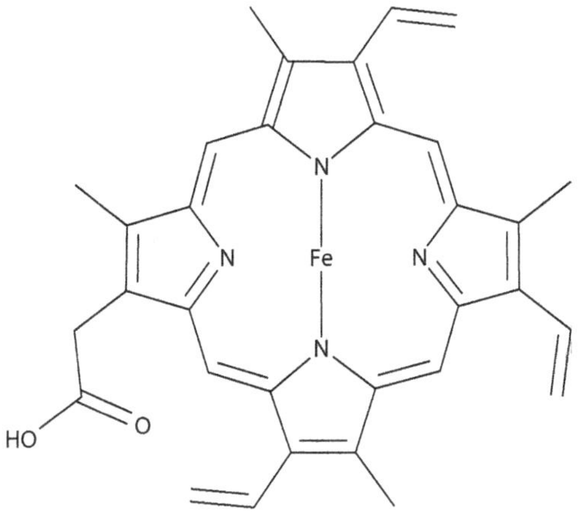

Heme

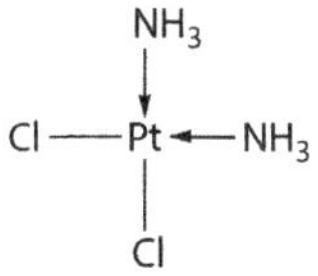

Cisplatin

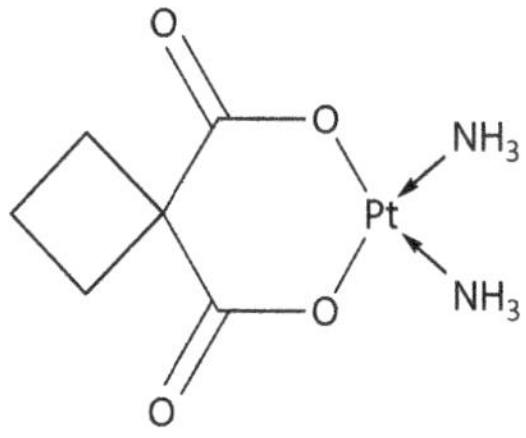

Carboplatin

Figure 6.2 **Examples of drugs that exist as complexes and/or have a high propensity for forming complexes.**   *(Continued)*

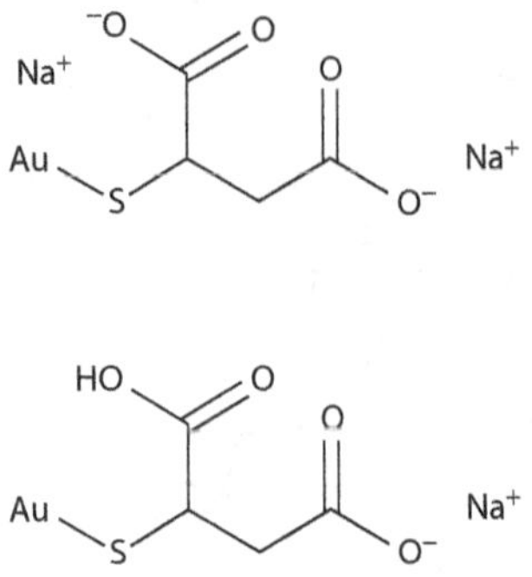

Solganol

Myocrisin

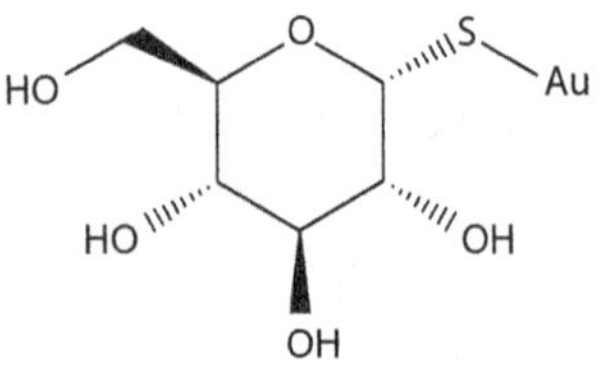

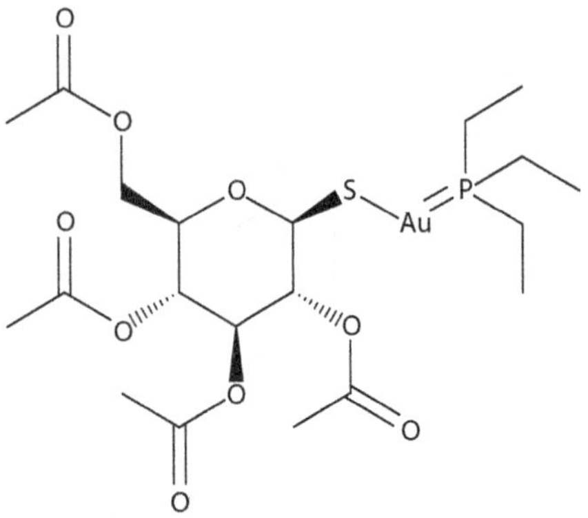

Ridaura

Figure 6.2 (Continued) Examples of drugs that exist as complexes and/or have a high propensity for forming complexes.

such as simvastatin, leading to a drug–drug interaction. Thus, coadministration of a drug that displaces warfarin from its protein-binding sites can cause high free-drug concentration, leading to toxicity of this low therapeutic index drug.

6.2 TYPES OF COMPLEXES

Depending on the type of interactions involved in complexation, ligand–substrate complexes are classified as follows:

- *Coordination complexes*: These are covalent complexes that form as a result of multiple Lewis acid–base reactions in which multiple neutral or anionic ligands bind a central, cationic substrate through multiple coordinate covalent bonds. Thus, the ionic covalent bonds are formed when an electron-rich atom on the ligand bonds with an electropositive atom of or on the substrate by donating its pair of electrons. Tetracycline complexation with divalent heavy metal cations is an example of a coordination complex.

- *Molecular complexes*: These are noncovalent complexes formed by multiple attractive interactions between two molecules, such as hydrogen bonding, electrostatic attraction, van der Waals forces, and hydrophobic interactions.

6.2.1 Coordination complexes

Metal complexes are the most common coordination complexes. Their structure involves one or more central metal atom or cation, surrounded by a number of substrates with negatively charged ions (such as carboxylate groups) or neutral molecules possessing lone pair of electrons (such as on nitrogen atoms of amine groups). The ions or molecules surrounding the metal are called ligands. The number of bonds formed between the metal ion and the ligand(s) is called the coordination number of the complex. Ligands are generally bound to a metal ion by a coordinate covalent bond (i.e., donating electrons from a lone electron pair into an empty metal orbital) and are thus said to be coordinated to the ion.

The interaction between the metal ion and the ligand is a Lewis acid–base reaction, in which the ligand (a base) donates a pair of electrons (:) to the metal ion (an acid) to form the coordinate covalent bond. For example,

$$Ag^+ + 2(: NH_3) \rightarrow [Ag(NH_3)_2]^+$$

Where, silver ion (Ag^+) is the central metal ion interacting with ammonia (NH_3) to form the silver–ammonia $[Ag(NH_3)_2]^+$ coordination complex. Ligands, such as $H2O$:, NC^-:, and Cl^-: donate a pair of electron in forming a complex. For example, silver–ammonia complexes can be neutralized with Cl^- to form $[Ag(NH_3)_2]Cl$.

Several enzymes involve coordination complexation of their amino acids to one or more heavy metal atoms. Coordination complexes play a critical role in controlling the structure and function of many enzymes. Heavy metal ions present in physiological proteins and enzymes facilitate the formation of coordination complexes that result in the functionality of the protein or the enzyme. For example, copper ion is present in proteins and enzymes, including hemocyanin, superoxide dismutase, and cytochrome oxidase. Zinc is present in many proteins and confers structure and stability, such as crystalline insulin. When present in deoxyribonucleic acid (DNA)-binding proteins, Zn^{2+} binds tetrahedrally with the two histidine and two cysteine residues of the protein to form a loop (zinc finger), which can fit into the major groove of genomic double-helical DNA (Figure 6.3).

Several nonenzymatic molecules of biological significance are coordination compounds. For example, vitamin B_{12} (cyanocobalamin) is a coordination complex of cobalt (Figure 6.2), and heme is a coordination complex of iron with the nitrogens of histidine residues of the protein (Figure 6.2). Heme proteins of myoglobin and hemoglobin are iron

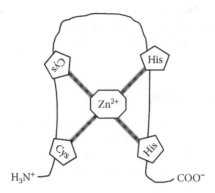

Figure 6.3 Formation of zinc finger due to zinc binding to histidine and cysteine residues in a peptide chain.

complexes that are essential for the transport of oxygen in the blood and tissues. Each heme residue contains one central iron atom in the ferrous oxidation state (Fe^{2+}) in coordinate bonds with a heterocyclic organic compound called porphyrin. The oxygen carried by heme proteins is bound directly to Fe^{2+} atom of the heme group. Oxidation of the iron to the ferric oxidation state (Fe^{3+}) renders the molecule incapable of binding oxygen.

Among drugs, anticancer drugs cisplatin and carboplatin are platinum (II) complexes (Figure 6.2). Rheumatoid arthritis drugs aurothiomalate (Myocrisin®), aurothioglucose (Solganol®), aurothiopropanol sulfonate (Allocrysin®), and nuranofin (Ridaura®) are gold complexes (Figure 6.2).

6.2.2 Molecular complexes

Molecular complexes involve noncovalent interactions between the ligand and the substrate, such as electrostatic attraction between oppositely charged ions, van der Waals forces, hydrogen bonding, and hydrophobic interactions. Molecular complexes can be subdivided based on the substrate and the ligand involved in complexation.

6.2.2.1 Small molecule–small molecule complexes

- Molecules bearing functional groups with opposite polarity can interact with each other in solution. For example, benzocaine interacts with caffeine as a result of a dipole–dipole interaction between the nucleophilic carbonyl oxygen of benzocaine and the electrophilic nitrogen of caffeine.
- Self-association complexes form when drug molecules in solution interact with one another to form dimers, trimers, or higher-order-association structures, including micelles. For example, daunomycin, mitoxantrone, and brivanib alaninate are known to self-associate in aqueous solution.

6.2.2.2 Small molecule–large molecule complexes

- Drugs often interact with macromolecules *in vitro*. For example, cationically charged drugs may interact with anionically charged excipients and polymers in the dosage form, such as tablet, to form a complex. Commonly encountered anionic hydrophilic polymers in the dosage form include the superdisintegrants croscarmellose sodium and sodium starch glycolate.
- Drugs can also form complexes with ion-exchange resins. Such complexation can lead to incomplete drug release from the dosage form. Examples of drugs that can form such complexes include basic drugs amitriptyline, verapamil, diphenhydramine, alprenolol, and atenolol. Ion-exchange resins that strongly bind drugs are also used to make sustained-release dosage forms. For example, ion-exchange resins carboxylic acid and sulfonic acid can bind cationic drugs and those with quaternary ammonium groups can bind anionic drugs.
- Several water-soluble pharmaceutical polymers, including polyethylene glycols (PEGs), polyvinylpyrrolidone (PVP), polystyrene, carboxymethylcellulose (CMC), and similar polymers containing nucleophilic oxygen, can form complexes with drugs in solution.
- Plasma protein binding. Drug–protein complexation between small-molecule drugs and large protein molecules in the plasma is mediated by reversible molecular interactions.
- Enzyme–substrate interactions. Enzyme–substrate interactions involve very specific noncovalent bonds between various amino acids of the enzyme folded into the substrate-recognition site. The requirement of formation of multiple specific bonds in specific orientation and location within the substrate-binding site for enzyme action ensures substrate recognition for activation of the enzyme.
- Inclusion/occlusion complexes. These complexes involve the entrapment of one compound in the molecular framework of another. Inclusion complexes are exemplified by the complexation of hydrophobic drugs by cyclodextrin molecules, which totally enclose the substrate. Occlusion complexes are exemplified by a specific case of cyclodextrin complexes where only the hydrophobic portion of an amphiphilic molecule is complexed by cyclodextrin.

Cyclodextrins are donut-shaped molecules of β-D-glucopyranose with 6, 7, or 8 cyclic residues of D-glucose, known as α-, β-, or γ-cyclodextrins, respectively (Figure 6.4). The cavity size ranges from 5Å for α-cyclodextrin to 8 Å for γ-cyclodextrin. In addition, several cyclodextrin derivatives, such as methyl-, dimethyl-, 2-hydroxypropyl, and sulfobutyl ether substitutions on the hydroxyl groups of the cyclodextrin, lead to different physicochemical properties. For example, Figure 6.5 shows the ampicillin–cyclodextrin occlusion complex.

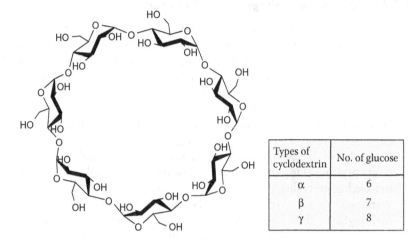

Types of cyclodextrin	No. of glucose
α	6
β	7
γ	8

Figure 6.4 Chemical structure of cyclodextrin.

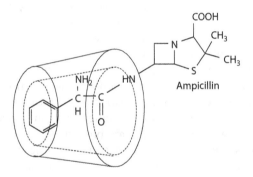

Ampicillin

Figure 6.5 Example of complexation of ampicillin by cyclodextrin.

Cyclodextrins are used to complex hydrophobic molecules or hydrophobic portions of a molecule. Complexation is mediated primarily by van der Waals force of attraction and hydrophobic interaction. The surface of cyclodextrins is highly hydrophilic because of the multiple hydroxyl (−OH) functional groups that can hydrogen bond with water. Thus, cyclodextrins can form reversible water-soluble inclusion or occlusion complexes of hydrophobic compounds. Cyclodextrins are nontoxic and do not illicit immune response. Cyclodextrin complexation can, therefore, serve as an effective means of increasing the aqueous solubility, stability, absorption, and bioavailability of hydrophobic drugs. Cyclodextrins have been used to complex and increase the solubility of various hydrophobic drugs, such as paclitaxel and hydrocortisone.

6.2.2.3 Large molecule–large molecule complexes

- Large molecule–large molecule complexes are exemplified by polyacids, which can form hydrogen-bonded complexes with PEGs (Figure 6.6). In addition, PVP can form complexes with poly(acrylic acids).
- Base–base interactions in DNA helix through interactions between the nucleotide bases involve hydrogen bonding and are responsible for the unique double-helical structure of the double-stranded DNA. The DNA double helix is stabilized by the hydrogen bond interactions among nucleotides. Adenine (A) forms two hydrogen bonds with thymine (T), and guanine (G) forms three hydrogen bonds with cytosine (C) (Figure 6.7).

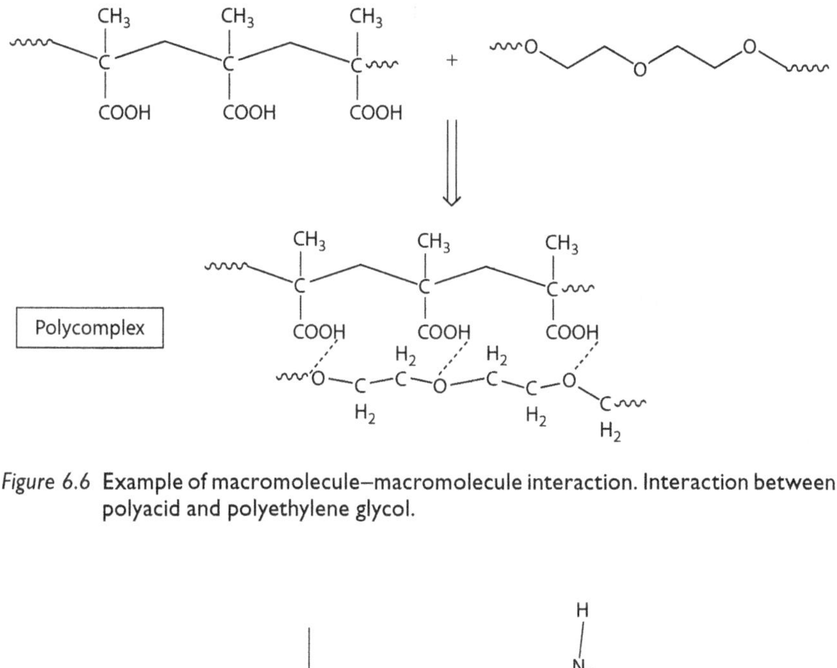

Figure 6.6 Example of macromolecule–macromolecule interaction. Interaction between polyacid and polyethylene glycol.

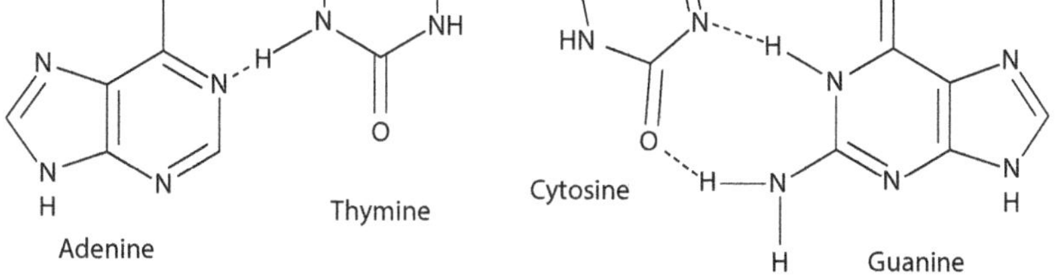

Figure 6.7 Complexation between bases in DNA molecules.

6.3 PROTEIN BINDING

A molecule (drug) that binds the protein is known as a ligand, and the protein with which it binds is called the substrate.

Protein binding is involved in the following:

- Plasma protein binding of drugs in the central or plasma pharmacokinetic compartment after administration.
- Drug–receptor interactions (when the receptor is a protein) leading to drug action.
- Substrate–enzyme interactions leading to enzyme action or inhibition.

Physical parameters of protein–ligand binding interaction include the kinetics of binding and its thermodynamics.

6.3.1 Kinetics of ligand–protein binding

Binding of a ligand (L) to a protein (P) to form a protein–ligand complex (PL) can be expressed as:

$$P + L \underset{k_d}{\overset{k_a}{\rightleftharpoons}} PL \qquad (6.1)$$

where, k_a and k_d are the equilibrium rate constants known as the association constant and the dissociation constant, respectively.

Their rate expressions can be written as:

$$k_a = \frac{[PL]}{[P][L]} \qquad (6.2)$$

$$k_d = \frac{[P][L]}{[PL]} \qquad (6.3)$$

The dissociation constant (k_d) has a unit of concentration (such as M), while the association constant (k_a) has the unit of inverse concentration (such as M−1).

Thus,

$$k_d = \frac{1}{k_a} \qquad (6.4)$$

6.3.1.1 Parameters of interest

Biopharmaceutical applications of protein binding require the determination of two key parameters:

1. Binding affinity (defined as the association constant, k_a)
2. Binding capacity (maximum number of ligand molecules that can be bound per molecule of protein, y_{max})

6.3.1.2 Experimental setup

Protein–ligand binding studies are usually carried out with fixed protein concentration and varying ligand concentration, or vice versa. At each concentration, the amount of ligand bound is separated from free ligand by techniques such as centrifugation and filtration. Free ligand concentration is then determined by analytical methods such as ultraviolet-visible spectroscopy (UV-VIS). The measurement of free ligand concentration as a function of total ligand concentration enables the determination of both the affinity and the capacity of ligand binding of the substrate.

The amount of ligand bound to the substrate in each experiment (y) can be expressed as a fraction of maximum concentration that can be bound (y_{max}), as

$$\theta = \frac{y}{y_{max}} \tag{6.5}$$

Where, y represents the molar concentration, the amount of ligand bound per unit molar concentration, or the amount of protein, and y_{max} represents the maximum binding capacity.

6.3.1.3 Determining k_a and y_{max}

6.3.1.3.1 Nonlinear regression I

Average number of ligand molecules bound per molecule of protein is expressed as the molar concentration of ligand bound to the protein per molar concentration of the protein. For the case of single binding site on the protein, molar concentration of ligand bound to the protein is given by [PL] and the total protein concentration is given by [P] + [PL]. Thus,

$$n = \frac{[PL]}{[P]+[PL]} \tag{6.6}$$

From the expression for the dissociation constant, k_d,

$$[PL] = \frac{[P][L]}{k_d} \tag{6.7}$$

Combining these two equations,

$$n = \frac{[P][L]/k_d}{[P]+[P][L]/k_d} = \frac{[L]/k_d}{1+[L]/k_d} = \frac{[L]}{k_d+[L]} \tag{6.8}$$

For a single ligand-binding site per protein,

$$n = \theta \tag{6.9}$$

Thus, the amount of ligand bound to the protein as a fraction of saturation concentration (θ), which is experimentally determined, can be written as:

$$\theta = \frac{[L]}{k_d+[L]} \tag{6.10}$$

Directly plotting θ against the free ligand concentration [L] gives a saturation curve (Figure 6.8a), and the data can be fitted by nonlinear regression to solve for y_{max} and k_d as parameters.

$$\theta = \frac{[L]}{k_d+[L]} \tag{6.11}$$

Or,

$$y = y_{max}\left(\frac{[L]}{k_d+[L]}\right) \tag{6.12}$$

However, nonlinear regression is a computationally intensive parameter-estimation method that uses algorithms for adjusting the equation parameters to best fit the data. Thus, it suffers the drawbacks of requiring software support, being dependent on the initial values of parameters chosen, and the possibility of coming up with incorrect parameters due to minimization of sum-of-square errors in a local region. Therefore, linearization of this equation followed by simple linear regression is traditionally preferred. Two methods for linearization are the double-reciprocal plot and the Scatchard plot.

6.3.1.3.2 Linear regression I: Double-reciprocal (Hughes–Klotz) plot

Inverting Equation 6.11,

$$\frac{1}{\theta} = 1 + \frac{k_d}{[L]} \tag{6.13}$$

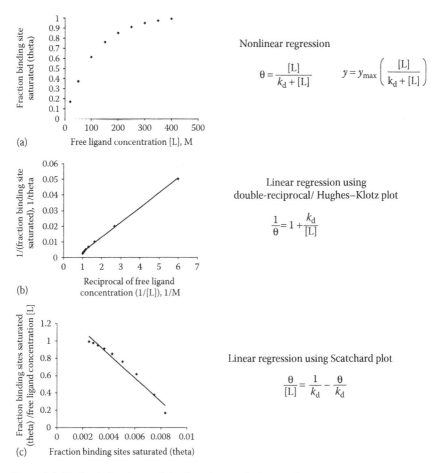

(a)

Nonlinear regression

$$\theta = \frac{[L]}{k_d + [L]} \qquad y = y_{max}\left(\frac{[L]}{k_d + [L]}\right)$$

(b)

Linear regression using
double-reciprocal/ Hughes–Klotz plot

$$\frac{1}{\theta} = 1 + \frac{k_d}{[L]}$$

(c)

Linear regression using Scatchard plot

$$\frac{\theta}{[L]} = \frac{1}{k_d} - \frac{\theta}{k_d}$$

Figure 6.8 Methods for determining ligand–protein interaction parameters.

In this equation, [L] represents the free ligand concentration, which is also experimentally determined. This is a linear form of the equation, whereby plotting $1/\theta$ against $1/[L]$ gives a straight line, with slope as k_d. This plot is known as the double-reciprocal plot, Lineweaver–Burk plot, Benesi–Hildebrand binding curve, or the Hughes–Klotz plot (Figure 6.8b).

As seen in Figure 6.8b, graphical treatment of data using Klotz reciprocal plot heavily weighs those experimental points obtained at low concentrations of free ligand and may, therefore, lead to misinterpretations regarding the protein-binding behavior at high concentrations of free ligand. The Scatchard plot (Figure 6.8c)—discussed in the next section—does not have this disadvantage and is, therefore, preferred for plotting data.

6.3.1.3.3 Linear regression II: Scatchard plot

The equation for θ can also be converted into:

$$\frac{\theta}{[L]} = \frac{1}{k_d + [L]} \tag{6.14}$$

Adding and subtracting $1/k_d$ from this equation:

$$\frac{\theta}{[L]} = \frac{1}{k_d + [L]} + \frac{1}{k_d} - \frac{1}{k_d} = \frac{1}{k_d} - \left(\frac{1}{k_d} - \frac{1}{k_d + [L]} \right)$$

$$= \frac{1}{k_d} - \left(\frac{k_d + [L] - k_d}{k_d(k_d + [L])} \right) = \frac{1}{k_d} - \left(\frac{[L]}{k_d(k_d + [L])} \right)$$

Which gives:

$$\frac{\theta}{[L]} = \frac{1}{k_d} - \frac{\theta}{k_d} \tag{6.15}$$

Or,

$$\frac{y/y_{max}}{[L]} = \frac{1}{k_d} - \frac{y/y_{max}}{k_d}$$

$$\frac{y}{[L]} = \frac{y_{max}}{k_d} - \frac{y}{k_d} \tag{6.16}$$

Thus, given that both θ and [L] are experimentally determined, plotting $\theta/[L]$ against θ would give a slope of $-1/k_d$ and an intercept of y_{max}/k_d. This linear plot is known as the Scatchard plot (Figure 6.8c). Interchanging the x- and y-axis of the Scatchard plot results in the Eadie–Hofstee plot.

Although the Scatchard plot is widely used for protein–ligand binding data analyses, it suffers from mathematical limitations. As seen in Figure 6.8c, the Scatchard transformation distorts experimental error, resulting in violation of the underlying assumptions of linear regression, viz., Gaussian distribution of error and standard deviations being the same for every value of the known variable. In addition, plotting $\theta/[L]$ against θ leads to the unknown variables being a part of both x- and y-axis, while linear regression assumes that y-axis is unknown and x-axis is precisely known.

6.3.2 Thermodynamics of ligand–protein binding

Binding affinity can also be inferred from the thermodynamics of binding. A binding interaction, where more stable bonds are formed than are broken, involves release of energy as heat. The amount of heat released can be precisely measured in carefully controlled experiments by a technique generally known as calorimetry. For example, isothermal titration calorimetry (ITC) involves the titration of one binding partner (ligand) into another (protein) while measuring the heat (enthalpy) change per unit volume of the ligand added to the protein. These data are integrated to yield enthalpy change per mole of the injectant and plotted against the molar ratio of ligand to protein (Figure 6.9). In this plot, the enthalpy difference between the starting value and the saturated value indicates enthalpy (ΔH) of binding, the slope of the transition indicates binding affinity, and the ligand/protein molar ratio at the inflexion point indicates the stoichiometry of binding, that is, the number of ligand molecules binding per protein molecule.

Thus, ITC can be used to determine the thermodynamic parameters associated with a physical or a chemical change. These parameters include the free energy (ΔG), enthalpy (ΔH), and entropy (ΔS) change, which are related to each other as:

$$\Delta G = \Delta H - T\Delta S$$

Spontaneous processes must have favorable overall free energy of reaction (negative ΔG). The ITC helps determine whether the ligand–protein binding is enthalpically driven (negative ΔH) or entropically driven (positive ΔS).

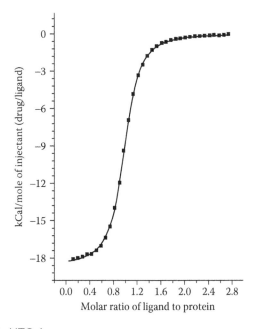

Figure 6.9 A typical ITC thermogram.

An entropically driven process is likely to be significantly influenced by the liquid medium. In the case of an enthalpically driven process, the binding constant and the enthalpy change associated with the binding are indicative of the strength of binding. Complexation is a binding process whereby the degrees of freedom of two or more molecules are reduced as they bind each other. Thus, complexation is an entropically unfavorable process (i.e., has a negative entropy).

An ITC experiment can also help determine the dissociation constant (k_d).

$$\Delta G = -RT \ln k_d \tag{6.17}$$

where:
　　R is the gas constant
　　T is the absolute temperature

6.3.3 Factors influencing protein binding

The physicochemical characteristics and concentration of the drug, the protein, and the characteristics of the liquid medium in which binding takes place influence drug (ligand)–protein binding.

6.3.3.1 Physicochemical characteristics and concentration of the drug

The extent of protein binding of many drugs is a linear function of their oil–water partition coefficient (Figure 6.10), which is a measure of their hydrophobicity. Thus, protein binding generally increases with an increase in drug lipophilicity. This indicates involvement of drug–protein hydrophobic interactions. This phenomenon can be used to predict the biological activity of a drug's analogs. For example, an increase in the

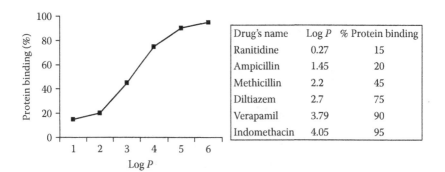

Drug's name	Log P	% Protein binding
Ranitidine	0.27	15
Ampicillin	1.45	20
Methicillin	2.2	45
Diltiazem	2.7	75
Verapamil	3.79	90
Indomethacin	4.05	95

Figure 6.10 Effect of lipophilicity (log P) on plasma protein binding of drugs.

lipophilicity of penicillins results in decreased activity. The hydrophobic binding of penicillin in serum proteins reduces their potency *in vivo*, by decreasing their free plasma concentration.

Increasing the concentration of the drug would generally increase the extent of binding. However, if the concentration is increased beyond the saturation concentration, saturation of some or all binding sites can occur and the proportion of drug bound would actually decrease, as the absolute amount of bound drug remains constant.

6.3.3.2 Physicochemical characteristics and concentration of the protein

In a dilute solution, increasing protein concentration is expected to increase the proportion of the drug bound. However, at high protein concentrations, the protein may agglomerate or self-associate, leading to shielding of the hydrophobic region(s), which can reduce drug binding if the drug—protein interaction is driven by hydrophobic interactions.

Physicochemical characteristics of the protein, such as the density distribution of hydrophobic groups on its surface, significantly influence the extent of drug–protein interaction. Thus, binding affinity of a drug toward different proteins can be markedly different.

6.3.3.3 Physicochemical characteristics of the medium

Binding interaction between the drug and the protein involves disruption of drug–solvent and protein–solvent bonds with the formation of drug–protein bonds. Thus, solvent medium that strongly interacts with either or both of the drug and protein can lead to thermodynamically unfavorable outcome of drug–protein interactions. In addition, change in the dielectric constant of the medium, such as in the presence of alcohol in aqueous solutions, can lead to altered forces of attraction and bonding between the drug and the protein. For example, salt concentration and dielectric constant of the solvent medium can significantly influence drug–protein interactions.

6.3.4 Plasma protein binding

Systemically administered drugs reach target organs and tissues through blood, which is a mixture of several substances, including proteins. In pharmacokinetic terms, the blood or the plasma is called the central compartment. In this compartment, drugs often bind plasma proteins. The drug exits the central compartment as it partitions into organs and tissues, called the peripheral compartment. Plasma protein binding of drugs is generally reversible, so that protein-bound drug molecules are released as the level of free drug in blood declines.

6.3.4.1 Plasma proteins involved in binding

Blood plasma normally contains about 6.72 g of protein per 100 cm³, the protein comprising 4.0 g of albumin, 2.3 g globulin, and 0.24 g of fibrinogen. Albumin (commonly called human serum albumin [HSA]) is the most abundant protein in plasma and interstitial fluid. Plasma albumin is a globular protein consisting of a single polypeptide chain of molecular weight 67 kDa. It has an isoelectric point of 4.9 and, therefore, a net negative charge at pH 7.4. Nevertheless, albumin is amphoteric and capable of binding both acidic and basic drugs. Physiologically, it binds relatively insoluble endogenous compounds, including unesterified fatty acids, bilirubin, and bile acids. Human serum albumin has two sites for drug binding:

1. Site I (warfarin site) binds bilirubin, phenytoin, and warfarin.
2. Site II (diazepam site) binds benzodiazepines, probenecid, and ibuprofen.

Plasma proteins other than albumin are sometimes the major binding partners of drugs. For example, dicoumarol is bound to β- and α-globulins, and certain steroid hormones are specifically and preferentially bound to particular globulin fractions. Among other proteins, α1-acid glycoprotein (AAG) binds to lipophilic cations, including promethazine, amitriptyline, and dipyridamole.

6.3.4.2 Factors affecting plasma–protein binding

The amount of a drug that is bound to plasma proteins depends on three factors:

1. Concentration of free drug
2. Drug's affinity for the protein-binding sites
3. Concentration of protein

6.3.4.3 Consequences of plasma–protein binding

The binding of drugs to plasma proteins can influence their action in a number of ways:

1. Reduce free drug concentration. Protein binding affects antibiotic effectiveness, as only the free antibiotic has antibacterial activity. For example, penicillin and cephalosporins bind reversibly to albumin, thus affecting their free concentrations in plasma.
2. Reduce drug diffusion. The bound drug assumes the diffusional and other transport characteristics of the protein molecules.

3. Reduce volume of distribution. Only free drug is able to cross the pores of the capillary endothelium. Protein binding will affect drug transport into other tissues. When binding occurs with high affinity, the drug is preferentially localized in the plasma or the central compartment. In pharmacokinetic measurements, this reflects as a low volume of distribution of the drug.

However, some drugs (e.g., warfarin and tricyclic antidepressants) may exhibit both a high degree of PPP and a large volume of distribution. Although drug bound to plasma proteins is not able to cross biological membranes, binding of drugs to plasma proteins is in a dynamic equilibrium with the drug bound to plasma proteins. If the unbound (or free) drug is able to cross biological membranes and has a greater affinity and capacity for binding to the tissue biomolecules, compared with the plasma proteins, the drug may exhibit high volume of distribution, despite also exhibiting high PPP. As free drug moves across membranes and out of vascular space, the equilibrium shifts, drawing drug off the plasma protein to *replenish* the free drug lost from vascular space. This free drug is now also able to traverse membranes and leave vascular space. In this way, a drug with a very low free fraction (i.e., a high degree of PPP) can exhibit a large volume of distribution.

4. Reduce elimination. Protein binding retards the metabolism and renal excretion of the drug. Proteins are not filtered through glomerular filtration. Thus, protein-bound drugs have reduced rate of filtration in the kidneys and metabolism in the liver.

5. Increase risk of fluctuation in plasma free drug concentration.

   a. In cases where a drug is highly protein-bound (around 90%), small changes in binding, protein concentration, or displacement of the drug by another coadministered drug (drug–drug interaction) can lead to drastic changes in the concentration of free drug in the body, thus affecting efficacy and/or toxicity.

      However, a plasma protein may have multiple binding sites. Thus, if drugs bind to different sites on a protein, there will not be a competitive binding interaction between them. Thus, some drugs that are highly bound to albumin exhibit competitive interactions, while others do not.

   b. Sometimes, drug administration may also cause displacement of body hormones that are physiologically bound to the protein, thus increasing free hormone concentration in the blood.

   c. Disease states that alter plasma protein concentration may alter the protein binding of drugs. If the concentration of protein in plasma is reduced, there may be an increase in the free fraction of the drugs bound to that protein. Similarly, if pathological changes in binding proteins reduce the affinity of drug for the protein, there will be an increase in the free fraction of drug.

6.3.4.3.1 Effect on dosing regimen

Plasma protein binding can affect dosing regimen of a drug in several ways.

1. Lower metabolism and elimination of a plasma-protein-bound drug can lead to longer plasma half-life, compared with an unbound drug. Thus, the protein-bound drug may serve as a reservoir of drug within the body, maintaining free drug concentration through equilibrium dissociation process. This leads to long half-life and sustained plasma concentrations. Thus, dosing frequency would need to be adjusted in cases where drug's PPP or the concentration of plasma proteins is affected (such as burns).
2. Dose adjustments are frequently required in the case of disease states that affect the protein to which the administered drug is bound. Certain disease states increase AAG concentration while reducing albumin concentration. For example, acute burns reduce the concentration of circulating albumin, resulting in an increase in the free fraction of drugs that are normally bound to albumin. On the other hand, AAG concentration is substantially increased after an acute burn, resulting in a decrease in the free fraction of drugs that are normally bound to this plasma protein.
3. Age-based dose adjustments often have to account for PPP of drugs. For example, newborns have selectively lower plasma protein levels than adults. Thus, although the neonatal HSA concentration at birth is 75%–80% of adult levels, AAG concentration is only ~50%. Thus, dose adjustment may be needed for drugs that bind AAG.
4. Drug–drug interactions. Drugs that compete for the same plasma-protein-binding site can displace one another. This can lead to increased free level of a drug. Minor perturbation in PPP can have a significant influence on free drug concentration. Thus, coadministration of certain drugs may be contraindicated or require dose adjustment.

6.3.5 Drug receptor binding

Target protein (receptor) binding is routinely utilized in drug discovery, with the goal of maximizing binding affinity and specificity. This is expected to result in a drug molecule that is highly potent and has low off-target activity and, thus, toxicity. The principles involved in delineating the kinetics of drug–receptor binding are same as discussed earlier for ligand–protein binding.

6.3.6 Substrate enzyme binding

Binding of a ligand, which serves as a substrate for an enzyme, to an enzyme is a part of a continuous process involving conversion of the substrate into the product(s) by the enzyme. This process involves continuous recycling of the enyzme's binding sites for fresh substrate, as each molecule of the substrate is converted into product(s). Thus, substrate–enzyme binding kinetics are represented in terms of the rate of binding, and the saturation of the binding kinetics is considered in terms of the maximum rate of binding.

The rate of binding kinetics for substrate–enzyme reactions follows a hyperbolic function, described by the Michaelis–Menten equation.

$$v = v_{max} \frac{[S]}{k_M + [S]} \tag{6.18}$$

where:
   v is the initial reaction rate
   v_{max} is the maximum reaction rate
   [S] is the substrate concentration
   k_M is the Michaelis–Menten constant, which represents the ratio of the rate of dissociation of the enzyme–substrate complex to its rate of formation.

The similarity of this equation to Equation 6.7 indicates similar basic principles involved in their derivation.

$$y = y_{max} \left(\frac{[L]}{k_d + [L]} \right) \tag{6.12}$$

REVIEW QUESTIONS

6.1 Name the following coordination compounds?
   A. $[CoBr(NH_3)_5]SO_4$
   B. $[Fe(NH3)_6][Cr(CN)_6]$
   C. $[Co(NH_3)5Cl]SO_4$
   D. $[Fe(OH)(H_2O)5]^{2+}$
   E. $(C_5H_5)Fe(CO)_2CH_3$
6.2 Write the molecular formulas of the following coordination compounds?
   A. Hexaammineiron(III) nitrate
   B. Ammonium tetrachlorocuprate(II)
   C. Sodium monochloropentacyanoferrate(III)
   D. Potassium hexafluorocobaltate(III)
6.3 Identify the most prominent human plasma protein?
   A. α1-acid glycoprotein (AAG)
   B. Human serum albumin (HSA)
   C. Globulin
   D. Insulin

6.4 Which of the following forces are involved in molecular complexes?
A. Hydrogen bonding
B. Hydrophobic interactions
C. Van der Waals forces
D. Covalent bonding
E. Ionic bonding

6.5 Which of the following forces are involved in coordination complexes?
A. Hydrogen bonding
B. Hydrophobic interactions
C. Van der Waals forces
D. Covalent bonding
E. Ionic bonding

6.6 What are the important parameters for characterizing drug–plasma protein binding?
A. Protein concentration
B. Drug concentration
C. Binding affinity
D. Binding capacity
E. Rate of binding

6.7 Explain the factors affecting plasma protein binding of drugs.

6.8 What is the effect of plasma protein binding on the dosing regimen of a drug?

FURTHER READING

Amiji MA (2003) Complexation and protein binding. In Amiji AM and Sandman BJ (Eds.) *Applied Physical Pharmacy*, New York: McGraw-Hill, pp. 199–229.

Sadler PJ and Guo Z (1998) Metal complexes in medicine: Design and mechanism of action. *Pure Appl. Chem.* 70: 863–871.

Florence AT and Atwood D (2006) *Physicochemical Principles of Pharmacy*, 4th ed., London: Pharmaceutical Press.

Connors K (2000) Complex formation. In Gennaro AR (Ed) *Remington: The Science and Practice of Pharmacy*, 20th ed., pp. 183–197.

Higuchi T and Lach JL (1954) Investigation of some complexes formed in solution by caffeine. IV. Interactions between caffeine and sulfathiazole, sulfadiazine, p-aminobenzoic acid, benzocaine, phenobarbital, and barbital. *J Am Pharm Assoc* 43: 349–354.

Davis ME and Brewster ME (2004) Cyclodextrin-based pharmaceutics: Past, present and future. *Nat Rev Drug Discov* 3: 1023–1035.

Vallner JJ (1977) Binding of drugs by albumin and plasma protein. *J Pharm Sci* 66: 447–465.

Chapter 7

Chemical kinetics and stability

LEARNING OBJECTIVES

On completion of this chapter, the students should be able to

1. Discuss the importance of kinetics of a reaction.
2. Differentiate between the rate and the order of a reaction.
3. Write and explain zero- and first-order rate equations and half-life expressions.
4. Use Arrhenius equation to determine the effect of temperature on reaction rates.
5. Compute shelf life (t_{90}) of drugs and expiration time.
6. Describe the log k versus pH profile of drugs and identify the pH of maximum stability.
7. Describe main drug degradation pathways.
8. Name some stabilization strategies for drug products.

7.1 INTRODUCTION

Drug substances and drug products are required to be physically and chemically stable under recommended storage conditions to maintain their identity, potency, and safety throughout the shelf life. Most drugs are susceptible to chemical decomposition in their dosage forms. Degradation can lead to loss of the drug's potency and generation of impurities in drug products. Regulatory agencies, such as the Food and Drug Administration (FDA), require identification, quantitation, and/or toxicological evaluation of impurities in drug products when they exceed a given threshold, which depends on the drug's daily dose. In addition to the time and cost associated with these investigations, if an impurity is found to be significantly toxic, it can compromise a drug development program. Understanding of the rates and mechanisms of drug degradation reactions and implementing stabilization strategies help identify strategies to control the extent of formation of impurities in drug products. For example, the knowledge of the rate at which a drug deteriorates under various conditions of pH, temperature, humidity, and light allows formulators to choose a vehicle that will retard or prevent drug degradation.

Chemical kinetics deals with rates of chemical reactions. A knowledge of reaction kinetics under various conditions helps identify mechanisms of drug degradation and stabilization.

7.2 REACTION RATE AND ORDER

The *rate* of a reaction is the amount or concentration of a degradation product formed or the reactant lost per unit time. The rate of a reaction is described by a *rate equation*. For example, for a hypothetical reaction,

$$a\mathrm{A} + b\mathrm{B} \rightarrow m\mathrm{M} + n\mathrm{N} \tag{7.1}$$

where:
A and B are the reactants
M and N are the products

a, b, m, and n are the *stoichiometric coefficients* (number of moles participating in the reaction) for the corresponding reactant or product

The rate of this reaction can be described in terms of rate of disappearance of A or B or the rate of appearance of M or N, which are all interrelated. Thus,

$$-\frac{1}{a}\frac{dC_A}{dt} = -\frac{1}{b}\frac{dC_B}{dt} = \frac{1}{m}\frac{dC_M}{dt} = \frac{1}{n}\frac{dC_N}{dt} \tag{7.2}$$

Where, dC_A is the change in the concentration of reactant A over a period of time dt, and respectively, for all reactants and products in the equation.

The rate equation, describing the rate of formation of a product or the rate of disappearance of a reactant, for this reaction can be written as:

$$\frac{dC_A}{dt} = \frac{dC_B}{dt} = -kC_A^a C_B^b \tag{7.3}$$

Or

$$\frac{dC_M}{dt} = \frac{dC_N}{dt} = kC_A^a C_B^b \tag{7.4}$$

where k is the rate constant.

The negative sign associated with the reactants indicates their rate of disappearance from the system; that is, the concentration after time period, t, is lower than the starting concentration. The rate equation for the products does not carry the negative sign, since it indicates the rate of formation or appearance of products in the system; that is, the concentration after time period, t, is higher than the starting concentration. The rate constant, k, is positive in both rate equations.

The *order* of a reaction is the sum of powers to which the concentration terms of the reactants are raised in the rate equation. The order of a reaction can also be defined with respect to a given reactant. Hence, in the aforementioned example, the order of the reaction with respect to reactant A is a, order with respect to reactant B is b, and the overall order of the reaction is $a + b$.

Identifying the order of a reaction helps understand the dependence of the reaction on the concentrations of starting materials. Thus, reactions can be of zero order (indicating independence to reactant concentrations), first order (indicating that reaction rate is proportional to the first power of one of the reactants), second order (indicating that reaction rate is proportional to the first power of two of the reactants or the second power of one of the reactants), or higher order.

For example, for the base-catalyzed hydrolysis of an ester in the reaction,

$$CH_3COOC_2H_5 + NaOH \rightarrow CH_3COONa + C_2H_5OH$$

Reaction rate is defined as:

$$Rate = -\frac{d[CH_3COOC_2H_5]}{dt} = -\frac{d[NaOH]}{dt}$$
$$= \frac{d[CH_3COONa]}{dt} = \frac{d[C_2H_5OH]}{dt} \tag{7.5}$$

The reaction rate equation is:

$$Rate = k[CH_3COOC_2H_5][NaOH] \tag{7.6}$$

The order of this equation is $1 + 1 = 2$, since both reactants are raised to the power of one in the rate equation.

The number of molecules taking part in a reaction is called the *molecularity* of a reaction. The molecularity of a reaction is determined by the mechanism of a reaction and is expressed in the reaction equation. The order of a reaction may or may not be same as the *molecularity* of a reaction. In the aforementioned example, the molecularity of the reaction is 2, which is same as the order of the reaction.

7.2.1 Pseudo-nth order reactions

The observed order of a reaction may sometimes be different than the sum of stoichiometric coefficients of reactants. The rate of a reaction may sometimes be independent of the concentration of one of the reactants, even though this reactant is consumed during the reaction. For example, if one of the two reactants is the solvent in which the other reactant is dissolved at low concentration, such as an aqueous solution of a hydrolytically sensitive drug, the order of the reaction may be independent of the solvent concentration—the reactant present in a significantly higher concentration. Such reactions are termed as *pseudo-nth order* reactions. Thus, a truly second-order reaction, such as equimolar reaction of an ester compound with water in an aqueous solution, that presents itself as a first-order reaction is termed a *pseudo-1st order* reaction.

For example, for the hydrolysis of a dilute solution of ethyl acetate,

$$CH_3COOCH_2CH_3 + H_2O \rightarrow CH_3COOH + CH_3CH_2OH$$

Reaction rate is defined as:

$$\text{Rate} = -\frac{d[CH_3COOC_2H_5]}{dt} = \frac{d[CH_3COOH]}{dt} = \frac{d[CH_3CH_2OH]}{dt} \quad (7.7)$$

The reaction rate equation is:

$$\text{Rate} = k[CH_3COOC_2H_5] \quad (7.8)$$

The order of this equation is 1, since only one of the reactants is involved in the rate equation. Nevertheless, the molecularity of this reaction is 2, since two molecules are involved in the mechanism of the reaction and are expressed in the reaction equation. Pseudo-order reactions are typically the cases where the molecularity of a reaction is not the same as the order of a reaction.

7.2.2 Determining the order of a reaction

The order of a reaction is determined experimentally, while molecularity of a reaction–which determines the equation–is determined by a thorough understanding of the reaction mechanism. The order of a reaction can be experimentally determined by one of the several methods:

1. Initial rate method. Initial rate of a reaction is measured for a series of reactions with varying concentrations of reactants to determine the power to which the reaction rate depends on the concentration of each reactant. Only the initial rate is measured to ensure that the reactant's concentration is the predominant influence on the reaction rate.

As a reaction proceeds, the reaction rate can be influenced by changes in reaction conditions, such as accumulation of product or by-product of a reaction. Thus, the measurement of only the initial rate of a reaction provides a robust way to quantitate the dependence of reaction rate on the concentration of reactant(s).

2. Integrated rate law method. The concentration–time data of a reaction can be used to assess how the rate of a reaction changes as a function of reactant concentration. This plot is compared to theoretical predictions made by integrated rate equations, discussed later in this chapter, to infer reaction order.

3. Graph method. Similar to the integrated rate law method, this method plots the concentration–time profile of a reaction graphically to check fit to different reaction order kinetics.

4. Half-life method. The dependence of half-life, the time it takes for the reactant concentration to reach half of the measured initial concentration, on the initial concentration of the reactant is different for reactions of different orders. Half-lives of reactants can be determined experimentally and compared to theoretical predictions to determine reaction order.

7.2.3 Zero-order reactions

A zero-order reaction is one in which the reaction rate is independent of the concentration(s) of the reactant(s). The rate of change of concentration of reactant(s) or product(s) in a zero-order reaction is constant and independent of the reactant concentration. Many decomposition reactions in the solid phase or in suspensions follow zero-order kinetics.

The reaction rate for most zero-order reactions depends on some other factor, such as absorption of light for photochemical reactions or the interfacial surface area for heterogeneous reactions (i.e., reactions that happen at the solid–liquid, liquid–gas, or solid–gas interface). Thus, the slowest or the rate-determining factor of the reaction is different than the concentration(s) of reactant(s) for zero-order reactions.

7.2.3.1 Rate equation

Reaction rate of a zero-order reaction is constant and independent of the reactant concentration. Thus, the rate expression for the change in reactant concentration, C, with time, t, for a zero-order reaction (Figure 7.1) can be written as:

$$-\frac{dC}{dt} = k_0 \tag{7.9}$$

Or

$$dC = -k_0 dt \tag{7.10}$$

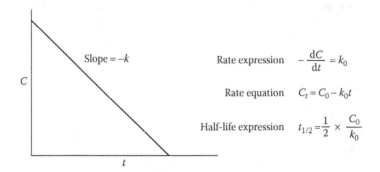

Figure 7.1 Zero-order kinetics. Plot of concentration, C, versus time, t.

where k_0 is the rate constant for a zero-order reaction. Thus, the change in the concentration of the reactant depends only on the time multiplied by a constant value, k_0.

Integrating this equation from concentration C_0 at time = 0 to concentration = C_t at time = t,

$$\int_{C_0}^{C_t} dC = -k_0 \int_0^t dt \tag{7.11}$$

$$C_t - C_0 = -k_0(t - 0) = -k_0 t \tag{7.12}$$

Thus, the rate equation (Figure 7.1) is:

$$C_t = C_0 - k_0 t \tag{7.13}$$

This is a linear equation of the form, $y = mx + c$. Thus, a plot of concentration, C_t, on the y-axis against time, t, on the x-axis (Figure 5.1) is linear, with a slope of $-k_0$ (downward slope of the line is indicated by the negative sign) and y-axis intercept of C_0.

7.2.3.2 Half-life

Scientists are frequently interested in the time required for the reduction of a given proportion of starting drug concentration. For example, a drug's shelf life is defined in terms of the time taken for the reduction of labeled drug concentration to its 90% level. The half-life ($t_{1/2}$) of a reaction is defined as the time required for one-half of the material to decompose. Thus, concentration of a reactant at its half-life ($C_{t1/2}$) is defined as half of the initial concentration (C_0); that is:

$$C_{t1/2} = \frac{C_0}{2} \tag{7.14}$$

Thus, for a zero-order reaction,

$$C_{t1/2} = C_0 - k_0 t_{1/2} \tag{7.15}$$

$$\frac{C_0}{2} = C_0 - k_0 t_{1/2} \tag{7.16}$$

After rearranging, the half-life expression (Figure 7.1) for a zero-order reaction is:

$$t_{1/2} = \frac{1}{2} \times \frac{C_0}{k_0} \tag{7.17}$$

7.2.4 First-order reactions

A first-order reaction is one in which the rate of reaction is directly proportional to the concentration of one of the reactants. Many decomposition reactions in the solid phase or in suspensions follow first-order kinetics. In a first-order reaction, concentration decreases exponentially with time, with the reaction rate slowing down progressively as the reactant is consumed in the reaction.

7.2.4.1 Rate equation

The rate of disappearance of the reactant is the concentration of the reactant multiplied by a constant. Thus, the first-order rate equation (Figure 7.2) is:

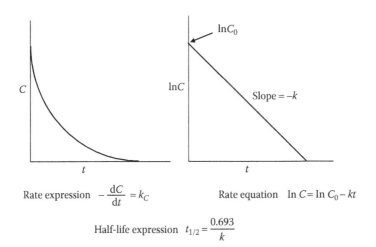

Rate expression $-\dfrac{dC}{dt} = kC$ Rate equation $\ln C = \ln C_0 - kt$

Half-life expression $t_{1/2} = \dfrac{0.693}{k}$

Figure 7.2 First-order kinetics. Plot of concentration, C, against time, t (A), and plot of natural logarithm of the concentration, C, against time, t.

$$-\frac{dC}{dt} = kC \tag{7.18}$$

Where, C is the reactant concentration at time t, and k is the first-order rate constant.

This equation can be rearranged as:

$$-\frac{dC}{C} = kdt \tag{7.19}$$

Integrating this equation from concentration C_0 at time $= 0$ to concentration $= C_t$ at time $= t$,

$$\int_{C_0}^{C} \frac{dC}{C} = -k\int_{0}^{t} dt \tag{7.20}$$

Solving this integral,

$$\ln C - \ln C_0 = -k(t - 0) = -kt \tag{7.21}$$

This equation can also be expressed as:

$$\log C - \log C_0 = \frac{-kt}{2.303} \tag{7.25}$$

Thus, the rate equation (Figure 7.2) is:

$$\ln C = \ln C_0 - kt \tag{7.22}$$

Or, in the exponential form, the rate equation can be expressed as:

$$C = C_0 e^{-kt} \tag{7.23}$$

Alternatively, by converting ln to the log base 10 ($\log_{10}$),

$$C = C_0 10^{-\frac{kt}{2.303}} \tag{7.24}$$

Thus, in a first-order reaction, the concentration decreases exponentially with time (Figure 7.2). A plot of log or ln of concentration against time is a straight line, whose slope provides the rate constant, k.

7.2.4.2 Half-life

Half-life, $t_{1/2}$, is defined as the time for the drug concentration to get to half of its original concentration (C_0); that is, $C_t = C_0/2$. This can be derived from the rate equation as:

$$\log C - \log C_0 = \frac{-kt}{2.303} \tag{7.25}$$

$$\log \frac{C}{C_0} = \frac{-kt}{2.303} \tag{7.26}$$

$$\log \frac{C_0}{C} = \frac{kt}{2.303} \tag{7.27}$$

$$k = \frac{2.303}{t} \log \frac{C_0}{C} \tag{7.28}$$

$$t = \frac{2.303}{k} \log \frac{C_0}{C} \tag{7.29}$$

Thus, the half-life of the reactant in a first-order reaction is given by:

$$t_{1/2} = \frac{2.303}{k} \log \frac{C_0}{C_0/2} \tag{7.30}$$

$$t_{1/2} = \frac{2.303}{k} \log 2 \tag{7.31}$$

Hence, the half-life expression (Figure 7.2) is:

$$t_{1/2} = \frac{0.693}{k} \tag{7.32}$$

For a reaction observing first-order kinetics, the half-life, $t_{1/2}$, or the time to any proportional reduction in concentration (e.g., $t_{0.9}$, i.e., time to 90% of initial concentration), is a constant number and independent of the initial reactant concentration, C_0.

7.2.5 Second-order reactions

Bimolecular reactions, reactions involving two different molecules A and B, involve reactions of two molecules.

$$A + B \rightarrow Products$$

The rates of bimolecular reactions are frequently described by a second-order equation. The rate of change in the concentrations of products and reactants in second-order reactions is proportional either to the second power of the concentration of a single reactant or to the first powers of the concentrations of two reactants.

When the speed of the reaction depends on the concentrations of A and B, with each term raised to the first power, the rate of decomposition of A is equal to the rate of decomposition of B and both are proportional to the product of the concentrations of the reactants. This can be expressed as the rate equation:

$$-\frac{d[A]}{dt} = -\frac{d[B]}{dt} = k[A][B]$$
(7.33)

7.2.5.1 Rate equation

Assuming that the initial concentrations of A and B are same, that is, C_0, and their concentration after time, t, is C, the rate equation can be written as:

$$-\frac{d[A]}{dt} = -\frac{d[B]}{dt} = k[A]^2 = k[B]^2$$
(7.34)

Or, using their concentration value, C, the *rate expression* (Figure 7.3) is:

$$-\frac{dC}{dt} = kC^2$$
(7.35)

$$\frac{dC}{C^2} = -kdt$$
(7.36)

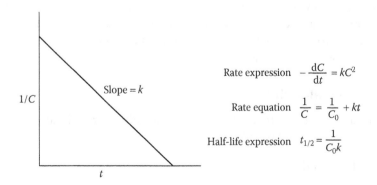

Rate expression $-\dfrac{dC}{dt} = kC^2$

Rate equation $\dfrac{1}{C} = \dfrac{1}{C_0} + kt$

Half-life expression $t_{1/2} = \dfrac{1}{C_0 k}$

Figure 7.3 Second-order kinetics. Plot of the reciprocal of the concentration, C, against time, t.

Integrating,

$$\int_{C_0}^{C} \frac{dC}{C^2} = -k \int_{0}^{t} dt \tag{7.37}$$

$$\frac{1}{C_0} - \frac{1}{C} = -kt \tag{7.38}$$

Or, the *rate equation* (Figure 7.3) is:

$$\frac{1}{C} = \frac{1}{C_0} + kt \tag{7.39}$$

7.2.5.2 Half-life

In a second-order reaction, the time to reach a certain fraction of the initial concentration (such as $t_{1/2}$ or $t_{0.90}$) is dependent on the initial concentration. The half-life is defined as:

$$C = \frac{C_0}{2} \tag{7.40}$$

Hence,

$$\frac{1}{C_0} - \frac{1}{C_0/2} = -kt_{1/2} \tag{7.41}$$

Or,

$$\frac{1}{C_0} = kt_{1/2} \tag{7.42}$$

Thus, the *half-life expression* (Figure 7.3) is:

$$t_{1/2} = \frac{1}{C_0 k} \tag{7.43}$$

Hence, for a second-order reaction, $t_{1/2}$ decreases with increasing initial concentration of the two reactants. This is consistent with the molecular mechanism of intermolecular reactions, where the two molecules must collide and react with each other for the reaction to happen. Thus, higher initial concentration increases the probability of collision and reaction between the molecules of two different types, increasing the probability and rate of the reaction.

7.2.6 Complex reactions

Often, a drug undergoes more than one chemical reaction or a series of reactions in the same environment. Such complex reactions can be exemplified by reversible, parallel, or consecutive reactions. In such cases, the experimental methods for detection of reaction rates usually have limitations in that each reaction intermediate and product may not be detected or accurately quantitated. Thus, one may be quantitating a product whose concentration is impacted by the starting material or drug concentration in a complex manner. Understanding how such complexities might be linked to the kinetics of reactions can help delineate the mechanisms of degradation of drugs. This section describes the predicted kinetics of complex reactions.

7.2.6.1 Reversible reactions

Reversible reactions are bidirectional; that is, the product can convert back to the reactant. The rate constant of the forward reaction can be designated by k_1, and the rate constant of the reverse reaction can be designated by k_{-1}.

$$A \underset{k_{-1}}{\overset{k_1}{\rightleftharpoons}} B$$

Assuming first-order reaction in either direction, the rate of the forward reaction at equilibrium is described by:

$$\text{Rate} = -\frac{d[A]}{dt} = \frac{d[B]}{dt} = k_1[A] - k_{-1}[B] \tag{7.44}$$

The rate of the forward reaction may not equal the rate of the reverse reaction. In fact, the rate of the forward and the reverse reactions may be affected by different environmental conditions. Equilibrium reactions are characterized by a constant ratio of the concentration of reactants and products, without regard to their absolute concentration.

7.2.6.2 Parallel reactions

Parallel reactions involve two or more simultaneous reaction pathways for a reactant. For example,

Assuming first-order kinetics for both reactions, the individual reaction rates and the rates of formation of individual products are defined by the individual rate equations:

$$\frac{d[B]}{dt} = k_1[A] \tag{7.45}$$

$$\frac{d[C]}{dt} = k_2[A] \tag{7.46}$$

The overall rate of degradation of a reactant is given by:

$$\text{Rate} = -\frac{d[A]}{dt} = k_1[A] + k_2[A] = (k_1 + k_2)[A] = k_{obs}[A] \tag{7.47}$$

where k_{obs} is the observed rate of degradation of the reactant A.

The concentration of reactant A at any time t is given by the exponential first-order equation:

$$[A] = [A_0]e^{-k_{obs}t} \tag{7.48}$$

Thus, the overall or observed rate of degradation is a sum of the rates of degradation of all individual parallel reactions that occur simultaneously.

7.2.6.3 Consecutive reactions

Consecutive reactions involve the formation of an intermediate, which is transformed into the final product.

$$A \xrightarrow{\;k_1\;} B \xrightarrow{\;k_2\;} C$$

The rate equations for this mechanism can be written as:

$$-\frac{d[A]}{dt} = k_1[A] \tag{7.49}$$

$$\frac{d[B]}{dt} = k_1[A] - k_2[B] \tag{7.50}$$

$$\frac{d[C]}{dt} = k_2[B] \tag{7.51}$$

The concentration time profiles for all species in this reaction can be obtained by simultaneously solving the above differential equations.

7.3 FACTORS AFFECTING REACTION KINETICS

To determine ways to prevent degradation of drugs in pharmaceutical formulations, it is important to identify the mechanism of drug degradation and the factors that affect the degradation rate or reaction kinetics. Once the route and kinetics of degradation have been identified, stabilization strategies that minimize reaction rates and maximize the shelf life of the drug product can be adopted.

7.3.1 Temperature

If a chemical reaction is endothermic (takes heat from the environment to react), increase in temperature generally accelerates the reaction. If a reaction is exothermic (gives out heat to the environment as it proceeds), temperature is generally inversely proportional to reaction rate. Most chemical reactions of pharmaceutical relevance in a dosage form are endothermic. Thus, an increase in temperature generally accelerates the reaction rate.

7.3.1.1 Arrhenius equation

The effect of temperature on the rate of drug degradation is expressed in terms of the effect of temperature on the reaction rate constant, k, by the *Arrhenius equation* (Figure 7.4):

$$k = Ae^{-E_a/RT} \tag{7.52}$$

where:
E_a is the activation energy
A is a preexponential constant
R is the gas constant (1.987 calories/degree.mole)
T is the absolute temperature (in Kelvin)

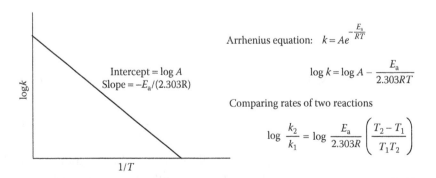

Arrhenius equation: $k = Ae^{-\frac{E_a}{RT}}$

$$\log k = \log A - \frac{E_a}{2.303RT}$$

Comparing rates of two reactions

$$\log \frac{k_2}{k_1} = \log \frac{E_a}{2.303R} \left(\frac{T_2 - T_1}{T_1 T_2} \right)$$

Intercept = log A
Slope = $-E_a/(2.303R)$

Figure 7.4 Arrhenius plot. Plot of the variation of the rate constant, k, versus reciprocal of the absolute temperature, T.

The Arrhenius expression can also be written as (Figure 7.4):

$$\ln k = \ln A - \frac{E_a}{RT} \tag{7.53}$$

Or

$$\log k = \log A - \frac{E_a}{2.303\,RT} \tag{7.54}$$

This equation is of the form $y = mx + c$ for a straight-line plot. Thus, an Arrhenius plot of $\log k$ on the y-axis against reciprocal of the absolute temperature $(1/T)$ on the x-axis yields E_a from the slope of the straight line (Figure 7.4). This equation is not amenable for direct application for the measurement of reaction rates, since A and E_a are unknown. Nevertheless, activation energy, E_a, can be calculated by comparing reaction rates at two different temperatures.

Thus, for temperatures T_1 and T_2,

$$k_1 = Ae^{-E_a/RT_1} \tag{7.55}$$

$$k_2 = Ae^{-E_a/RT_2} \tag{7.56}$$

Thus,

$$\frac{k_2}{k_1} = \frac{Ae^{-E_a/RT_2}}{Ae^{-E_a/RT_1}} = e^{E_a/RT_1 - E_a/RT_2} = e^{E_a/R(1/T_1 - 1/T_2)} = e^{E_a/R((T_2 - T_1)/T_1 T_2)} \tag{7.57}$$

Which is same as:

$$\ln \frac{k_2}{k_1} = \ln \frac{E_a}{R}\left(\frac{T_2 - T_1}{T_1 T_2}\right) \tag{7.58}$$

Or, as in Figure 7.4,

$$\log \frac{k_2}{k_1} = \log \frac{E_a}{2.303R}\left(\frac{T_2 - T_1}{T_1 T_2}\right) \tag{7.59}$$

Thus, measurement of the reaction rate constant, k, at two different temperatures allows the calculation of the activation energy, E_a, for a given reaction.

7.3.1.2 Shelf life

The Arrhenius plot can be used to determine the shelf life of the drug. The half-life ($t_{1/2}$) and shelf life ($t_{0.90}$) expressions from the reaction order can be substituted for the reaction rate constants, k, in the above equations to directly infer product's shelf life at a given temperature. These calculations allow the calculation of temperature of optimum drug stability over its shelf life. If a drug is stable at room temperature ($25°C$), it is usually labeled for storage at controlled room temperature (range $15°C–30°C$). If a drug is unstable at room temperature but stable at lower refrigerated temperature ($5°C$), it is usually labeled for storage under refrigerated conditions (range $2°C–8°C$). This is the case, for example, with various injectables such as penicillin, insulin, oxytocin, and vasopressin. Typically, a shelf life of 24 months or more is desired for all commercial products to allow enough time for manufacture, storage, distribution, and consumption by the patient. Appropriate temperature, packaging configuration, and drug product storage conditions are determined to achieve the desired shelf life. Stabilization strategies for drugs against degradation during storage are often required for achieving and extending the desired shelf life.

7.3.1.3 Thermodynamics of reactions

Arrhenius equation provides a mathematical basis of connecting reaction kinetics to the collision theory and the transition state theory of chemical reactions. The collision theory states that reactions happen by collisions that happen among reacting molecules in a favorable configuration. It highlights the need for several random collisions for effective collisions, which lead to the reaction, to occur. The collision theory predicts an increase in intermolecular collisions as a function of temperature, thus leading to higher number of effective collisions and higher reactivity at higher temperatures.

The transition state theory states that an activation energy barrier must be surpassed for a reaction to become spontaneous. This activation energy barrier can be understood as the energy of collisions required for them to overcome the intermolecular repulsions at close contact for effective intermolecular reactions to occur. Arrhenius equation relates the rate of a reaction, k, with the activation energy barrier, E_a.

$$\ln k = \ln A - \frac{E_a}{RT}$$

Once the activation energy barrier is overcome, the free energy difference between the reactants and the products, ΔG, determines the reaction rate. This is given by the equation:

$$\Delta G = -RT \ln k \tag{7.60}$$

In addition, the free energy difference between the reactants and the products is a measure of the difference in the enthalpy, ΔH, and entropy, ΔS, between the reactants and products. This is represented by the equation:

$$\Delta G = \Delta H - T\Delta S \tag{7.61}$$

Greater the free energy difference, that is, lower the free energy of the products than that of the reactants, the faster the reaction. Thus, a negative ΔG facilitates a forward reaction. This can be achieved by lower enthalpy of the products than that of the reactants, achieving a negative ΔH, or higher entropy of the products than the reactants, achieving a high ΔS and a negative $T\Delta S$.

These equations can be used in conjunction with each other to connect a reaction's thermodynamic parameters to reaction rates, which can be determined experimentally. This allows the determination of thermodynamic parameters, such as free energy and entropy change, of various reactions.

7.3.2 Humidity

Water can influence reaction kinetics by acting as a reactant, a solvent (i.e., a reaction medium), or a plasticizer.

7.3.2.1 Water as a reactant

For hydrolytically sensitive drugs, water acts as a *reactant* and increases the drug degradation rate directly by participating in a bimolecular reaction. Such reactions may follow second-order or pseudo first-order kinetics, depending on whether water is available in the reaction medium in limited (such as contamination in a solvent or adsorbed water in a solid-state excipient) or ample (such as the solvent) quantity.

7.3.2.2 Water as a plasticizer

A plasticizer is a substance that is used as an additive to promote fluidity in a solid state. For example, polyethylene glycol (PEG) is commonly used as a plasticizer in tablet film-coating applications to allow the formation of a flexible film that wraps around the tablet core. Small amounts of free water (e.g., adsorbed on the surface) present in the solid particles can promote localized dissolution and fluidity or flow of reacting molecular species. Thus, for drugs that are not hydrolytically sensitive, water can increase reaction rates by acting as a *plasticizer* in the solid dosage forms, by increasing the molecular mobility and diffusion rates of the reactive components. This is commonly seen in solid dosage forms such as tablets and capsules, where the reaction rates are dependent on the humidity during storage.

7.3.2.3 Water as a solvent

In addition to the role of water as a solvent in the liquid dosage forms, small amounts of adsorbed water can also act as a *solvent* in the microenvironment within a solid dosage form. This can affect reaction rates by the following:

1. Solubilizing reacting components and increasing their mobility
2. Affecting the disproportionation of the salt form of the drug to its free acid or free base form, which may have different reactivity compared with the salt form of the drug
3. Removing the product away from the reacting species, so that equilibrium reactions proceed more rapidly toward the formation of the product

Disproportionation of the salt form of a drug in a solid dosage form to its constituent free acid or free base form of the active pharmaceutical ingredient (API) is commonly attributed to the dissolution of the salt in the free water in the dosage form.

7.3.2.4 Determination and modeling the effect of water/humidity

Experimentally, the effect of water or humidity on the stability of a dosage form is determined by determining drug degradation kinetics at different temperature and humidity storage conditions. This is accomplished by storing the drug product under open-dish conditions at different controlled temperature and humidity conditions for different time periods, followed by analysis of the degradation products. These studies are called *isothermal degradation rate studies*, since the temperature is kept constant throughout the study.

The effect of relative humidity (RH) at a fixed temperature on drug's degradation rate constant, k, can be incorporated using an empirically determined humidity effect constant, B, as:

$$k = e^{B(RH)} \tag{7.62}$$

This equation may be combined with Arrhenius equation for the effect of temperature on reaction rate:

$$k = A e^{-\frac{E_a}{RT}} \tag{7.63}$$

To obtain,

$$k = A e^{B(RH) - E_a/RT} \tag{7.64}$$

This combined equation predicts reaction rate as a function of both temperature and humidity.

The effect of humidity on reaction rate constant is an empirically fitted model. Hence, this model can take different forms, depending on the experimental system under investigation. For example, some systems may be better described by the following equation:

$$k = Ae^{E_a/RT + B(RH)} \tag{7.65}$$

Nonetheless, combining the humidity effect with the temperature effect on reaction rate constant provides a better estimation of the extent of drug degradation over its shelf life.

7.3.3 pH

7.3.3.1 Disproportionation effect

The pH of the drug solution in a liquid dosage form and the microenvironmental pH in a solid dosage form can significantly influence drug stability by affecting the proportion of ionized versus unionized species of a weakly acidic or a weakly basic drug. The proportion of free acid or free base form of a drug at a given pH is modeled by the *Henderson–Hasselbalch equation*.

$$pH = pK_a + \log \frac{[\text{salt}]}{[\text{acid}]} \tag{7.66}$$

for a drug that is the salt of a weak acid, or

$$pH = pK_a + \log \frac{[\text{base}]}{[\text{salt}]} \tag{7.67}$$

for a drug that is the salt of a weak base.

Disproportionation of a salt into its free acid or base form can influence reactivity by changing the concentration of the reacting species. Generally, the free acid or the free base form of a drug is more reactive. Thus, drugs that are salts of free acids are unionized in greater proportion and, consequently, are more reactive at acidic pH, and drugs that are salts of free bases are unionized in greater proportion and, consequently, are more reactive at basic pH.

7.3.3.2 Acid–base catalysis

Acid (H^+) and base (OH^-) can catalyze several reactions directly. For example, the rate of an ester hydrolysis reaction catalyzed by hydrogen or hydroxyl ions can vary considerably with pH. The H⁺ ion catalysis predomi-nates at a lower pH and the OH⁻ ion catalysis operates at a higher pH.

Acids and bases can affect reaction kinetics by *specific or general catalysis*. For example, in specific catalysis, the reaction rate depends only on the pH of the system and not on the concentration of actual acid or base salts (such as HCl vs. HF or NaOH vs. KOH) contributing the H⁺ or the OH⁻ ions. In general catalysis, all species capable of donating or sequestering protons contribute to the reaction rate, and proton transfer from an acid to the solvent or from the solvent to a base is the rate-limiting step. General catalysis is usually evident by changing reaction rates with changing buffer concentration at a constant pH.

7.3.3.3 pH–rate profile

Rates of chemical reactions are often determined at different pH values to identify the pH of optimal drug stability. The pH–rate profiles are two-dimensional plots of observed reaction rate constant (k_{obs}) on the y-axis against pH on the x-axis. The shape of a pH–rate profile reflects on the mechanism of the reaction. For example, Figure 7.5 shows representative pH–rate profiles that indicate, for the corresponding subfigures, (A) base catalysis and stability at acidic pH, (B) acid catalysis and stability at basic pH, (C) a continuum of acid and base catalysis with a narrow pH region of maximum drug stability, and (D) acid and base catalysis under extreme ionization conditions and a wide pH region of maximum drug stability.

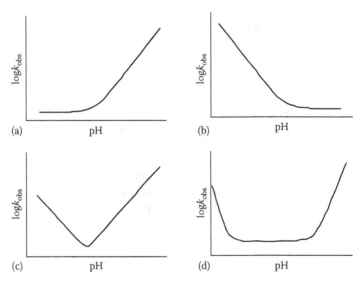

Figure 7.5 Typical pH stability profiles. Examples of pH—stability profiles for a drug that degrades under basic (a), acidic (b), or both acidic and basic conditions (c and d).

Proteins are particularly sensitive to changes in pH, particularly with respect to the conformation of the secondary and tertiary structures. Changes in the ionization of amino acid side chains in proteins with changes in pH can lead to folding or unfolding to varying degrees. Proteins exhibit overall charge neutrality at their *isoelectric point*, where the proportion of the positively charged groups within the protein is the same as that of the negatively charged groups. Proteins tend to be most stable in their most folded state, called the native state, which is generally obtained by appropriate balance of charged and uncharged groups on the surface. The *pH of optimal stability* can be determined by plotting log k against pH. For example, recombinant α-antitrypsin (rAAT) has a V-shaped pH—stability profile, with optimal stability at pH 7.5.

7.3.4 Cosolvent and additives

For liquid dosage forms, cosolvents are frequently used to improve drug solubility and stability in the vehicle. These cosolvents are commonly one or more of *PEG, propyleneglycol (PG), and ethanol*. In addition, water-miscible surfactants, such as polysorbate 80, and polymers, such as polyvinyl alcohol (PVA), may be used. Other common components of liquid dosage forms include buffers to maintain desired pH, ionic components for isotonicity of parenteral solutions, preservatives, sweeteners, flavors, and colorants.

These additives in liquid formulations lead to simultaneous changes in physicochemical conditions of the reaction medium, such as *dielectric constant, ionic strength, surface tension, and viscosity*, all of which may affect rates of chemical reactions. The effect of ionic strength and dielectric constant depends on the relative hydrophilicity of the reactants and products. If, for example, products achieve greater solubilization in the reaction medium, the rate of the reaction would be higher. This is due to the ability of the products to diffuse away from the reaction site, leading to shift in the equilibrium of the reaction toward the formation of the products. Similarly, if the reacting species have opposite charges, a solvent with a low dielectric constant accelerates the reaction rate. This could be attributed to lower solubilization of the reacting species, which also have affinity with each other, thus increasing the propensity for the reaction. On the other hand, if the reacting species have the same charge, a solvent with high dielectric constant will accelerate the reaction by forming bonds and dissolving both species, thus reducing intermolecular repulsion due to like charges.

7.3.4.1 Drug–excipient interactions

Chemical interaction between components in solid dosage forms can impact, often increasing, the rate of drug degradation.

Buffer salts are often added to maintain a formulation at optimal pH. These salts may often affect the degradation rate. For example, the hydrolysis rate of codeine is almost 20 times higher in phosphate buffer at neutral pH than in an unbuffered solution at this pH. At neutral pH, the major buffer species are $H_2PO_4^-$ and HPO_4^{2-}, either of which may act as a general base catalyst for codeine degradation.

Excipients that have specific functional groups such as the carboxylate group on croscarmellose sodium or the sulfate group on sodium lauryl sulfate can exhibit specific interactions with the drug substance that can destabilize a drug. These interactions can be direct reaction, salt disproportionation, or facilitation of a reaction by surface adsorption. In addition, excipients often contain small quantities of reactive substances, called reactive impurities. These reactive impurities in excipients can react with low-dose drugs in the dosage form to cause drug degradation. For example, PEG and polyvinyl-pyrrolidone (PVP) commonly have peroxide impurities that can cause oxidative degradation of sensitive drugs.

The effect of excipients on drug stability is usually assessed early in drug development through *excipient compatibility studies*. Drug degradation rate is determined in physical mixtures of a drug with individual or a combination of excipients. Excipient compatibility studies are also useful later in drug development when unexpected impurities are observed. Mechanistic investigation of the reaction pathway leading to the formation of these impurities becomes a cornerstone of drug product stabilization strategies.

7.3.4.2 Catalysis

Components of a dosage form can frequently act as, or bring in species that act as, reaction *catalysts*. A catalyst affects the rate of change in the concentrations of products and reactants in a chemical reaction but not the equilibrium concentration of reactants and products in the reaction. As seen in Figure 7.6,

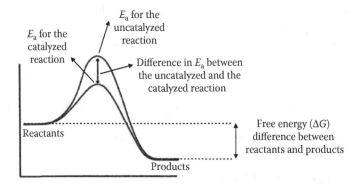

Figure 7.6 Effect of catalyst. Transition state during reaction progress (on the x-axis from left to right) with the energetics (on the y-axis) is indicated by the peak in the energy requirement for the reactants to convert to products. The presence of a catalyst changes the reaction pathway such that the height of this peak is lowered.

a catalyst may change the reaction pathway and lower the energy of activation required for a reaction, thus accelerating the reaction. However, the thermodynamic driver for a reaction, that is, free energy difference between the reactants and the products, remains the same for an uncatalyzed versus a catalyzed reaction. Thus, *a catalyst influences the speed but not the extent of a reaction*. In addition, a catalyst does not get chemically altered itself.

In pharmaceutical dosage forms, heavy metal contaminants in excipients and drug substances often act as catalysts.

7.4 DRUG DEGRADATION PATHWAYS

Major degradation pathways include hydrolysis, oxidation, and photolysis.

7.4.1 Hydrolysis

Hydrolysis is the common degradation pathway of carboxylic acid derivatives, such as esters, amides, lactams, lactones, imides, and oximes (Figure 7.7).

Chemical class	Structures
Ester	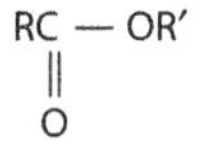 $RC - OR'$, with $\parallel O$
Amide	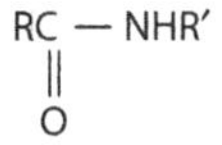 $RC - NHR'$, with $\parallel O$
Lactam, cyclic amide	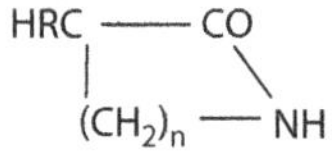 $HRC - CO$, $(CH_2)_n - NH$
Lactone, cyclic ester	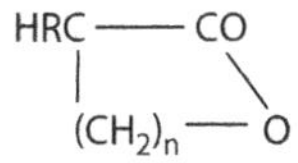 $HRC - CO$, $(CH_2)_n - O$
Imide	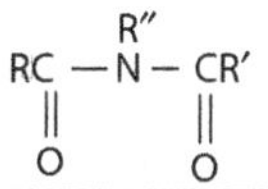 $RC - \overset{R''}{N} - CR'$, with $\parallel O \quad \parallel O$
Oximes	$R_2C = NOR$

Figure 7.7 Chemical groups susceptible to hydrolysis.

7.4.1.1 Ester hydrolysis

An ester hydrolysis pathway may involve, for example, nucleophilic attack of hydroxyl oxygen on the electropositive carbon, followed by breakage of the labile bond in the parent compound.

Hydrolysis is generally acid- and/or base-catalyzed, which becomes evident when pH–rate profile is constructed. Drugs that contain ester linkages include acetylsalicylic acid (aspirin), physostigmine, methyldopa, tetracycline, and procaine. Hydrolysis of the ester linkage in atropine and aspirin are shown in Figures 7.8 and 7.9, respectively, with their typical pH–rate profiles. In case of atropine, below pH 3, the main reaction is hydrogen-ion-catalyzed hydrolysis of the protonated form of atropine. Above pH 5, the main reaction is hydroxide-ion-catalyzed hydrolysis of the same species. The pH of maximum stability of atropine is 3.7.

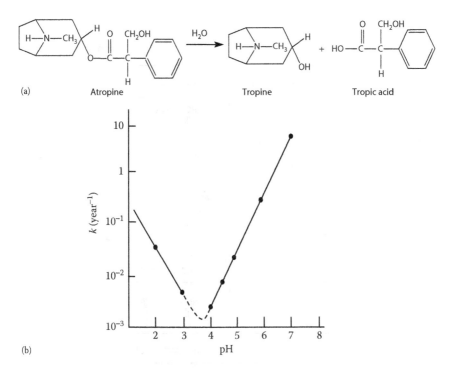

Figure 7.8 Hydrolysis of atropine. (a) Hydrolytic reaction scheme and (b) hydrolysis rate of atropine as a function of pH.

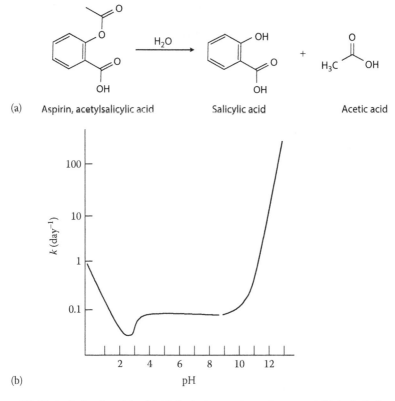

(a) Aspirin, acetylsalicylic acid   Salicylic acid   Acetic acid

(b)

Figure 7.9 Hydrolysis of aspirin. (a) Hydrolytic reaction scheme and (b) hydrolysis rate of aspirin as a function of pH.

7.4.1.2 Amide hydrolysis

Another chemical structure commonly found in pharmaceuticals is the amide group. It is considerably more stable than the ester group to hydrolysis under normal physiological conditions but can be broken down at extreme pH. The greater stability of the amide group, compared with the ester group, is due to the lower positive-charge density on the electropositive carbon. Hydrolysis of the amide group can be represented as:

Chloramphenicol decomposition below pH 7 proceeds primarily through hydrolytic cleavage of the amide function. Antibiotics possessing the β-lactam structure, which is a cyclic amide, are hydrolyzed rapidly by ring opening of the β-lactam group. Penicillins and cephalosporins belong to this category. The decomposition of these compounds in aqueous solution is catalyzed by hydrogen ion, solvent, hydroxide ion, and sugars. Deamidation and isomerization of asparaginyl residues are the major hydrolytic degradation reactions in proteins.

7.4.1.3 Control of drug hydrolysis

Hydrolysis is frequently catalyzed by hydrogen ions (specific-acid catalysis) or hydroxyl ions (specific-base catalysis) or both (specific-acid–base catalysis). Hydrolysis can be minimized by determining the pH of maximal stability and then formulating the drug product at this pH.

For solid formulations, minimizing the exposure of the drug product to moisture during manufacture and shelf life storage can minimize hydrolytic drug degradation. Moisture content should be as minimal as possible in solid dosage forms containing drugs susceptible to hydrolysis. In addition, *desiccants* may be used in drug product packages, such as bottles, for storage over the product shelf life.

For liquid formulations, reduction of the dielectric constant of the vehicle by the addition of nonaqueous *cosolvents* such as alcohol, glycerin, and propylene glycol may reduce hydrolysis. Another strategy to suppress hydrolysis is to make the drug *less soluble*. For example, the stability of penicillin in procaine–penicillin suspensions was significantly increased by reducing its solubility by using additives such as citrates, dextrose, sorbitol, and gluconate. *Complexation* of drugs with excipients, such as cyclodextrins, may also reduce hydrolysis. For example, the addition of caffeine to the aqueous solutions of benzocaine, procaine, and tetracaine was shown to decrease their base-catalyzed hydrolysis.

7.4.2 Oxidation

After hydrolysis, oxidation is the next most common pathway for drug degradation. Oxidation usually involves a reaction with oxygen. As illustrated below and in Figure 7.10, oxygen exists in two states: the ground or the triplet state, which contains two unpaired electrons in the outer molecular orbitals, and the singlet state, which contains all paired electrons. The * in the following molecular orbital notation and Figure 7.10 indicates antibonding molecular orbitals.

Oxygen atom: O (total electrons = 16) $1s^2, 2s^2, 2p_x^2, 2p_y^1, 2p_z^1$

Oxygen molecule: O_2 (total electrons = 16)

Triplet/ground state: $1s^2, 1s^{*2}, 2s^2, 2s^{*2}, 2p_x^2, 2p_y^2, 2p_z^2, 2p_x^{*1}, 2p_y^{*1}, 2p_z^{*0}$

Singlet/excited state: $1s^2, 1s^{*2}, 2s^2, 2s^{*2}, 2p_x^2, 2p_y^2, 2p_z^2, 2p_x^{*2}, 2p_y^{*0}, 2p_z^{*0}$

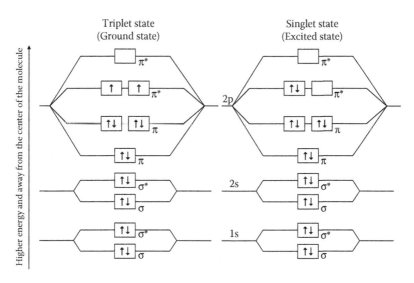

Figure 7.10 Molecular orbital illustration of the triplet and singlet states of oxygen molecule.

Most organic compounds are in the singlet state (with paired electrons). Most organic molecules are in the paired singlet state, which is their ground state. According to the molecular orbital theory of conservation of spin angular momentum of electrons, reactions between two molecules in the singlet state are favored, but not of a molecule in the singlet state with another molecule in the triplet state. Therefore, the atmospheric oxygen (triplet state) is unreactive. However, oxygen can be excited to singlet state both chemically and photochemically, leading to its higher reactivity, leading to oxidation reactions.

Oxidation in small-molecule drugs often involves free radical-mediated autocatalytic reaction initiated by the abstraction of hydrogen from the carbon next to a heteroatom, followed by reaction with oxygen to form a peroxide free radical. In addition, direct nucleophilic attack of the lone pair of electrons on the nitrogen can lead to N-oxide formation. Steroids and sterols represent an important class of drugs that are subject to oxidative degradation through the carbon–carbon double bonds, to which peroxyl radicals can readily add. Similarly, polyunsaturated fatty acids are susceptible to oxidation. Polyene antibiotics, such as amphotericin B, which contains seven conjugated double bonds, are subject to attack by peroxyl radicals, leading to aggregation and loss of activity. In proteins, several electron-rich functional groups are susceptible to oxidation, such as sulfhydryl in cysteine, imidazole in histidine, thioether in methionine, phenol in tyrosine, and indole in tryptophan.

Oxidation can involve coordination of the lone pair of electron on the nitrogen to oxygen to form N-oxide or free-radical autoxidation mechanism. The electron transfer or nucleophilic reactions are exemplified by peroxide anion reactions under basic conditions. Free radical-mediated oxidation reactions tend to be self-propagating until the substrate is depleted. These reactions could be initiated by the presence of an initiator, such as heavy metal, peroxides, and oxygen, along with environmental stresses such as heat and light. Termination of free radical-mediated oxidative reactions involves bimolecular reactions of radicals with another species, such as another free radical or a stabilizing conjugated system, to produce nonreactive products. Free radical reactions are characterized by a delay or *lag time* in their detection, which corresponds to the time required for the gradual build-up of free radicals in the system.

Oxidation is frequently catalyzed by transition metal contaminants (e.g., Fe^{2+}/Fe^{3+} and Cu^+/Cu^{2+}). The reacting metal species is regenerated in these reaction systems, commonly known as *Fenton's systems*.

The free radical formed can react with oxygen to produce a peroxide radical, and the reaction propagates as:

Peroxides (ROOR') and hydroperoxides (ROOH) are photolabile, breaking down into hydroxyl (HO•) and/or alkoxyl (RO•) radicals, which are themselves highly oxidizing species. The free radical reaction continues until all the free radicals are consumed or destroyed by inhibitors or by side reactions, which eventually break the chain. Reaction termination involves reactions of two free radicals to form nonfree radical end products.

7.4.2.1 Control of drug oxidation

Oxidation reaction proceeds until the substrate is consumed and/or the free radicals are destroyed by inhibitors or by side reactions, which eventually break the chain. The stabilization strategies for free radical-mediated oxidative degradation involve either or both:

- Inhibiting the initiation and/or propagation phases
- Promoting chain termination

Antioxidants are commonly used in formulations of susceptible compounds to stabilize drug products. Antioxidants can be categorized into three general categories based on their mechanism of action:

1. *Inhibitors of initiation*: Compounds that prevent the initiation phase of the free radical-mediated chain reaction and/or remove catalytic initiators. These are exemplified by the chelating agents, such as ethylenediaminetetraacetic acid (EDTA).
2. *Free radical terminators*: Compounds that react with free radicals and inhibit the propagation phase of the free radical chain reaction. These are exemplified by butylated hydroxyanisole (BHA) and butylated hydroxytoluene (BHT).
3. *Reducing agents*: Compounds that possess lower redox potential than the oxidation substrate in the formulation, thereby acting as a reducing agent by getting preferentially oxidized. These are exemplified by ascorbic acid, thiols (such as thioglycerol and thioglycolic acid), and polyphenols (such as propyl gallate).

In addition to the use of antioxidants, replacement of headspace oxygen in pharmaceutical containers with an inert gas, such as nitrogen, can minimize oxidation. The use of chelating agents, such as EDTA, and minimized content of heavy metal ions, such as iron, cobalt, and nickel, can prevent metal-catalyzed oxidation. Other approaches to minimize oxidation include the use of opaque or amber containers when light-induced photooxidation is involved.

7.4.3 Photolysis

Many pharmaceutical compounds, including phenothiazine tranquilizers, hydrocortisone, prednisolone, riboflavin, ascorbic acid, and folic acid, degrade on exposure to light. Some light-sensitive functional groups such as indole in tryptophan can adsorb energy from light illumination and form electronically excited species with high reactivity. The most common photodegradation of proteins is photo-induced autoxidation. The amino acids susceptible to photooxidation are His, Trp, Met, and Cys.

Light is a form of electromagnetic radiation, with energy given by:

$$E = h\nu = \frac{hc}{\lambda} \tag{7.68}$$

where:

E is the energy

h is the Plank's constant

c is the speed of light $(3 \times 10^8 \, m/sec)$

ν is the frequency of light

λ is the wavelength of light

Lower the wavelength, higher the frequency and more the energy in the radiation. Absorption of electromagnetic radiation by a molecule causes excitation of electrons and thus higher reactivity of the molecule. As this molecule loses energy to come back to the ground state, it can transfer that energy to another molecule in its vicinity. This process is called *photosensitization*. Thus, a molecule that does not directly absorb light (but acts as an acceptor of energy quanta) can be excited in the presence of a light-absorbing molecule (which acts as a donor of energy quanta). The acceptor molecule, thus, is frequently termed a *quencher*, since it relaxes the excited state of the donor molecule.

Colored compounds absorb light of lower wavelength and emit it at higher wavelength. Thus, colored compounds are usually susceptible to photolytic degradation. In addition, photolysis of a drug substance frequently leads to discoloration, in addition to chemical degradation.

Light also causes electronic transition of the low-reactive triplet state of oxygen to the higher-reactive singlet state. In addition, the excited triplet state of organic molecules can react with the ground triplet state of oxygen. Thus, oxidation very often accompanies photooxidation in the presence of oxygen and light.

Photooxidation processes can be of two types. *Type I photooxidation*, also called an electron transfer or free radical process, involves transfer of an electron or a proton by the light-absorbing donor to the acceptor, thus converting the acceptor to a reactive anion or neutral radical. The reactive acceptor then reacts with triplet-state oxygen. In *Type II photooxidation*, ground-state triplet molecular oxygen acts as a quencher of the excited singlet or triplet states of organic molecules, thus absorbing energy to convert itself to the excited-state singlet molecular oxygen. The singlet molecular oxygen is more reactive, since it has similar spin state as ground-state organic molecules.

7.4.3.1 Control of photodegradation of drugs

Control strategies to prevent photodegradation of drugs can include the use of amber-colored glass containers and storage in the dark. Amber glass excludes light of wavelength less than 470 nm and protects drugs sensitive to ultraviolet light. In addition, application of primary barrier on the dosage form, such as film coating of tablets, can prevent drug degradation. For example, film coating of tablets with vinyl acetate containing oxybenzone prevents discoloration and photolytic degradation of sulfasomidine tablets.

7.5 STABILIZATION STRATEGIES

Some of the factors may not be easily modifiable. For example, drug's pK_a, salt/counterion, propensity for disproportionation (i.e., separation of the counterion from the free base or the free acid form of the drug) in the dosage form, crystalline form, concentration of the drug in the dosage form, and intrinsic solubility are determined by the chemistry and clinical needs of the compound.

Other external factors, such as temperature, humidity, pH, light, and additives that may act as reaction catalysts or quenchers, may be readily controlled to achieve desired drug product's shelf life stability.

Drug product stabilization is a constantly evolving field with new information and insights, helping stabilize a wide variety of molecules. For example,

- Wet granulation manufacturing process for solid dosage forms results in better distribution of the stabilizing agent within the powder matrix than the dry granulation process. Thus, improved product stabilization can be obtained with the wet granulation process.
- Polyionic surfactants can promote disproportionation of the salt form of a drug in the dosage form to its free acid or free base form, which is generally more reactive than the salt form. Stabilization strategies in such cases consist of avoiding highly ionized excipient or creating a microenvironment in the drug product that disfavors disproportionation of the drug.
- Reduction of particle size of the drug, though good for improving the dissolution rate and bioavailability of the drug from the dosage form, may be counterproductive to solid-state reactions that are limited by the surface area of the compound. Use of larger-particle-size drug or physical separation of the drug from the reactive excipient in the dosage form is an effective strategy in such cases.

Identifying degradation mechanisms/pathways, delineating reaction kinetics, and adopting stabilization strategies that can support drug product stability for clinical studies and shelf life during commercial manufacture are the key roles of pharmaceutical scientists in modern drug development. The science of drug product stabilization keeps evolving with the discovery of ever-newer chemistries, with newer molecules in the discovery pipeline.

REVIEW QUESTIONS

7.1 Use one or more of the choices given below for answering the following set of questions. These questions can be answered by referring to Figures 7.1 through 7.3.
     i.   Zero-order reaction

    ii. First-order reaction
   iii. Second-order reaction

      A. For which of these reaction kinetic models is the drug concentration at any time, t, independent of the initial drug concentration?

      B. For which of these reaction kinetic models is the half-life independent of the initial drug concentration?

      C. For which of these reaction kinetic models is the rate of drug degradation independent of the initial drug concentration?

      D. Which of these reaction kinetic models would show an exponential decline in drug concentration over a period of time?

      E. Which of these reaction kinetic models would give a straight line when concentration is plotted against time?

      F. Which of these reaction kinetic models would give a straight line when the inverse of concentration is plotted against time?

      G. Which of these reaction kinetic models would give a straight line when logarithm of concentration is plotted against time?

      H. For which of these reaction kinetic models is the drug concentration at any time t (and half-life) would be greater when higher initial concentration is used (i.e., directly proportional to the initial drug concentration)?

      I. For which of these reaction kinetic models is the drug concentration at any time t (and half-life) would be lower when higher initial concentration is used (i.e., inversely proportional to the initial drug concentration)?

7.2 Which equation is used to predict the stability of a drug at room temperature from experiments at increased temperatures?

    A. Stokes equation
    B. Arrhenius equation
    C. Michaelis–Menten equation
    D. Henderson–Hasselbalch equation
    E. Noyes–Whitney equation

7.3 Which equation is used to predict disproportionation of a weakly acidic or a weakly basic drug as a function of pH?

    A. Stokes equation
    B. Arrhenius equation
    C. Michaelis–Menten equation
    D. Henderson–Hasselbalch equation
    E. Noyes–Whitney equation

7.4 When an acid catalyzed reaction is affected by the concentration and strength of the buffer species, it is known as:

    A. Specific-acid catalysis
    B. Specific-base catalysis
    C. General-acid catalysis
    D. General-base catalysis

7.5   In a first-order reaction involving the decomposition of hydrogen peroxide for a period of 50 min, the concentration expressed in volume was found to be 10.6 mL from an initial concentration of 72.6 mL.
   A.   Calculate k.
   B.   Calculate the amount of hydrogen peroxide not decomposed after 30 min.

7.6   For a second-order reaction,

$$C_2H_5COOC_2H_5 + KOH \rightarrow C_2H_5COOK + C_2H_5OH$$

   Diethyl acetate and potassium hydroxide were at a concentration of 0.05 M. Concentration of potassium hydroxide was observed to change by 0.0088 mol/L over a period of 35 min. Determine the rate constant, k, for the reaction and the half-life.

7.7   A formulation for an analgesic is found to degrade at 110°C (383°K), with a rate constant of $k_1 = 2.0$ h^{-1} and k_2 at 150°C (383°K) of 3.8 h^{-1}. Calculate the activation energy and the frequency factor, A ($R = 1.987$ cal deg^{-1}. mol^{-1}).

7.8   The shelf life of a liquid drug is 21 days at 5°C. For approximately how long will the drug be stable at 37°C?

FURTHER READING

Carstensen JT (1995) *Drug Stability: Principles and Practices*, 2nd ed., New York: Marcel Dekker.

Eley JG (2003) Reaction kinetics. In Amiji MM and Sandamann BJ (Eds.), *Applied Physical Pharmacy*, New York: McGraw-Hill, pp. 231–284.

Florence AT and Attwood D (2006) *Physicochemical Principles of Pharmacy*, 4th ed., London: Pharmaceutical Press.

Gosh TK (2005) Chemical kinetics and stability. In Ghosh TK and Jasti BR (Eds.) *Theory and Practice of Contemporary Pharmaceutics*, Boca Raton, FL: CRC Press, pp. 217–256.

Hovorka SW and Schöneich C (2001) Oxidative degradation of pharmaceuticals: Theory, mechanisms and inhibition. *J Pharm Sci* 90: 253–269.

Kim C (2004) *Advanced Pharmaceutics: Physical Principles*, Boca Raton, FL: CRC Press, pp. 257–239.

Lu ZR (2005) Stability of proteins and nucleic acids. In Mahato RI (Ed.) *Biomaterials for Delivery and Targeting of Proteins and Nucleic Acids*, Boca Raton, FL: CRC Press, pp. 352–374.

Narang AS et al. (2010) Excipient compatibility. In Qiu Y, Chen Y, and Zhang GGZ (Eds.) *Developing Solid Oral Dosage Forms*, New York: Elsevier, pp. 125–146.

Tønnesen HH (2001) Formulation and stability testing of photolabile drugs. *Int J Pharm* 225: 1–14.

Zhou D et al. (2010) Drug stability and degradation studies. In Qiu Y, Chen Y, and Zhang GGZ (Eds.) *Developing Solid Oral Dosage Forms*, New York: Elsevier, pp. 87–124.

Chapter 8

Interfacial phenomena

LEARNING OBJECTIVES

On completion of this chapter, the students should be able to

1. Identify different types of interfaces based on physical states of the system.
2. Describe examples where interfacial phenomena are important in biological and pharmaceutical systems.
3. Define and differentiate between surface tension and interfacial tension.
4. Describe the importance of interfacial tension in pharmaceutical formulation.
5. Compare physical adsorption and chemisorption.
6. Describe the differences between Langmuir, Freundlich, and Brenner, Emmett, and Teller's (BET) adsorption isotherms.
7. List the factors affecting adsorption of a solute from solution to an undissolved solid.

8.1 INTRODUCTION

A boundary between two phases (a phase being one of the three states of matter—gas, liquid, or solid) is termed an *interface*. An interface between solid and gas or liquid and gas is typically called a *surface*. Liquid–liquid interfaces result from the contact of mutually immiscible liquids. Liquid–solid interfaces result from the contact of an insoluble solid with the liquid. The phenomena resulting from and at the boundary of the two phases are termed *interfacial phenomena*. Interfacial phenomena result from the different environment (at the molecular level) faced by the molecules of both phases at the interface, as compared with the bulk of each phase.

8.2 LIQUID–LIQUID AND LIQUID–GAS INTERFACES

A liquid or a solid phase can be defined as a conglomeration of like molecules, held together by intermolecular bonds that hold the molecules in association and proximity with each other. The two phases—liquid and solid—differ in the degree of order in the association of the molecules, with the solid phase being more ordered than the liquid phase. Within the solid phase, the crystalline phases are more ordered than the amorphous phases. The gas phase, on the other hand, is the least ordered, with the molecules undergoing random Brownian motion, independent of other molecules.

The bonds that hold a phase together are van der Waals force, ionic, dipole, and hydrogen bonds—depending on the atomic structure of the molecules of a phase. For example, water molecules are held together predominantly by hydrogen bond and dipole forces, whereas octane molecules are held together by weak van der Waals forces. The strength of intermolecular forces of attraction and the proximity of the molecules follow the general trend: solids > liquids > gases. In the bulk of a phase, a molecule is surrounded by other molecules of the same type and encounters similar forces in all directions, which tend to neutralize each other. At the interface, a molecule encounters directionally different forces (Figure 8.1). Forces of attraction between the molecules of the same type within a phase can be termed *cohesive* forces, and the resulting phenomenon is termed *cohesion*. Similarly, forces between the molecules of different types at the interface can be termed *adhesive forces*, and the resulting phenomenon is termed *adhesion*.

At the liquid–gas interface, cohesive forces are generally greater than adhesive forces, leading to an inward pull on the molecules toward the bulk. This force pulls and keeps the molecules of the interface together and tends to contract the surface, resulting in minimization of the exposed surface area. Thus, a liquid droplet tends to be spherical, since this shape

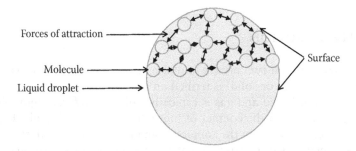

Figure 8.1 A liquid droplet depicted with some molecules (small spheres) with mutual forces of attraction (depicted with arrows). The molecules at the surface experience attractive forces from all directions, except at the interface, leading to a pull toward the bulk of the liquid.

can contain the maximum volume per unit surface area. Expansion of surface requires application of force. This force can be expressed in terms of surface or interfacial tension.

8.2.1 Surface tension

Surface tension (γ) is the force per unit length that must be applied in *parallel* to the surface to expand the surface, counterbalancing the net *inward* pull. It has units of force per unit length, for example, dyne/cm. Surface tension of a liquid film is commonly determined by creating a film of the liquid in a horizontal bar apparatus (Figure 8.2) and pulling the film using standard weights until the film breaks. Surface tension of the solution forming the film is a function of the force that must be applied to break the film over the length of a movable bar in contact with the film. Since the film has two liquid–gas interfaces (one above and one below the plane of the bar), the total length of the contact is equal to twice the length of the bar.

Thus,

$$\gamma = \frac{f_b}{2L} \qquad (8.1)$$

where:
   f_b is the force required for breaking the film
   L is the length of the film or the movable bar

Surface tension of a liquid is constant. Thus, this equation indicates that the amount of force required to break the film is directly proportional to the

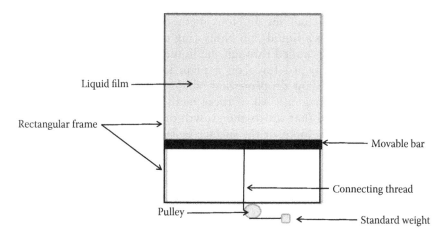

Liquid film

Rectangular frame

Movable bar

Connecting thread

Pulley

Standard weight

Figure 8.2 A simplistic representation of a rectangular block apparatus for determining the surface tension of a liquid.

length of the film. In other words, the amount of force required, or work done, to create additional surface is directly proportional to the amount of new surface being created.

8.2.2 Interfacial tension

Interfacial tension is the force per unit length that must be applied in *parallel* to the interface to expand the interface, counterbalancing the net *inward* pull of the two phases. While the term *surface tension* is reserved for liquid–gas and solid–gas interfaces, the term *interfacial tension* is commonly used for liquid–liquid interfaces. Interfacial tension has the same symbol (γ) and units (dyne/cm) as surface tension and is derived similarly from the amount of force required to create new interface. Subscripts are commonly used to distinguish between different interfacial tensions. For example, $\gamma_{L/L}$ is the interfacial tension between two liquids (designated "L"), and $\gamma_{L/V}$ is the surface tension between a liquid and its vapor (designated "V") in the gas phase.

Usually, the *interfacial tension* (liquid–liquid) of a hydrophilic liquid is less than its *surface tension* (liquid–vapor). This is because the adhesive forces between two liquid phases forming an interface are generally higher than those between a liquid and a gas phase. For example, at ~20°C, the interfacial tension between water and carbon tetrachloride is 45 mN/m, while the surface tension of water is 72.8 mN/m.

8.2.3 Factors affecting surface tension

Surface tension is measured with devices known as *tensiometers*. These devices measure the force by which a surface is held together while the force is applied on the surface to expand it. The methods for surface tension measurement include the du Nouy method (maximum pull on a rod or plate immersed in a liquid), du Nouy ring method (maximum downward force on a ring pulled through the liquid–air interface), Wilhelmy plate method (downward force on a plate lowered to the surface of the liquid), and pendant drop method (shape of the drop at the tip of needle by optical imaging). All of these methods measure the inherent force within a liquid that resists the growth or expansion of its surface. Factors affecting this force, or the surface tension, of a liquid include the following:

- *Nature of the liquid*: Greater the cohesive forces between the molecules of a liquid, higher its surface tension. Thus, the surface tension of water (72.8 mN/m at 20°C) is higher than that of methanol (22.7 mN/m). Mixing of the two miscible solvents leads to an intermediate surface tension. For example, a 7.5% solution of methanol in water has a surface tension of 60.9 mN/m.

- *Temperature*: Surface tension of most liquids decreases linearly with an increase in temperature. This is because of greater Brownian motion of individual molecules that leads to reduction in the intermolecular attractive forces and, thus, the reduced *inward pull* of the molecules on the surface.

8.2.4 Surface free energy

Surface free energy of a liquid is defined as the work required for increasing the surface area. Surface free energy (W) and surface tension (γ) are related by:

$$W = \gamma \Delta A \tag{8.2}$$

Where, W is the work done, or the surface free energy (ergs) input, required to increase the surface by an area ΔA (cm^2) for a liquid that has the surface tension γ (dynes/cm).

Surface free energy represents the amount of energy put into the system per unit increase in surface area. Thermodynamically, surface free energy represents the Gibbs free energy at constant temperature and pressure.

$$W = \Delta G = \gamma \Delta A \tag{8.3}$$

Thus, surface tension (γ) can be represented as the increment in Gibbs free energy per unit area.

$$\gamma = \frac{\partial G}{\partial A} \tag{8.4}$$

> **Example 1:** If the length of the bar (Figure 8.2) is 5 cm and the mass required to break a liquid film is 0.5 g, what is the surface tension of the soap solution? What is the work required to pull the wire down 1 cm?
>
> Since $\gamma = \dfrac{f_b}{2L}$
>
> $\therefore \gamma = (0.50 \text{ g} \times 981 \text{ cm/s}^2)/10 \text{ cm} = 49 \text{ dyn/cm}$
>   In addition,
>
> $$W = \gamma \Delta A$$
>
> $\therefore W = 49 \text{ dyn/cm} \times 10 \text{ cm}^2 = 490 \text{ ergs}.$

8.3 SOLID–GAS INTERFACE

8.3.1 Adsorption

If a solid comes in contact with a gas or a liquid, there is an accumulation of gas or liquid molecules at the interface. This phenomenon is known as *adsorption*. Adsorption refers to the surface binding of a liquid or gas molecule (*adsorbate*) onto a solid surface (*adsorbent*). Examples of adsorbents are highly porous solids, such as charcoal and silica gel, and finely divided powders, such as talc. Adsorbate could be any molecule, such as a drug compound.

Removal of the adsorbate from the adsorbent is known as *desorption*. A physically adsorbed gas may be desorbed from a solid by increasing the temperature and reducing the pressure. *Adsorption* is a surface phenomenon, distinct from *absorption*, which implies the penetration through the solid surface into the core of the solid.

8.3.2 Factors affecting adsorption

The degree of adsorption depends on the following:

- The chemical nature of the adsorbent and the adsorbate. Since adsorption is a result of an adhesive process, whereby two types of molecules interact with one another, the nature of the two types of molecules will determine their attractive interactions.
- Surface area of the adsorbent. Greater the surface area of the adsorbent, more the absolute amount of adsorbate that can be adsorbed. In modeling the adsorption phenomenon, the amount of adsorbate per unit adsorbent is usually calculated. In this scenario, the specific surface area (surface area per unit mass) of the adsorbent plays a role in determining the amount of adsorbate per unit mass of the adsorbent. This phenomenon indicates that a finely divided solid (of the same mass as a coarse particulate solid) would adsorb greater amount of adsorbate.
- Temperature. Temperature increases molecular motion, and its effect on adsorption depends on the relative change in the intermolecular forces of attraction between the molecules of the two phases. Generally, an increase in Brownian motion with increasing temperature reduces adsorption.
- Partial pressure (gas) or concentration (liquid) of the adsorbate. Generally, greater the solute (adsorbate) partial pressure or concentration, greater the rate of adsorption.

8.3.3 Types of adsorption

Adsorption can be physical or chemical in nature. Table 8.1 compares the characteristics of physical and chemical adsorption.

8.3.3.1 Physical adsorption

Physical adsorption is rapid, nonspecific, and relatively weak. It is typically mediated by weak noncovalent forces of attraction, such as van der Waals

Table 8.1 Characteristics of physical and chemical adsorption

Properties	Physical adsorption	Chemical absorption
Adsorption forces	Weak van der Waals forces Heat of adsorption <50 kJ/mol	Involves transfer or sharing of electrons between adsorbent and adsorbed molecules. Heat of adsorption is about 60–420 kJ/mol
Specificity	Nonspecific, will occur to some degree in any system	Specific, that is, occurs only when reaction is possible between adsorbent and adsorbate
Reversibility	Reversible, that is, adsorbate can be removed easily from surface in an unchanged form	Irreversible, that is, adsorbate is removed with difficulty in a changed form. For example, oxygen adsorbed by carbon is removed as carbon dioxide
Number of adsorbed layers	Monomolecular layer formed at low pressure, followed by an additional layer as pressure increases (multilayer)	Restricted to formation of monolayer
Rate of adsorption	Rapid at all temperature	Proceeds at a finite rate, which increases rapidly with rise in temperature

forces, and is reversible. Physical adsorption is an exothermic process, since heat is released with the formation of attractive interactions between molecules of the two phases. Physical adsorption may be associated with three phenomena:

- *Monolayer formation*: Adsorption of a solute on a solid surface leads to a monolayer formation, as the solute occupies the available surface in a single layer.
- *Multilayer formation*: Surface adsorption may continue into multilayer formation if the adsorption is facilitated by the interactions of solute molecules with other solute molecules (that are already adsorbed on the solid surface). Once the monolayer formation is complete and the conditions (such as solute concentration in the liquid or partial pressure of the gas) are supportive, multimolecular adsorption may take place.
- *Condensation*: The adsorbate may condense in the pores or capillaries of the adsorbent, leading to changes in the kinetics of the rate and the extent of adsorption.

8.3.3.2 Chemical adsorption (chemisorption)

Chemical adsorption or *chemisorption* is an irreversible process in which the adsorbent gets covalently linked to the adsorbate by chemical bonds.

Chemisorption is specific and may require activation energy. Therefore, this process is slow, and only a monolayer may be formed.

8.3.4 Adsorption isotherms

An adsorption isotherm is a graph that shows the amount of solute/adsorbate adsorbed per unit mass of a solid/adsorbent as a function of the equilibrium partial pressure (P) of the gaseous solute or the concentration (c) of the solute in the liquid at a constant temperature (thus, the term *isotherm*).

8.3.4.1 Type of isotherms

The isotherms can generally be classified into five types [UK34](Figure 8.3):

- Type I isotherms (e.g., ammonia on charcoal at 273 K) show a fairly rapid rise in the amount of solute adsorbed with increasing pressure to a limiting value. This phenomenon is due to the adsorption being restricted to a monolayer.
- Type II isotherms (e.g., nitrogen on silica gel at 77 K) are frequently encountered and represent multilayer physical adsorption on nonporous solids. They are often referred to as *sigmoid isotherms*. These isotherms are characterized by rapid solute adsorption to a limiting value, which sustains for certain increase in the partial pressure of the solute. Thereafter, multilayer adsorption initiates at an exponentially increasing rate.
- Type III adsorption isotherm (e.g., bromide at 760°C or iodine at 790°C on silica gel) shows large deviation from Langmuir model, no flattish portion in the curve, and the formation of multilayer films.
- Isotherm IV is typical of adsorption onto porous solids and involves the formation of a monolayer, which is followed by multilayer formation. An asymptote toward a limiting value is observed after each additional layer formation.
- Type V isotherm is similar to a type III isotherm in terms of the initial rate of solute adsorption increasing exponentially with solute

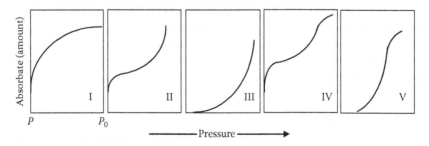

Figure 8.3 Types of adsorption isotherms.

concentration or partial pressure. This behavior is seen in relatively few instances in which the heat of adsorption of the solute in the first layer is less than the latent heat of condensation of successive layers. This promotes more rapid deposition of subsequent layers of adsorbed solute over the previous layer. Type III isotherm does not involve an eventual asymptote toward a limiting value, while type V isotherm does.

8.3.4.2 Modeling isothermal adsorption

Adsorption of a solute on a solid substrate at constant temperature (i.e., *isothermal* conditions) is a kinetic and a thermodynamic equilibrium phenomenon that can be described with the help of empirical or semiempirical equations. Modeling adsorption helps us understand a system and builds predictive ability to interpret the implications of changing system variables on the amount of free versus adsorbed solute. For example, in the case of drug adsorption on activated charcoal for preventing drug absorption into the systemic circulation after an oral overdose, the modeling of adsorption isotherm enables simulation of absorption and pharmacokinetics of the drug in the presence and absence of charcoal and the effect of different quantities of drug and charcoal. This can help determine the required dose of charcoal for a given drug overdose. In addition, modeling the adsorption data can be used to generate information about the system that would otherwise be unavailable. For example, gas adsorption on a solid substrate is used to quantify the specific surface area of a solid.

Isothermal adsorption can be modeled by using Freundlich, Langmuir, or BET equations.

8.3.4.2.1 Freundlich adsorption isotherm

Some cases of isothermal adsorption of a gas on a solid can be explained by the empirical Freundlich equation (Figure 8.4a).

$$Y = \frac{x}{m} = kp^{1/n} \tag{8.5}$$

where:
  Y is the mass ratio of the adsorbent on the adsorbate, given by the ratio of the mass of gas (x) adsorbed per unit mass (m) of adsorbent at the partial pressure of gas (p).
  The k and n are constants for a particular system at a constant temperature.

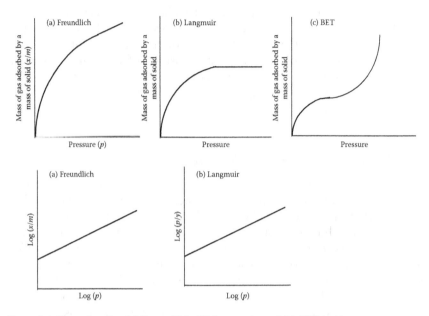

Figure 8.4 Plots showing (a) Freundlich, (b) Langmuir, and (c) BET isotherms.

The Freundlich isotherm, thus, states dependence of the mass of gas adsorbed on the partial pressure of gas with nonlinear kinetics, which depends on the specific combination of the adsorbent, the adsorbate, and the environment. Thus, the constants k and n depend not only on the substrate (adsorbate) and the gas (adsorbent) but also on the system (environment, such as other constituents).

The above equation can be written logarithmically as:

$$\log\left(\frac{x}{m}\right) = \log k + \frac{1}{n}\log p \tag{8.6}$$

A plot of log (x/m) against $log\ p$ yields a straight line, with slope $1/n$ and intercept $log\ k$. This allows the experimental determination of the constants for a given system.

Freundlich isotherm models multilayer adsorption and mostly represents physical adsorption that does not reach saturation.

8.3.4.2.2 Langmuir adsorption isotherm

Langmuir developed an equation based on the theory that the molecules or atoms of gas are adsorbed on active sites of the solid to form a layer one-molecule thick (monolayer) (Figure 8.4b). Langmuir adsorption isotherm predicts not only a dependence on the partial pressure of gas (p) but also saturable kinetics of the overall rate of adsorption (K), which is defined as the ratio of the forward (adsorption) reaction rate constant (k_a) to the reverse (desorption) reaction rate constant (k_d). Thus, for the adsorption reaction,

$$A + B \underset{k_d}{\overset{k_a}{\underset{\leftarrow}{\rightarrow}}} AB$$

$$K = \frac{\overset{k_a}{\rightarrow}}{\underset{k_d}{\leftarrow}}$$

Langmuir adsorption isotherm predicts that:

$$\frac{y}{y_{max}} = \frac{KP}{1 + KP}$$

where, y is the number of available surface adsorption or binding sites occupied by the adsorbent, also expressed as the mass of gas adsorbed per unit mass of adsorbent, and y_{max} is the total number of surface adsorption or binding sites on the adsorbate, also expressed as the maximum mass of gas that a unit mass of adsorbent can absorb when monolayer is complete.

Therefore, the Langmuir adsorption isotherm predicts that y never exceeds y_{max}, even as the rates of forward, adsorption, reaction reach but never exceed 1.

The simplified equation of Langmuir isotherm is:

$$\frac{p}{y} = \frac{1}{Ky_{max}} + \frac{p}{y_{max}} \tag{8.7}$$

A plot of p/y against p yields a straight line, with $1/y_{max}$ as the slope and $1/Ky_{max}$ as the intercept. This allows the experimental estimation of the values of y_{max} and K.

Langmuir adsorption isotherm is often indicative of chemisorption and has the following characteristics:

- Adsorption is localized to the active regions on the surface, and only monolayer adsorption takes place.
- Heat of adsorption is independent of surface coverage, indicating that all molecules being adsorbed experience the same attractive force, independent of the neighboring adsorbed molecules.

8.3.4.2.3 BET adsorption isotherm

The BET adsorption isotherm models multilayer gas adsorption and assumes that the forces involved in physical adsorption are the same as those responsible for the condensation of the adsorbate.

The BET equation relates the partial pressure of gas (P) with the relative proportion of the adsorbed molecules (Y/Y_m) by the equation:

$$\frac{P}{Y(P_0 - P)} = \frac{1}{Y_m b} + \frac{b-1}{Y_m b}\frac{P}{P_0} \tag{8.8}$$

where:

p is the partial pressure of adsorbate

y is the mass of adsorbate per unit mass of adsorbent

P_0 is the vapor pressure of adsorbate when the adsorbent is saturated with adsorbate molecules

Y_m is the maximum quantity of adsorbate adsorbed per unit mass of the adsorbent

b is the constant proportional to the difference between the heat of adsorption of the gas in the first layer and the latent heat of condensation in the successive layers

The BET isotherms occur when gases undergo physical adsorption onto nonporous solids to form a monolayer, followed multilayer formation. The BET isotherms have a sigmoidal shape (Figure 8.4c) and represent type II isotherms.

8.4 SOLID–LIQUID INTERFACE

Many pharmaceutical systems deal with the adsorption of solutes from solutions onto solid surfaces. These can be exemplified by the adsorption of drug or hydrophilic polymer on suspended drug particles in a suspension or the adsorption of drug on activated charcoal administered in the case of oral drug overdose.

8.4.1 Modeling solute adsorption

The adsorption of solute molecules from solution may be treated in a manner analogous to the adsorption of gas molecules on the solid surface. Isothermal adsorption can be expressed by Langmuir equation in the following form:

$$\frac{c}{y} = \frac{1}{K y_{max}} + \frac{c}{y_{max}} \tag{8.9}$$

where, c is the equilibrium concentration of the solute in the solution and replaces p, the partial pressure of the gas. A plot of c/y against c yields a straight line, and y_{max} and K can be obtained from the slope and intercept of this plot.

The Langmuir binding isotherm was utilized in determining the affinity and extent of interaction of drug and excipients in the dosage form and the impact of this interaction on oral bioavailability of drugs.

8.4.2 Factors affecting adsorption from solution

Adsorption from solution depends on the following factors:

1. *Solubility of adsorbate/solute*: The rate of adsorption of a solute is inversely proportional to its solubility in the solvent from which adsorption occurs. For adsorption to occur, solute–solvent bonds must first be broken. The greater the solubility, the stronger are these bonds and, hence, the lower the rate of adsorption. Conversely, the lower the solubility of the solute in the solvent, the higher its rate of adsorption onto the solid adsorbent.

2. *Solute concentration*: An increase in the solute concentration increases the rate of adsorption that occurs at equilibrium until a limiting value is reached.

3. *Temperature*: Adsorption is an exothermic process, that is, heat is released when stronger adsorbate–adsorbent bonds are formed. Thus, increase in temperature reduces adsorption. This can also be understood as increased Brownian motion of the solute molecules at higher temperature.

4. *pH*: The pK_a value(s) of the solute determines the relative proportion of ionized and unionized species of the solute and solute solubility in solution as a function of pH. The pH of the solution may also influence surface polarity of the solid substrate by changing the ionization, ion adsorption, or selective dissolution, as discussed earlier. The effect of pH on adsorption depends on the nature of intermolecular forces between solute and solute, solute and solvent, and solute and solid substrate as a function of the ionization status of an ionizable solute. The pH of the solution would also affect the solubility of the solute.

   Adsorption generally increases as the ionization of the drug is suppressed; that is, the extent of adsorption reaches a maximum when the drug is completely unionized. This is related to higher aqueous solubility of the ionized form. For amphoteric compounds, adsorption is at a maximum at the isoelectric point.

5. *Nature of adsorbent/solid substrate*: The physicochemical nature of the adsorbent can affect the rate and extent of adsorption by changes in the molecular forces of attraction between the adsorbate and the

adsorbent. In addition, the extent of adsorption is proportional to the surface area of the adsorbent. Thus, an increased surface area, achieved by a reduction in particle size or the use of a finely divide or porous adsorbing material, increases the extent of adsorption.

8.4.3 Wettability and wetting agents

Adsorption of the solvent, water, onto a solid substrate is termed wetting. The *wettability* of a material can be ascertained by observing the contact angle that water makes with the surface. *Contact angle* is the angle between a liquid droplet and the surface of the solid over which it spreads. As shown in Figure 8.5, the lower the contact angle (θ), the higher the wetting. Contact angle can range from 0° to 180°. For example, mercury does not wet most solid surfaces and its contact angle is well above 120° for most surfaces.

The balance of intermolecular forces involved in determining the adsorption of solute on a solid surface is the same for the adsorption/wetting of solvent/water on a solid surface. Powders, such as sulfur, charcoal, and magnesium stearate, that are not easily wetted by water are called *hydrophobic*. Powders, such as zinc oxide, talc, and magnesium carbonate, that are readily wetted by water are called *hydrophilic*.

A *wetting agent* lowers the contact angle and aids in displacing an air phase at the surface and replacing it with a liquid phase. Wetting agents could be of the following types:

1. *Surfactants*: Surfactants with hydrophile-lipophile balance (HLB) values between 7 and 9 are used as wetting agents, generally in the concentration of about 0.1% w/v. Surfactants reduce the interfacial tension between solid particles and a vehicle. As a result of the lowered interfacial tension, air is displaced from the surface of particles, and wetting and deflocculation of dispersed solid particles are promoted. Examples of surfactants used as wetting agents include polysorbates (Tweens) and sorbitan esters (Spans), as well as sodium lauryl sulfate.
2. *Hydrophilic colloids*: Acacia, bentonite, tragacanth, alginates, and cellulose derivatives act as protective colloids by coating hydrophobic particles with a multimolecular layer. This changes the surface properties of the solid, making it more hydrophilic, and promotes wetting.

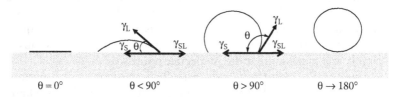

Figure 8.5 Contact angles from 0° to 180°.

3. *Solvents*: Water-miscible solvents, such as alcohol, glycerol, and glycols, can act as wetting agents by getting adsorbed on the solid surface, which makes the surface more hydrophilic, and reducing the dielectric constant of water, which can alter the balance of solute solubility in the bulk of the solvent versus surface adsorption.

8.5 BIOLOGICAL AND PHARMACEUTICAL APPLICATIONS

Interfacial phenomena are important in the following biological and pharmaceutical applications:

- *Physical stability of biphasic dosage forms*, such as suspensions and emulsions, are affected by the stabilization of the solid–liquid and the liquid–liquid interfaces, respectively.
- *Gas exchange in the lung*: Biological surfactants in the lung lower the surface tension of the alveolar membrane. Thus, alveoli can expand easily with inspiration and do not collapse at the end of expiration. If there is little or no surfactant in the lungs to assist these processes, the alveoli collapse, leading to respiratory distress syndrome.
- *Preventing absorption after oral overdose and poisoning*: Activated charcoal, magnesium oxide, and tannic acid are administered to reduce the absorption of an oral overdose of many drugs such as colchicines, phenytoin, aspirin, and chlorphenamine.
- *Hemoperfusion*: Many cases of severe drug overdoses can be treated by direct perfusion of the blood over charcoal granules. Although activated charcoal granules are very effective in adsorbing many toxic materials, they are not safe to use because they tend to embolize particles and remove blood platelets. Charcoal-induced embolism was reduced by microencapsulation of activated charcoal granules in biocompatible membranes, such as acrylic hydrogels.
- *Adsorption in drug formulation*: Some drugs tend to adsorb onto solid surfaces, which may reduce the rate and/or extent of drug release from the dosage form. This is exemplified by ionic interactions of ionizable drugs with ion-exchange resins. This phenomenon is used to create sustained- or extended-release dosage forms and in the use of resins for oral overdose.
- *Adsorption to packaging components*: Adsorption of medicaments onto the container and closure material can reduce the potency of the drug product.
- *Improving drug dissolution*: The dissolution rate of poorly soluble drugs can be improved by adsorption of a small amount of surfactant on the surface of drug particles.

- *Protein adsorption*: Adsorption of proteins onto surfaces is a fast process and depends on concentration, charge, temperature, and hydrophobicity. Adsorption of protein on hydrophobic surfaces can catalyze its unfolding and aggregation, leading to physical instability in drug product formulation. Thus, containers and closures for the storage and administration of protein therapeutics, including intravenous infusion sets, need to be carefully evaluated for protein—surface interaction.

REVIEW QUESTIONS

8.1 Which of the following is NOT true for gas adsorption on a solid?
   A. Chemical adsorption is reversible
   B. Physical adsorption is based on weak van der Waals forces
   C. Chemical adsorption may require activation energy
   D. Chemical adsorption is specific to the substrate
   E. All of the above
8.2 What is the difference between absorption and adsorption? Compare physical and chemical adsorption.
8.3 What is adsorption isotherm? What are the types of adsorption isotherms? What is the BET equation used for? What are its inherent assumptions in terms of nature of adsorption (physical or chemical) and molecules adsorbed (monomolecular or multimolecular)?
8.4 Why it is easy to measure the amount of adsorption of a pure gas but difficult to measure the adsorption of a pure liquid?
8.5 What is a wetting agent? What are the types of wetting agents used for formulation of pharmaceutical suspension?
8.6 Calculate the surface tension of a 2% w/v solution of a wetting agent that has a density of 1.008 g/cm^3 and that rises 6.60 cm in a capillary tube having an inside radius of 0.02 cm.
8.7 The surface tension of an organic liquid is 25 ergs/cm^2, the surface tension of water is 72.8 ergs/cm^2, and the interfacial tension between the two liquids is 30 ergs/cm^2 at 20°C. What is the work of cohesion of the organic liquid and the work of adhesion between the liquid and water at 20°C?

REFERENCE

Narang et al. (2011) Reversible and pH-dependent weak drug-excipient binding does not affect oral bioavailability of high dose drugs. *J Pharm Pharmacol* **64**: 553–565.

FURTHER READING

Bummer PM (2000) Interfacial phenomena. In: Gennaro A (Ed.) *Reminton's The Science and Practice of Pharmacy*, 20th ed., Easton, PA: Lippincott Williams & Wilkins, pp. 275–287.

Fell JT (1988) Surface and interfacial phenomena. In: Aulton ME (Ed.) *Pharmaceutics: The Science of Dosage Form Design*, Edinburgh: Churchill Livingstone, pp. 50–61.

Florence AT and Attwood D (2006) *Physicohemical Principles of Pharmacy*, London: Pharmacuetical Press.

Lambros MP and Nicolau SL (2003) Interfacial phenomena. In: Amiji MM and Sadamann BJ (Eds.) *Applied Physical Pharmacy*, New York: McGraw-Hill, pp. 327–363.

Rosen MJ (1989) *Surfactants and Interfacial Phenomena*, New York: Wiley.

Chapter 9

Colloidal dispersions

LEARNING OBJECTIVES

On completion of this chapter, the students should be able to

1. Define the size range for colloidal dispersions and how it is different from molecular and coarse dispersions.
2. Define and differentiate between lyophilic, lyophobic, and association colloids.
3. Identify two methods of preparation of hydrophobic colloids.
4. Differentiate between the stabilization strategies for hydrophilic and hydrophobic colloids.
5. Describe the electrical, kinetic, and colligative properties of colloids.
6. Discuss how the electrical properties of colloids can be used for improving their physical stability.

9.1 INTRODUCTION

Dispersed systems consist of one phase, known as the *dispersed phase*, distributed throughout a *continuous phase* or *dispersion medium*. The dispersed systems range in size from particles of atomic and molecular dimensions to visible particles, which can be up to several millimeters in diameter. On the basis of the size of the dispersed phase, dispersed systems are classified into the following types:

1. *Molecular dispersions* (<1 nm): Molecular dispersions are true solutions of one component in another. They are visibly homogeneous. The size of dispersed phase of molecular dispersions is typically less than 1 nm in diameter. True solutions do not scatter light and are clear or colored.

221

2. *Colloidal dispersions* (1 nm to 0.5 μm): Colloidal dispersions have the dispersed-phase size larger than the molecular dimensions of the dispersed phase in true solutions, while being much smaller than the particles that would be visible to the naked eye. The size of dispersed phase of molecular dispersions is typically 1 nm to 0.5 μm in diameter. Colloidal dispersions scatter light and appear turbid. Many natural systems, such as suspensions of microorganisms, blood, and isolated cells in culture, are colloids. Some hydrophilic colloids can be used as blood plasma substitutes to maintain osmotic pressure.

3. *Coarse dispersions* (>0.5 μm): Coarse dispersions have a particle size significantly larger than molecular and colloidal dispersions, such that the dispersed phase rapidly and spontaneously segregates if appropriate stabilization strategies are not utilized. The size of the dispersed phase of coarse dispersions is typically greater than 0.5 μm in diameter. Coarse dispersions scatter light and are visually cloudy/milky. Emulsions and suspensions are examples of coarse dispersions.

Colloidal solutions are preferred for pharmaceutical applications, where maximizing the surface area of the dispersed phase is important. Some examples of colloids used as pharmaceuticals are as follows:

- Colloidal kaolin is used for toxin absorption in the gastrointestinal (GI) tract.
- Colloidal aluminum hydroxide is used for neutralizing excess acid in stomach.
- Colloidal dispersion of amphotericin B and sodium cholesteryl sulfate (Amphocil®) is used as an antifungal agent.
- Colloidal silver chloride, silver iodide, and silver protein are effective germicides. They do not cause irritation, which is characteristic of ionic silver salts.
- Colloidal copper has been used in the treatment of cancer.
- Colloidal gold as a diagnostic agent for paresis and colloidal mercury for syphilis.
- Psyllium hydrophilic colloid (Metamucil®) is used as an oral laxative.

9.2 TYPES OF COLLOIDAL SYSTEMS

On the basis of the type and extent of molecular interactions of the dispersed phase with the dispersion medium, colloidal systems can be classified into three groups: lyophilic, lyophobic, and association colloids.

9.2.1 Lyophilic colloids

A *lyophilic colloid* (*solvent loving*) is a system in which the dispersed phase
has an affinity for the dispersion medium. Depending on the type of dispersion medium (solvent), both lipophilic (i.e., lipid-loving, which represents
the same characteristics as hydrophobic or water-hating) and hydrophilic
(i.e., water-loving, which represents the same characteristics as lipophobic
or lipid-hating) colloids can be lyophilic (solvent-loving). Thus, *lipophilic
colloids* are a dispersion of the lipophilic or hydrophobic material in an
organic solvent. *Hydrophilic colloids* are a dispersion of hydrophilic material in an aqueous medium.

Owing to their affinity for the dispersion medium, lyophilic materials
form colloidal dispersions with relative ease. Examples of lyophilic colloids
include gelatin, acacia, proteins (such as insulin), nucleic acids, albumin,
rubber, and polystyrene. Of these, the first five produce lyophilic colloids
in aqueous dispersion medium and are called hydrophilic colloids. Rubber
and polystyrene form lyophilic colloids in organic solvents and are thus
referred to as lipophilic colloids.

9.2.2 Lyophobic colloids

Lyophobic (*solvent-hating*) colloids are composed of materials that have
little attraction, if any, for the dispersion medium. Lyophobic colloids are
intrinsically physically unstable. These are formed by the mismatch of preferred molecular interactions of the dispersed phase and the dispersion
medium. For example, water and hydrophilic molecules prefer stronger
hydrogen bonding, dipole–dipole interactions, and electrostatic interactions, while lipids and hydrophobic molecules prefer weaker van der Waals
interactions. Examples of lyophobic colloids are gold, silver, arsenous sulfate, and silver iodide. Thus, dispersion of hydrophobic molecules, particles, or material in an aqueous medium results in *hydrophobic colloids*.
Special methods and energy input are required to prepare stable lyophobic
colloids, as they do not form spontaneously.

Differences in the properties of hydrophilic and hydrophobic colloids are
summarized in Table 9.1.

9.2.3 Association colloids

Association, or *amphiphilic colloids* are formed by the grouping or self-
association of the dispersed phase, which is amphiphilic (e.g., surface-active
agents). These molecules exhibit both lyophilic and lyophobic properties.
At low concentrations, amphiphiles exist separately as molecular dispersions
or true solutions and do not form a colloid. However, at higher concentrations, self-association of several monomers, or individual molecules occurs

Table 9.1 Differences in properties of hydrophilic and hydrophobic colloids

Property	Hydrophilic colloids	Hydrophobic colloids
Ease of dispersion of materials in dispersion medium	Usually occurs spontaneously	Special treatment necessary
Stability toward electrolytes	High concentrations of soluble electrolytes necessary to cause precipitation	Relatively low concentrations of electrolytes will cause precipitation
Stability toward prolonged dialysis	Stable	Unstable if the ions that are necessary for colloid stability get removed in dialysis
Reversibility after precipitation	Reversible (i.e., easily redispersible)	Irreversible
Viscosity	Usually higher than that of the dispersion medium	Similar to that of the dispersion medium
Protective ability	Capable of acting as protective colloids	Incapable of acting as protective colloids

leading to *micelle* formation. The concentration at which micelles are formed is known as the *critical micelle concentration* (CMC). The number of monomers that aggregate to form a micelle is called the *aggregation number*. As with lyophilic colloids, formation of association colloids is spontaneous, provided that the concentration of the amphiphile in solution exceeds the CMC.

9.3 PREPARATION OF COLLOIDAL SOLUTIONS

Lyophilic and association colloids are formed spontaneously by simple mixing of the dispersed-phase ingredients with the dispersion medium.

The preparative methods of hydrophobic colloids may be divided into methods that involve the breakdown of larger particles into colloidal dimensions (*dispersion method*) and that in which the colloidal particles are formed by the aggregation of smaller particles, such as molecules (*condensation methods*).

- *Dispersion methods* involve the reduction of particle size of coarse particles by input of energy, which can be done using ultrasonic methods, electrical methods, or shearing.
- *Condensation methods* involve the aggregation of subcolloidal-sized dispersed phase into colloidal particles, which usually involves supersaturation of the dispersed-phase concentration, followed by spontaneous formation and growth of dispersed-phase nuclei. Supersaturation may be brought about by the addition of solute, change in solvent,

or reduction in temperature. For example, if sulfur is dissolved in alcohol and the concentrated solution is then poured into an excess of water, many small nuclei form in the supersaturated solution. They grow rapidly to form a colloidal solution.

Other condensation methods involve a chemical reaction, such as reduction, oxidation, or hydrolysis. For example, *colloidal sulfur* may be obtained by passing hydrogen sulfide through a solution of sulfur dioxide.

9.4 PROPERTIES OF COLLOIDAL SOLUTIONS

9.4.1 Kinetic properties

Properties of colloidal systems that arise from the motion of particles with respect to the dispersion medium are known as kinetic properties. These include Brownian motion, diffusion, sedimentation, and osmosis.

9.4.1.1 Brownian movement

Brownian motion results from asymmetry in the force of collisions of molecules of the dispersion medium on the dispersed phase. This results in random dispersed-phase particle motion, called Brownian movement. Since the speed of motion of the molecules of the dispersion medium increases with temperature, Brownian motion is a function of temperature. Increase in temperature generally increases Brownian motion of dispersed-phase particles. The velocity of the particles also increases with decreasing particle size, which can be attributed to lower inertia of smaller particles. Similarly, increasing the viscosity of the medium decreases Brownian movement due to greater resistance to movement of the dispersed-phase particles.

9.4.1.2 Diffusion

Colloidal particles are subject to random collisions with other dispersed-phase particles, usually with a greater force, in addition to the molecules of the dispersion medium. This leads to the overall movement, called diffusion, of the dispersed-phase particles from a region of high concentration to a region of low concentration. The rate of diffusion of the dispersed-phase particles is given by *Fick's first-law equation*:

$$\frac{dM}{dt} = -DS\frac{dC}{dx} \tag{9.1}$$

where:
dM is the mass of substance diffusing in time dt across a cross-sectional area S
dC/dx is a concentration gradient

dC over the diffusion distance dx
D is the diffusion coefficient

The diffusion coefficient of the dispersed phase, D, is related to the frictional coefficient, f, of the particles, which quantitates the resistance to the movement of particles in the dispersion medium. The diffusion coefficient, D, and the frictional coefficient, f, are inversely related to each other and are linearly dependent on temperature, T, as explained by the *Einstein's law of diffusion*:

$$Df = kT \qquad (9.2)$$

or

$$D = \frac{kT}{f}$$

where:
   k is the Boltzmann constant
   T is the absolute temperature
   f is the frictional coefficient

The Boltzmann constant is a physical constant that relates the average kinetic energy of particles in a gas with the temperature of the gas and is obtained by dividing the gas constant, R, by the Avogadro's number, N, that is, the number of molecules per mole of a substance. The Boltzmann constant has the dimensions of energy over temperature, same as entropy, and is quantitatively $1.38064852(79) \times 10^{-23}$ J/K. Thus,

$$k = \frac{R}{N}$$

The frictional coefficient is dependent on the size of particles and the viscosity of the dispersion medium by the equation:

$$f = 6\pi\eta r \qquad (9.3)$$

where:
   η is the viscosity of the medium
   r the radius of the particle

Thus, diffusion coefficient depends on the viscosity and temperature of the dispersion medium and the size of the dispersed phase by the equation:

$$D = \frac{kT}{6\pi r} \qquad (9.4)$$

This equation indicates that the diffusion coefficient is inversely proportional to the viscosity of the medium and the radius of the diffusing particles, while it is directly proportional to the temperature.

Expressing the Boltzmann constant in terms of the gas constant and the Avogadro's number yields the *Stokes–Einstein equation*:

$$D = \frac{RT}{6\pi\eta r N} \tag{9.5}$$

However, this equation assumes spherical particles and does not take into account particle shape effects, which can be important in the case of complex molecules, such as proteins, and linear polymers that can entangle during movement. In addition, greater the asymmetry or deviation from sphericity, greater the resistance to flow.

9.4.1.3 Sedimentation

When stored undisturbed, the dispersed phase tends to separate out from the dispersion medium and concentrate in one region of the dispersion. When the dispersed-phase density is higher than that of the dispersion medium, the dispersed phase accumulates at the bottom, or sediments, and this process is called sedimentation. This is the case for most aqueous suspensions. When the dispersed-phase density is lower than that of the dispersion medium, such as in the case of aqueous emulsions, the dispersed phase accumulates toward the top of the container, or creams, and this process is called creaming. Both these phenomena are governed by the same physics, and for simplicity; this section will focus on sedimentation.

The rate of settling of particles, that is, the velocity (v) of sedimentation, is given by the *Stokes' law* equation:

$$v = \frac{2r^2(\rho - \rho_0)g}{9\eta_0} \tag{9.6}$$

where:
   ρ is the density of the particles
   ρ_0 is the density of the dispersion medium
   η_0 is the viscosity of the dispersion medium
   g is the acceleration due to gravity

In a centrifugation experiment, g is replaced by angular acceleration $\omega 2x$, where ω is the angular velocity and x is the distance of the particle from the center.

Stokes' law was derived for dilute dispersions of spherical particles. It does not take into consideration deviation of particle shape from sphericity and interparticulate interactions, especially at high dispersed-phase concentration. Thus, Stokes' law may not be quantitatively exactly applicable to the concentrated dispersions. However, the qualitative, or rank-order, effects of the factors indicated by the Stokes equation still hold true. For example, an increase in the mean particle size or in the difference between the densities of the solid and liquid phases increases the rate of sedimentation. Using the Stokes equation, creaming of an emulsion or sedimentation of a given suspension can be reduced by forming smaller particles, increasing the viscosity of continuous phase and/or decreasing the density difference between two phases.

9.4.2 Electrical properties

Electrical properties of the dispersed phase refer to the electrostatic charge on the surface of the particles and its impact on the interaction of the dispersed-phase particles with each other and with the dispersion medium.

9.4.2.1 Surface charge

Surface charge on the dispersed phase plays an important role in the following:

- Physical stability of colloids. Greater the electrostatic repulsion among the dispersed-phase particles, greater the physical separation and uniform appearance of the colloidal dispersion. However, when settled, the cake formed may not be easily redispersible. Therefore, a balance of electrostatic charge on the particles is sought that promotes formation of uniform dispersion and also allows easy redispersibility on settling.
- Filtration efficiency of submicron particles, which can be diminished considerably by particle aggregation or particle affinity for the filtration membrane.
- Determining the conformation of macromolecules such as polymers, polyelectrolytes, and proteins by influencing macromolecule–solvent interactions and intramolecular interactions within the polymers and macromolecules.

Most substances acquire a surface electric charge when brought in contact with an aqueous medium by ionization, ion adsorption, and/or ion dissolution.

9.4.2.1.1 Ionization

Surface charge arising from ionization on the particles is the function of the pH of the environment and the pK_a of the particle's surface functional groups. For example, proteins and peptides acquire charge through the ionization of surface carboxyl and amino groups to COO^- and NH_3^+

ions, respectively. The state and extent of ionization of these groups and the net molecular charge depend on the pH of the medium and the pK_a of the functional groups, as determined by the Henderson–Hasselbalch equation.

Macromolecules such as proteins have many ionizable groups. Thus, at pH below its isoelectric point (PI), the protein molecule bears an overall positively charge, and at pH above its PI, the protein molecule bears an overall negative charge—even though there may be domains within the protein structure that would be uncharged or bear the opposite charge. At the PI of a protein, the total number of positive charges equals the total number of negative charges in the protein, resulting in the net charge being zero.

As an illustration for the amino acid alanine, which has one amino and one carboxylate group, this phenomenon may be represented as:

Ionized molecule has stronger electrostatic, dipole, and hydrogen bond interactions than unionized functional group or molecule. Thus, ionization increases dispersed-phase–aqueous-solvent interactions and generally stabilizes the dispersion. Addition of salt to solutions of ionized proteins can reduce protein–solvent interactions and protein solubility. Thus, addition of salt to precipitate the protein of interest is a common procedure in experimental sciences. In solutions of multiple proteins, increasing salt concentration can sequentially precipitate proteins in the increasing order of their aqueous solubility.

However, at the isolectric point, proteins with multiple functional groups can self-associate through interactions of oppositely charged functional groups. Thus, often, a protein is least soluble at its isoelectric point due to the attractive interactions between different protein molecules. At the isolectric point, water-soluble salts such as ammonium sulfate, which partially neutralize surface charges and reduce interparticle attractions, may increase protein solubility.

9.4.2.1.2 Ion adsorption

Surfaces that are already charged usually show a tendency to adsorb counterions from solution. For example, a positively charged surface selectively adsorbs chloride (Cl^-) ions from a salt (NaCl) solution. This results in an excess of the countercharge (i.e., negative charge in the example of Cl^- ions) on the surface of the dispersed phase, compared with the bulk solution. A second layer of charged coions concentrate above the countercharged surface. In the aforementioned example, the free Na^+ ions in solution form a second layer over the Cl^- ions on the surface. These two layers of electrical charge on a charged surface are together called *electrical double layer.* Figure 9.1 shows the electrical double layer on a surface, with the first layer of negatively charged counterions and the second layer of positively charged

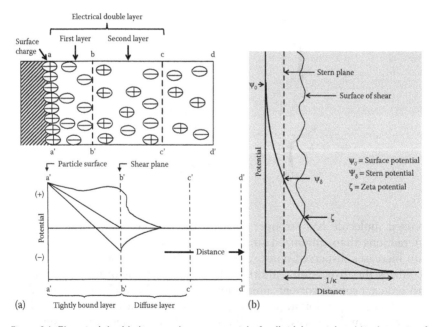

Figure 9.1 Electrical double layer and zeta potential of colloidal particles. (a) schematic of electrical double layer at the separation between two phases, showing distribution of ions and (b) changes in potential with distance from particle surface.

coions. A lower net surface charge results on charged surfaces that form electrical double layer due to unequal adsorption of oppositely charged ions, resulting in only partial neutralization of particle surface charge.

This phenomenon can also enable nonpolar surfaces to develop charge by adsorption of charged solutes from solution. For example, surfactants strongly adsorb by the hydrophobic effect and determine the surface charge when adsorbed.

9.4.2.1.3 Ion dissolution

Ionic substances can acquire a surface charge by unequal dissociation of the oppositely charged ions. For example, in a dispersion of silver iodide particles with excess [I⁻] in solution, the dispersed particles carry a negative charge. This is due to the suppression of dissociation of the I⁻ ions on the surface of particles by the common-ion effect.

$$AgI \rightleftharpoons Ag^+ + I^-$$

Similarly, the net charge on AgI particles is positive if excess Ag^+ ions are present in the solution. In this case, therefore, the silver and iodide ions are referred to as *potential-determining ions*, since their concentrations determine the electric potential at the particle surface.

9.4.2.2 Electrical double layer

As explained in Section 9.4.2.1.2, the surface charges of dispersed-phase particles influence the distribution of the nearby ions in the polar dispersion medium. Ions with opposite charge (known as *counterions*) are attracted toward the surface, and ions with like charges (known as *coions*) form a second layer on the concentrated layer of counterions. This leads to the formation of *an electric double layer* made up of a neutralizing excess of counterions close to the charged surface and coions. The *electrical double-layer theory* explains the distribution of ions with the changing magnitude of the electric potentials, which occur in the vicinity of the charged surface.

At a particular distance from the surface, the concentration of anions and cations is equal; that is, conditions of electrical neutrality prevail in bulk solution. The system *as a whole* is electrically neutral, even though there are regions of unequal distribution of anions and cations. This is illustrated in Figure 9.1. The first layer extends from aa' to bb' and is tightly bound to the surface. This rigid layer attached to the particle surface is called the *stern* layer. The second layer extends from bb' to cc' and is more diffuse. Hypothetical *planes* are defined as boundaries around the surface of particles that define true hydrated particle size of electrically double-layered dispersed-phase particles (called *stern plane*) and the plane that defines the movement of these particles in solution (called *shear plane*). The stern plane is at the center of the first layer of hydrated ions from the surface. The shear plane is the boundary of the first layer of hydrated ions from the surface.

The thickness of the electrical double layer is defined by the *Debye–Huckel radius or length parameter*, which characterizes the distance from surface at which the particle charge is completely screened by other charges in solution. This parameter is dependent on the electrolyte concentration of the aqueous media. The thickness of the electrical double layer shrinks with increase in electrolyte concentration in solution.

9.4.2.2.1 Nerst and zeta potentials

Electrothermodynamic or *Nerst potential* (E) is defined as the difference in potential between the actual surface and the electroneutral region of the solution. This is the potential at the particle surface (aa' in Figure 9.1). However, when the particles are set in motion by electrical forces, such as electrophoresis, a small layer of solvent with oppositely charged ions moves concurrently with the particles. The boundary of this layer is termed the *shear plane* (bb'), since this distinguishes the moving from the stationary part of the solvent. The electrical potential at the shear plane bb' is known as the *electrokinetic* or *zeta potential*, ζ. The ζ potential is defined as the difference in potential between the surface of the tightly bound layer (shear plane) and the electroneutral region of the solution.

The ζ potential, rather than the Nerst potential, truly governs the degree of repulsion between the adjacent, similarly charged, dispersed particles. Therefore, measurement and optimization of ζ potential is needed for the stability of dispersed systems. The ζ potential can be impacted by all three mechanisms discussed in the previous section, viz., ionization, ion dissolution, and ion adsorption. In addition, surfactant molecules that adsorb by the hydrophobic effect on the surface of the dispersed phase can affect the ζ potential.

9.4.2.2.2 DLVO theory

DLVO theory is named in honor of Russian physicists B. Derjaguin and L. Landau and Dutch pioneers in colloid chemistry, E. Verwey and J. Overbreek. These scientists independently formulated the theories of interaction forces between colloidal particles in the 1940s to help predict colloidal stability of charged particles in dispersion. This theory explains the stability of dispersed colloids in aqueous suspensions on the basis of the balance of two opposite forces between the dispersed-phase particles: electrostatic force of repulsion and van der Waals force of attraction.

DLVO theory of colloidal stability states that the only interactions involved in determining the stability of colloidal dispersed particles are electric repulsion (V_R) and van der Waals attraction (V_A) and that these interactions are additive. Therefore, the total potential energy of interaction (V_T) is given by:

$$V_T = V_A + V_R \tag{9.7}$$

Thus, a stable dispersion is obtained when the repulsive forces dominate, while a physically unstable dispersion is obtained when attractive forces dominate.

9.4.2.3 Electrophoresis

Electrophoresis is the movement of charged particles (with the attached ions and the solvent in the tightly attached first electrical layer) relative to the stationary liquid dispersion medium, under the influence of an applied electric field. Migration of particles in an electric field occurs due to the motion of the particle and its counterion cloud away from the electrode of the same charge and toward the electrode of the opposite charge.

Electrophoretic mobility (μ) of a molecule is a function of its net charge (Q) and size (radius, r). Thus,

$$\mu = \frac{Q}{r}$$

Experimentally, the electrophoretic mobility is determined as the particle velocity (v) per unit electrical field (E). Thus,

$$\mu = \frac{v}{E} \tag{9.8}$$

Hence, electrophoresis experiments can be used to determine the net surface charge on the particles.

9.4.3 Colligative properties

Colligative properties are the properties that depend only on the *number* of nonvolatile molecules in solution, without regard to their size or molecular weight, or the solute–solute or solute–solvent interactions.

9.4.3.1 Lowering of vapor pressure

Addition of a nonvolatile solute to a solvent lowers its vapor pressure, since solute occupies some of the surface of the solvent. This, therefore, reduces the rate of evaporation of the solvent.

The extent of decrease in the vapor pressure with the addition of solute to a solvent is given by *Raoult's law*, which states that the vapor pressure of an ideal solution is dependent on the vapor pressure of each individual component, weighted by the mole fraction of that component in solution. Thus,

$$P_A = X_A P_A^0 \tag{9.9}$$

where:

P_A is the vapor pressure of the colloidal solution
X_A is the mole fraction of solute in the solvent
P_A^0 is the vapor pressure of the pure solvent

Thus, the reduction of vapor pressure of a solvent is directly proportional to the concentration of solute in that solvent.

9.4.3.2 Elevation of boiling point

Addition of a nonvolatile solute leads to the elevation of boiling point due to the nonvolatile solute displacing the corresponding number of solvent molecules from the surface and, consequently, reducing the number of solvent molecules that are able to escape into the vapor phase from the solution. The extent of increase in the boiling point by the addition of a nonvolatile solute is given by:

$$\Delta T_b = K_b m \tag{9.10}$$

where:

K_b is the molal boiling point elevation constant
ΔT_b is the elevation of boiling point
m is the molal amount of solute in solution

Thus, the extent of elevation of boiling point is directly proportional to the concentration of nonvolatile solute in solution. The extent of this effect is different for each solvent. The value of K_b for different solvents is available in literature.

9.4.3.3 Depression of freezing point

Addition of a nonvolatile solute results in reduction in solvent–solvent interactions; This leads to depression of freezing point of the solvent.
The extent of decrease in the freezing point is given by:

$$\Delta T_f = K_f m \tag{9.11}$$

where:

K_f is the molal freezing point depression constant
ΔT_f is the depression of freezing point
m is the molal amount of solute in solution

Similar to the case with the elevation of boiling point, the extent of depression of freezing point is directly proportional to the concentration of nonvolatile solute in solution. The extent of this effect is different for each solvent. The value of K_f for different solvents is available in literature.

9.4.3.4 Osmotic pressure

Osmosis involves flow of solvent molecules through a membrane toward its concentration gradient, which is opposite of the concentration gradient of the solute in solution. The use of a membrane with a well-defined pore size leads to its semipermeable nature; that is, only molecules below a certain size or molecular weight are able to pass through the membrane. Thus, the use of a membrane through which the colloidal solutes are not able to diffuse, with solutes at different solution concentration on either side of the membrane, promotes solvent flow from a solution of low solute concentration to a solution of high solute concentration. This process of solvent flow is called *osmosis,* and the relative difference in the pressure of solvent generated by the concentration gradient on either side of the membrane is called *osmotic pressure,* π. The osmotic pressure is given by:

$$\pi = \left(\frac{n}{v}\right)RT = MRT \tag{9.12}$$

where:
    R is the gas constant
    T is temperature in Kelvin
    M is the difference in the molar concentration of solute in solution, which is defined as the number of moles of solute, n, per unit volume of solution, v

Thus, osmotic pressure of a solution is directly proportional to its solute concentration and temperature, through their impact on the motion of solvent molecules. The higher the solute concentration and the temperature, the higher the osmotic pressure.

9.4.4 Optical properties

Colloidal solutions scatter light, since their particle diameter is within the range of wavelength of visible light. This phenomenon is known as *Tyndall effect.* Thus, light passing through a colloidal solution with particle diameter of ~200 nm leads to scattering, resulting in turbid or milky appearance. This property is utilized in quantifying the number of suspended particulates in a liquid or gas colloidal solution by using a turbidimeter or *nephelometer* by calibrating the amount of turbidity at different concentrations of a colloidal solution.

9.5 PHYSICAL STABILITY OF COLLOIDS

Physical stability of colloidal dispersions depends on the balance of the following forces:

1. Electrical forces of repulsion between dispersed-phase particles
2. Forces of attraction between dispersed-phase particles
3. Forces of attraction between the dispersed phase and the dispersion medium

Accordingly, colloidal dispersions can be stabilized by the following:

1. Modulating the electric charge on the dispersed particles. The presence and magnitude, or the absence, of a charge on a colloidal particle are important determinants of the stability of colloidal systems. This can be done through ion adsorption, dissociation of ionizable functional groups, and ion dissolution. In addition, ionized species added to the aqueous solution, such as salt, can influence the overall zeta potential on the surface of the dispersed phase.
2. Surface coating of the particles to minimize adherence on collisions. This effect is significant for hydrophilic colloids. Thus, addition of soluble hydrophilic polymers to colloidal dispersions can entangle dispersed-phase particles, minimizing the speed and impact of interparticle collisions.

The stabilizations strategy depends on the type of colloid and the specific properties of a colloidal system.

9.5.1 Stabilization of hydrophilic colloids

Hydrophilic and association colloids are thermodynamically stable and exist in a true solution, so that the system constitutes a single phase and is visually clear. When negatively and positively charged hydrophilic colloids are mixed, the particles may separate from the dispersion to form a layer rich in the colloidal aggregates. The colloid-rich layer is known as a *coacervate*, and the phenomenon by which macromolecular solutions separate into two liquid layers is referred to *coacervation*. For example, when the solutions of gelatin and acacia are mixed in a certain proportion, coacervation results. Gelatin at a pH below 4.7 (its isoelectric point) is positively charged, whereas acacia carries a negative charge that is relatively unaffected by pH in the acid range. The viscosity of the outer layer is markedly decreased below that of the coacervate, which is considered as *incompatibility*. Coacervation need not involve the interaction of charged particles. Coacervation of gelatin may also be brought about by the addition of alcohol, sodium sulfate, or a macromolecular substance such as starch.

In colloidal dispersions, frequent interparticle collisions due to Brownian movement can destabilize the system. Thus, increase in temperature often compromises the physical stability of these systems.

9.5.2 Stabilization of hydrophobic colloids

In contrast to hydrophilic colloids, lyophobic or hydrophobic colloids are thermodynamically unstable but can be stabilized by imparting electric charge on the dispersed particles, which can prevent aggregation by increasing the repulsion between like particles. Addition of a small amount of electrolyte to a hydrophobic colloid tends to stabilize the system by imparting a charge to the particles. Addition of excess amount of electrolyte may result in the accumulation of opposite ions and reduce the ζ potential below its critical value, leading to destabilization. The critical potential for finely dispersed oil droplets in water is about 40 mV. This high value indicates relative instability and the need for significant electrostatic charge repulsion for stabilization. In comparison, the critical ζ potential of colloidal gold is nearly zero, which suggests that the particles require only a minute charge for stabilization.

REVIEW QUESTIONS

9.1 Which of the following statements about lyophilic colloidal dispersions is TRUE?
   A. They tend to be more sensitive to the addition of electrolytes than lyophobic systems
   B. They tend to be more viscous than lyophobic systems
   C. They can be precipitated by prolonged dialysis
   D. They separate rapidly
   E. All of the above
   F. None of the above

9.2 Compounds that tend to accumulate at interface and reduce surface or interfacial tension are known as:
   A. Antifoaming agents
   B. Detergents
   C. Wetting agents
   D. Surfactants
   E. Interfacial agents

9.3 Indicate which of the following statements is TRUE and which is FALSE:
   A. Particle size of molecular dispersions is larger than a colloidal dispersion.
   B. Zeta potential influences colloidal stability.
   C. Nernst potential is higher than zeta potential.

   D. Zeta potential is electrothermodynamic in nature.
   E. Hydrophilic colloids form turbid solutions.
9.4 Classify disperse systems based on the particle size of their dispersed phase. Which of these systems are not visible to the naked eye?
9.5 A. List three mechanisms involved in acquisition of surface charge in a molecule.
   B. Formulation of amino acids as solutions for parenteral administration requires careful consideration of the isoelectric point and the ionization status of the amino acids. Consider that your laboratory is given the amino acid alanine (structure given below) to be formulated into a solution.

   i. Which chemical groups in alanine will affect its ionization.
   ii. Assign either of the two pK values (2.35 and 9.69) of alanine to each group.
   iii. Predict the structure of L-alanine at pH 2, 7, and 10.
9.6 A. What is zeta potential?
   B. Zeta potential of the particles is routinely used for assessing the stability of pharmaceutical emulsions and suspensions. Suggest a reason why the surface charge of the particles is not used for this purpose?
9.7 Define and differentiate aggregation and coagulation in a colloidal system.
9.8 A. Define Stokes' law.
   B. Using the Stokes' law equation, explain how we can minimize the sedimentation and creaming phenomena.
   D. Sedimentation by ultracentrifugation is often utilized to determine the particle size of submicron particles. Suggest the principle behind this application.
   E. Suggest two reasons why this method is more suited to water-insoluble compounds than to water-soluble molecules.
9.9 A lyophilic colloid can be:
   A. Hydrophilic
   B. Hydrophobic
   C. Lyophobic
   D. All of the above
   E. None of the above

FURTHER READING

Mahato RI (2004) Dosage forms and drug delivery systems. In Gourley DR (Ed.) *APh's Complete Review for Pharmacy*, New York: Castle Connelly Graduate Publishing, pp. 37–64.

Florence AT and Attwood D (2006) *Physicochemical Principles of Pharmacy*, 4th ed., London: Pharmaceutical Press.

Sinko PJ (2005) *Martin's Physical Pharmacy and Pharmaceutical Sciences*, 5th ed., Philadelphia, PA: Lippincott Williams & Wilkins, pp. 561–583.

Im-Emsap W, Siepman J and Paeratakul O (2002) Disperse system. In Baker and Rhodes (Eds.) *Modern Pharmaceutics*, 4th ed., New York: Marcel Dekker, pp. 237–280.

Li X and Jasti B (2005) Theory and applications of diffusion and dissolution. In Ghosh TK and Jasti BR (Eds.) *Theory and Practice of Contemporary Pharmaceutics*, Boca Raton, FL: CRC Press, pp. 197–215.

Aulton ME, (Ed.) (1988) *Pharmaceutics: The Science of Dosage Form Design*, New York: Churchill Livingstone.

Chapter 10

Surfactants and micelles

LEARNING OBJECTIVES

On completion of this chapter, the students should be able to

1. Define surfactants and exemplify their applications in pharmaceutical dosage forms.
2. Describe micelles, types of micelles, critical micelle concentration (CMC), and the factors that affect the size and CMC of micelles.
3. Differentiate between micelles and liposomes.
4. Define the mechanism, factors affecting, and the benefits of micellar solubilization.

10.1 INTRODUCTION

Surface-active agents, or surfactants, are substances that preferentially localize or adsorb to surfaces or interfaces and reduce surface or interfacial tension. Common interfaces of pharmaceutical relevance are those between two insoluble liquids or the air–water interface. The interfacial tension between two surfaces results from lower forces of attractive interaction between the two materials (~adhesion) than within the two materials (~cohesion), which arise from the differences in the types of molecular interactions in a material. For example, hydrocarbon/oil molecules predominantly bind by hydrophobic interactions, whereas water molecules bond by hydrogen bonding and polar/dipole interactions. Thus, in an oil–water system, the water–water interactions and the oil–oil interactions are stronger than the oil–water interactions. This leads to a thermodynamic

241

propensity of the system to minimize the interfacial area, the extent of which may be expressed in terms of interfacial tension. Surface tension is a special case of interfacial tension, when one of the materials is air.

A surfactant preferentially adsorbs to the interface due to its molecular characteristics. Adsorption of surfactant at the interface results in changes in the nature of the interface and reduces interfacial tension between the two liquids. This phenomenon is of considerable influence in pharmaceutical formulations. For example, the lowering of the interfacial tension between oil and water phases facilitates emulsion formation. The adsorption of surfactants on insoluble particles reduces solid–liquid interfacial tension and enables drug particles to be dispersed in a suspension.

When a surfactant is added to a liquid in excess of what is needed to completely cover the surface, the surfactant forms self-associating structures within the liquid. These structures are called micelles. When formed in water, these micelles have a hydrophobic core and a hydrophilic shell. The incorporation of insoluble compounds within micelles of the surfactants in an aqueous solution can solubilize these insoluble drugs. Therefore, surfactants are commonly used as *emulsifying agents, solubilizing agents, detergents*, and *wetting agents*.

10.2 SURFACTANTS

A surfactant molecule has two distinct regions—hydrophilic (*water-liking*) and hydrophobic (*water-hating*). The existence of such two regions in a molecule is known as *amphipathy*, and the molecules are consequently referred to as *amphipathic* molecules or *amphiphiles*. The hydrophilic portions are typically the functional groups that bear electronegative atoms that can form hydrogen bonds with water and can participate in dipole–dipole interactions. The hydrophobic portions are usually saturated or unsaturated hydrocarbon chains or, less commonly, a heterocyclic or aromatic ring system. Depending on the number and nature of the polar and nonpolar functional groups present, the amphiphile may be predominantly hydrophilic, predominantly lipophilic, or almost equal in hydrophilic and lipophilic characters. For example, straight-chain alcohols, amines, and acids are amphiphiles that change from being predominantly hydrophilic to predominantly lipophilic as the number of carbon atoms in the alkyl chain is increased.

Surfactants are usually depicted with a circle representing a polar (hydrophilic) head group and a wiggly chain or a rectangular box depicting a nonpolar (lipophilic) region.

The surface activity (ability to reduce surface/interfacial tension) of a surfactant depends on its ability to preferentially partition into the interface, which, in turn, depends on the balance between its hydrophilic and hydrophobic properties. The surfactant molecules localize at the surface, with the hydrophobic regions pointing toward and bonding the hydrophobic liquid (or air), while the hydrophilic regions pointing toward and bonding the aqueous or hydrophilic liquid. Thus, the surfactant molecules replace the bulk liquid molecules on the surface with molecules that show mutual attraction for both sides of the surface, which reduces surface tension. For air–water surfaces, an increase in the length of the hydrocarbon chain of a surfactant results in an increased surface activity. Conversely, an increase in the hydrophilicity results in a decreased surface activity.

10.2.1 Types of surfactants

Surfactants are generally classified according to the nature of the hydrophilic group (Table 10.1). The hydrophilic regions can be *anionic* (negatively charged at certain pH values), *cationic* (positively charged at certain pH values), or *nonionic* (not charged at all pH values). In addition, some surfactants possess both positively and negatively charged groups. These surfactants can exist in either or both anionic or cationic states, depending on the pH of the solution and the pK_a of the ionizable groups on the surfactants. Such surfactants are known as *ampholytic* compounds.

Table 10.1 Classification of surfactants

Anionic surfactants
- Sodium stearate
- Sodium dodecyl sulfate (SDS)
- Sodium dodecyl benzene sulfonate
- Sodium cholate

Cationic surfactants
- Hexadecyltrimethylammonium bromide
- Dodecyl pyridinium chloride

Nonionic surfactants
- Heptaoxyethylene monohexadecyl ether

Ampholytic (Zwitterionic) surfactants
- N-dodecyl alanine
- Lecithin

10.2.1.1 Anionic surfactants

The hydrophilic group of anionic surfactants carries a negative charge, such as R-COO⁻, RSO$_4^-$, or RSO$_3^-$, where R represents an organic group. Anionic surfactants have high hydrophilicity and are used as detergents and foaming agents, such as in shampoos. Examples of anionic surfactants include soap (sodium salt of fatty acids, R-COO⁻Na⁺), sodium dodecyl sulfate ($C_{12}H_{25}SO_4Na^+$) (SDS), alkylpolyoxyethylene sulfate (R-[CH$_2$CH$_2$O]$_n$SO$_4^-$), and alkylbenzene sulfonate (R-C$_6$H$_5$-SO$_3^-$). Some of these surfactants, such as *SDS, also known as sodium lauryl sulfate (SLS)* (Figure 10.1), are used to create sink conditions during *in vitro* drug-release studies for new drug product development. It is very water soluble and has bacteriostatic action against gram-positive bacteria. Therefore, SLS also finds use as a preoperative skin cleaner and in medicated shampoos.

10.2.1.2 Cationic surfactants

Cationic surfactants have a cationic group, a functional group that can be positively charged at certain pH values, as the hydrophilic portion of the molecule. For example, primary (RNH$_2$), secondary (R$_2$NH), or tertiary amines (R$_3$N) are positively charged at low pH values. However,

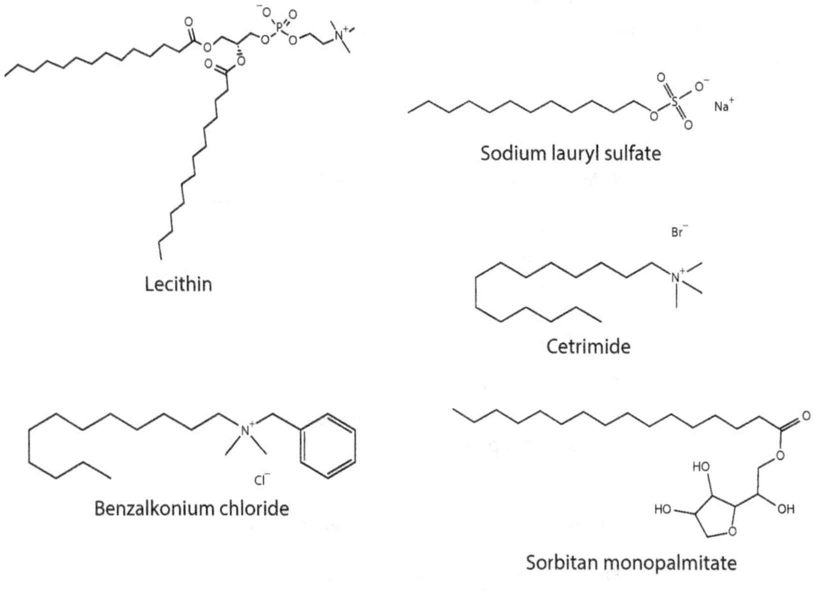

Figure 10.1 Structures of some surfactants.

the quaternary amines (R_4N^+) are permanently positively charged irrespective of the solution's pH. Most cationic surfactants are quaternary derivatives of alkylamines, for example, alkyl trimethyl ammonium salts, dialkyl dimethyl ammonium salts, and alkyl benzyl dimethyl ammonium salts.

Cationic surfactants are used in fabric softeners and hair conditioners. In addition, cationic surfactants can destabilize biological membranes due to the interaction of their cationic groups with the negatively charged phospholipids on the cell membranes. This results in their germicidal activity. Thus, the quaternary ammonium and pyridinium cationic surfactants have bactericidal activity against a wide range of gram-positive and some gram-negative organisms and are commonly used as preservatives in pharmaceutical formulations. They may also be used on the skin for cleansing of wounds. For example, solutions containing 0.1%–1% *cetrimide* (Figure 10.1) are used for cleaning the skin, wounds, and burns, as well as for cleaning contaminated vessels. *Benzalkonium chloride* (Figure 10.1) is a mixture of alkyl benzyl dimethyl ammonium chlorides. Its dilute solution may be used for the preoperative disinfection of the skin and mucous membranes, for application to burns and wounds, and for cleaning polyethylene tubing and catheters. Benzalkonium chloride is also used as a preservative in eye drops.

10.2.1.3 Nonionic surfactants

Nonionic surfactants contain ether $[-(CH_2CH_2O)_nOH]$ and/or hydroxyl $[-OH]$ hydrophilic groups. Thus, these surfactants are nonelectrolytes; that is, their hydrophilic groups do not ionize at any pH value. Nonionic surfactants are commonly used for stabilizing oil-in-water (o/w) and water-in-oil (w/o) emulsions. Since the nonionic surfactants do not contain an ionizable group, their properties are much less sensitive to changes in the pH of the medium and the presence of electrolytes. In addition, they have fewer interactions with cell membranes compared with the anionic and cationic surfactants. Thus, nonionic surfactants are preferred for oral and parenteral formulations because of their low tissue irritation and toxicity.

Most commonly used nonionic surfactants include *Spans and Tweens*. Sorbitan fatty acid esters (Spans), such as sorbitan monopalmitate (Figure 10.1), are oil-soluble emulsifiers that promote the formation of w/o emulsions. Polyethylene glycol sorbitan fatty acid esters (Tweens) are water-soluble emulsifiers that promote the formation of o/w emulsions. The Spans and the Tweens come in different molecular weight or size ranges, which differ in their physical properties.

10.2.1.4 Ampholytic surfactants

Ampholytic surfactants possess both cationic and anionic groups in the same molecule. Their ionization state in solution is dependent on the pH of the medium and the pK_a of ionizable groups. For example, the acidic functional groups, such as carboxylate, sulfate, and sulfonate, are negatively charged (ionized) at pH > pK_a, while the basic functional groups, such as amines, are positively charged (ionized) at pH < pK_a. The extent of ionization of functional groups, that is, the proportion of molecules in solution that bear the positive or the negative charge, at a given pH is governed by the Henderson–Hasselbalch equation, discussed elsewhere in this book. *Lecithin* (Figure 10.1), for example, is an ampholytic surfactant and is used for parenteral emulsions.

10.2.2 Hydrophile–lipophile balance system

In 1949, Griffin devised an arbitrary scale of values to serve as a measure of relative contributions of the hydrophilic and lipophilic regions of a surfactant to its overall hydrophilic/lipophilic character, which could be used to select emulsifying agents for a given application. This system is now widely known as the hydrophile–lipophile balance (HLB) system. The higher the HLB value of an emulsifying agent, the more hydrophilic it is. The emulsifying agents with lower HLB values are less polar and more lipophilic.

The HLB values of some commonly used surfactants are listed in Table 10.2. The Spans, that is, sorbitan esters, are lipophilic and have low HLB values (1.8–8.6); the Tweens, poly(oxyethylene) derivatives of the Spans, are hydrophilic and have high HLB values (9.6–16.7). Figure 10.2 illustrates

Table 10.2 HLB values of commonly used surfactants

Names of surfactants	HLB
Sorbitan laurate (Span 20)	8.6
Sorbitan palmitate (Span 40)	6.7
Sorbitan stearate (Span 60)	4.7
Sorbitan oleate (Span 80)	4.3
Sorbitan trioleate (Span 85)	1.8
Polyoxyethylene sorbitan laurate (Tween 20)	16.7
Polyoxyethylene sorbitan palmitate (Tween 40)	15.6
Polyoxyethylene sorbitan stearate (Tween 60)	14.9
Polyoxyethylene sorbitan oleate (Tween 80)	15.0
Polyoxyethylene sorbitan trioleate (Tween 85)	11.0
Brij 30	9.5
Brij 35	16.9
Sodium oleate	18.0
Potassium oleate	20.0

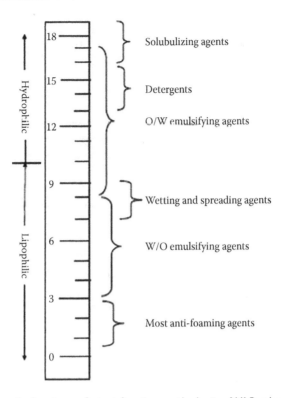

Figure 10.2 A scale showing surfactant function on the basis of HLB values.

a scale showing surfactant function on the basis of HLB values. Utilizing this numbering system, it is possible to establish an HLB range of optimum efficiency for each application of surfactants.

10.2.2.1 Type of emulsion formed

Surfactants with the proper balance of hydrophilic and lipophilic affinities are effective emulsifying agents, since they concentrate at the oil–water interface, while being present in the two phases (oil and water) in different concentrations. Thus, a lipophilic surfactant would have higher concentration in oil, while a hydrophilic surfactant would have higher concentration in water. The phase with higher surfactant concentration tends to become the external phase in an emulsion. Thus, the HLB of a surfactant, or a combination of surfactants, determines whether an o/w or w/o emulsion results. An emulsifying agent with high HLB is preferentially soluble in water and results in the formation of an o/w emulsion. The reverse situation is true with surfactants of low HLB value, which tend to form w/o emulsions. In general, o/w emulsions are formed when the HLB of the emulsifier is ~9–12, and w/o emulsions are formed when the HLB is ~3–6.

10.2.2.2 Required HLB of a lipid

The lipid phase used in an emulsion can be assigned a *required HLB* (or *RHLB*) value. The RHLB *for* the lipid phase of an emulsion is the HLB value *of* the surfactant that provides the lowest interfacial tension between the two phases to form an o/w or a w/o emulsion. The RHLB, thus, provides guidance to surfactant selection for a specified lipid phase for the formation of a stable emulsion. The RHLB may be experimentally determined by preparing a series of emulsions with surfactants of different HLB values and selecting the HLB value that resulted in the physically most stable emulsion, as assessed, for example, by the separation of phases on undisturbed storage of the emulsion. A list of RHLB values for common emulsifying agents is usually available in the literature (Table 10.3).

Table 10.3 Required HLB for some oil-phase ingredients for making o/w and w/o emulsions

Oil phase	w/o emulsion	o/w emulsion
Acetophenone		14
Cottonseed oil	6–7	
Lauric acid		16
Linoleic acid		17
Oleic acid		16
Ricinoleic acid		17
Stearic acid		15
Cetyl alcohol		14
Decyl alcohol		14
Lauryl alcohol		14
Tridecyl alcohol		14
Benzene		16
Carbon tetrachloride		14
Castor oil		8
Chlorinated paraffin		14
Kerosene		12
Lanolin, anhydrous	8	12
Aromatic mineral oil	4	
Paraffinic mineral oil	4	
Mineral spirits	4	
Petrolatum		
Beeswax	5	9
Candelilla		14–15
Carnauba		12
Paraffin	4	10

This value is utilized in the HLB concept to prepare an emulsion by selecting an emulsifier that has the same, or nearly the same, HLB value as the RHLB of the oleaginous phase of the intended emulsion. For example, mineral oil has an assigned RHLB value of 4 if a w/o emulsion is desired, and it has a value of 10.5 if an o/w emulsion is desired.

10.2.2.3 Required HLB of a formulation

The HLB values are additive. Therefore, calculation of the required HLB of a formulation is done by weighting the RHLB of each oil-phase ingredient (excluding any emulsifiers) as a weight percentage of the total oil-phase ingredients. For example, if the oil-phase ingredients of an o/w emulsion consist of 10% mineral oil, 3% capric/caprylic triglyceride, 2.5% isopropyl myristate, 4% cetyl alcohol, and the remaining emulsifiers, water, preservative, sweeteners, flavors, and colorants, the percentage oil-phase ingredients in the formulation would be calculated as $10 + 3 + 2.5 + 4 = 19.5\%$. The RHLB of the oil phase for a desired o/w emulsion would be calculated as follows:

Oil-phase ingredient	% In the formulation	% Contribution to the oil phase	RHLB of the ingredient	RHLB contribution to the formulation
Mineral oil	10	$10/19.5 \times 100 = 51.3$	10.5	$51.3/100 \times 10.5 = 5.4$
Capric/caprylic triglyceride	3	$3/19.5 \times 100 = 15.4$	5.0	$15.4/100 \times 5.0 = 0.8$
Isopropyl myristate	2.5	$2.5/19.5 \times 100 = 12.8$	11.5	$12.8/100 \times 11.5 = 1.5$
Cetyl alcohol	4	$4/19.5 \times 100 = 20.5$	15.5	$20.5/100 \times 15.5 = 3.2$

Thus, the RHLB of the oil phase is $5.4 + 0.8 + 1.5 + 3.2 = 10.9$. Hence, this formulation would require the use of an emulsifier, or a combination of emulsifiers, which should have the HLB of 10.9 to make an optimum physically stable o/w emulsion.

10.2.2.4 Assigning a hydrophile–lipophile balance value to a surfactant

The HLB value of a surfactant reflects a one-fifth fraction of the hydrophilic portion of the surfactant on a molecular-weight basis. For example, for calculating the HLB value of 22 moles of ethoxylate of oleyl alcohol, the molecular weight of 22 moles of ethylene oxide [$-CH_2-O-CH_2-$], with the repeating unit molecular weight of 44, is calculated to represent the hydrophilic portion of a surfactant. Thus, $22 \times 44 = 968$. This mass is added to the molecular weight of oleyl alcohol, 270, to get the total molecular weight of the surfactant. Thus, $968 + 270 = 1238$. The percentage molecular weight of the surfactant that is hydrophilic is, therefore, $968/1238 \times 100 = 77\%$. Taking a one-fifth fraction, the HLB of this surfactant would be $77/5 = 15.4$.

The HLB values are assigned only to nonionic surfactants. Thus, the HLB values are generally in the range of 0.5–19.5. Nevertheless, the HLB values of ionic surfactants are provided in the literature as an indication of their relative hydrophilicity. Some HLB values are listed in Table 10.2. Thus, an ionic surfactant with an HLB value of 40 simply indicates that it is highly hydrophilic.

10.2.2.5 Selection of surfactant combination for a target hydrophile–lipophile balance value

Frequently, a combination of two or more surfactants, usually with different molecular weight/size, is used instead of just one surfactant. A combination of surfactants ensures better packing at the interface and greater physical stability of the emulsion. In using a combination of surfactants, their HLB values are additive. Thus, the HLB value of a combination of surfactants is the weighted average of the HLB of each surfactant. For example, if 50% each of Span 20 and Span 80 were mixed together, the HLB of their combination would be $50/100 \times 8.6 + 50/100 \times 4.3 = 6.45$. Similarly, the use of 90% Span 80 and 10% Span 20 would give combined HLB value of 4.7, which is the same HLB value as that of Span 60. However, the use of Span 20 + Span 80 is expected to give a more stable emulsion than Span 60 in the same quantities.

10.3 MICELLES

At low concentrations in solutions, amphiphiles exist as monomers and predominantly occupy the surface or interface. As the concentration is increased above the level required to completely occupy the surface (known as the critical micelle concentration or the critical micellization concentration and abbreviated as *CMC*), subvisible self-association structures form in solution. These soluble aggregates, which may contain up to 50 or more monomers, are called *micelles*. Therefore, micelles are small, generally spherical structures composed of both hydrophilic and hydrophobic regions of surfactant molecules. In an aqueous bulk solution environment, the hydrophobic region is embedded on the inside (Figure 10.3). Conversely, in a hydrophobic, lipid, or lipophilic bulk solution, the hydrophilic region is embedded on the inside.

The surfactant monomers in micelles are in dynamic equilibrium with free molecules (monomers) in solution, resulting in a continuous flux of monomers between the solution and the micellar phase.

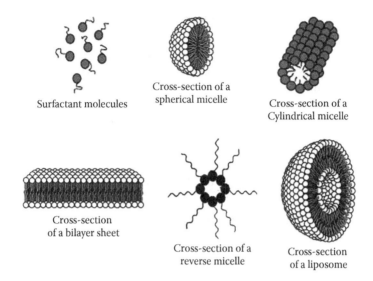

Surfactant molecules

Cross-section of a spherical micelle

Cross-section of a Cylindrical micelle

Cross-section of a bilayer sheet

Cross-section of a reverse micelle

Cross-section of a liposome

Figure 10.3 Types of micelles. Spherical micelles are formed when the concentration of monomers in the aqueous solution reaches the critical micelle concentration (CMC). Elongation of spherical micelles at high concentration leads to the formation of a cylindrical micelle. Reverse micelles are formed in a nonpolar solvent.

10.3.1 Types of micelles

The shape of micelles formed by a particular surfactant is greatly influenced by the geometry of the surfactant molecules. At higher surfactant concentrations, micelles may become asymmetric and eventually assume *cylindrical* or *lamellar* structures (Figure 10.3). Thus, spherical micelles exist at concentrations relatively close to the CMC. Oil-soluble surfactants have a tendency to self-associate into *reverse micelles* in nonpolar solvents, with their polar groups oriented away from the solvent and toward the center, which may also enclose some water (Figure 10.3).

10.3.2 Micelles versus liposomes

Micelles are unilayer structures of surfactants, whereas liposomes have a lipid bilayer structure that encloses the solvent medium (water) (Figure 10.3). Although both micelles and liposomes are formed from amphiphilic monomers, the structure and properties of the monomers play a role in determining which of these structures forms. In addition, liposomes are not formed spontaneously—they require an input of energy and are typically formed by the application of one or more of agitation, ultrasonication, heating, and extrusion.

10.3.3 Colloidal properties of micellar solutions

Micellar solutions are different from other types of colloidal solutions (such as colloidal suspensions of particles), since micelles are *association colloids;* that is, the associated surfactant molecules are colloidal in size in solution. The micelles are formed by reversible self-association of monomers. The minimum concentration of a monomer at which micelles are formed is called the *critical micelle concentration* or the *critical micellization concentration* (CMC).

The number of monomers that aggregate to form a micelle is known as the *aggregation number* of the micelle. The size of micelles depends on the number of monomers per micelle and the size and molecular shape of the individual monomers. For example, the longer the hydrophobic chain or the lower the polarity of the polar group, the greater the tendency for monomers to *escape* from water to form micelles and, hence, lower the CMC. The CMC and number of monomers per micelle differ for different types of surfactants. Some examples are listed in Table 10.4.

As the surfactant concentration in a solution is progressively increased, the properties of the solution change gradually. Not all surfactants form micelles. In the case of surfactants that form micelles, a sharp inflection point in the physical properties of the solution is observed at the CMC. The properties that are affected include the following:

Surface tension: As illustrated in Figure 10.4, surface tension of a surfactant solution decreases steadily up to the CMC but remains constant above

Table 10.4 Critical micellization concentration and number of surfactant molecules per micelle

Name	Molecular formula	CMC (mM)	Surfactant molecules/micelle
Sodium octant sulfonate	$n\text{-}C_8H_{17}\text{-}SO_3Na$	150	28
Sodium decane sulfonate	$n\text{-}C_{10}H_{21}SO_3Na$	40	40
Sodium dodecane sulfonate	$n\text{-}C_{12}H_{25}SO_3Na$	9	54
Sodium lauryl sulfate	$n\text{-}C_{12}H_{25}OSO_3Na$	8	62
Decyltrimethylammonium bromide	$n\text{-}C_{10}H_{21}N(CH_3)_3Br$	63	36
Dodecyltrimethylammonium bromide	$n\text{-}C_{12}H_{25}N(CH_3)_3Br$	14	50
Tetradecyltrimethylammonium bromide	$n\text{-}C_{12}H_{29}N(CH_3)_3Br$	3	75
Octaoxyethylene glycol monododecyl ether	$n\text{-}C_{12}H_{25}O(CH_2CH_2O)_8H$	0.13	132
Dodecaoxyethylene glycol monododecyl ether	$n\text{-}C_{12}H_{25}(CH_2CH_2O)_{12}H$	0.14	78

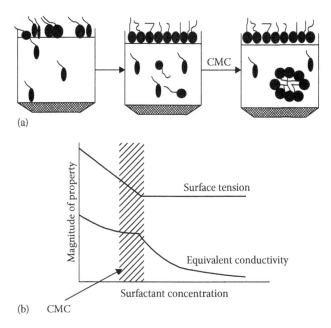

Figure 10.4 Micellization of an ionic surfactant (a) and its effect on conductivity and surface tension (b).

the CMC. This is attributed to the saturation of surface occupation of a surfactant above the CMC. Below the CMC, as the surfactant concentration in the solution is increased, more and more surfactant molecules partition into the surface or interface, leading to a steady reduction in surface tension. Above the CMC, the surface or interface is already completely full or saturated with the surfactant. Thus, further addition of the surfactant leads to minimal changes in surface tension. The excess surfactant added to the solution forms micelles in the bulk of the liquid.

Conductivity: The conductivity of a solution due to the presence of monovalent inorganic ions is affected by the surfactant's concentration, since the polar head group of the surfactant can bind the ions, leading to reduced number of free ions available for conductance. As a surfactant is added to the solution, some of the surfactant occupies surface and some is available in the bulk of the solution, binding the counterions. Thus, solution conductivity reduces steadily as a function of the surfactant's concentration. As shown in Figure 10.4, this change is much more rapid above the CMC, following a sharp inflection point at the CMC. This is attributed to most of the added surfactant (above the CMC) being available in solution for binding with the ions.

Solubility: Solubility of a hydrophobic molecule in an aqueous solution increases slightly with the surfactant concentration below the CMC but shows significant and sharp increase above the CMC. Below the CMC, an increase in the solubility of a hydrophobic drug results from changes in the characteristics of the solvent medium (such as dielectric constant) and drug–surfactant interaction. Above the CMC, additional drug solubilization results from the hydrophobic drug getting incorporated into the micelles.

Osmotic pressure: Micelles, formed above the CMC, act as association colloids, leading to an increase in the osmotic pressure of the colloidal solution.

Light-scattering intensity: Light scattering shows a sharp increase above the CMC due to the formation of colloidal micelles that scatter light.

10.3.4 Factors affecting critical micelle concentration and micellar size

- *Size and structure of hydrophobic group*: An increase in the hydrocarbon chain length causes a logarithmic decrease in the CMC. This is because an increase in hydrophobicity reduces aqueous solubility of the surfactant and increases its partitioning into the micelles. Micellar size increases with an increase in the hydrocarbon chain length, owing to an increase in the volume occupied per surfactant in the micelle.
- *Nature of hydrophilic group*: An increase in hydrophilicity increases the CMC due to increased surfactant solubility in the aqueous medium and reduced partitioning into the interface. As the proportion of surface/interface to bulk surfactant concentration reduces, more of added surfactant is required to achieve saturation of the surface before micelles can form. Thus, nonionic surfactants have very lower hydrophilicity and CMC values compared with ionic surfactants with similar hydrocarbon chains.
- *Nature of counterions*: About 70%–80% of the counterions of an ionic surfactant (e.g., Na^+ is a counterion for carboxylate and sulfonate groups, and Cl^- is a counterion for quaternary amine groups) are bound to the micelles. The nature of the counterion influences the properties of these micelles. For example, size of micelles formed with a cationic surfactant increases according to the series $Cl^- < Br^- < I^-$ and with an anionic surfactant according to the series $Na^+ < K^+ < Cs^+$. This is a function of not only the size and electronegativity of the counterion but also the size of the hydration layer around the counterion. The weakly hydrated (smaller, highly electronegative) ions are adsorbed more closely to the micellar surface and neutralize the charge on the surfactant more effectively, leading to the formation of smaller micelles.
- *Addition of electrolytes*: Addition of electrolytes, such as salt, to solutions of ionic surfactants decreases the CMC and increases the size of

the micelles. This is due to a reduction in the effective charge on the hydrophilic headgroups of the surfactants. This tips the hydrophilic lipophilic balance toward greater lipophilicity, increases the proportion of surface/interface to bulk surfactant concentration below the CMC, and promotes the formation of micelles in the bulk liquid. In contrast, micellar properties of nonionic surfactants are only minimally affected by the addition of electrolytes.

- *Effect of temperature*: Size of micelles increases and CMC decreases with increasing temperature up to the cloud point for many nonionic surfactants due to increased Brownian motion of the monomers. Temperature has little effect on ionic surfactants. This is due to stronger hydrogen bonding and electrical forces governing the hydrophilic interactions of ionic surfactants than nonionic surfactants.
- *Alcohol*: Addition of alcohol to an aqueous solution reduces the dielectric constant and increases the capacity of the solution to solubilize amphiphilic (surfactant) and hydrophobic molecules. Thus, greater surfactant solubility in the hydroalcoholic solutions decreases the surface/interface to bulk solution concentration of the surfactant, thus increasing the CMC.

10.3.5 Krafft point

Krafft point (K_t), also known as the critical micelle temperature or Krafft temperature, is the minimum temperature at which surfactants form micelles, irrespective of the surfactant concentration. Below the Krafft point, surfactants maintain their crystalline molecular orientation form even in an aqueous solution and are not distributed as freely tumbling random monomers that are able to self-associate to form micelles. The International Union of Pure and Applied Chemistry's *Gold Book* (http://goldbook.iupac.org) defines Krafft point as the temperature at which the solubility of a surfactant rises sharply to that at the CMC, the highest concentration of free monomers in solution. The Krafft point is determined by locating the abrupt change in slope of a graph of the logarithm of the solubility against temperature (T), or $1/T$. Below K_t, the surfactant has a limited solubility, which is insufficient for micellization. As the temperature increases, solubility increases slowly. At the Krafft point, surfactant crystals melt and the surfactant molecules are released in solution as monomers, which can also get incorporated into micelles. Above the Krafft point, micelles form and, due to their high solubility, contribute to a dramatic increase in the surfactant solubility.

10.3.6 Cloud point

Cloud point is the temperature *at which some surfactants begin to precipitate* and the solution becomes cloudy. The appearance of turbidity at the cloud point is due to separation of the solution into two phases. For nonionic surfactants, aqueous solubility is at least partially attributed to the hydration of their hydrophilic regions by water molecules. Increasing solution temperatures up to the cloud point leads to an increase in micellar size. Increasing temperature above the cloud point imparts sufficient kinetic energy to the hydrating water molecules to effectively dissociate from the surfactant and bond exclusively with the bulk water. This produces a sufficient overall drop in the solubility of the surfactant to cause surfactant precipitation and cloudiness of solution. At elevated temperatures, the surfactant separates as a precipitate. When in high concentration, it separates as a gel. This phenomenon is commonly seen with many nonionic polyoxyethylate surfactants in solution.

Organic solubilized molecules or solution additives, such as ethanol, generally decrease the cloud point of nonionic surfactants. Addition of aliphatic hydrocarbons increases the cloud point. Aromatic hydrocarbons or alkanols may increase or decrease the cloud point, depending on the concentration.

10.3.7 Micellar solubilization

Micelles can be used to increase the solubility of materials that are normally insoluble or poorly soluble in the dispersion medium used. This phenomenon is known as *solubilization*, and the incorporated substance is referred to as the *solubilizate*. For example, surfactants are often used to increase the solubility of poorly soluble steroids. The location, distribution, and orientation of solubilized drugs in the micelles influence the kinetics and extent of drug solubilization. These parameters are determined by the molecular location of the interaction of drugs with the structural elements or functional groups of the surfactant in the micelles.

10.3.7.1 Factors affecting the extent of solubilization

Factors affecting the rate and extent of micellar solubilization include the nature of surfactants, the nature of solubilizates, temperature, and pH.

1. *Nature of surfactants*: Structural characteristics of a surfactant affect its solubilizing capacity because of its effect on the solubilization site within the micelle. In cases where the solubilizate is located within the core or deep within the micelle structure, the solubilization capacity increases with increase in alkyl chain length. For example, there was an increase in the solubilizing capacity of a series of polysorbates for selected barbiturates as the alkyl chain length was increased from C_{12} (polysorbate 20) to C_{18} (polysorbate 80).

An increase in the alkyl chain length increases the hydrophobicity of the core and micellar radius, reduces pressure inside the micelle, and increases the diffusive entry of the hydrophobic drug into the micelle. In addition, the solubilization of the poorly soluble drug tropicamide increased with increase in the oxyethylene content of poloxamer. On the other hand, an increase in the ethylene oxide chain length of a polyoxyethylated nonionic surfactant led to an increase in the total amount solubilized per mole of surfactant because of the increasing number of micelles. Thus, the effect of increase in the number of micelles of the same (smaller) size can be very different than increase in the size of micelles.

2. *Nature of solubilizate (drug being solubilized)*: The location of solubilizates in the micelles is closely related to the chemical nature of the solubilizate. In general, nonpolar, hydrophobic solubilizates are localized in the micellar core. Compounds that have both hydrophobic and hydrophilic regions are oriented with the hydrophobic group facing or in the core and the hydrophilic or polar groups facing toward the surface. For a hydrophobic drug solubilized in a micelle core, an increase in the lipophilicity or the lipophilic region or surface area of the drug leads to solubilization near the core of the micelle and enhances drug solubility.

   Unsaturated compounds are generally more soluble than their saturated counterparts. Solubilizates that are located within micellar core tend to increase the size of the micelles. Micelles become larger not only because their core is enlarged by the solubilizate but also because the number of surfactant molecules per micelle increases in an attempt to cover the swollen core.

3. *Effect of temperature*: In general, the amount of the drug solubilized increases with an increase in temperature (Figure 10.5). The effect is particularly pronounced with some nonionic surfactants, where it is a consequence of an increase in the micellar size with increasing temperature.

4. *Effect of pH*: The main effect of pH on solubilizing ability of nonionic surfactants is to alter the equilibrium between ionized and unionized drugs. The overall effect of pH on drug solubilization is a function of proportion of ionized and unionized forms of the drug in solution and in micelles, which is determined by (1) the pK_a value of the ionizable functional group(s), (2) the solubility of the ionized and unionized forms in the solution, and (3) the solubilization capacity of the micelles for the ionized and unionized forms. Generally, the unionized form is the more hydrophobic form and is solubilized to a greater extent in the micelles than the ionized form.

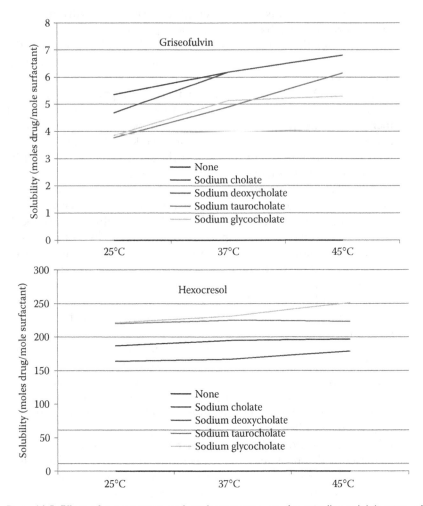

Figure 10.5 Effect of temperature and surfactant type on the micellar solubilization of griseofulvin and hexocresol. (Modified from Bates, T.R, Gilbaldi, M. and Kanig, J.I. *J. Pharm. Sci.*, 55, 191, 1966. With Permission.)

10.3.7.2 Pharmaceutical applications

Several insoluble drugs have been formulated by using micellar solubilization. For example:

- Phenolic compounds, such as cresol, chlorocresol, and chloroxylenol, are solubilized with soap to form clear solutions for use as disinfectants.
- Polysorbates have been used to solubilize steroids in ophthalmic formulations.

- Polysorbate are used to prepare aqueous injections of the water-insoluble vitamins A, D, E, and K.
- Nonionic surfactants are efficient solubilizers of iodine.

10.3.7.3 Thermodynamics/spontaneity

Micellar solubilization involves partitioning of the drug between the micellar phase and the aqueous solvent. Thus, the standard free energy of solubilization, ΔG_s, can be computed from the partition coefficient, K, of the drug between the micelle and the aqueous medium:

$$\Delta G_s = -RT \ln K \tag{10.1}$$

where:
 R is the gas constant
 T is the absolute temperature

Change in free energy with micellization can be expressed in terms of the change in enthalpy (ΔH_s) and entropy (ΔS_s) as:

$$\Delta G_s = \Delta H_s - T\Delta S_s \tag{10.2}$$

Thus,

$$\Delta H_s - T\Delta S_s = -RT \ln K$$

Or,

$$\ln K = -\frac{\Delta H_s}{R} \cdot \frac{1}{T} + \text{constant}$$

where the constant is $\Delta S_s/R$, assuming that the change in entropy from micellization is constant. Thus, experimental determination of enthalpy of micellization can be a useful tool to predict ΔG_s, which, in turn, indicates whether micellar incorporation of a drug would be spontaneous. When ΔG_s is negative, solubilization process is spontaneous. When ΔG_s is positive, solubilization does not occur.

> Example 1: Given $\Delta H_s = 2830$ cal/mol and $\Delta S_s = -26.3$ cal/K mol, does ammonium chloride spontaneously transfer from water to micelles?
>
> $$\Delta G_s = \Delta H_s - T\Delta S_s = 2830 \text{ cal/mol} - (298\text{K})(-26.3 \text{ cal/kmol})$$
>
> which is positive, indicating that micellar solubilization (transfer) would not occur.

Example 2: Given $\Delta H_s = -1700$ cal/mol and $\Delta S_s = 2.1$ cal/K mol, does amobarbital spontaneously transfer from water to a micellar solution (sodium lauryl sulfate, 0.06 mol/L)?

$$\Delta G_s = \Delta H_s - T\Delta S_s = 1700 \text{ cal/mol} - (298 K)(-2.1 \text{ cal/kmol}) = -2326 \text{ cal/mol}$$

which is negative, indicating that micellar solubilization (transfer) would indeed spontaneously occur.

REVIEW QUESTIONS

10.1 Which of the following dosage forms may utilize surface-active agents in their formulations?
   A. Emulsions
   B. Suspensions
   C. Colloidal dosage forms
   D. Creams
   E. All of the above

10.2 Increasing the surfactant concentration above the critical micellar concentration will result in:
   A. An increase in surface tension
   B. A decrease in surface tension
   C. No change in surface tension
   D. All of the above

10.3 Which of the following surfactants is incompatible with anionic bile salts?
   A. Polysorbate 80
   B. Potassium stearate
   C. Sodium lauryl sulfate
   D. Benzalkonium chloride
   E. All of the above

10.4 Which of the following statements is TRUE?
   A. Most substances acquire a surface charge by ionization, ion adsorption, and ion dissolution
   B. The term *surface tension* is used for liquid–vapor and solid–vapor interfaces
   C. At the isoelectric point, the total number of positive charges is equal to the total number of negative charges on a molecule
   D. All of the above
   E. None of the above

10.5 A. Enlist three pharmaceutical applications of surfactants.
   B. Enlist three different types of surfactants.
   C. You formulated an emulsion by using a surfactant with an HLB value of 18. The emulsion was highly unstable. Explain why there was a problem with emulsion stability.

10.6   Define micelles, critical micellization concentration, and aggregation numbers. What are the types of micelles and how do they form? Describe with the help of a diagram.

10.7   Draw a diagram to illustrate the change in surface tension and conductivity with increasing the concentrations of sodium lauryl sulfate in water versus that of glucose. Why should the profile be different in two cases?

10.8   Define and differentiate between cloud point and Krafft point. What are the three factors affecting cloud point?

10.9   Define micellar solubilization with the aid of a diagram. Mention any three properties of the drug that are affected by this phenomenon. How does the alkyl chain length of the surfactant affect the solubilization of a hydrophobic drug?

10.10 Which of the following is a (i) cationic, (ii) anionic, (iii) nonionic, or (iv) ampholytic surfactant?

   A.   Cetrimide
   B.   Benzalkonium chloride
   C.   Sodium lauryl sulfate
   D.   Lecithin
   E.   Span 20
   F.   Tween 80

FURTHER READING

Aulton ME (Ed.) (1988) *Pharmaceutics: The Science of Dosage Form Design*, New York: Churchill Livingstone.

Bates TR, Gibaldi M, and Kanig JL (1966) Solubilizing properties of bile salt solutions. I. Effect of temperature and bile salt concentration on solubilization of glutethimide, griseofulvin, and hexestrol. *J Pharm Sci* 55: 191–199.

Florence AT and Attwood D (2006) *Physicochemical Principles of Pharmacy*, 4th ed., London: Pharmaceutical Press.

Rangel-Yagui CO, Pessoa A Jr, and Tavares LC (2005) Micellar solubilization of drugs. *J Pharm Pharm Sci* 8: 147–165.

Saettone MF, Giannoccini B, Delmonte G, Campigli V, Tota G, and La Marca F (1988) Solubilization of tropicamide by poloxamers: Physicochemical data and activity data in rabbits and in humans. *Int J Pharm* 43: 67–76.

Sinko PJ (2005) *Martin's Physical Pharmacy and Pharmaceutical Sciences*, 5th ed., Philadelphia, PA: Lippincott Williams & Wilkins.

Chapter 11

Pharmaceutical polymers

LEARNING OBJECTIVES

On completion of this chapter, the students should be able to

1. Define monomers, oligomers, polymers, and repeating units.
2. Classify polymers based on type and arrangement of repeating units.
3. Describe different types of polymerization processes.
4. Describe the mechanistic basis of the behavior of polymers in solutions.
5. Identify and define the key features of polymers—their structures, molecular weight distribution, crystallinity, and solubility.

11.1 INTRODUCTION

Polymers are widely used in pharmaceutical dosage forms. Based on the solubility of polymers in water, they can be classified as water-soluble polymers, water-insoluble polymers, and hydrogels.

Water-soluble polymers: These polymers are used in many different ways, so to increase the viscosity of the aqueous solutions; to improve and maintain the physical stability of suspensions; to promote the adhesion of solid particles of different types, leading to granulation in wet processes; to form a flexible film on tablets during the coating of tablets; as adhesives for buccal and bioadhesive drug delivery systems; as emulsifying agents; as flocculating agents; and as components of sustained and site-specific drug delivery systems. Water-soluble polymers used in the coating of tablets include hydroxypropyl methylcellulose (HPMC) or polyvinyl alcohol (PVA) as a film former, and polyethylene glycol (PEG) as a plasticizer. Water-soluble polymers used for the stabilization of suspensions and emulsions include carrageenan, hydroxypropyl cellulose (HPC), and xanthan gum.

Water-soluble polymers can be cross-linked to give *hydrogels*. For example, crospovidone is cross-linked polyvinylpyrrolidone (PVP), and croscarmellose sodium is cross-linked carboxymethyl cellulose (CMC) sodium. Crospovidone and croscarmellose sodium are used as superdisintegrants in oral solid dosage forms. The rapid and high water absorption capability of these water-insoluble cross-linked polymers aids in the disintegration of compressed dosage forms.

Water-insoluble polymers: These polymers are used to form membranes and matrices for sustained-release and localized drug delivery systems. Being water-insoluble, these polymers help delay, slow, or sustain the rate of drug release. An example is the coating of a tablet with a sustained-release water-insoluble polymer. For such an application, water-insoluble polymer is mixed with a limited quantity of a water-soluble polymer, which dissolves in contact with aqueous fluid, leading to the formation of pores in the membrane through which the drug can diffuse out of the dosage form. Factors influencing drug release from these systems include membrane thickness, drug solubility in the membrane, and the porosity of the polymer matrix. Water-insoluble polymers used in the coating of tablets include polymers that dissolve at basic pH but not at acidic pH. Such polymers are called enteric polymers, and such coating on the tablet is called enteric coating. Polymers used for the *enteric coating* of tablets include cellulose acetate phthalate (CAP), hydroxypropyl methylcellulose phthalate (HPMCP), and methacrylic acid—methyl methacrylate copolymers (Eudragit®).

11.2 DEFINITIONS AND ARCHITECTURES OF POLYMERS

Polymers are high molecular weight natural or synthetic molecules made up of small *repeating units*, the connected molecular structure that repeats over and over again in a polymer. The structures of common polymers, their repeating units, and their monomers are shown in Figure 11.1. The structural unit enclosed in brackets or parentheses is referred to as the *repeating unit*. To indicate the repetition, a subscript n is frequently placed after the closing bracket, for example, $-[-CH_2CH_2-]_n-$. For polymers of a well-defined and known number of repeating units, the number of monomeric units constituting a polymer replaces the subscript "n."

Polymers are synthesized from simple molecules called *monomers* by a process called *polymerization*. The structure and molecular formula of the monomer and the repeating unit are very similar but not exactly the same.

If only a few monomer units are joined together, the resulting low molecular weight polymer is called an *oligomer*. For example, dimer, trimer, and tetramer are structures formed with two, three, or four monomer units, respectively.

End groups: There are the structural units that terminate polymer chains. Where end groups are specified, they are shown outside the brackets, for example:

$$CH_3CH_2 - [-CH_2CH_2-]_n - CH_2CH_3$$

Homopolymers are composed of single, identical repeating units forming the polymer chain or backbone. *Heteropolymers* or *heterochain polymers* contain more than one type of repeating unit in their backbone. When two or more monomers combine in specific repeating pattern to make a hetero-polymer, the polymer is called a *copolymer*. In copolymers, the monomeric units may be distributed randomly (random copolymer), in an alternating fashion (alternating copolymer), or in blocks (block copolymer). A graft copolymer consists of one polymer branching from the backbone of the other. Polymer molecules may be linear or branched, and separate linear or branched chains may be joined by cross-links.

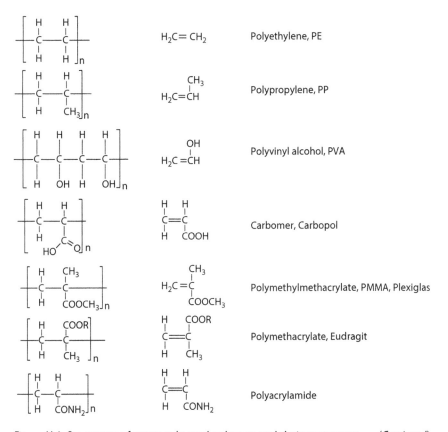

Figure 11.1 Structures of commonly used polymers and their monomers.   *(Continued)*

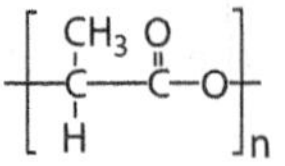

	Poly(d, l-lactide), polylactide
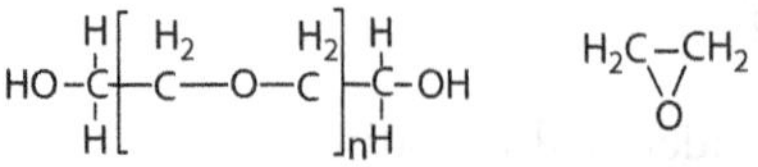	Polyethylene glycol, polyoxyethylene glycol, PEG
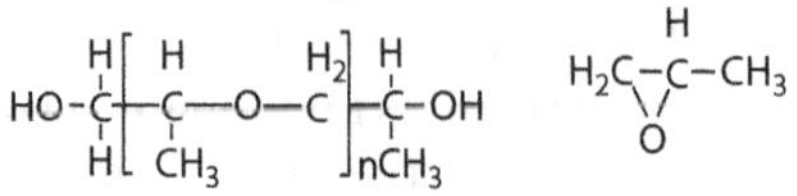	Polypropylene oxide, PPO
	Polystyrene
	Poly(vinyl chloride)
	Polytetrafluoroethylene
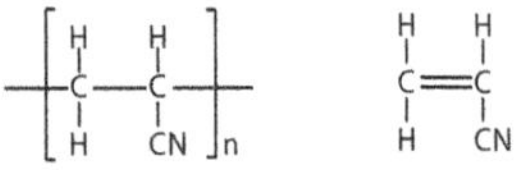	Polyacrylonitrile
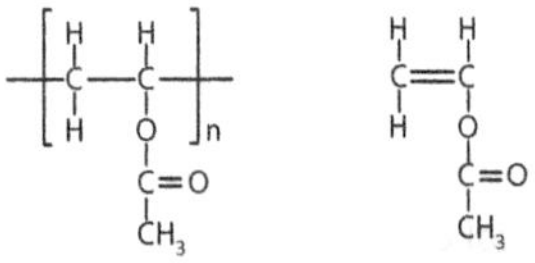	Poly(vinyl acetate)
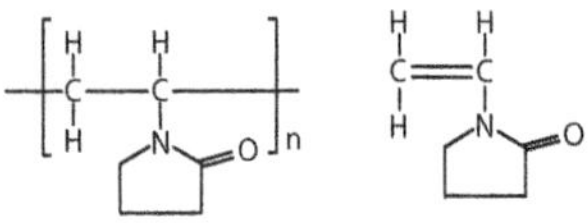	Polyvinylpyrrolidone
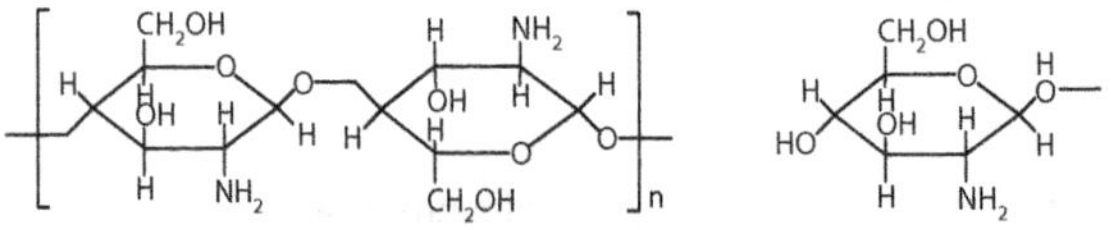	Poly(D-glucosamine), chitosans
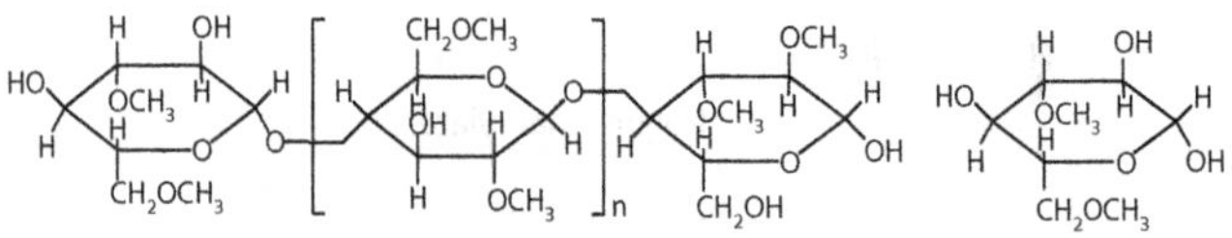	Methylcellulose, Methocel A

Figure 11.1 (Continued) Structures of commonly used polymers and their monomers.

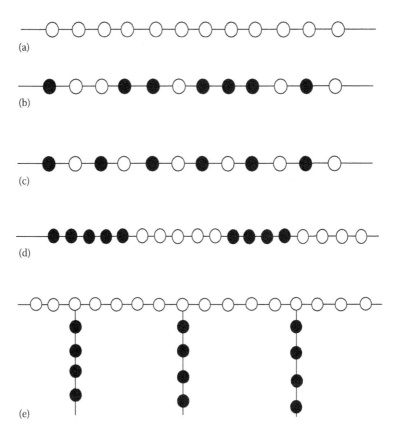

Figure 11.2 Polymer architectures: (a) linear homopolymer, (b) random copolymer, (c) alternating copolymer, (d) block copolymer, and (e) graft copolymer.

Figure 11.2 shows various arrangements of the hypothetical monomers A and B in the copolymer. Where blocks of A (○) and B (●) alternate in the backbone, the polymer is designated an -[-AB-]- *multiblock* copolymer. If the backbone consists of a single block of each, it is an AB (○●) *diblock* copolymer. Other possibilities include ABA (○●○) or BAB (●○●) *triblock* copolymers. For example, when vinyl pyrrolidone, a monomer, is polymerized, it forms the linear polymer PVP, also known as povidone. Polyvinyl pyrrolidone is a commonly used polymer in pharmaceutical processing and products, such as artificial tears. It is a protective colloid capable of forming complex with molecular iodine and is thus used in iodine tincture. Polypropylene sulfone is an alternating copolymer synthesized by copolymerization of propylene and sulfur dioxide.

Polymers can be linear, star-shaped, or branched, including the so-called star block copolymers. A branched polymer is not necessarily a graft polymer. *Star polymers* contain three or more polymer chains emanating from

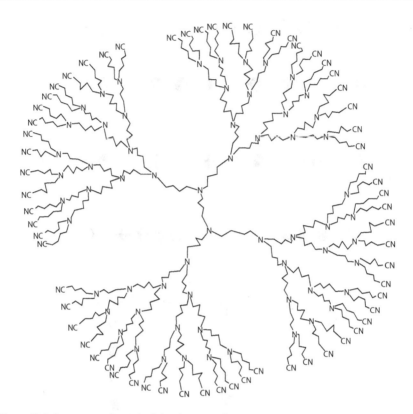

Figure 11.3 Structure of a typical dendrimer polymer.

a core structural unit. *Comb polymers* contain pendant chains (which may or may not be of equal length) and are related structurally to graft copolymers. *Dendrimers,* also known as *starburst* or *cascade polymers*, resemble star polymers, except that each leg of the star exhibits repetitive branching in the manner of a tree. Dendrimers are highly branched polymer constructs formed from a central core, which defines their initial geometry (Figure 11.3). Their branch-like structure leads to a spherical shape, which can become as large as the size of micelles or nanospheres, depending on the size of the polymer.

11.3 POLYMER MOLECULAR WEIGHT AND WEIGHT DISTRIBUTION

Both synthetic and natural polymers exist in a range of sizes, defined by the number of monomeric units or their molecular weights. The polymerization process produces polymers of different sizes. Thus, any given batch or quantity of a polymer is a mixture of polymers of different sizes. The nominal (or labeled) molecular weight of a polymer is an average molecular weight, which is inferred by the bulk property of the polymer such as chemical analysis, osmotic pressure, or light scattering. When determined by chemical analysis or osmotic pressure measurement, the reported molecular weight is the *number average molecular weight*, M_n, since these analytical tools are sensitive to the number of polymer chains of different sizes. Thus, for a mixture containing n_1, n_2, n_3, ... moles of polymers with molecular weights M_1, M_2, M_3, ..., respectively, the number average molecular weight is defined by:

$$M_n = \frac{n_1 M_1 + n_2 M_2 + n_3 M_3 + \cdots}{n_1 + n_2 + n_3 + \cdots} = \frac{\sum n_i M_i}{\sum n_i}$$

Thus, the number average molecular weight is the arithmetic mean of the molecular weight of all the polymer chains in the sample.

On the other hand, measurement techniques such as light scattering produce a response that depends on the molecular weight of the polymer chain. Thus, larger molecules produce greater light scattering. The molecular weight of the polymer chain or species of each size carries greater weightage in generating the measured response. Thus, the molecular weight is weighted in the inference of molecular weight by using such techniques. Such a molecular weight is defined as the weight average molecular weight, M_w. The weight average molecular weight for the same sample would be defined as:

$$M_w = \frac{n_1 M_1^2 + n_2 M_2^2 + n_3 M_3^2 + \cdots}{n_1 M_1 + n_2 M_2 + n_3 M_3 + \cdots} = \frac{\sum n_i M_i^2}{\sum n_i M_i}$$

The weight average molecular weight is generally greater than the number average molecular weight. Thus, the average polymer molecular weight measured by light scattering is greater than the polymer molecular weight obtained by osmotic pressure measurement.

11.4 BIODEGRADABLITY AND BIOCOMPATIBILITY

Biodegradability refers to the ability of the biological systems, such as the human body, to degrade and eliminate the polymer. This is important to ensure that any polymer administered as a part of a dosage form can break down into smaller pieces and be eliminated by the body, without causing undue toxicity of accumulation.

Biocompatibility of a polymer refers to the tolerance of the biological system, such as the human body, to the polymer. A biocompatible polymer is not sensed as foreign or harmful by the biological system, and the body does not mount an immune response against the polymer. This is critical, for example, for administration of polymers as a part of implantable devices to prevent any adverse reactions. Thus, biocompatible polymers avoid chronic inflammation and long-term complications.

Most biodegradable polymers have hydrolysable linkages, namely ester, orthoester, anhydride, carbonate, amide, urea, and urethane, in their backbones. Such linkages allow the biodegradable polymer to break down into metabolic products by hydrolysis or enzymatic action. Biodegradable polymers are reduced to soluble fragments that are either excretable or metabolized under physiological conditions.

Biodegradable biocompatible polymers are used to deliver a wide range of drugs to diseased tissues, often in a sustained-release dosage form for drug release and action over a prolonged period. Commercially available products that use such polymers include Decaptyl®, Lupron Depot®, Zoladex®, Adriamycin® and Capronor®.

11.5 PHYSICOCHEMICAL PROPERTIES AND SOLUBILITY

The chemical reactivity of polymers depends on the chemistry of their monomer units. However, their physicochemical properties depend, to a large extent, on the way the monomeric units are put together. This contributes to the versatility of synthetic polymers. Various arrangements of monomers, shown as ○ and ● in Figure 11.2, can be produced with consequent effects on the physical properties of the resulting polymer. For example, Eudragit polymers are commercially available with a wide range of physicochemical properties, including the effect of pH on solubility.

Depending on the polymer type and the types of cosolvents used, the overall solubility of a single polymer or a mixture of polymers can be altered in a formulation. Polymers that are sufficiently polar interact with the water through hydrogen bonding to provide the favorable negative energy of solvation; that is, energy is released on dissolution of such polymers in water. Water-soluble polymers tend to increase the viscosity of their solutions as a function of polymer concentration due to entanglement and bonding of solvent molecules and reduction of slip planes in the colloidal polymer solutions. Hydrophilic polymers also tend to swell by absorbing water in solution. In contrast, water-insoluble or hydrophobic polymers tend to form thin films or matrices.

Figure 11.4 represents polymer morphologies in solution, gel, and solid states. In solution, the polymer conformation depends on the interaction between the polymer and the solvent and whether the polymer chains associate to form micelles. Covalent cross-linking, hydrogen bonding, entanglement, or hydrophobic interactions among polymer chains in solution form

"Good" solvent (↑solubility)   "Poor" solvent (↓solubility)

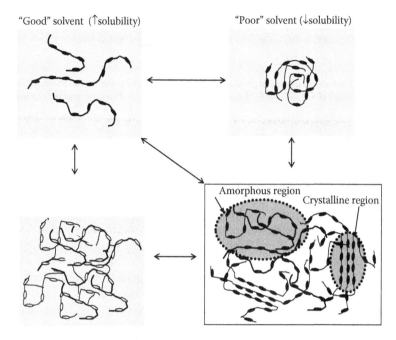

Amorphous region   Crystalline region

Figure 11.4 Representation of polymer morphologies in solution, gel (hydrogel or lipogel), and solid states. In solution, the polymer conformation depends on the polymer-solvent interaction and whether the polymer chains associate to form micelles. Gels can be formed by covalent cross-linking, hydrogen bonding, or hydrophobic interactions.

gels. Agarose gel, edible Gello, and polyacrylamide gel are classic examples of aqueous gels. Aqueous gels tend to hold a significant amount of water in a solid-like structure but are immensely pliable and fragile.

11.6 BLOCK COPOLYMERS

Block copolymers consist of two or more repeating units in a specific pattern. The different repeating units of a block copolymer differ in chemical structure and physicochemical properties. For example, in a block copolymer of the type AAABBBAAA, in which A is water-soluble repeating unit and B is water-insoluble repeating unit, the insoluble parts tend to colocalize and aggregate in solution. Poly(oxyethylene)-poly(oxypropylene)-poly(oxyethylene) (PEO-PPO-PEO) block copolymers (Figure 11.5), commercially known as *Pluronic®* or *Poloxamer*, exhibit such properties and are used as nonionic surfactants. In addition, aqueous solutions of some Poloxamers exhibit temperature-induced phase transitions from solution to gel, when the polymer concentration is above a critical value.

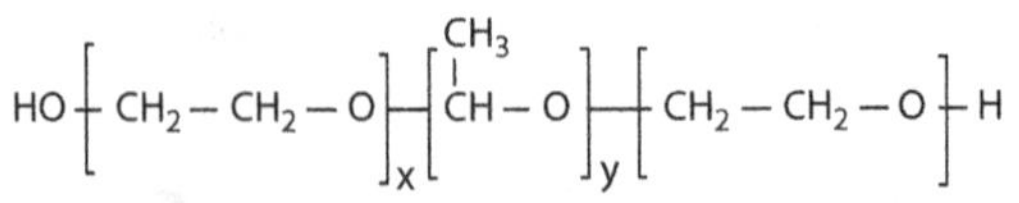

$$\text{HO} \left[CH_2 - CH_2 - O \right]_x \left[\begin{array}{c} CH_3 \\ | \\ CH - O \end{array} \right]_y \left[CH_2 - CH_2 - O \right] H$$

Figure 11.5 Chemical structure of poly(ethylene oxide-co-propylene oxide-co-polyethylene oxide) (PEO-PPO-PEO) (commercially known as Pluronic and poloxamer).

Block copolymer micelles are of great interest due to the following reasons:

- Hydrophobic drugs can be physically entrapped in the core of block copolymer micelles and transported at concentrations that exceed their intrinsic water solubility. An important property of micelles is their ability to increase the solubility of materials that are normally insoluble or only slightly soluble in the dispersion medium used.
- Hydrophilic blocks, which are often composed of PEO or PEG, can form a tight shell, or corona, around the micellar core. Diblock copolymer micelles with a PEO corona resist protein adsorption and cell adhesion. This helps prevent recognition by the phagocytotic reticuloendothelial system (RES) cells, which is a rapid metabolism and elimination pathway for sensitive drug delivery systems.

11.7 INTELLIGENT OR STIMULI-SENSITIVE POLYMERS

The terms *intelligent, stimuli-sensitive,* or *stimuli-responsive* polymers refer to polymers that exhibit relatively large and sharp changes in physical or chemical properties in response to a small change in the environment, such as pH and temperature. Changes in the environment that affect polymer properties are termed stimuli, while the resulting changes in the polymer and the system (such as dissolved state of the polymer in a solvent) are termed the responses. The mechanistic basis of changes in the physical properties of stimuli-responsive polymers is generally a modification in the structure of the polymer in solution. For example, water-soluble polymers and copolymers can undergo conformational change or phase transition in response to environmental stimuli. These changes may exhibit as swelling, change in solubility and conformation of polymer matrix or chain, or polymer precipitation. When a soluble polymer is stimulated to precipitate, it will be selectively removed from the solution. When such polymers are grafted or coated onto a solid support, then one may reversibly change the water adsorption into the polymer-coated surface of the solid, thus changing the wettability of the surface. When a

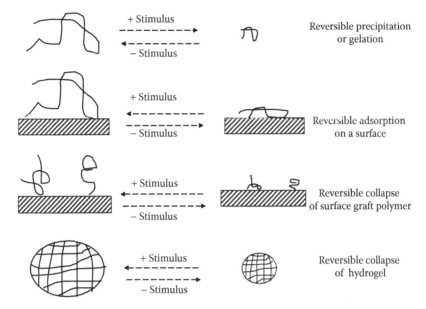

Figure 11.6 Schematic representations of stimuli-sensitive polymers in solutions, on surface, and as hydrogels.

hydrogel is stimulated to collapse, it shrinks in size, squeezes out water from its pores, turns opaque, and becomes stiffer. Figure 11.6 shows the schematic representations of stimuli-sensitive polymers in solutions, on surface, and as hydrogels.

The stimuli responsiveness of the stimuli-sensitive polymers originates in the bulk and surface chemistry and the architecture of these organic compounds. Physicochemical properties of organic compounds are a result of their surface chemistry (i.e., elemental and functional group composition) and architecture (i.e., surface exposure of functional groups and molecular domains). The surface exposure or display of functional groups could be different within the different forms of the same molecule. For small molecules, this is exemplified by the existence of different crystalline forms (polymorphism) and different morphologies of the same crystalline form. These result in different surface properties (such as solubility and dissolution rate) and bulk properties (such as powder adhesion) of crystals, depending on the differences in the functional groups and molecular domains exposed on the surface.[1] For large molecules, this is exemplified by protein structure where surface exposure of functional groups of a peptide chain can lead to a protein being hydrophilic and globular or hydrophobic and membrane-embedded, with many potential secondary and tertiary structure possibilities.

In addition to the changes in these physicochemical properties, for relatively large-molecular-weight organic compounds (such as polymers and proteins), the sheer multitude of functional groups and presence of molecular domains (such as hydrophobic regions) impart special characteristics to the interactions of these molecules with the external environment. These interactions include solute–solute and solute–solvent interactions in the dissolved state. Such interactions are responsible for phenomena such as micellization and swelling/collapse of a cross-linked scaffold.

At a molecular level, stimuli responsiveness of polymers is typically based on changes in polymer–polymer and polymer–solvent interactions. For example, polymers that bear multiple ionizable weakly acidic or weakly basic functional groups undergo ionization as a function of pH, resulting in changes in the strength and extent of polymer interactions with the solvent. Polymer structure, number and positioning of the functional groups, and the strength of their interactions determine the macroscopic response of the polymer system to the environmental stimulus. For example, some polymeric systems can undergo reversible or irreversible phase transformation from a single-phase solution to biphasic precipitated or aggregated state. However, certain polymer solutions may display gelling or micellization without physical phase separation. In addition, colloidal or biphasic polymeric systems that contain cross-linked polymers can show polymer swelling or shrinkage with changes in the type and strength of polymer–solvent interactions.

11.7.1 pH-Responsive polymers

Polymers that bear ionizable functional groups can undergo reversible (e.g., by hydration) or irreversible (e.g., by hydrolysis) changes in response to changes in the environmental pH. Hydration can lead to reversible polymer swelling and deswelling or collapse. Hydrolysis, on the other hand, can lead to irreversible polymer-chain degradation or breakage, resulting in changes in the overall molecular weight and monomer content of the polymer. The reversible pH-sensitive polymers are typically polyelectrolytes that contain a multiple weakly acidic or weakly basic (ionizable) functional groups, whose ionization status can change in response to the environmental pH. The weak acid, HA, is ionized at basic pH to H^+ and A^-, while the weak base, B, is ionized at acidic pH to BH^+. In the ionized state, increased polymer–water interaction through hydrogen and electrostatic bond interactions leads to higher proportion of polymer-associated water of hydration. In addition, electrostatic repulsion between functional groups bearing the same charge on the polymer backbone can lead to polymer-chain expansion. In the unionized state, weak dipole–dipole and hydrophobic interactions within and between the polymer chains can lead to polymer collapse, solvent exclusion, and reduced hydrodynamic volume, eventually causing polymer aggregation or precipitation.

Changes in polymer properties, such as water solubility, water absorption, and polymer degradation by hydrolysis, as a function of solution pH can be utilized in various ways. For example:

- *Sustained drug delivery* can be achieved by forming a well-mixed matrix or a core–shell structure of a water-soluble drug in a water-insoluble polymer that degrades in the appropriate pH environment. Hydrolytic degradation of the polymer leads to slow and sustained drug release.
- *pH-triggered drug release* in the target tissue can be achieved using polymers linked to drugs through hydrolyzable functional groups. For example, polymeric prodrugs of 5-amino salicylic acid (5-ASA) such as methacryloyloxyethyl-5-amino salicylate (MOES) and N-methacryloylaminoethyl-5-amino salicylamide (MAES) utilize hydrolyzable ester linkages that release the drug in the colon on cleavage of the biodegradable linkers.
- *Tumor tissue targeting* of a cytotoxic drug can be achieved by covalent conjugation with or entrapment in the drug delivery system composed of a pH-sensitive polymer. Tumor tissues have a slightly lower pH compared to nontumor tissues. The polymers that show significant change in their physicochemical properties as a result of decrease in the environmental pH, usually due to the ionization of basic functional groups, can be utilized for targeted drug delivery to the tumors.
- *Intracellular drug release* in specific organelles or regions of the cell can be achieved based on the mechanism of cellular uptake and the pH-responsive properties of the polymer utilized in a drug delivery system. The pH responsiveness of polymers has been utilized in nonviral gene and antisense drug delivery by utilizing polycationic polymers, such as poly(ethyleneimine) (PEI), poly(L-lysine) (PLL), and poly(L-histidine) (PLH). These polymers possess multiple amine functional groups that are cationically (positively) charged and ionized at acidic pH but unionized at basic and neutral pH. On cellular uptake through the endosomal pathway, as the pH of the endosomes becomes more and more acidic toward lysosomal pH, the ionization of these polymers leads to water retention and increase in osmotic pressure of the endosomal vesicles, causing their disruption. These carriers, then, are able to release the drug cargo intracellularly before the endosomes become the lysosomes.
- *Enteric- and colon-targeted oral drug delivery* can be achieved by coating a drug delivery system using a polymer that exhibits pH-dependent solubility or degradation. Gastrointestinal (GI) fluid's pH changes progressively from acidic to basic from the stomach through the intestines to the colon. The changes in the GI fluid's pH can be utilized to release a drug at a particular physiological location in the GI tract. In particular, dosage forms of drugs that are sensitive to

the acidic environment of the stomach, or whose release in the stomach is otherwise undesirable, can be coated with a polymer that would be insoluble at the acidic stomach pH and soluble at the basic intestinal pH. Such polymers are known as enteric polymers, and such a coating on the dosage form is termed enteric coating. Enteric polymers are exemplified by methacrylic acid polymer (Eudragit® L100), methyl methacrylate polymer (Eudragit® S100), CAP, HPMCP, and carboxymethyl ethylcellulose etc.

11.7.2 Thermosensitive polymers

Thermosensitive polymers show changes in physicochemical properties with pharmaceutically and physiologically relevant changes in temperature. An aqueous polymer solution might show phase transition above or below a certain temperature, called the *critical solution temperature* (CST). Solutions may exhibit upper or lower CST, depending on their temperature. Solutions that exhibit upper CST (UCST) are monophasic and isotropic above a certain temperature (but biphasic below that temperature), while solutions that exhibit a lower CST (LCST) are monophasic below a certain temperature (and biphasic above that temperature). The CST is also known as the *cloud point*, since phase separation occurs at this temperature, leading to cloudy appearance of the solution.

Injectable depot formulations, tissue engineering, and temperature-induced tumor drug delivery exemplify biopharmaceutical applications of thermoresponsive polymers. Temperature-responsive biodegradable polymers that show sol–gel transition between room temperature and physiological temperature can be utilized to prepare a drug formulation as a solution that precipitates or forms a gel on injection. The *in vivo* gel serves as a drug reservoir for sustained or controlled drug release over a period of time. Injectable gel-forming polymers are exemplified by block copolymers of poly(ethylene oxide) (PEO) and poly(propylene oxide) (PPO). Commercially available injectable polymers that undergo sol–gel transition include BST-Gel® (BioSyntech) and ReGel® (Macromed) etc.

11.7.3 Other stimuli-responsive polymers

Light-responsive polymers can be designed to respond to either visible or ultraviolet (UV) light. This behavior is typically exhibited in a polymer solution or hydrogel. Hydrophilic polymers can form hydrogels in water, which are three-dimensional polymer chain networks with unique physical properties such as higher viscosity and reduced fluidity than liquid water or dilute polymer solutions. The mechanism of stimuli responsiveness of light-responsive hydrogels may involve chemical bond cleavage with the higher-energy UV irradiation or may involve energy transfer to the polymer network with lower-energy visible radiation.

Electrically responsive polymers respond to the application of electrical field for triggering or controlling drug release. These allow user control over the amplitude of current, duration of impulses, and the time interval between electrical impulses. Drug delivery systems that respond to electrical signal are typically polyelectrolyte polymers with multiple ionizable functional groups. The application of electric field aligns the dipoles in polar or charged molecules, resulting in a change in the polymer conformation. This conformational change could involve polymer swelling, shrinkage, or a change in shape. A small change in the electric potential across a polyelectrolyte gel can lead to significant (up to several 100-fold) reversible change in the volume of the gel.[2]

Magnetically responsive drug delivery systems have been used to target the location of drug release to a particular tissue and/or trigger the drug release from a system based on the application of an external magnetic field. This is usually accomplished by the incorporation of ferromagnetic micro- or nanoparticles in the drug delivery system. Magnetically responsive drug delivery systems might require exposure to strong magnetic fields for prolonged period of time. This exposure is likely to generate localized heating, which could be a part of the drug-release mechanism. For example, Katagiri et al. showed drug release from magnetically responsive lipid bilayer capsules prepared with magnetite, iron (II, III) oxide (Fe_3O_4), nanoparticles.[3] The lipid bilayer was deposited on top of the magnetite nanoparticles. On application of alternating magnetic field, heating of the magnetite resulted in phase transition of the bilayer membrane, leading to release of the drug incorporated in the bilayers.

Ultrasound-responsive drug delivery systems commonly use ultrasound as a permeation enhancer through biological membranes such as skin, lungs, intestinal wall, and blood vessels. The use of ultrasound for increasing transdermal drug delivery is known as *sonophoresis* or *phonophoresis*. Ultrasound increases skin permeability through formation of bubbles caused by acoustic cavitation. The use of ultrasound for drug delivery can be based on energy transfer from the ultrasonic waves, leading to chemical degradation of the polymer.

II.8 WATER-SOLUBLE POLYMERS

Polymers that have sufficient number of electronegative atoms and/or functional groups that can form hydrogen bonds with water tend to dissolve in water and are called water-soluble polymers. Water-soluble polymers have an ability to increase the viscosity of solvents at low concentrations, to swell or change shape in solution, and to adsorb at surfaces. The rate of dissolution of a water-soluble polymer depends on its molecular weight. Larger the molecules, stronger the forces holding the chains together and lower the rate of dissolution. Greater the degree of crystallinity of the polymer in the solid state, lower the rate of dissolution. This combination of slow dissolution rate and formation of viscous surface layer makes high-molecular-weight hydrophilic polymers suitable for use in controlling the release rate of soluble drugs. For example, high-molecular-weight HPMC is used as a matrix controlled-release drug delivery carrier.

Examples of commonly used water-soluble polymers include the following:

11.8.1 Carboxypolymethylene (carbomer, carbopol)

Carboxypolymethylene, also known as carbomer, carbopol, or carboxyvinyl polymer, is a high molecular weight polymer of acrylic acid, containing a high proportion of carboxyl groups. This polymer is used as a binding agent in tablets and a suspending agent in other pharmaceutical preparations. The carboxylic groups impart it an acidic character. Thus, its aqueous solutions are acidic. On neutralization with a base, the carboxylic groups become ionized and form stronger hydrogen-bond associations with other polymer chains and the solvent, water. Consequently, carboxypolymethylene solutions become very viscous, with a maximum viscosity at pH between 6 and 11.

11.8.2 Cellulose derivatives

Cellulose itself is insoluble in water. Its partial aqueous solubility is attributed to substitutions, such as methylation and carboxymethylation. Ethyl methylcellulose is soluble in hot and cold water and does not form a gel. Methylcellulose is poorly soluble in water and forms a gel on heating. Sodium carboxymethylcellulose, being an ionized carboxylic acid salt, is soluble in water at all temperatures.

11.8.3 Natural gum (acacia)

Acacia gum, also known as gum arabica, is a complex arabinogalactan-type polysaccharide exuded by acacia trees. Acacia solutions are highly viscous in water. It is one of the most widely used emulsifiers and thickeners.

11.8.4 Alginates

Alginates, also called align or alginic acid, is an anionic polysaccharide in the cell walls of brown algae. It forms a viscous gum on binding with water. Alginate solutions are less readily gelled than acacia gum and are used as stabilizers and thickening agents.

II.8.5 Dextran

Dextran is a complex, branched polymer or polysaccharide composed of glucose molecules. Hence, it is also called glucan. Partially hydrolyzed dextran reduces blood viscosity and is used as a plasma substitute and a volume expander. It exerts an osmotic pressure comparable with that of plasma. Thus, it is used to restore or maintain blood volume in severe trauma.

II.8.6 Polyvinylpyrrolidone

Polyvinylpyrrolidone (PVP), also known as povidone, is a homopolymer of N-vinyl pyrrolidone (Figure 11.5). It is commonly used as a suspending and dispersing agent. It is also used as binding and granulating agent for tablets and as a vehicle for drugs such as penicillin, cortisone, procaine, and insulin to delay their absorption and prolong their action.

II.8.7 Polyethylene glycol

Polyethylene glycol (PEG) is a polyether compound with repeating units of ethylene oxide and a terminal hydroxyl group. The electronegative oxygen confers water solubility on this polymer. Polyethylene glycols have different physical states, depending on their molecular weight, with low-molecular-weight PEGs being liquid at room temperature, while high-molecular-weight PEGs being crystalline solids. Polyethylene glycols are water-soluble and miscible and can dissolve drugs that are not soluble in water. Thus, PEGs are commonly used to increase drug solubility. Polyethylene glycols are also used as plasticizers in coating suspensions to form an elastic film during tablet coating.

II.9 BIOADHESIVE/MUCOADHESIVE POLYMERS

A bioadhesive polymer can adhere to a biological substance (usually the surface of an anatomical location) and remain there for an extended period of time, compared with a nonbioadhesive polymer or material. When the adhering surface or the biological substance is a mucous membrane, then the bioadhesive polymer is referred to as a *mucoadhesive* polymer. All drug delivery systems come in physical contact with an anatomical location of the body. For drugs that are systemically absorbed, the drug passes through that anatomical location into the systemic circulation. Such an anatomical location has been called the *site of drug absorption*. The duration of time for which a drug delivery system remains in contact with the site of absorption is termed *residence time*. The rate of drug absorption combined with the residence time of the drug delivery system at the site of drug absorption determines the total amount of drug absorbed. Thus, increasing the residence time of the drug delivery system at the site of drug absorption, through the use of bioadhesive or mucoadhesive polymers, can increase the bioavailability of a drug.

Physiological processes usually limit the residence time of a drug delivery system. For example, bronchiolar cilia and mucosal clearance limit the duration of contact of a foreign material with the bronchiolar tissue due to the ciliary motion and the mucosal clearance rate. Similarly, duration of time for which a drug stays in the gastric compartment is a function of the gastric emptying time.

The residence time of a drug at the site of drug absorption can be increased by the following actions:

1. Altering the physiological processes governing the normal residence time, for example, by slowing down normal mucosal clearance rate/ciliary motion for bronchiolar drug delivery and increasing the gastric emptying time for gastric drug delivery.
2. Introducing another rate-limiting process that would govern the residence time, for example, by using a gastroretentive drug delivery system that has lower density and remains in the stomach for a prolonged period of time or by incorporating a bioadhesive or a mucoadhesive polymer in the drug delivery system.

The mucus is a highly viscous aqueous fluid that serves to protect the epithelial cell lining of various organs and organ systems such as respiratory, GI, urogenital, visual, and auditory pathways. The mucosal fluid, secreted by the cells in the mucosal membranes, is typically rich in glycoproteins and may contain other ingredients such as immunoglobulins and inorganic salts. Glycoproteins are natural hydrophilic polymers that consist of a protein or polypeptide backbone with covalently attached oligosaccharide (i.e., glycan) side chains. This composition of the mucus indicates its high hydrophilicity and polymeric nature. Accordingly, the polymeric materials that have strong hydrogen-bonding groups display mucoadhesive properties. In addition, linear long chains in high-molecular-weight polymers tend to entangle in the glycoproteins, enhancing mucoadhesion. Common hydrogen-bonding groups in polymers include hydroxyl, carboxyl, amines, and sulfates. Polymers that exhibit such functional groups, such as several polyacrylic acid and cellulose derivatives, are bioadhesive in nature. Examples of polyacrylic acid-based polymers are carbopol, polycarbophil, polyacrylic acid, polyacrylate, poly(methylvinylether-co-methacrylic acid), poly(2-hydroxyethyl methacrylate), and poly(methacrylate). Cellulose derivatives are exemplified by CMC, hydroxyethyl cellulose (HEC), HPC, methyl cellulose (MC), and methyl hydroxyethyl cellulose. Some other bioadhesive polymers include chitosan, gums, PVP, and PVA.

11.9.1 Advantages of mucoadhesive drug delivery systems

Mucoadhesive drug delivery systems can be utilized for both local and systemic drug delivery applications. In the case of local drug delivery, such as vaginal drug delivery, the use of mucoadhesive polymer in the drug delivery system can lead to higher residence time and prolonged duration of local action of the medication. In the case of systemic drug delivery, such as oral administration into the GI tract, localization of the drug delivery system to a particular site (e.g., the site that has a high rate of drug permeability) can lead to (a) more intimate contact between the dosage form and the site of drug absorption, which can increase local drug concentration and the rate of drug absorption or flux and/or (b) higher residence time at the site of drug absorption, leading to an increase in the total amount of drug absorbed (bioavailability).

11.9.2 Mechanism of mucoadhesion

Mucoadhesive polymers interact with a mucosal surface in two stages: (i) contact and (ii) consolidation. The first contact of the mucoadhesive polymer with the mucosal surface leads to surface adhesion due to multiple favorable hydrogen-bond and electrostatic interactions and polymer expansion due to water uptake and plasticization of the drug delivery system. The polymer and the drug delivery system expand and spread on the mucosal surface. The subsequent strong bonding (adhesion) between the polymer and the mucus is a function of polymer chain diffusion, hydration and plasticization, and interlocking bond formation. Attractive interactions between the hydrophilic mucoadhesive polymer and the hydrophilic polymeric glycoproteins in the mucus lead to mutual entanglement and interpenetration of the polymeric chains. This facilitates the formation of more and deeper electrostatic and hydrogen-bond interactions, which promote bioadhesion or mucoadhesion. Thus, mucoadhesion is facilitated by the presence of hydrogen-bond-forming groups in the polymeric chain, flexibility of the polymer chains, and the surface activity of the drug delivery system. Mechanical forces at the site of adhesion can help in deeper penetration and mechanical interaction of the polymers.

11.9.3 Quantitation of mucoadhesion

The strength of mucoadhesion, S_m, is the force, F, required to separate two surfaces after adhesion has been established, per unit surface area (A). Thus,

$$S_m = \frac{F}{A}$$

The S_m can be calculated *in vitro* on the isolated mucus immobilized on an artificial surface or *ex vivo,* using a biological surface, such as an isolated intestinal lumen. *In vivo* assessment of bioadhesion is usually done by measuring the residence time of the dosage form at the site of bioadhesion by an imaging technique.

11.9.4 Sites of application

Mucoadhesive drug delivery systems can be used to deliver a drug to and/or through several anatomical sites in the human physiology, including oral cavity, vagina, nasal cavity, skin (transdermal), conjunctiva of the eye, and the GI tract.

Buccal drug delivery systems seek to deliver drug locally into the oral cavity for local treatment of oral lesions. When used to deliver a drug to the systemic circulation, such as by sublingual administration, bypassing the hepatic first-pass metabolism can contribute to higher bioavailability. Similarly, drug delivery to the nasal mucosa and the vaginal tissue is utilized for local drug action or rapid drug absorption in the systemic circulation, bypassing the hepatic first-pass metabolism.

Ocular drug delivery using mucoadhesive polymers seeks to address the problem of *excessive drainage of the drug via the lachrymal glands* before adequate absorption can take place. Prolonged retention of the drug on the cornea reduces precorneal drainage loss of the drug and increases the duration of drug absorption, thus improving ocular bioavailability. Mucoadhesive polymers adhere to the mucin coat covering the conjunctiva and the corneal surface of the eye. Ocular mucoadhesion markedly prolongs the residence time of a drug in the conjunctival sac, since clearance of a mucoadhesive dosage form is controlled by the much slower rate of mucus turnover rather than the tear turnover rate.

Oral mucoadhesive drug delivery systems have been utilized to effect adhesion of particulate insoluble drugs to the GI mucosal surface. Incorporation of mucoadhesive polymers, such as chitosan, poly(acrylic acid), alginate, poly(methacrylic acid), and sodium carboxymethyl cellulose, into the oral solid drug delivery systems can increase the residence time of the drug and adhesion of particulate drug to the mucosal surface, leading to higher local concentration at the site of drug absorption.

11.9.5 Dosage forms

Mucoadhesive dosage forms include tablets, granules, films, patches, solutions, gels, and ointments. The selection of dosage form depends on the route of drug administration as well as the desired characteristics of the drug delivery system. For example, while tablet and granules are suitable for administration through the oral route, solutions are more suitable for ocular and nasal drug delivery, patches for transdermal drug delivery, films for buccal drug delivery, and gels and ointments for vaginal drug delivery.

11.9.6 Transdermal patches

Transdermal patches deliver drugs through the skin. Percutaneous absorption of a drug generally results from direct penetration of the drug through the stratum corneum, deeper epidermal tissues, and the dermis. When the drug reaches the vascularized dermal layer, it becomes available for absorption into the general circulation.

Among the *factors* influencing percutaneous absorption are the physicochemical properties of the drug, including its molecular weight, solubility, partition coefficient, nature of vehicle, and condition of the skin. Chemical permeation enhancers, iontophoresis, or both are often used to *enhance* the percutaneous absorption of a drug.

In general, patches are *composed* of three key compartments: a protective seal that forms the external surface and protects it from damage, a compartment that holds the medication itself and has an adhesive backing to hold the entire patch on the skin surface, and a release liner that protects the adhesive layer during storage and is removed just before application.

Most patches belong to one of the two general *types*—the reservoir system and the matrix system. The reservoir system incorporates the drug in a compartment of the patch, which is separated from the adhesion surface. Drug transport from the patch to the skin in channelized and controlled through a rate-limiting surface layer. The matrix system, on the other hand, incorporates the drug uniformly across the patch in a polymer matrix. Diffusion of the drug through the polymer matrix and the bioadhesive properties of the polymer determine the rate of drug absorption.

Marketed transdermal patches are *exemplified* by Estraderm® (estradiol), Testoderm® (testosterone), Alora® (estradiol), Androderm® (testosterone), and Transderm-Scop® (scopolamine). Transderm® relies on rate-limiting polymeric membranes to control drug release. Nicoderm® is a nicotine patch, which releases nicotine over 16 h, continuously suppressing the smoker's craving for a cigarette.

REVIEW QUESTIONS

11.1 Which property is NOT TRUE for poly(oxyethylene)-poly(oxypropylene)-poly(oxyethylene) block copolymers?
   A. Surfactant
   B. Forms micelles
   C. Biodegradable
   D. Thermosensitive
   E. All of the above
   F. None of the above

11.2 Which of the following is TRUE for biomaterials?

  A. The greater the degree of crystallinity of the polymer, the lower the rate of dissolution.

  B. Molecular weight and molecular weight distribution affect solvent penetration and crystallinity.

  C. Increase in the main-chain polarity increases the glass transition temperature of a polymer.

  D. Modification of biomaterial surfaces with polyethylene glycol minimizes protein adsorption and/or platelet adhesion.

  E. All of the above.

  F. None of the above.

11.3 Define the following nomenclatures using chemical structures of commonly used polymers:

  A. Biomaterials and biocompatibility

  B. Block and graft copolymers

  C. Repeating unit and end group

  D. Monomer and oligomer

11.4 What is the degree of polymerization (DP) of (a) a sample of poly(methyl methacrylate) with average molecular weight of 50,000, and (b) a sample of poly(tetramethylene-m-benzenesulfonamide) with average molecular weight of 26,000?

11.5 Suppose we have a polymer sample consisting of 9 moles with molecular weight 15,000 and 5 moles with molecular weight 25,000. What is the number average molecular weight (M_n) and the weight average molecular weight (M_w)?

REFERENCES

1. Waknis V, Chu E, Schlam R, Sidorenko A, Badawy S, Yin S, and Narang AS (2014) Molecular basis of crystal morphology-dependent adhesion behavior of mefenamic Acid during tableting. *Pharm Res* **31**(1): 160–172.
2. Tanaka T, Nishio I, Sun ST, and Ueno-Nishio S (1982) Collapse of gels in an electric field. *Science* **218**(4571): 467–469.
3. Katagiri K, Nakamura M, and Koumoto K (2010) Magnetoresponsive smart capsules formed with polyelectrolytes, lipid bilayers and magnetic nanoparticles. *ACS Appl Mater Interfaces* **2**(3): 768–773.

FURTHER READING

El-Sayed ME, Hoffman AS, and Stayton PS (2005) Smart polymeric carriers for enhanced intracellular delivery of therapeutic macromolecules. *Expert Opin Biol Ther* **5**: 23–32.

Florence AT and Attwood D (2006) *Physicochemical Principles of Pharmacy* 4th ed., London: Pharmaceutical Press.

Mahato RI (Ed.) (2005) *Biomaterials for Delivery and Targeting of Proteins and Nuclec Acids*, Boston, MA: CRC Press.

Na K and Bae YH (2005) pH sensitive polymers for drug delivery. In Kwon GS (Ed.) *Polymeric Drug Delivery Systems*, London: Taylor & Francis, pp. 129–194.

Sinko PJ (Ed.) (2006) *Martin's Physical Pharmacy and Pharmaceutical Sciences*, 5th ed., New York: Lippincott Williams & Wilkins, pp. 585–627.

Stevens MP (1999) *Polymer Chemistry: An Introduction*, 3rd ed., New York: Oxford University Press.

Tomalia DA (1995) Dendrite molecules. *Sci Am* **272**: 62–66.

Chapter 12

Rheology

LEARNING OBJECTIVES

On completion of this chapter, the students should be able to

1. Define rheology and describe its applications in pharmaceutical sciences.
2. Describe Newtonian and non-Newtonian types of flow.
3. Discuss pseudoplastic and dilatant rheograms and identify shear-thinning and shear-thickening phenomena.
4. Discuss the importance of thixotropy in semisolid pharmaceutical dosage forms.

12.1 INTRODUCTION

Rheology is the study of flow properties of liquids. It addresses the viscosity characteristics of solution and colloidal systems. The flow of simple liquids can be described by viscosity, an expression of the resistance to flow. Liquids and solutions for which the flow characteristics can be adequately described by viscosity are called Newtonian fluids, and they are said to exhibit Newtonian flow characteristics. However, the flow of complex dispersions cannot be adequately described by viscosity. These fluids are termed non-Newtonian, and they are said to exhibit non-Newtonian flow characteristics.

Rheological properties are important considerations in the manufacturing, analysis, and use of several dosage forms, including solutions, emulsions, suspensions, pastes, lotions, suppositories, parenteral injectable drug products, and intravenous infusions. Viscosity is important for the mixing and flow of materials, their packaging into containers, and their removal before use—whether achieved by pouring from a bottle, extrusion from a tube, or passage through a syringe needle. For example, pourability, spreadability, and syringeability of an emulsion are determined by its rheological properties.

In addition, viscosity is an important consideration as a critical material attribute (CMA) and critical (in-process) material attribute (CiMA) for pharmaceutical manufacturing. For example, viscosity of solutions of polymers and liquid ingredients such as polyethylene glycol is a CMA used for the quality control of incoming raw materials, while viscosity of suspension of polymer, colorant, and opacifier used for tablet coating is a CiMA during product manufacturing.

12.2 NEWTONIAN FLOW

Newton's law of flow states that the application of stress on a liquid leads to flow in the direct proportion to the amount of stress applied. The constant that relates the flow of a liquid to the applied stress is called viscosity, η. Thus,

$$t = \eta \bullet D$$

where, D is the rate of flow and t is the applied stress.

Fluids that obey Newton's law of flow are referred to as *Newtonian fluids*, and fluids that deviate from the linear stress–flow proportionality expressed by the Newton's law are called *non-Newtonian fluids*.

Viscosity can also be expressed as the constant that relates the shearing stress (force per unit area applied parallel to the direction of flow of liquid) and shear rate (volume of fluid flow per unit area of the liquid). Thus,

$$F = \frac{F'}{A} = \eta \frac{dv}{dr} \tag{12.1}$$

where:
  force per unit area (F'/A) required to bring about flow is called the
    shearing stress (F)
  dv/dr is the rate of shear representing the velocity of fluid movement per
    unit distance from the plane of shear stress

The units of F'/A are dynes per cm^2 or Pascal. The units of dv/dr are (cm/sec)/cm = sec^{-1}. Thus, the Standard International (SI) unit of viscosity is Pascal*second (Pa*s). The more commonly used unit of viscosity is dyne*second/cm^2, which is named Poise (P).

10 Poise = 1 Pascal*second

10 centiPoise = 1 Poise

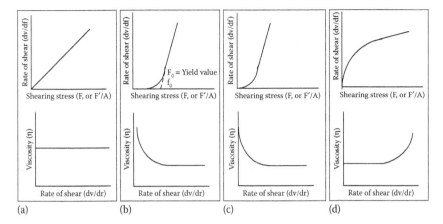

Figure 12.1 Plots of rate of shear and viscosity as a function of shearing stress for (a) Newtonian, (b) plastic, (c) pseudoplastic, and (d) dilatant flows.

The higher the viscosity of a liquid, the greater the shearing stress required to produce a certain rate of shear. A plot of the rate of shear against shearing stress yields a *rheogram*. A Newtonian fluid will plot a straight line, with the slope of the line being η (Figure 12.1a).

Another term, fluidity, ϕ, is defined as the reciprocal of viscosity:

$$\phi = 1/\eta \qquad (12.2)$$

In the case of Newtonian fluids, viscosity is a constant for a fluid at a given temperature and pressure. It does not change with increasing shear rate. Water and dilute solutions typically exhibit Newtonian flow properties.

12.2.1 Temperature dependence and viscosity of liquids

Viscosity of a liquid generally decreases as the temperature is raised. Increased Brownian motion, rapid movement of liquid molecules among each other, at higher temperatures reduces energy expense from the shearing force in overcoming intermolecular cohesion forces, leading to a greater proportion of energy utilized in influencing the direction of flow. In other words, fluidity increases with an increase in temperature.

The relationship between viscosity and temperature can be represented by the following equation, which is analogous to the Arrhenius equation of chemical kinetics:

$$\eta = Ae^{E_v/RT} \qquad (12.3)$$

where:

η is the viscosity (centiPoise)

A is the constant, depending on molecular weight and molar volume of liquid

E_v is the activation energy required to initiate flow between molecules (cal/mol)

R is the gas constant (1.987 cal*K^{-1}*mol^{-1})

T is the temperature in Kelvin

This equation predicts that the decrease in viscosity with an increase in temperature is nonlinear and follows an exponential profile. Thus, from the temperature of highest viscosity of the fluid, slight changes in temperature lead to much more change in fluid viscosity compared with the effect of temperature on viscosity close to the lowest viscosity of a fluid. The temperature–viscosity relationships are critical for routine handling of liquids in pharmaceutical manufacturing under ambient temperature conditions and handling of drug products by patients. For example, an injectable solution stored in a refrigerator may have much higher viscosity and may be difficult to inject than the same solution brought to room temperature before injection.

12.3 NON-NEWTONIAN FLOW

Most pharmaceutical fluids, such as colloidal dispersions, emulsions, liquid suspensions, and ointments, do not follow Newton's law of flow. The viscosity of the fluid varies with the rate of shear. Depending on how viscosity changes with shear, there are three general types of non-Newtonian flow behaviors: plastic, pseudoplastic, and dilatant (Figure 12.1b–d).

12.3.1 Plastic flow

Substances that undergo plastic flow are called Bingham bodies; they are defined as substances that exhibit a yield value as the point at which plastic flow curve intersects shearing stress axis (Figure 12.1b). Plastic flow is associated with, for example, the presence of flocculated particles in concentrated suspensions. Flocculated solids are light, fluffy conglomerates of adjacent particles held together by weak van der Waals forces. A certain shearing stress must be exceeded in order to break up van der Waals forces, which is the yield value, f_0. The yield value is an indicator of flocculation. Higher the yield value, greater the degree of flocculation.

A plastic system resembles a Newtonian system at shear stresses below the yield value. The characteristics of plastic flow can be summarized as follows:

- Plastic flow does not begin until a shearing stress, corresponding to a yield value, f, is exceeded.

- The curve intersects the shearing stress axis but does not cross through the origin.
- The materials are said to be *elastic* at shear stresses below the yield value.

12.3.2 Pseudoplastic (shear-thinning) flow

Pseudoplastic flow is characterized by decrease in viscosity with increasing shear stress. This leads to increasing rate of shear (flow) for the same amount of change in the shear stress as the shear stress levels are increased. Thus, these fluids tend to flow more easily with increasing shear stress and flow and are thus called *shear-thinning* fluids. The molecular origin of the shear-thinning behavior of fluids is in preferential alignment of the molecules of solution during flow, such that intermolecular cohesion forces are reduced as the rate of shear and the rate of flow increase. Thus, shear thinning occurs when molecules align themselves during flow, such that they slip and slide past each other.

Linear polymers in solution exhibit pseudoplastic flow. A large number of pharmaceutical products, including natural and synthetic gums (e.g., liquid dispersions of tragacanth, sodium alginate, methyl cellulose, and sodium carboxymethylcellulose), exhibit pseudoplastic flow properties.

The characteristics of pseudoplastic flow materials can be summarized as follows:

- Pseudoplastic substances begin to flow when a shearing stress is applied; therefore, they exhibit no, or very low, yield value. Thus, the shear stress—shear rate profile does cross the origin (Figure 12.1c).
- Viscosity of a pseudoplastic substance decreases with increasing shear stress and shear rate.

12.3.3 Dilatant (shear-thickening) flow

Dilatant flow is characterized by an increase in viscosity with increasing shear stress. This leads to decreasing rate of shear (i.e., reduced flow and higher viscosity) for the same amount of change in the shear stress as the shear stress levels are increased. Thus, these fluids tend to get more viscous and *thicker* and flow with greater difficulty with increasing shear stress and flow; they are thus called *shear-thickening* fluids. The molecular origin of the shear-thickening behavior of fluids is in entanglement with increasing attractive intermolecular interactions among the molecules of solution during flow, such that the intermolecular cohesion forces increase as the rate of shear and the rate of flow increase. Thus, shear thickening occurs when molecules get entangled, swell, or otherwise align themselves during flow, such that the intermolecular forces of attraction are higher as the flow increases.

Generally, dilatant solutions are those that exhibit an increase in solute volume when sheared. This leads to the reduction in *free*, or unbound, solvent volume between the solute molecules. Thus, the viscosity increases with shear rate. Dilatant systems are usually suspensions with a high percentage ($\geq$ 50% w/w) of dispersed small, deflocculated particles. These systems exhibit an increase in resistance to flow with increasing rates of shear.

The characteristics of dilatant flow materials can be summarized as follows:

- Solutes in dilatant solutions increase in volume when sheared. When the stress is removed, the dilatant system returns to its original state of fluidity.
- Viscosity increases with increasing shear rate.

In concentrated suspensions, particles are closely packed, with the inter-particle void volume being at a minimum at rest. Nevertheless, the amount of vehicle in the suspension is sufficient to fill this volume and to allow the particles to move relative to one another at low rates of shear. Dilatant suspensions can be poured from a bottle, since these are reasonably fluid under these conditions. The bulk of the system dilates (expands) with increase in shear stress. The particles in a deflocculated suspension take an open form of packing; that is, each particle is detached from every other particle, occupying its own space and interacting with solvent molecules to its full surface. Thus, the solvent-filled interparticle void volume is lower for deflocculated suspensions. Accordingly, resistance to flow increases with an increase in flow, because the particles no longer completely get wetted or lubricated by the vehicle. Eventually, the suspension will set up as a firm paste.

The Newton's flow equation described earlier, $t = \eta * D$, where, D is the rate of flow and t is the applied stress, can be modified to:

$$t = \eta * D * e^N$$

For Newtonian fluids, $N = 0$ and the expression $e^N = 1$, thus yielding the expression for Newtonian flow. For dilatant or shear-thickening fluids, $0 < N < 1$. As dilatancy increases, the shear rate response to shear stress decreases exponentially; thus, N decreases as dilatancy increases for shear-thickening fluids.

12.4 THIXOTROPY

Thixotropy is the property of some non-Newtonian fluids to show a time-dependent change in viscosity. For dilatant (shear-thickening) thixotropic fluids, the longer the fluid undergoes shear, the more its viscosity. For pseudoplastic (shear-thinning) thixotropic fluids, the longer the fluid undergoes shear, the lower its viscosity.

Many gels and colloids are pseudoplastic thixotropic materials, exhibiting a stable form at rest but becoming fluid when agitated with a reversible gel–sol transformation phenomenon. When sheared by mixing, such as simple shaking, the matrix relaxes and forms a solution with the characteristics of a liquid dosage form for ease of use. Although pseudoplastic fluids show thinning with only increasing rate of shear, thixotropic fluids show increasing flow and thinning as the duration of mixing increases, even at the same shearing forces. On setting, the higher-viscosity plastic state resumes as a network gel forms and provides a rigid matrix that stabilizes suspensions and gels.

The molecular basis of thixotropic behavior of fluids lies in changes in intermolecular interactions of solute on persistence of shear. Thus, pseudoplastic thixotropic systems show thinning behavior not only with increasing shear but also with increasing duration of shear, owing to changes in the alignment of solute molecules that reduce their intermolecular interactions. Conversely, dilatant thixotropic systems show thickening behavior not only with increasing shear but also with increasing duration of shear, owing to swelling and/or intermolecular entanglement or interparticle interactions, which cause increased interparticle bonding or intermolecular attractive interactions with time.

The term thixotropic fluid is sometimes used to refer to pseudoplastic (shear-thinning) thixotropic behavior. Conversely, the term negative thixotropy or antithixotropy is sometimes used to refer to a dilatant (shear-thickening) thixotropic behavior, that is, a time-dependent increase rather than a decrease in apparent viscosity on application of a shearing stress.

The main advantage of pseudoplastic thixotropic preparations is that the particles remain in suspension during storage, but when required for use, the pastes are readily made fluid by tapping or shaking. This is true for both pseudoplastic and thixotropic fluids. For thixotropic fluids, the duration of shearing or shaking, even at the same shear stress, also impacts shear thinning to the fluid. For example, concentrated parenteral suspensions containing from 40% to 70% w/v of procaine penicillin G in water show high inherent pseudoplastic thixotropic behavior.

12.4.1 Hysteresis loop

Hysteresis loop represents different path of response (shear rate) to the experimental parameter (shear stress) when the experimental parameter is increased or decreased. Thixotropic systems show an up–down curve, called hysteresis loop, such that for a given shear stress, the flow response is a function of the history of the sample—increasing the shear stress (up curve) leads to different flow behavior than if the shear stress were decreasing (down curve). Typical rheograms for pseudoplastic and dilatant systems exhibiting this behavior are shown in Figure 12.2.

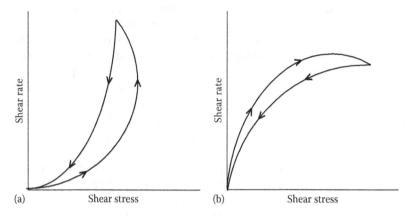

Figure 12.2 Thixotropy in pseudoplastic and dilatant flow systems: (a) thixotropy in pseudoplastic material and (b) thixotropy in dilatant material.

Rheograms of pseudoplastic thixotropic materials are highly dependent on the rate at which shear is increased or decreased and the length of time for which a sample is subjected to any one rate of shear or shear stress. As shown in Figure 12.3 for a pseudoplastic thixotropic system shown in Figure 12.2a, the shear rate (or flow) increases from point "a" to point "b" with an increase in shear stress and decreases from "b" to "e" with a decrease in the shear stress. This forms the hysteresis loop "abe." However, if the sample was taken to point "b" and the shear rate was held constant for a certain period of time (say, t_1 seconds), the rate of shear (and hence the consistency and

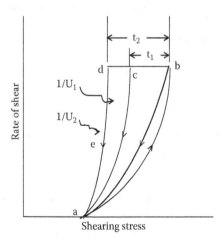

Figure 12.3 Relationship between shearing stress and rate of shear for a plastic system possessing thixotropy.

the flow) increases for the same shear stress (vertical upward movement of line from point "b," not shown in the figure). Consequently, to maintain the same desired rate of shear (as in point "b"), the amount of shear stress required is progressively lower as time passes ($t_2 > t_1$)—represented by the horizontal line from "b" to "c" for t_1 and "b" to "d" for t_2 in Figure 12.3. This decrease in the required shear stress to maintain the same rate of shear is attributed to reduction in the degree or amount of structure in the sample.

Decreasing the rate of shear to zero after having reached the state "b," "c," or "d" (depending on the time for which shear stress is applied at constant shear rate) would create hysteresis loops aba, aca, or ada, respectively. Therefore, in contrast to pseudoplastic or dilatant materials, the rheogram of a thixotropic material is not unique but depends on the rheologic history of the sample and the approach used to obtain the rheogram. For example, keeping the constant shear stress at point "b" for times t_1 and t_2 would result in very different rheograms than the rheogram shown in Figure 12.3, where the shear rate was kept constant at point "b." Student exercise: draw the rheograms for the case where shear stress is kept constant for times t_1 and t_2 at point "b" and compare them with that in Figure 12.3. This property of thixotropic systems is important to bear in mind when attempting to obtain a quantitative measure of viscosity and flow of thixotropic systems.

12.5 PHARMACEUTICAL APPLICATIONS OF RHEOLOGY

Thixotropy is a desirable property in liquid pharmaceutical preparations. A well-formulated thixotropic suspension does not settle out readily in the container and becomes fluid on shaking, with increasing duration of shaking resulting in greater reduction of viscosity or increase in flow. A similar pattern of behavior is desirable with emulsions, lotions, creams, ointments, and parenteral suspensions to be used for intramuscular depot therapy. With regard to suspension stability, there is a relationship between the degree of thixotropy and the rate of sedimentation—the greater the thixotropy, the lower the rate of settling. Importantly, the degree of thixotropy of a system may change over time (e.g., storage during shelf life) and result in an inadequate formulation.

Examples of pharmaceutical applications of rheology can be exemplified by the following:

- Surfactants, poly(oxyethylene)-poly(oxypropylene)-poly(oxyethylene) block copolymers, also known as Pluronic® and poloxamers, exhibit Newtonian behavior in the liquid state, at low concentrations and low temperatures. Poloxamer vehicles are used in dermatological bases or topical ophthalmics, since they are nontoxic and form clear water-based gels.

- Polymer solutions are used as wetting agents for contact lenses or tear substitutes for the dry eye syndrome. These solutions should ideally exhibit pseudoplastic behavior; that is, low viscosity at high shear rates produces lubrication during blinking, and high viscosity at low or zero shear rate prevents fluid from flowing away from the cornea when eyelids are not blinking. Both natural (e.g., dextran) and synthetic (e.g., polyvinyl alcohol) ones are used with the addition of various preservatives. High-molecular-weight sodium hyaluronate with a concentration of 0.1%–0.2% is used for the dry eye condition.
- Rheologic properties of suppositories at rectal temperatures can influence the release and bioabsorption of drugs from suppositories, particularly those having a fatty base. Some fatty acid bases exhibit either Newtonian or plastic flow at rectal temperatures.
- When polymer, opacifier, and surfactant suspensions are prepared for tablet coating, the rheologic behavior of the suspension determines the ability to spray the suspension through a thin nozzle for tablet coating.
- Stability and pourability of liquid preparations is determined by their viscosity and thixotropic behavior.

REVIEW QUESTIONS

12.1 Indicate which statement is TRUE and which one is FALSE.
   A. Pseudoplastic flow is shear-thinning type and dilatant is shear-thickening type.
   B. Flocculated systems exhibit negative thixotropy, while deflocculated system with more than 50% by volume of solid dispersed particles exhibits dilatant flow behavior.
12.2 Which of the following is the most desired behavior in high-concentration suspension formulations sold in a bottle for use by the patient? Explain why.
   A. Pseudoplastic flow
   B. Dilatant flow
   C. Thixotropic flow
   D. Antithixotropic flow
   E. Newtonian flow
12.3 Define Newton's law of flow and draw the diagrams to illustrate the effect of shear stress on the rate of flow and viscosity of fluids obeying Newton's law. Draw flow diagrams to illustrate the effect of shear stress on the rate of flow and viscosity for three types of non-Newtonian fluids.
12.4 Define thixotropy and draw a hysteresis loop to explain the thixotropic phenomenon. Explain why thixotropic phenomenon is desirable for pharmaceutical formulations.

FURTHER READING

Amiji MM (2003) Rheology. In Amiji MM and Sandmann BJ (Eds.) *Applied Physical Pharmacy,* New York: McGraw-Hill, pp. 365–395.

Briceno MI (2000) Rheology of suspensions and emulsions. In Nielloud F and Marti-Mestres G (Eds.) *Pharmaceutical Emulsions and Suspensions,* New York: Marcel Dekker.

Mahato RI (2004) Dosage forms and drug delivery systems. In Gourley DR (Ed.) *APh's Complete Review for Pharmacy,* New York: Castle Connelly Graduate Publishing, pp. 37–64.

Schott H (2000) Rheology, In Gennaro AR (Ed.) *Remington's The Science and Practice of Pharmacy,* 20th ed., Philadelphia: Lippincott Williams & Wilkins, pp. 335–355.

Sinko PJ (2005) *Martin's Physical Pharmacy and Pharmaceutical Sciences,* 5th ed., Philadelphia, PA: Lippincott Williams & Wilkins, pp. 561–583.

Chapter 13

Radiopharmaceuticals

LEARNING OBJECTIVES

On the completion of this chapter, the students should be able to

1. Define radioactivity and types of radiation.
2. Name units of measurement of radioactivity.
3. Identify key radiation safety health hazards.
4. Define four key elements of radiation safety.
5. Identify types of applications of different types of radioactive compounds.
6. Identify salient elements in the formulation and dispensing of radiopharmaceuticals.
7. Calculate activity and dose of a radioactive material.

13.1 INTRODUCTION

Radioisotopes, also called radionuclides, are usually artificially produced unstable atoms of a naturally occurring element. These isotopes have the same number of electrons and protons as the naturally occurring element, but different number of neutrons. More than 1000 radioisotopes are known to occur. Of these, only about 50 are naturally occurring. Most radioisotopes are produced by bombarding the atoms of the stable, naturally occurring element with fast-moving neutrons produced in a nuclear reactor or particle accelerator. These isotopes tend to revert to the natural, stable elements at a rate that is specific to each isotope of each element. The rate of conversion of an isotope to its stable elemental composition determines its time in existence and is measured by half-life, the time it takes for half of the radioisotope population to convert.

In a hospital setting, radiopharmaceuticals are typically handled by the nuclear pharmacy or radiopharmacy, involved in the preparation of radioactive materials for diagnosis and/or treatment of specific diseases. In diagnostic applications, radiopharmaceuticals accumulate in specific tissues or cells and emit radiation, which can be collected and processed into images, showing the location of the accumulation in the body, for diagnostic purposes. In therapeutic applications, the high-energy radiation released by radiopharmaceuticals destroys undesired local cells and tissue.

13.1.1 Types of radiation

The unstable nuclei of radioisotopes dissipate energy, in the form of specific types of radiation, as they spontaneously convert to the stable parent isotopes. These radiations are commonly known as alpha, beta, or gamma rays.

1. Alpha radiation is a result of excess energy dissipation by unstable nuclei in the form of alpha particles. The alpha particles have two positive charges and a total mass of four units. This is exemplified by polonium $^{210}Po_{84}$ decaying to $^{206}Po_{82}$, in a notation where superscript before the element's symbol represents the atomic mass and the subscript after the element's symbol represents the atomic number. The alpha particles, being heavy, are ejected at about 1/10th the speed of light and are not very penetrating. They can travel about 1–4 inches in the air.

2. Beta radiation is produced through beta decay of unstable nuclei and can follow either of the three processes: electron emission, positron emission, and electron capture.

    • Negative beta decay involves the emission of an energetic electron and an antineutrino (which does not have a resting mass). In the resulting nucleus, a neutron becomes a proton and stays in the nucleus. Thus, the proton number (atomic number) of the resulting nucleus increases by one, while the mass number (total number of protons and neutrons in the nucleus) does not change. For example, this process occurs for tritium (^3H) decay to radioactive helium (^3He).

    • Positive beta decay involves the emission of a positron, similar to an electron in all aspects but with opposite charge, and a neutrino. In the resulting nucleus, a proton converts to a neutron. Thus, the atomic number of the daughter nucleus is one less than the parent, while the atomic mass remains the same.

    • Electron capture is a process whereby an orbiting electron combines with a nuclear proton to form a neutron (which remains in the nucleus) and a neutrino (which is emitted). In the resulting nucleus, the atomic number reduces by one, while the atomic mass stays the same.

- Beta decay is usually a slower process compared with alpha or gamma decay. Most beta particles are emitted at the speed of light.
3. Gamma rays are the most penetrating electromagnetic radiation of shortest wavelength and highest energy, just above the X-ray region of the electromagnetic spectrum. Gamma rays can be produced by the decay of the radioactive nuclei or of certain subatomic particles. The mechanism of formation of high-energy gamma ray photons is currently not well understood. Characteristic features of these types of radiations emitted by the radioisotopes are summarized in Table 13.1.

13.1.2 Units of radioactivity

The quantity of radioactive material is measured in terms of activity rather than mass. The amount of radioactivity is typically expressed in the units of Curie (Ci), which is a measure of radioactivity per unit mass of material. The international system of units (SI system) recommends becquerel (Bq) as a unit of radioactivity. One Bq represents the amount of radiation produced from one disintegration per second (dps). One Ci is 37-billion Bq or 37 GBq.

While Ci is the unit of measurement of radioactivity, the absorbed dose of ionizing radiation is expressed in rad, the dose equivalent (when radiation is applied to humans) is expressed in rem, and the exposure to radiation is quantitated in *roentgen* (R). One rad represents the amount of radiation that releases energy of 100 ergs per gram of matter. Erg is a unit of energy or work that equals 10^{-7} Joules. Rem is the dosage in rads that causes the same amount of biological injury as 1 rad of X-rays or gamma rays.

Table 13.1 Characteristic features of different types of radiation emitted by radioisotopes

Property	Alpha rays	Beta rays	Gamma rays
Composition	Identical to the nucleus of Helium-4 (^{4}He), with two protons and two neutrons	Electrons or positrons	High-energy photons
Charge	Positive (2+)	Negative (1−) electron or positive (1+) positron	None
Mass	Four	Zero	Zero
Penetration depth	Least	Medium	Highest
Principal application	High-energy bombardment sources	Radiotherapy and radioimmunotherapy agents	Tracers and diagnostic agents

In the SI system of units, where Bq is the unit of radioactivity, gray (Gy) is the unit of expression of absorbed dose, Sievert (Sv) is the dose equivalent unit, and exposure is expressed in coulomb per kilogram body weight (C/kg). One rad is 0.01 Gy and one rem is 0.01 Sv.

13.1.3 Radiation safety

Radiation exposure can lead to several side effects that can be understood as the impact of radiation on rapidly dividing cells. The following side effects are commonly observed in patients undergoing radiation therapy of cancer:

- Hair loss
- Gastrointestinal irritation becoming evident as nausea, vomiting, diarrhea, and stomach upset
- Low white blood cell count (leucopenia).
- Local side effects such as reddening and itchiness of the skin, if applied
- Oral mucositis, leading to sore mouth or oral ulcers

Generally, doses higher than 30 μCi are administered in a hospital setting to ensure adequate safety monitoring.

Avoiding unintended exposure to radiation in a laboratory setting is a key function of the organizational environmental, health, and safety (EHS) organizations. These are done through careful inventory control, engineering controls when handling radioactive materials (such as the use of fume hoods), and proper storage and disposal of radioactive material and contaminated waste. In addition, the following protection guidelines are recommended for the users:

1. *Time*: The shorter the time of potential use of a radioactive material, the shorter the duration of exposure. Thus, quick and efficient work with minimal time of exposure of the radioactive material to the ambient laboratory environment is recommended.
2. *Distance*: The farther a person is from a source of radiation, the lower the dose of radiation exposure. In addition, physical contact with the radioactive material is generally avoided with the use of devices to manipulate or move stored containers of radioactive material.
3. *Shielding*: Radioisotopes are typically handled in lead containers, since lead absorbs and is impervious to all radiation. X-ray technicians and laboratory personnel wear lead-coated aprons to block potential direct exposure to radiation.
4. *Quantity*: The amount of radioactive material in the working area and inventory is generally minimized. Multiple procurements of small quantities are preferred over purchasing and storing one large quantity.

Note that temperature and pressure are not included in the list of radiation safety considerations. In other words, handling a radioactive compound under refrigerated conditions does not provide any lower exposure to radiation than handling the same compound at the room temperature. The decay rate of radionuclides is insensitive to temperature and pressure under normal usual laboratory operating conditions.

13.2 BIOMEDICAL APPLICATIONS OF RADIOPHARMACEUTICALS

There are about 200 radioisotopes that are in current medical use. Radioisotopes can be employed as either radiation sources, for radiotherapy applications, or as radioactive tracers, which are commonly used as diagnostic agents. They are also used to determine the biodistribution of particular compounds (in which those radioisotopes are incorporated).

13.2.1 Radioactive tracers and diagnosis

Use of radioisotopes as tracers and diagnostic agents depends on the ease of detection of the radiation emitted by the isotope and the ability of the isotopic element to be incorporated into the molecule that is being traced (such as during the biodistribution studies of new drug candidates). As diagnostic agents, radioactive elements are typically adsorbed or incorporated on a carrier. Thus, chemical identity, ability to use during synthesis, and the form of the radioisotopes are important for their applications as tracers and diagnostic agents. Diagnostic uses of radioisotopes can be exemplified by thyroid function studies using low dose ^{131}I, erythrocyte tagging for identification of type of anemia using ^{51}Cr, and metabolic studies using ^{14}C. The ^{14}C radioisotope detection in breath can be used to detect the presence of ulcer-causing bacteria *Helicobacter pylori*. There is an increasing preference for the use of nonradioactive methods of analyses, wherever possible, due to the handling risks associated with radioactive isotopes.

For the use of radioisotopes as diagnostic agents and tracers, the dose of radiation administered to the patients or normal healthy volunteers should be as low as possible, while maintaining accuracy and sensitivity of analytical detection. Thus, the radioisotopes for diagnostic use should ideally be compounds with low half-life that exhibit rapid elimination kinetics and are administered in low doses. Typically, radioisotopes that emit gamma rays are used for diagnostic use, since gamma rays are the most penetrating; the radiation does not stay in the body and is quickly received by the detector. Specialized analytical methods are often developed to analyze low concentration of radioisotopes in plasma and tissue samples.

Technetium-99m (^{99m}Tc)—a metastable nuclear isomer of technetium-99 (^{99}Tc), with a half-life of about 6 hours and biological elimination half-life of about 1 day, is the most common radioisotope used in medicine. ^{99}Tc is sourced at the hospitals from its more stable and easily transportable parent isotope, molybdenum-99 (^{99}Mo), with a half-life of about 66 hours, in lead containers. The hospital extracts and uses the needed quantities of ^{99m}Tc, as ^{99}Mo degrades to ^{99m}Tc.

13.2.2 Radiotherapy

Use of radioisotopes as radiation sources for radiotherapy aims to utilize the tissue damage that results from radiation to, for example, reduce the amount of cancerous tissue. Selection of radioisotopes as radiotherapy agents depends mainly on the type and energy of radiation emitted by the isotope and its depth of tissue penetration. Chemical identity and reactivity are of relatively less importance for radiotherapy applications.

Radioisotopes used for therapy can be applied to the target tissue either from an external source or on administration to the patient as a drug.

1. External source application of radiation has the advantages of duration and amount of dose titration, with direct observation of the target tissue, and of being able to remove the radiation source—and terminate treatment—at any time. Radioisotopes used for external therapy are exemplified by cobalt (^{60}Co) and cesium (^{137}Cs). They have been used for the treatment of undesired lesions.

2. Internal application, or administration of the radiotherapy agent to the patient, has a limitation that the source of radiation cannot be removed once administered. Therefore, the amount of radioisotope administered to the patient must be carefully controlled. Radioisotopes that have been used for internal therapy include gold (^{198}Au), iridium (^{192}Ir), phosphorus (^{32}P as sodium phosphate), yttrium (^{90}Y), iodine (^{131}I as sodium iodide), and palladium (^{103}P).

   • Colloidal gold (^{198}Au) suspensions have been used in the cases of fluid accumulation in the abdomen (peritoneal cavity) or chest (plural cavity), associated with malignant tumors. The colloidal suspension diffuses throughout the fluid and, over time, tends to aggregate at the surface of the cavity.

   • Nylon ribbons containing iridium (^{192}Ir) seeds at periodic intervals can be implanted into the interstitial cavity, such as abdominal, for the treatment of tumors. These ribbons are surgically removable.

   • Radiophosphorus (^{32}P) can be injected parenterally as a solution of a highly soluble sodium salt. The phosphorus tends to accumulate in rapidly proliferating cells and tissues. Accordingly, it has been used for the treatment of polycythemia vera (too many red blood cells produced by the bone marrow) and chronic granulocytic or myeloid leukemia (too many blood cells produced by the bone marrow). At relatively high doses (1.5–5 mCi), ^{32}P accumulates in the bone marrow and can suppress the production of blood cells.

- Yttrium (^{90}Y) has a strong affinity for chelating agents, which can be used for targeting carriers. Yttrium chelate with pentetic acid or diethylenetriaminepentaacetic acid (DTPA) can be used for localization to the lymphatics, and its chelate with ethylenediaminetetraacetic acid (EDTA) can lead to localization in the bone.

- Iodine (^{131}I), used as a water-soluble salt, sodium iodide, is perhaps the most commonly known radioisotope used for the treatment of goiter, Graves' disease, and thyroid cancer. Iodine is selectively taken up by the thyroid gland in the neck. The uptake of radioactive iodine can cause localized tissue destruction by radiation produced within the gland. Targeted uptake of ^{131}I by select tissues can be achieved by incorporation into compounds such as metaiodobenzylguanidine (mIBG). The ^{131}I-labeled mIBG is selectively taken up by the adrenal medullary tissues and can be used to treat carcinomas of or metastases from the adrenal medullary glands.

- Radioimmunotherapy is the targeting of radioisotopes to specific cells, tissues, and tumor types by covalent conjugation of a radioisotope to monoclonal antibodies or their antigen-binding fragments. For example, ^{131}I can be conjugated to antibodies by using N-hydroxysuccinimide (NHS) to produce radiolabeled antibodies. Copper isotope, ^{67}Cu, can be conjugated to antibodies by using the chelating agent [6-p-nitrobenzyl]-1,4,8,11-tetraazacyclotetradecane-N, N', N'', N''' tetraacetate (TETA) for radioimmunotherapy.

13.3 RADIATION DETECTION EQUIPMENT

The most common radiation detection equipment includes the following:

1. Scintillation detector for the detection of gamma-radiation emitting probes. A scintillation probe can be suitably modified to detect the localization of isotopes in the organs of interest. For example, detection of ^{131}I uptake by the thyroid and the uptake of red blood cells labeled with ^{51}Cr by the spleen require appropriately modified gamma scintigraphy equipment.

2. Scintillation counters are typical laboratory equipment used for the detection and measurement of ionizing radiation. These counters can

be used to test samples of *in vitro* testing (such as drug release or dissolution) and *in vivo* samples after digestion into a homogeneous liquid (such as biodistribution studies). The operating principle of a scintillation counter is the excitation of a scintillating material, typically a transparent crystal, with the high-energy photons of the incident ionizing radiation. A scintillating material is a luminescent material that absorbs incoming high-energy radiation and reemits the absorbed energy in the form of light. The scintillating material could be, for example, cesium iodide, to detect protons and alpha particles, sodium iodide containing small amounts of thallium to detect gamma radiation, zinc sulfide to detect alpha particles, and lithium iodide to detect neutrons. This instrumentation also contains a sensitive photomultiplier tube that converts light energy to electrical signal and the needed electronics to quantitatively process and display this signal.

3. Positron emission tomography (PET) is a functional imaging technique applied in nuclear medicine to measure whole body metabolism. It detects gamma rays emitted by a positron-emitting radionuclide ^{18}F fluorodeoxyglucose that is administered to the patient as a tracer before the procedure. A computerized tomography (CT) X-ray scan is concurrently performed on the patient to construct a three-dimensional (3D) image of the patient, which is then utilized to construct a 3D location of the radioisotope in the body in what is known as the CT-PET scan. This scan can detect regional metabolic activity, as indicated by regional glucose uptake. This scan is commonly used to detect cancer metastases.

   A similar nuclear medicine tomographic (providing 3D information in 2D cross-sectional slices) imaging technique is the single-photon emission computed tomography (SPECT or SPET). This technique is often applied in an organ-specific way, with the administration of a specific radionuclide or its conjugate with a targeting ligand.

4. Geiger counter, also known as the Geiger–Muller counter, is a typical name for a handheld device for measuring ionizing radiation most commonly used by the laboratory safety personnel. It detects ionizing radiation, including alpha particles, beta particles, and gamma rays, using the ionization effect produced by the radiation in a Geiger–Muller tube. The Geiger–Muller tube is filled with an inert, unionized gas (He, Ne, or Ar) at low pressure and is equipped with an anode and a cathode under high voltage (400–600 V). However, there is no flow of current, since the gas in the chamber is unionized. Ionization of the inert gas with incident radiation leads to the flow of current in direct proportion to the amount of incident radiation, which is detected and reported.

13.4 FORMULATION OF RADIOPHARMACEUTICALS

Preparation of radiopharmaceuticals for clinical administration necessitates several considerations. For example:

1. *Sterility*: All formulations intended for parenteral administration need to be manufactured under aseptic conditions that provide a reasonable assurance of sterility. In addition, absence of microbial contamination needs to be shown periodically to validate such a working environment. Aseptic filtration as the final step in the process and/or terminal sterilization after packaging in the final container, for example, by autoclaving, are the preferred practices that ensure sterility of the formulation in the final product container.

2. *Pyrogens and endotoxin limit*: Endotoxins are bacterial cell-wall components that can elicit fever response in humans and are therefore called pyrogenic. Endotoxin content of the formulations must be within the limits that would ensure that the rate and total daily amount of endotoxin intake in the patients stay below acceptable regulatory limits. The endotoxin limit for human administration is limited to five endotoxin units (as defined by the United States Pharmacopeia [USP]) per kilogram per hour (in the case of infusion) for intravenous (IV) administration. All incoming raw materials, including glassware contact surfaces and packaging materials, and processing techniques can contribute to the endotoxin content of the final formulation.

3. Adsorption of the radiopharmaceutical to the container of storage or dispensing can lead to reduced dose or potency administered to the patient. This can be obviated by using silicone-coated low-adsorption containers as—well as by using inert carrier-loaded radiopharmaceuticals, which have lower diffusivity and adsorption potential.

Certain radioisotopes have short half-life, which introduces additional complexity in their use. For example, ^{99m}Tc and ^{113m}In have a short half-life. The parenteral formulations of these radioisotopes must be prepared in the clinic under aseptic conditions, ideally immediately before administration to the patient.

Institutions, such as hospitals, that handle radioisotopes have a medical isotope committee to carefully guide, monitor, and control the handling of the radioisotopes to maintain patient and user safety. Generally, the institutional medical isotope committee is charged with the responsibility to define the specific details of radioisotope use and disposal, such as facilities, storage requirements, inventory requirements, and personnel procedures.

13.5 CALCULATIONS

Calculations for radioactive materials are important for dispensing or administering the right dose to a patient has also for the laboratory researcher to use the right amount of the radioactive material. In addition, the phenomenon of radioactive decay necessitates an understanding of changing radioactivity of a preparation over a period of time.

Half-life of a radionuclide is the time taken for half of the radionuclide's atoms to decay. Radioactive decay is the process by which an atomic nucleus of an unstable atom loses energy by emitting ionizing particles. If λ be the rate of decay of a radionuclide (number of disintegrations per unit time) and τ be the average lifetime of an atom before it decays, the number N of radioactive atoms available in a sample at any point in time, t, represented as $N(t)$, is a function of the number of radioactive atoms present in the same amount of sample at the initial time point, $t = 0$, by first-order rate of degradation of the radionuclide (since the rate of degradation depends on the number of available radioactive nuclei) by the equation:

$$N(t) = N(0)e^{-\lambda t}$$

Note that this is a typical first-order equation, where a measure of the rate of a phenomenon $[N(t)/N(0)]$ is an exponential function of the rate of that phenomenon, λ, times the elapsed time. The minus in the exponent represents the reduction of radioactivity with time.

Moreover, since,

$$\lambda = \frac{1}{\tau}$$

$$N(t) = N(0)e^{-\frac{t}{\tau}}$$

Thus, radioactive decay is exponential with a constant probability.

Typical of the first-order reaction kinetics, at half-life, $t = t_{1/2}$ and $N(t_{1/2}) = \frac{N(0)}{2}$.

Thus,

$$t_{1/2} = \tau \ln 2 = \frac{\ln 2}{\lambda} = \frac{0.693}{\lambda}$$

or, the rate of decay, λ, can be calculated from $t_{1/2}$ as:

$$\lambda = \frac{\ln 2}{t_{1/2}} = \frac{0.693}{t_{1/2}}$$

As seen in this equation, highly radioactive materials (high λ) have a short half-life, whereas low-radioactivity emitting materials last longer.

Example

Ten mCi of a radioactive isotope was received in the laboratory on February 15. This isotope has a half-life of 60 days. How much is the remaining radioactivity on March 15?

Solution:
Representing radioactivity as the number of available radionuclides,

$N(0) = 10$ mCi
$t = 30$ days
$t_{1/2} = 60$ days
$N(t) = ?$

Now, $N(t) = N(0)e^{-\lambda t}$
To use this equation, we need to calculate λ, the rate of decay.

$$\lambda = \frac{\ln 2}{t_{1/2}} = \frac{0.693}{t_{1/2}} = \frac{0.693}{60} = 0.01155 \ \text{day}^{-1}$$

Now,

$$N(t) = N(0)e^{-\lambda t} = 10 \ \text{mCi} \times e^{-0.01155 \times 30} = 7.07 \ \text{mCi}$$

Note that the initial radioactivity of 10 mCi would become 5 mCi in 60 days, since the $t_{1/2}$ is 60 days. However, after 30 days of storage, on day 30, the activity of the radioactive compound is more than a quarter of the initial radioactivity. This highlights that the rate of decay of a radioactive compound follows a nonlinear first-order exponential rate kinetics.

REVIEW QUESTIONS

13.1 Scintigraphy is used to detect radioisotopes that emit which kind of radiation?
   A. Alpha
   B. Beta
   C. Gamma
   D. Delta
13.2 Which radiation is the most penetrating?
   A. Alpha
   B. Beta
   C. Gamma
   D. Delta

13.3 Which radiation is the least penetrating?
  A. Alpha
  B. Beta
  C. Gamma
  D. Delta

13.4 Artificial radioisotopes are produced by bombarding atoms of stable, naturally occurring elements with fast-moving ...
  A. Electrons
  B. Protons
  C. Neutrons
  D. Atoms

13.5 Which one of the following is the most commonly used radioisotope in medical applications today?
  A. Technetium
  B. Molybdenum
  C. Iodine
  D. Carbon

13.6 Containers that are used to safely store and transport radioisotopes must be made of which metal?
  A. Iron
  B. Aluminum
  C. Copper
  D. Lead

13.7 Which of the following statement is not correct?
  A. Isotope has the same proton number but different neutron number
  B. Isotope has different mass
  C. All the isotopes are radioactive
  D. None of the above

13.8 An archeologist finds that an artifact containing 14C was found to contain 3 disintegrations per minute (dpm) of radioactivity. Knowing from literature that the half-life of 14C is 5700 years and that the decay rate of a fresh 14C sample is 15 dpm, how much age would the archeologist assign to the artifact?

13.9 The doctor prescribed a dose of 30 mCi of a particular radioisotope to a patient. The hospital has this radioisotope in inventory, purchased 9 days ago at 600 mCi/g. Knowing from literature that this radioactive element has a half-life of 6 days, how much dose in grams should the pharmacist dispense for the patient?

13.10 Which one of the following is not a key consideration to improve radiation safety?
  A. Time
  B. Temperature
  C. Distance
  D. Shielding

13.11 Which one of the following is a handheld device used for the detection of ionizing radiation by laboratory safety personnel?

A. PET scanner

B. SPECT scanner

C. Geiger–Muller counter

D. Scintillation counter

13.12 Radionuclides typically decay by α-particle, β-particle, or γ-ray emission. For diagnostic radiopharmaceuticals, which type of decay is preferable?

A. α-particle emission

B. β-particle emission

C. γ-ray emission

D. All of the above

Chapter 14

Drug delivery systems

LEARNING OBJECTIVES

On completion of this chapter, the students should be able to

1. Describe the effect of particle size and zeta potential on drug delivery and targeting.
2. Describe different routes of drug administration and discuss specific characteristics of drug delivery systems for each specific route.
3. Describe targeted delivery systems to specific organs, tissues, and cells.

14.1 INTRODUCTION

Drug delivery systems (DDSs) are polymeric or lipid carrier systems that transport drugs to their targets or receptor sites in a manner that provides their maximum therapeutic activity, prevent their degradation or inactivation during transit to the target site(s), and protect the body from adverse reactions due to inappropriate disposition. The goal of a DDS is to release the drug(s) to simultaneously provide maximal safety, effectiveness, and reliability (Figure 14.1). Design of an effective delivery system requires a thorough understanding of the drug, the disease, and the target site. Various physicochemical product properties that influence the quality features of plasma clearance kinetics, tissue distribution, metabolism, and cellular interactions of a drug can often be controlled by using a delivery system.

DDSs can broadly be classified into two groups: macromolecular drug carrier systems and particulate carrier systems (such as, microspheres, nanospheres, and liposomes). For site-specific delivery, the drug is released directly into a specific area, whereas in non-site-specific delivery, the drug is released and it enters the body systemically. Following administration, targeting of drugs to specific sites in the body can be achieved by linking particulate systems or macromolecular carriers to monoclonal antibodies or to

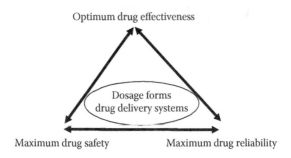

Figure 14.1 Objectives of a dosage form or a drug delivery system.

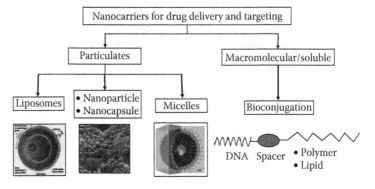

Figure 14.2 Commonly used nanocarriers for drug delivery and targeting.

cell-specific ligands (e.g., asialofetuin, glycoproteins, and immunoglobulins) or by alterations in the surface characteristics, so that they are not recognized by the reticuloendothelial system (RES). The ability of a macromolar or particulate carrier system to deliver a drug to a target site depends on the following characteristics: molecular weight/size, surface charge, surface hydrophobicity, and presence of targeting ligands. Figure 14.2 shows commonly used nanocarriers for drug delivery and targeting.

In this chapter, various DDSs will be described. Biological events and processes influencing drug targeting will also be discussed. This chapter will provide the reader with an insight into the rapid developments in the area of drug delivery and targeting.

14.2 TARGETED DRUG DELIVERY

Drug delivery in a selective manner to a biological target, such as organ, tissue, cells, or intracellular organelles, is called targeted drug delivery. Targeted drug delivery is differentiated from target-based drug development or drug targets, which are defined as the molecular targets that the drugs modulate for their pharmacological action. Drug therapy that aims to utilize drug molecules that target a specific protein or receptor for their action is called targeted drug therapy. Targeted drug delivery, on the other hand, refers to the science and technology of presenting a drug to its site of action. The overall goal of all drug-targeting strategies is the improvement of efficacy and/or safety profile of a drug substance.

Targeted drug delivery can involve either drug delivery to a specific organ or tissue or avoiding drug delivery to a specific organ, tissue, or cells. Targeted drug delivery to a particular physiological location can bring the drug to its primary site of action. Thus, it can help improve the efficacy of a drug or prevent its undesired toxicities in other tissues or organs. In addition, sometimes, targeted strategies are intended to avoid drug exposure to a specific organ or tissue.[1] This can help avoid specific drug-related toxicities in particular organs, such as the kidney. For example, intravenous (IV) injection of liposomal doxorubicin has lower nephrotoxicity and cardiotoxicity than IV injection of doxorubicin solution.

The most significant advantages of targeted drug delivery are realized in acute disease states, for example, targeting cytotoxic anticancer drugs to a specific organ (e.g., brain, lungs, liver, kidney, and colon) or the tumor tissue. For example, prodrugs of doxorubicin have been prepared with folate ligand conjugated through bovine serum albumin or polyethylene glycol (PEG), which enable targeting of the drug to the tumors that express folate receptors. Two important design elements of targeted DDSs are (a) the selection of the target organ or tissue and (b) the selection of the targeting strategy.

- The selection of target organ or tissue is governed by the pharmacological need of the disease state and the drug substance. For example, drugs are targeted to the blood–brain barrier (BBB) for drug delivery to the brain for neurodegenerative diseases such as Alzheimer's disease.
- The selection of the targeting strategy for the DDS is governed by the pathophysiology of the target tissue and how it can be utilized to impart stimuli-responsive physicochemical property changes in the DDS. For example, leaky vasculature of the tumor tissue can be utilized for passive drug targeting by designing a DDS that is smaller in particle size and thus can extravasate to the tumor site. In addition, expression of specific biochemical receptors on cell surface of tumor tissues can be utilized for active targeting of the DDS to tumor cells.

Several drug-targeting approaches have successfully transitioned from the proof of concept to the clinical application and have become a state of the art. Examples of targeted drug delivery platforms that have become well accepted in the clinical practice include the following:

- Enteric coating of oral solid dosage forms to overcome chemical instability against acidic pH of the gastrointestinal (GI) tract or adverse effects of the drug in the gastric environment

- Pulmonary drug delivery by dry powder inhalation
- Ocular inserts for drug delivery to the surface of the eye
- Transdermal and implantable DDSs for sustained systemic absorption or local drug delivery.[1]

In addition, several drug delivery strategies being explored are at different preclinical and clinical stages of advancement. Targeted delivery of small and macromolecular drugs has been discussed in depth in a recent book.[2] In this chapter, we will describe different drug-targeting strategies and the role of physicochemical properties of the DDS, combined with the disease mechanism and tissue physiology, in the identification of target as well as a targeting strategy and vehicle.

The design and development of targeted drug delivery agents are based on the biological principles of physiological differences in target tissues compared with other organs or tissues that can be utilized for targeting approaches. These biological differences are matched to the physicochemical principles of drug release from the DDS that targets its drug cargo to a specific organ or tissue.[3] Such a DDS can be exemplified by the drug carriers that include stimuli-responsive polymers that demonstrate a significant change in their properties with relatively minor change in an environmental physicochemical stimulus. The environmental physicochemical or mechanical stimuli that can cause response in the stimuli-responsive DDSs are exemplified by the following:

- pH
- Ionic strength or osmolarity
- Light
- Heat
- Electricity
- Ultrasound
- Oxidation–reduction (redox) potential

14.3 PRODRUGS

A prodrug is formed by chemical modification of a biologically active drug that will liberate the active compound *in vivo* by enzymatic or hydrolytic cleavage. The objective of employing a prodrug is to increase drug absorption and to reduce side effects. Therefore, a prodrug is often classified as a controlled release dosage form. Prodrugs which are more lipophilic than the parent drug can increase membrane penetration and thus drug absorption. For example, phenytoin 2-monoglycerides, a lipophilic phenytoin prodrug, afforded significant increase in oral absorption and bioavailability. The prodrug form can protect the parent compound from hydrolysis or enzymatic attack. A series of ester prodrugs of proprandol protected the drug from first-pass metabolism. An example of a prodrug is enala-pril maleate, which on oral administration is bioactivated by hydrolysis to enaprilat, an angiotensin-converting enzyme (ACE) inhibitor used in the treatment of hypertension.

14.4 SOLUBLE MACROMOLECULAR CARRIERS

Both natural and synthetic water-soluble polymers have been used as macromolecular drug carriers. Soluble carriers include antibodies and soluble polymers such as poly(hydroxypropyl methacrylate), poly(L-lysine), poly(aspartic acid), poly(vinylpyrrolidone), poly(N-vinyl-2-pyrrolidone-co-vinylamide), and poly(styrene co-maleic acid/anhydride). The drug can be attached to the polymer chain either directly or via a biodegradable spacer. Conjugation of a drug to a polymer ensures that this free drug is not available for diffusion to all body systems; hence reducing unintended effects (side effects) and reducing the dose of the drug required for achieving a given concentration at the target site. The spacer overcomes problems associated with the shielding of the drug moiety by the polymer backbone. The spacer allows greater exposure of the drug to the biological milieu, thereby facilitating drug release. Different components of soluble macro-molecular carrier systems are illustrated in Figure 14.3. An example of

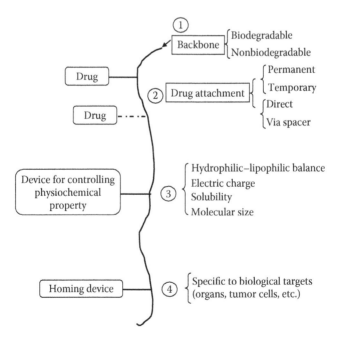

Figure 14.3 Components of a soluble macromolecular carrier system.

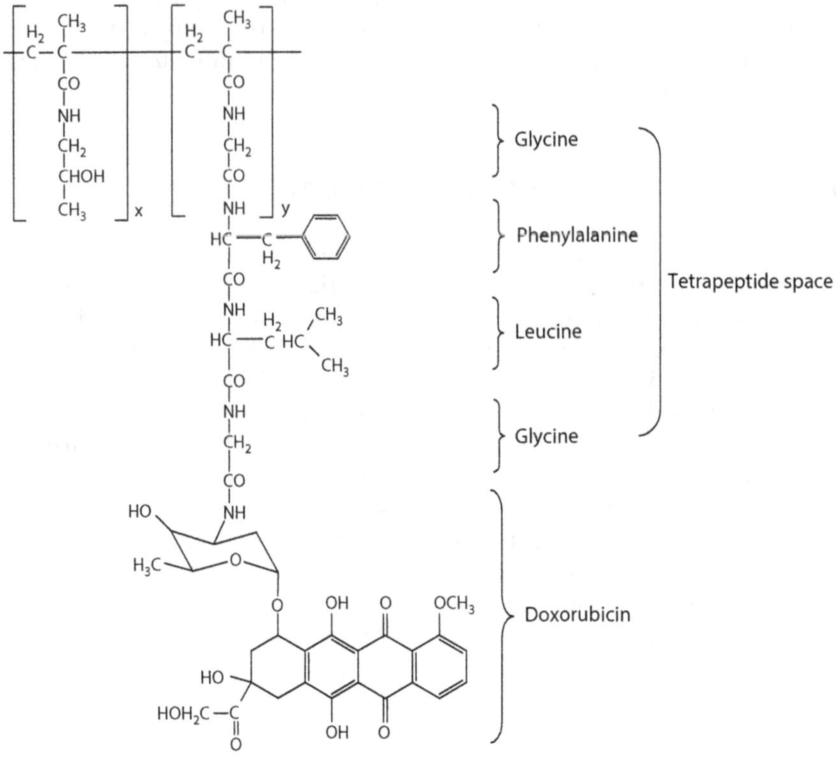

Figure 14.4 Chemical structure of N-(2-hydroxypropyl)methacrylamide (HPMA) copolymer–doxorubicin conjugate.

a biodegradable spacer is the tetrapeptide Gly-Phe-Leu-Gly, which is cleaved by cathepsin B in the lysosomal compartment of cells. For example, N-(2-hydroxypropyl)methacrylamide (HPMA) copolymer has been conjugated to doxorubicin by using this tetrapeptide as a spacer (Figure 14.4).

Attachment of polyethylene glycol (PEG) to proteins can protect them from rapid hydrolysis or degradation within the body and can increase blood circulation time and lower the immunogenicity of proteins. PEGylated forms of interferons, PEG-Intron™ and Pegasys™ (for treating hepatitis C and reducing dosing frequency from daily injections to once-a-week injection dosing), adenosine deaminase, and L-asparaginase are currently on the market. PEGylation improves the solubility and stability of macromolecule by minimizing the uptake by the cells of the RES. Since PEG drug conjugates are not well absorbed from the gut, they are mainly used as injectables.

The drug–polymer conjugate may also contain a receptor-specific ligand to achieve selective access to, and interaction with, the target cells, while decreasing adverse side effects to healthy cells. For example, galactose receptors are present on liver parenchymal cells, thus the inclusion of galactose residues on a drug carrier can deliver the drug to these cells. Similarly, monoclonal antibodies can be used for targeting drug–polymer conjugates to the tumors.

14.5 PARTICULATE CARRIER SYSTEMS

Many particulate carriers have been designed for drug delivery and targeting. These include liposomes, micelles, microspheres, and nanoparticles. In general, particulate carriers are phagocytosed by the macrophages of the mononuclear phagocyte system (MPS), thereby localizing predominantly in the liver and spleen. However, sterically stabilized particulate carriers have extended circulation times. The *in vivo* fate of particulate DDSs depends on the size, shape, charge, and surface hydrophobicity of the particles.

14.5.1 Liposomes

Liposomes are microscopic phospholipid vesicles. The phospholipid usually has a hydrophilic headgroup and two hydrophobic chains. Phosphatidylcholine (PC), a neutral phospholipid, has emerged as the major component used in the preparation of liposomes. Phosphatidylglycerol and phosphatidylethanolamine are also widely used. These lipid moieties spontaneously orient in water to give the hydrophilic headgroup facing out into the aqueous environment and the lipid chains orienting inward, thus avoiding the water phase; this gives rise to bilayer structures. In general, liposomes can be multilamellar vesicles (MLVs), which have diameters in the range of 1–5 μm. Extrusion or sonication of MLVs results in the production of small unilamellar vesicles (SUVs) with diameters in the range of 0.02–0.08 μm. Large unilamellar vesicles (LUVs) can also be made by evaporation under reduced pressure, resulting in liposomes with a diameter of 0.1–1 μm. The bilayer-forming lipid is the essential part of the lamellar structure, while the other compounds are added to impart certain characteristics to the vesicles.

Location of a drug entrapped in the liposomes depends on the drug's solubility characteristics. Water-soluble drugs can be entrapped in liposomes by intercalation in the aqueous bilayers, while lipid-soluble drugs can be entrapped within the hydrocarbon interiors of the lipid bilayers. Liposomes can encapsulate small molecular weight drugs, proteins, peptides, oligonucleotides, and genes. Examples of applications where liposomes have been successfully employed to provide therapeutic benefit include amphotericin B liposomes. The use of the antifungal agent amphotericin B formulated in liposomes has been approved by the Food and Drug Administration (FDA) for treating systemic mycoses.

The rigidity and permeability of the bilayer strongly depend on the type and quality of lipids used. The phospholipids bilayer can be two physical states based on their structural rigidity: gel state or a more fluid state known as the *liquid crystalline* state. Preference for either state depends on various characteristics of lipid components, including (1) the chain length of hydrophobic alkyl chain, (2) the degree of unsaturation of alkyl chain, and (3) the polar headgroup structure. For example, a C_{18} saturated alkyl chain produces rigid bilayers with low permeability at room temperature. The presence of cholesterol also tends to rigidify the bilayers. This state also depends on temperature. Thus, liposomes of a given lipid will assume gel state at a lower temperature, while they become liquid crystalline at a higher temperature.

14.5.1.1 Types of liposomes

Liposomes can be classified into the following categories: conventional liposomes, stealth liposomes, targeted liposomes, and cationic liposomes.

14.5.1.1.1 Conventional liposomes

These are neutral or negatively charged liposomes typically composed of only phospholipids, glycolipids, and/or cholesterol, without derivatization, to increase the circulation time. These liposomes are generally used for passive targeting to the phagocytic cells of the MPS, localizing predominantly in the liver and spleen. Conventional liposomes have also been used for antigen delivery.

14.5.1.1.2 Sterically stabilized (stealth) liposomes

Since conventional liposomes are recognized by the immune systems as foreign bodies, PEG–lipid (commonly known as PEGylated lipid), such as PEG–phosphatidylethanolamine (PEG–PE) is often included in the preparation of liposomes. The PEGylated lipid reduces the uptake of liposomes by MPS or the RES, resulting in prolonged circulation half-life. Simply persisting in the bloodstream, PEGylated liposomes can localize into tumors and most sites of inflammation. For an example, ALZA Corporation developed STEALTH™ liposomes, which evade recognition by the immune system because of their unique PEG coating. Doxil™ is a STEALTH™ liposome formulation of doxorubicin used for the treatment of AIDS-related Kaposi's sarcoma.

14.5.1.1.3 Targeted liposomes

In addition to a PEG coating, most stealth liposomes also have some sort of biological species attached as a ligand to the liposome to enable binding via a specific expression on the targeted drug delivery site. These targeting ligands could be monoclonal antibodies (making an *immunoliposome*), vitamins, or specific antigens. Targeted liposomes can target nearly any cell type in the body and deliver drugs that would naturally be systemically delivered. Naturally toxic drugs can be much less toxic if delivered only to diseased tissues.

14.5.1.1.4 Cationic liposomes

Cationic liposomes are positively charged liposomes and used for nucleic acid delivery. Cationic liposomes interact with the negatively charged phosphate backbone of DNA or RNA, leading to neutralization of the charge. Cationic liposomes are prepared using a cationic lipid and a colipid, such as dioleoylphosphatidylethanolamine (DOPE) or cholesterol. The cationic amphiphiles differ markedly and may contain single multiple charges (primary, secondary, tertiary, or quaternary). The three basic components of cationic lipids include (1) a hydrophobic lipid anchor group, which helps in forming liposomes and can interact with cell membranes; (2) a linker group; and (3) a positively charged headgroup, which interacts with nucleic acids, leading to nucleic acid condensation and charge neutralization. The linker group is an important component that determines the chemical stability and biodegradability of the lipid. The physicochemical properties of liposome/nucleic acid complexes are strongly influenced by the relative proportions of each component and the structure of the headgroup.

14.5.1.2 Fabrication of liposomes

All methods of making liposomes involve three to four basic stages: drying down of lipids from organic solvents (usually chloroform), dispersion of the lipid mixtures in aqueous media, purification of the resultant liposomes, and analysis of the final products. Figure 14.5 illustrates the stages common to different liposome preparation methods. The drying down of large volume of organic solutions is most easily carried out in rotary evaporator, fitted with a cooling coil and water bath. Rapid evaporation of solvents is carried out by gentle warming (20°C–40°C) under pressure (400–700 mmHg). The main difference between the various methods of preparation is in the way in which the membrane components are dispersed in aqueous media. The following methods are often used for dispersion of lipid membrane components on hydration and agitation: extrusion, mechanical dispersion, microfluidization, sonification, detergent dialysis, and ethanol injection. Hydrated lipid solutions will initially form large, MLVs. After the initial pass through an extrusion membrane, the particle size distribution will tend toward a bimodal distribution. After sufficient passes through the membrane, a unimodal, normal distribution is

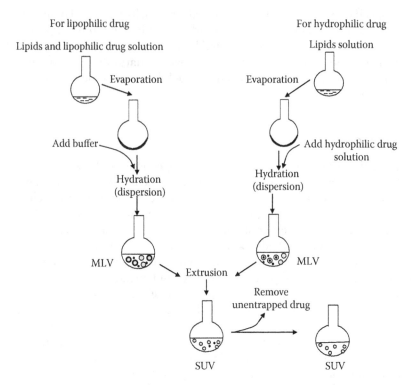

For lipophilic drug

Lipids and lipophilic drug solution

Evaporation

Add buffer

Hydration
(dispersion)

MLV

For hydrophilic drug

Lipids solution

Evaporation

Add hydrophilic drug
solution

Hydration
(dispersion)

MLV

Extrusion

Remove
unentrapped drug

SUV

SUV

Figure 14.5 Stages common to different liposome preparation methods.

obtained. Bath or probe ultrasonicators are also used to prepare liposomes from hydrated lipid films.

Microfluidizer is also used for preparation of liposomes from concentrated lipid suspensions. The microfluidizer is a machine that pumps fluid at very high pressure (10000 psi, which is 600–700 bar) through a 5-µm filter, after which it is forced along microchannels, which then direct the two streams of fluid to collide together at right angles at a very high velocity. The fluid collected can be recycled through the pump and interaction chamber until vesicles of the required dimensions are obtained.

In the *ethanol injection method of liposome preparation*, an ethanol solution of lipids is injected rapidly into an excess of saline or other aqueous medium, through a fine needle. This procedure can yield a high proportion of SUV of 25–50 nm; however, lipid aggregates and larger vesicles may form if the mixing is not thorough enough. The major limitations of ethanol injection method of liposome preparation are (1) poor solubility of lipids in ethanol (40 mM for PC), (2) volume of ethanol that can be introduced into the medium (7.5% v/v maximum), (3) poor drug encapsulation efficiency, and (4) difficulty in removal of ethanol from phospholipid membranes.

14.5.2 Microparticles and nanoparticles

A *microcapsule* has drug located centrally within the particle, where it is encased within a unique polymeric membrane. The core can be solid, liquid, or gas, and the envelope is made of a continuous, porous or nonporous polymeric phase. A drug can be dispersed inside the polymeric envelope as solid particulates or dissolved in solution, emulsion, suspension, or combination of both emulsion and suspension. In contrast, a *microsphere* has its drug dispersed throughout the particle; that is, the internal structure is a matrix of drug and polymeric excipient (Figure 14.6). Small molecular weight drugs, proteins, oligonucleotides, and genes can be encapsulated into microparticles to provide their sustained release at disease sites. A *microcapsule* is a reservoir-type system in which drug is located centrally within the particle, whereas a *microsphere* is a matrix-type system in which drug is dispersed throughout the particle. Microcapsules usually release their drug at a constant rate (zero-order release), whereas microspheres typically give a first-order release of drugs.

14.5.2.1 Fabrication of microparticulates

Microencapsulation is a technique that involves the encapsulation of small particles or solution of drugs in a polymer film or coat. Different methods of microencapsulation result in either microcapsules or microspheres. The most common methods of preparing microparticles and nanoparticles are emulsion and interfacial polymerization, and coacervation.

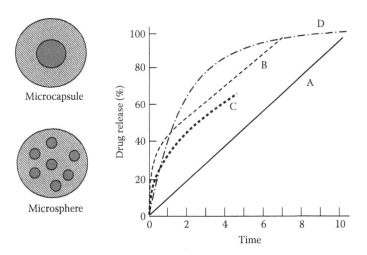

Figure 14.6 Schematic representation and drug-release profiles of microspheres and microcapsules: (a) microcapsules free of burst effect, (b) microcapsules with burst effect, (c) microspheres with square-root time release, and (d) microspheres with first-order release.

14.5.2.1.1 Emulsification

The first step in almost any microencapsulation technique involves the formation of an emulsion, usually of a polymeric solution inside a continuous phase. Similarly, to disperse nonsoluble drugs inside polymeric solution, emulsions must be created. Thus, a thorough understanding of emulsion formation and properties is extremely important. The emulsion formation determines the resulting particle size in the final process of encapsulation. An emulsion is achieved by applying mechanical force, which deforms the interface between the two phases to such an extent that droplets form. These droplets are typically large and are subsequently disrupted or broken into smaller ones. The ability to disrupt the larger droplets is a critical step in emulsification and in encapsulation, where an emulsion is prepared. The size of the oil-phase droplets obtained is determined by how rapidly the system is agitated when the oil phase is added to the aqueous phase; it also determines the size of the microparticles produced.

However, protein and nucleic acid drugs are fairly labile and can be destroyed due to the application of mechanical shear, and thus, preventive measures should be taken to stabilize these drugs during emulsification process. A suitable surfactant is needed to produce a stable emulsion, a result achieved by lowering the surface tension. Devices commonly used for production of emulsions are the following:

- Ultrasonicator
- Homogenizer
- Microfluidizer
- Injection
- Stirring
- Many more

Albumin and some other water-soluble proteins can be used to prepare microspheres, involving the formation of a water-and-oil (w/o) emulsion and stabilization of the protein by cross-linking, using glutaraldehyde or heat denaturation. A mixture of petroleum ether and cottonseed oil (60:40) containing 0.5% v/v Span® 80 can be used as a *continuous phase* (~100 mL). Serum albumin is dissolved in phosphate-buffered saline (PBS) containing 0.1% w/v sodium dodecyl sulfate (~1 mL). The albumin solution is added dropwise to the continuous phase, stirred with 2,500 rpm with a homogenizer. After 1 hour of mixing, *glutaraldehyde* solution (100 μL, 5%–12%) is added dropwise to the w/o emulsion, which is stirred for 1 hour at room temperature to allow cross-linking. Alternatively, microspheres can be stabilized by *heat denaturation* at 100°C–120°C. Following stabilization, microspheres are freed of oil by washing with petroleum ether (x3) and isopropanol (x2); they are then suspended in PBS and stored at 4°C until required. Biodegradation of albumin microspheres and drug-release rate are dependent on the concentration of glutaraldehyde concentration or degree of heat denaturation. Apart from albumin, other proteins such as *hyaluronidase* and chitosan can also be used for preparation of microspheres by using cross-linkers.

14.5.2.1.2 Solvent evaporation

Solvent evaporation is the most popular method of preparation of microparticles. A core material and capsule wall material are dissolved in a *water-immiscible, volatile organic solvent,* and the resulting solution is emulsified in an aqueous solution. The solvent is allowed to evaporate, thereby producing solid microcapsules or microparticles. Another way is to form a double emulsion, where an aqueous core material solution is emulsified in a polymer-volatile organic solvent solution. The resulting emulsion is emulsified in water, giving a double emulsion. Evaporation of the volatile solvent yields a solid microcapsule with an aqueous core. *Methylene chloride* (CH_2Cl_2) is a preferred solvent because of its volatility (boiling point, 41°C) and its capacity of dissolving broad range of polymers. Chloroform and ethyl acetate can also be used. A mixture of methylene chloride (a water-immiscible solvent) and acetone (a water-miscible solvent) can also be used.

The added drug may be completely dissolved in the polymer solution, or it may be completely insoluble and simply form a dispersion, suspension, or suspension–emulsion. In the latter case, the solid particles must be micronized, so that their mean diameter is much less than the desired mean microsphere size. To aid emulsification, a surfactant is normally dissolved in the water phase before the oil-in-water emulsion is formed. A good example is partially hydrolyzed (88%) poly(vinyl alcohol) (PVA).

After obtaining desired droplet size and emulsion stability, the system is stirred at constant rate, followed by solvent evaporation by using a rotary evaporator. Following solvent evaporation, the microparticles are separated from the suspending medium by filtration or centrifugation, washed, and dried. The maximum drying temperature must remain below the glass temperature of the polymer encapsulant or the microspheres fuse together. Although the solvent evaporation process is conceptually simple, the nature of the product can be affected by the following factors:

- Polymer molecular weight and concentration
- Polymer crystallization
- Type of drug and method of incorporation (solid, liquid, and suspension)
- Organic solvent used
- Type and concentration of surfactant used in the aqueous phase
- Ratio of organic phase to aqueous phase
- Rate of stirring
- Evaporation temperature

In general, semicrystalline polymers often give porous structures, with spherulites on the surface of the microspheres. Uniform, pore-free spheres are most readily obtained with amorphous polymers.

Biodegradable polylactide (PLA) and its copolymer with glycolide (polylactide-co-glycolide [PLGA]) are commonly used for preparing microparticles, from which the drug can be released slowly over a period of a month or so. Microspheres can be used in a wide variety of dosage forms, including tablets, capsules, and suspensions. Table 14.1 lists some of the FDA-approved commercial products of microspheres. *Lupron Depot* from TAP Pharmaceuticals is an FDA-approved preparation of PLGA microspheres for sustained release of a small-peptide luteinizing hormone-releasing hormone (LHRH) agonist. More recently, PLGA microspheres of recombinant human growth hormone have been developed and marked successfully by Genentech, Inc., under the trade name of Nutropin Depot. Polylactide-co-glycolide degrades into lactic and glycolic acids.

14.5.2.1.3 Interfacial (or in situ) polymerization

In *interfacial polymerization*, oil-soluble monomers and water-soluble monomers react at the water/oil interface of w/o or o/w dispersions, resulting in the formation of polymeric microcapsules. The process involves an initial emulsification step, in which an aqueous phase, containing a reactive monomer and a core material, is dispersed in a nonaqueous phase. This is followed by the addition of a second monomer to the continuous phase. Monomers in the two phases then diffuse and polymerize at the interface to form a thin film. The most widely used example of microcapsule preparation using this method is the interfacial polymerization of water-soluble alkyl-diamines with oil-soluble acid dichlorides to form polyamides. Examples of other polymeric wall materials include polyurethanes, polysulfonamides, polyphthalamides, and poly(phenyl esters). Interfacial polymerization of a monomer almost always produces microcapsules, whereas solvent evaporation may result in microspheres or microcapsules, depending on the amount of drug loading.

14.5.2.1.4 Complex coacervation

Complex coacervation uses the interaction of two oppositely charged polyelectrolytes in water to form a polymer-rich coating solution called a complex coacervate. This solution (or coacervate) engulfs the liquid or solid being encapsulated, thereby forming an embryo capsule. Cooling the system causes the coacervate (or coating solution) to gel via network formation. Gelatin and gum arabic are primary components of most complex coacervation systems. Coacervation uses the common phenomenon of *polymer–polymer incompatibility* to form microcapsules. The first step is to form a solution of gelatin in deionized water at 11 wt% and 45°C–55°C. Once the gelatin and gum arabic solutions are prepared, the drug is emulsified or dispersed in the 45°C–55°C gelatin solution. Once the drug–gelatin emulsion or dispersion is formed, it is diluted by addition of a known volume of the 45°C–55°C deionized water and 11wt% gum arabic solution (45°C–55°C). The pH of the resulting mixture is adjusted to 3.8–4.4 by addition of acetic acid. After the pH is adjusted, the system is allowed to cool down to room temperature and then to below 10°C, and at this point, glutaraldehyde is slowly added to cross-link the polymer. The system is stirred gently throughout this cooling period.

Alginates form gels on reaction with calcium salts: These gels consist of almost 99% of water and 1% or less of alginate. Cross-links are caused either by simple ionic bridging of two carboxyl groups on adjacent polymer chains via calcium ions or by chelation of single calcium ions by hydroxyl and carboxyl groups on each of the pair of polymer chains. Several types of viable cells (erythrocytes, sperm cells, hepatoma cells, and hepatocytes), tissues (pancreatic endocrine tissues and islets), and other labile biological substances are encapsulated within semipermeable alginate microspheres. The process involves suspending the living cells or tissues in sodium alginate solution, and the suspension is then extruded to produce microdroplets, which fall into a $CaCl_2$ solution and form gelled microbeads with the cells or tissues entrapped. These microbeads are next treated with polylysine solution, which displaces the surface layer of calcium ions and forms a permanent polysalt shell or membrane.

Porous microspheres are formed by gelation of the following:

- Sodium alginate and chitosan
- Sodium alginate and $CaCl_2$
- Sodium alginate and polylysine

14.5.2.1.5 Hot melt microencapsulation

In hot melt microencapsulation, melted polymers are mixed with drugs, and the mixture is then suspended in an immiscible solvent that is heated at 5°C above the melting point of the polymer and stirred continuously. Once the emulsion is stabilized, it is cooled until the core material has solidified. The solvents used in this process are usually silicon and olive oils. After cooling, microspheres are washed with petroleum ether to have free-flow powders. To avoid degradation of drugs due to heat, polymers with low melting points are used in this process. Robert Langer and his associates used this method to prepare *polyanhydride* microspheres.

14.5.2.1.6 Solvent removal

In the solvent-removal process, the fabrication occurs at the room temperature totally in organic solvents, which is good for hydrolytically labile polymers such as polyanhydrides and water-soluble drugs. Mathiowitz and Langer used this method to encapsulate zinc insulin into polyanhydride, such as poly(carboxyphenoxypropane-co-sebacic acid) (poly(CPP-SA), 50:50, microspheres. In this example, the polymer was dissolved in methylene chloride, the desired amount of the drug was added, and then the mixture was suspended in silicon oil containing Span 85. Petroleum ether was then added, and the mixture was stirred until methylene chloride was extracted into the oil solution and sufficient microcapsule hardening was achieved. The microspheres were isolated by filtration, washed with petroleum ether, and dried overnight under vacuum. The solvent-removal process is somewhat different from organic-phase separation (or coacervation) process. In solvent removal, the polymeric solution is introduced into the continuous phase, an emulsion is formed first, and then the organic solvent is extracted into the continuous phase. In the organic-phase separation, the polymer is dissolved in the continuous phase, a phase inducer is introduced, a coacervate is formed, and, finally, drug encapsulation occurs.

14.5.2.1.7 Spray drying

In this process, polymers are dissolved in a volatile solvent, such as methylene chloride. The spray-drying process involves dispersion of the core material in a solution of coating substance and spraying the mixture into an environment that causes the solvent to evaporate. This method is often used to encapsulate heat-sensitive drugs in polyanhydride microspheres.

14.5.3 Nanoparticles

Nanoparticles are solid colloidal particles ranging in size from 10 to 1,000 nm. Depending on the fabrication process, two different types of nanoparticles can be obtained, namely nanospheres and nanocapsules. Nanospheres have a matrix-type structure, in which a drug is dispersed, whereas nanocapsules exhibit a membrane-wall structure, with an oily phase containing the drug. Because these nanoparticles have very high surface areas, drugs may also be adsorbed on their surface. Biodegradable nanoparticles from poly(lactic acid)-poly(glycolic acid) (PLGA), polycaprolactone, and polyalkylcyanoacrylates have been widely studied. Gelatin nanoparticles are also used, which are prepared by desolvation of a gelatin solution containing drug bound to the gelatin. Hardening of the gelatin nanoparticles is achieved by glutaraldehyde, which cross-links with gelatin and is more efficient than formaldehyde.

14.6 ORAL DRUG DELIVERY

The preferred route of administration for pharmaceutical products has been oral ingestion. As a drug passes through the GI tract, it encounters different environments with respect to pH, enzymes, electrolytes, fluidity, and surface features, all of which can influence drug absorption. There is a great variation in the pH across the GI tract, which runs from the mouth to the anus. The interdigestive migration of a drug or a dosage form is governed by GI motility, wherein the drug is exposed to different pHs at different time periods. The stomach has an acidic pH, varying from 2 to 4. The acidic pH in the stomach increases up to a pH of 5.5 at the duodenum. The pH then increases progressively from the duodenum to the small intestine (a pH of 6–7) and reaches a pH of 7–8 in the distal ileum. After the ileocecal junction, the pH falls sharply to 5.6 and then climbs up to neutrality during transit through the colon. Owing to the pH variation in the GI tract, pH-sensitive polymers have been historically utilized as an enteric-coating material. Enteric-coated products featuring pH-sensitive polymers include tables, capsules, and pellets and are designed to keep an active substance intact in the stomach and then to release it into the upper intestine.

Apart from the pH, mucosal layer plays an important role in drug absorption from the lumen of the GI tract. Small intestine has large epithelial surface area, which consists of mucosa, villi, and microvilli. Drug must first diffuse through the unstirred aqueous layer, the mucous layer, and the glycocalyx (which is the coating of the mucous layer) in order to reach the microvilli, which is the apical cell membrane. The tight junction between the cell membranes of adjacent epithelial cells acts as a major barrier to the intercellular passage of drug molecules from the intestinal lumen to the lamina propria.

The low oral bioavailability of peptide and protein drugs is primarily due to their large molecular size and vulnerability to proteolytic degradation in the GI tract. Most protein and peptide drugs are susceptible to rapid degradation by digestive enzymes. Furthermore, most peptide and protein drugs are rather hydrophilic and thus are poorly partitioned into epithelial cell membranes, leading to their absorption across the GI tract through passive diffusion.

Various delivery systems have been proposed to increase drug absorption from the colon and ileum and minimize exposure of the drug to proteolytic enzymes. Enteric coatings that delay drug release for a sufficient period of time have been used to target both the ileum and colon. In addition, encapsulation into polymeric materials that are degraded by the human colonic microflora has been proposed as a method to increase drug absorption from the intestine. Coadministration of enzyme inhibitors and absorption enhancers has shown some promise. Encapsulation into erodible or biodegradable nanoparticles has been explained as a way of protecting drugs from enzymatic degradation. Submicron-size particles are absorbed through transcytosis by both enterocytes and M cells.

14.7 ALTERNATIVE ROUTES OF DELIVERY

For systemic action of drugs, the oral route has been the preferred route of administration. However, when administered by the oral route, many therapeutic agents are subjected to extensive presystemic elimination by GI degradation and/or hepatic metabolism. Delivery of drugs via the absorptive mucosa in various easily accessible body cavities (Figure 14.7), like the buccal, nasal, ocular, sublingual, rectal, and vaginal mucosae, offers distinct advantages over peroral administration for systemic drug delivery, since these alternative routes of drug delivery avoid the first-pass effect of drug clearance.

14.7.1 Buccal and sublingual drug delivery

The buccal and sublingual mucosae in the oral cavity provide an excellent alternative for the delivery of certain drugs. Oral transmucosal absorption is generally rapid because of the rich vascular supply to the mucosa. These routes provide improved delivery for certain drugs that are inactivated by first-pass intestinal/hepatic metabolism or by proteolytic enzymes in the GI tract.

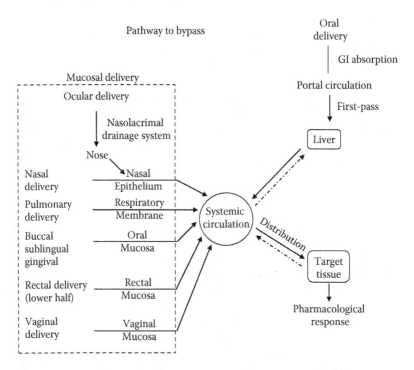

Figure 14.7 Various mucosal routes that bypass hepatic first-pass metabolism associated with oral administration.

The sublingual mucosa is relatively permeable and is suitable for delivery of low-molecular-weight lipophilic drugs when a rapid onset of action with infrequent dosing is required. Sublingual DDSs are generally of two different designs: (a) rapidly disintegrating tablets and (b) soft gelatin capsules filled with a drug in solution. Such systems create a very high drug concentration in the sublingual region before the drug is systemically absorbed across the mucosa. Therefore, rapidly disintegrating sublingual tablets are frequently used for prompt relief from an acute angina attack.

The buccal mucosa is considerably less permeable than the sublingual area and is generally not able to provide rapid absorption properties. The buccal mucosa has an expanse of smooth muscle and relatively immobile mucosa, which makes it a more desirable region for retentive systems used for oral transmucosal drug delivery. Thus, the buccal mucosa is suitable for sustained delivery applications, delivery of less permeable molecules, and perhaps peptide drugs. One of the major disadvantages associated with buccal drug is the low flux, which results in low drug bioavailability. Therefore, buccal DDSs usually include a penetration (permeability enhancer) to increase the flux of drugs through the mucosa. Another limitation associated with this route of administration is the poor drug retention at the site of absorption. Consequently, bioadhesive polymers have been extensively employed in buccal DDSs. The duration of mucosal adhesion depends on the type and viscosity of the polymer used. Nicotine in a gum vehicle when chewed is absorbed through the buccal mucosa. Glyceryl trinitrite has been found quite effective when administered through this route.

14.7.2 Nasal drug delivery

Although nasal route is traditionally used for locally acting drugs, this route is getting more attention for the systemic delivery of various peptide drugs that are poorly absorbed via the oral route. The major advantages of nasal administration include the fast absorption, rapid onset of action, and avoidance of hepatic and intestinal first-pass effects. There are three major barriers to drug absorption across nasal mucosa. These include a physical barrier composed of the mucus and epithelium, a temporal barrier controlling the mucosal clearance, and an enzymatic barrier acting principally on protein and peptide drugs. The physical barrier consists of a lipoidal pathway and an aqueous pore pathway. Nasally administered drugs have to pass through the epithelial cell layer to reach the systemic circulation. Nasal absorption of weak electrolytes is dependent on the degree of ionization, with higher nasal absorption of a drug at a pH lower than its pK_a.

Dosage forms for nasal absorption must deposit and remain in the nasal cavity long enough to allow effective absorption. Commonly used dosage forms administered through this route are nasal sprays and drops. The nasal spray deposits drug in the proximal part of the nasal atrium, whereas nasal drops are dispersed throughout the nasal cavity. A nasal spray requires that the particles have a diameter larger than 4 μm to be retained in the nose and to minimize the passage into the lungs. Nasal sprays are commercially available for buserelin, desmopressin, oxytocin, and calcitonin.

14.7.3 Pulmonary drug delivery

The respiratory tract includes the nasal mucosa, hypopharynx, and large and small airway structures (trachea, bronchi, bronchioles, and alveoli). This tract provides a large mucosal surface for drug absorption. Lung epithelium is highly permeable and has low metabolic activity compared with the liver and intestine. With a large surface area and highly permeable membrane, alveolar epithelium permits rapid absorption. This route of administration is useful for treating pulmonary conditions and for drug delivery to other organs via the circulatory systems. In general, lipid-soluble molecules are absorbed rapidly from the respiratory tract, and thus, an increasing number of drugs is being administered by this route, including bronchodilators (e.g., beclometasone), corticosteroids, antibiotics, antifungal agents, antiviral agents, and vasoactive drugs.

Since the lung has a large surface area and a highly permeable membrane, the lung is an ideal site for absorption of macromolecules, such as proteins, peptides, oligonucleotides, and genes. For example, DNase alpha (Pulmozyme®, Genentech), an enzyme used to reduce the viscosity of mucus in the airways of patients with cystic fibrosis, is most effective when inhaled. This protein is thus delivered directly to its site of action by nebulization. The recent approval of inhaled human insulin by the FDA for use in diabetes mellitus stands as a major advancement in the field of pulmonary delivery of macromolecules and systemically acting drugs.

14.7.4 Ocular drug delivery

Drugs are usually topically applied to the eyes in the form of drops or ointments for local action. Following topical administration, the drug is eliminated from the eye by nasolacrimal drainage, tear turnover, productive corneal absorption, and nonproductive conjunctival uptake. There are two barriers to ocular drug adsorption: (a) the blood–aqueous barrier and (b) the blood–retina barrier. The blood–aqueous barrier is composed of the ciliary epithelium, the epithelium of the posterior surface of the iris, and blood vessels within the iris. Drugs enter the aqueous humor at the ciliary epithelium and at blood vessels. Many substances are transported out of the vitreous humor at the retinal surface.

The cornea and the conjunctiva are covered with a thin film, the tear film, which protects the cornea from dehydration and infection. For drugs administered through the topical route, the cornea is the main barrier to drug absorption. The cornea consists of three parts: the epithelium, the stroma, and the endothelium. Both the endothelium and the epithelium have high lipid content and thus are penetrated by drugs in their union-ized lipid-soluble forms. The stroma lying between these two structures has a high water content, and thus, drugs that have to negotiate the corneal barrier successfully must be both lipid soluble and water soluble to some extent.

Ocular drug absorption depends on both drug ionization and tear turn-over. For example, the pH 5 solution induces more tear flow than the pH 8 solution, thus the concentration gradient is reduced, and transport of both ionized and nonionized drugs is less at pH 5. The duration of drug action in the eye can be extended by two approaches: (1) by reducing drainage using viscosity-enhancing agents, suspensions, emulsions, ointments, and polymeric matrices and (2) by improving corneal drug penetration using ionophores and liposomes.

Prodrug derivatization can be employed to overcome low corneal per-meability of water-soluble drugs. The drug molecules can be chemically modified to obtain suitable structural configuration and physicochemical properties to afford maximal corneal adsorption. However, a prodrug must be converted enzymatically or chemically to the parent drug in vivo to elicit its effect. Choline esterases, which are abundant in the corneal epithelium, can be used for delivery of more lipophilic esterified prodrugs of water-soluble compounds to the eye.

14.7.5 Rectal drug delivery

Rectal administration provides rapid absorption of many drugs and is an alternative when oral administration is inconvenient because of the inability to swallow or because of GI side effects such as nausea, vomit-ing, and irrigation. More importantly, rectal drug administration has the advantage of minimizing or avoiding hepatic first-pass metabolism. The rectal bioavailability of lidocaine in human is 65%, as compared with an oral bioavailability of 30%. Rectal route is used to administer diazepam to children who are suffering from epileptics, in whom it is difficult to estab-lish IV access. However, rectal administration of drugs is inconvenient and has irregular drug absorption. Moreover, rectal administration should be avoided in immunosuppressed patients in whom even minimal trauma could lead to the formation of an abscess.

14.7.6 Vaginal drug delivery

Vaginal epithelium is permeable to a wide range of substances, including steroids, prostaglandins, antibiotics, estrogens, and spermicidal agents. Most steroids are readily absorbed by vaginal epithelium, leading to their higher bioavailability compared with their oral administration, because of a reduced first-pass metabolism. For drugs with high membrane permeability, vaginal absorption is determined by permeability of the aqueous diffusion layer, whereas for drugs with low membrane permeability, such as testosterone and hydrocortisone, vaginal absorption is determined by membrane permeability. Vaginal ointments and creams contain drugs such as anti-infectives, estrogenic hormone substrates, and contraceptive agents. Contraceptive creams contain spermicidal agents and are used just before intercourse.

REVIEW QUESTIONS

14.1 The solution instilled as eye drops into the ocular cavity may disappear from the precorneal area of the eye by which of the following route(s):
A. Nasolacrimal drainage
B. Tear turnover
C. Corneal absorption
D. Conjunctival uptake

14.2 After oral drug delivery, drugs are absorbed in the gastrointestinal tract, and through the portal circulation, they enter the liver, where they are destroyed by so-called:
A. Second-pass metabolism
B. Drug efflux metabolism
C. First-pass metabolism
D. Drug decomposition
E. None of the above

14.3 The lung:
A. Has a highly permeable membrane
B. Has a membrane that provides an effective barrier to drug absorption
C. Provides easy access to the bloodstream
D. None of the above

14.4 Which layer is the major rate-limiting barrier for permeation of hydrophilic drugs across the cornea?
A. Endothelial layer
B. Stroma
C. Epithelial layer
D. A and C

14.5 Liposomes containing an anticancer drug are rapidly taken up by the cells of reticuloendothelial system on systemic administration. How can one extend the blood circulation time of this liposomal system?

14.6 Define polymeric micelles and liposomes. What is a common feature of these two carrier systems?

14.7 Why oral delivery of protein and peptide drugs is often not preferable?

14.8 How can a matrix system be differentiated from a reservoir system?

14.9 What are Peyer's patches? How can they be exploited in drug delivery and targeting?

REFERENCES

1 Narang AS, and Mahato RI (2010) Targeting colon and kidney: Pathophysiological determinants of design strategy. In Narang AS, and Mahato RI (Eds.) *Targeted Delivery of Small and Macromolecular Drugs*, Boca Raton, FL: CRC Press, pp. 351–370.

2. Narang AS, and Mahato RI, (Eds.) (2010) *Targeted Delivery of Small and Macromolecular Drugs*, Boca Raton, FL: CRC Press, p 614.

3. Narang AS, and Varia S (2011) Role of tumor vascular architecture in drug delivery. *Adv Drug Deliv Rev* **63**(8): 640–658.

FURTHER READING

Crommelin DJ and Storm G (2003) Liposomes: From the bench to the bed. *J Liposome Res*. **13**: 33–36.

Goldberg M and Gomez-Orellana I (2003) Challenges for the oral delivery of macromolecules. *Nature Rev Drug Discov* **2**: 289–295.

Hiller AM et al. (Eds.) (2001) *Drug Delivery and Targeting*, New York: Taylor and Francis

Kopecek J, Kopeckova P, Minko T, and Lu ZR (2000) HPMA copolymer-anticancer drug conjugates: Design, activity, and mechanism of action. *Eur J Pharm Biopharm* **50**: 61–81.

Kreuter J (1994) *Colloidal Drug Delivery Systems*. Marcel Dekker, New York.

Leach C (2005) Inhaled insulin gets a positive recommendation from the PDA advisory panel: The door opens wider for the future of inhaled drugs. *AAPS News Magazine*, December, p. 20.

Li VHK, Robinson JR, and Lee VHL (1987) Influence of drug properties and routes of drug administration on the design of sustained and controlled release systems. In: Robinson JR and Lee VHL (Eds.) *Controlled Drug Delivery*, 2nd ed., New York: Marcel Dekker, pp. 3–94.

Lu ZR, Kopeckova P, and Kopecek J (1999) Polymerizable Fab' antibody fragments for targeting of anticancer drugs. *Nat Biotechnol* **17**: 1101–1104.

Mathiowitz E, Kretz MR, and Bannon-Peppas L (1999) Microencapsulation. In *Encyclopedia of Controlled Drug Delivery*, New York: John Wiley & Sons, pp. 493–546.

Moghimi SM, Hunter AC, and Murray JC (2001) Long-circulating and target-specific nanoparticles: Theory to practice. *Pharmacol Rev* **53**: 283–318.

Poznansky MJ and Juliano RL (1984) Biological approaches to the controlled delivery of drugs: A critical review. *Pharmacol Rev* **36**: 277–336.

Tomlinson E (1987) Theorpy and practice of site-specific drug delivery. *Adv Drug Deliv Rev* **1**: 87–198.

Chapter 15

Organ-specific drug delivery

LEARNING OBJECTIVES

On the completion of this chapter, the students should be able to

1. Define targeted drug delivery and its advantages.
2. Describe how targeted drug delivery improves the therapeutic efficacy.
3. Discuss the physiological considerations for drug targeting to different organs and cells.
4. Describe the principles of targeted drug delivery to different organs.
5. Discuss the physicochemical principles involved in designing a targeted drug delivery system (DDS).
6. Describe the drug delivery platforms utilized for targeted drug delivery.

15.1 INTRODUCTION

Drug delivery in a selective manner to a biological target, such as an organ, tissue, cells, or intracellular organelles, is called *targeted drug delivery*. Targeted drug delivery is achieved by ensuring high drug concentration in a target organ or tissue through the systemic circulation. In other words, a drug is delivered systemically or is absorbed into the systemic circulation, first—before accumulating at the target site of action. This modality is exemplified by the intravenous administration of a liposomal delivery system of a cytotoxic drug such that the drug distribution into the kidney is avoided, thus minimizing the renal side effects of the drug.

This modality is distinguished from *localized drug therapy*, wherein the drug is not intended to reach systemic circulation. The drug is administered or applied locally for local action. This modality is exemplified by localized application of an antibiotic on a skin laceration or infection.

A third modality is the utilization of different organs or organ systems to enable drug absorption into the systemic circulation. For example, transdermal or sublingual drug delivery is intended for drug absorption into the systemic circulation. These aspects are generally covered under *alternate routes of drug delivery.*

Often, however, route of drug administration or organ for drug delivery may be utilized for drug delivery for localized action as well as drug delivery to the systemic circulation, depending on the pathological condition and the therapeutic need. For example, pulmonary drug administration of steroids is utilized for localized anti-inflammatory action, whereas pulmonary insulin administration is intended for systemic delivery. There are also cases where parenteral nanoparticulate drug administration is targeted for drug delivery to the lung.

Therefore, this chapter provides an integrated discussion of various organs and tissues utilized either as alternate routes for drug delivery to the systemic circulation or as organ or tissue drug targets, either by local or after systemic administration.

15.2 NEED FOR ALTERNATE, NONORAL ROUTES TO THE SYSTEMIC CIRCULATION

Delivery of drugs via the absorptive mucosa in various easily accessible body cavities, such as the buccal, nasal, ocular, sublingual, rectal, and vaginal mucosae are pursued when it offers advantages over peroral administration for systemic drug delivery, since the preferred route of administration for pharmaceutical product has been oral ingestion. The need for alternate routes of drug delivery into the systemic circulation originates with the challenges involved in the systemic delivery of drugs administered orally.

As a drug passes through the (GI) tract, it encounters different environments with respect to pH, enzymes, electrolytes, fluidity, and surface features, all of which can influence drug absorption. There is a great variation in the pH across the GI tract, which runs from the mouth to the anus. The interdigestive migration of a drug or a dosage form is governed by GI motility, wherein the drug is exposed to different pHs at different time periods. The stomach has an acidic pH varying from 2 to 4. The acidic pH in the stomach increases up to a pH of 5.5 in the duodenum. The pH then increases progressively from the duodenum to the small intestine (a pH of 6–7) and reaches a pH of 7–8 in the distal ileum. After the ileocecal junction, the pH falls sharply to 5.6 and then climbs up to neutrality during transit through the colon. Due to the pH variation in the GI tract, pH-sensitive polymers have been historically utilized as an enteric coating material. Enteric-coated products featuring pH-sensitive polymers include tablets, capsules, and pellets and are designed to keep an active substance intact in the stomach and tend to release it to the upper intestine.

Apart from the pH, mucosal layer plays an important role in drug absorption from the lumen of the GI tract. Small intestine has a large epithelial surface area, which consists of mucosa, villi, and microvilli. Drug must first diffuse through the unstirred aqueous layer, the mucus layer, and the glyco-calyx (which is the coating of the mucus layer) to reach the microvilli, which is the apical cell membrane. The tight junction between the cell membranes of adjacent epithelial cells acts as a major barrier to the intercellular passage of drug molecules from the intestinal lumen to the lamina propria.

The low oral bioavailability of peptide and protein drugs is primarily due to their large molecular size and vulnerability to proteolytic degradation in the GI tract. Most protein and peptide drugs are susceptible to rapid degradation by digestive enzymes. Furthermore, most peptide and protein drugs are rather hydrophilic, and thus are poorly partitioned into the epithelial cell membranes, leading to their absorption across the GI tract through passive diffusion.

Various delivery systems have been proposed to increase drug absorption from the colon and ileum and minimize exposure of the drug to proteolytic enzymes. Enteric coatings that delay drug release for a sufficient period of time have been used to target both the ileum and colon. In addition, encap-sulation into polymeric materials that are degraded by the human colonic microflora has been proposed as a method to increase drug absorption from the intestine. Coadministration of enzyme inhibitors and absorp-tion enhancers have shown some promise. Encapsulation into erodible or biodegradable nanoparticles have been explained as a way of protecting drugs from enzymatic degradation. Submicron size particles are absorbed through transcytosis by both enterocytes and M cells, which are epithelial cells of the mucosa-associated lymphoid tissues.

For systemic action of drugs, the oral route has been the preferred route of administration. When administered by the oral route, however, many therapeutic agents are subjected to extensive presystemic elimination by GI degradation and/or hepatic metabolism.

Several nonoral routes of drug delivery have been utilized to provide ade-quate drug concentrations in the systemic circulation, in addition to local-ized drug treatment. These include the rectal, vaginal, and the transdermal routes of drug administration.

15.2.1 Rectal drug delivery

Rectal administration provides rapid absorption of many drugs and is an alternative when oral administration is inconvenient because of inability to swallow or because of GI side effects such as nausea, vomiting, and irritation. More importantly, rectal drug administration has the advantage of minimizing or avoiding hepatic first-pass metabolism. The rectal bio-availability of lidocaine in human is 65%, as compared to an oral bioavail-ability of 30%. Rectal route is used to administer diazepam to children who are suffering from epilepsy and in whom it is difficult to establish intravenous access. However, rectal administration of drugs is inconvenient and has irregular drug absorption. Moreover, rectal administration should be avoided in immunosuppressed patients in whom even minimal trauma could lead to the formation of an abscess.

15.2.2 Vaginal drug delivery

Vaginal epithelium is permeable to a wide range of substances including steroids, prostaglandins, antibiotics, estrogens, and spermicidal agents. Most steroids are readily absorbed by vaginal epithelium, leading to their higher bioavailability compared to their oral administration because of a reduced first-pass metabolism. For drugs with high membrane permeability, vaginal absorption is determined by permeability of the aqueous diffusion layer, whereas for drugs with low membrane permeability, such as testosterone and hydrocortisone, vaginal absorption is determined by membrane permeability. Vaginal ointments and creams contain drugs such as anti-infectives, estrogenic hormone substrates, and contraceptive agents. Contraceptive creams contain spermicidal agents and are used just prior to sexual intercourse.

15.2.3 Transdermal drug delivery

Transdermal patches deliver drugs through the skin. Percutaneous absorption of a drug generally results from direct penetration of the drug through the *stratum corneum*, deeper epidermal tissues, and the dermis. When the drug reaches the vascularized dermal layer, it becomes available for absorption into the general circulation. Among the *factors* influencing percutaneous absorption are the physicochemical properties of the drug, including its molecular weight, solubility, partition coefficient, nature of the vehicle, and condition of the skin. Chemical permeation enhancers, iontophoresis, or both are often used to *enhance* the percutaneous absorption of a drug.

In general, patches are *composed* of three key compartments: (1) a protective seal that forms the external surface and protects it from damage, (2) a compartment that holds the medication itself and has an adhesive backing to hold the entire patch on the skin surface, and (3) a release liner that protects the adhesive layer during storage and is removed just prior to application.

Most patches belong to one of two general *types*—the reservoir system and the matrix system. The reservoir system incorporates the drug in a compartment of the patch, which is separated from the adhesion surface. Drug transport from the patch to the skin is channelized and controlled through a rate-limiting surface layer. The matrix system, on the other hand, incorporates the drug uniformly across the patch in a polymer matrix. Diffusion of the drug through the polymer matrix and the bioadhesive properties of the polymer determines the rate of drug absorption.

Marketed transdermal patches are *exemplified* by Estraderm® (estradiol), Testoderm® (testosterone), Alora® (estradiol), Androderm® (testosterone), and Transderm-Scop® (scopolamine). Nicoderm® is a nicotine patch, which releases nicotine over 16 h, continuously suppressing the smoker's craving for a cigarette. In addition, occlusive dressings are available, which have low water vapor permeability. These dressings help prevent water loss from the skin surface, resulting in increased hydration of the *stratum corneum.*

15.2.4 Buccal and sublingual drug delivery

The buccal and sublingual mucosae in the oral cavity provide an excellent alternative over oral tablets for certain drugs. Oral transmucosal absorption is generally rapid because of the rich vascular supply to the mucosa. These routes provide improved delivery for certain drugs that are inactivated by first-pass intestinal/hepatic metabolism or by proteolytic enzymes in the GI tract.

The sublingual mucosa is relatively permeable, and is suitable for delivery of low molecular weight lipophilic drugs when a rapid onset of action with infrequent dosing is required. Sublingual DDSs are generally of two different designs: (a) rapidly disintegrating tablets and (b) soft gelatin capsules filled with a drug in solution. Such systems create a very high drug concentration in the sublingual region before they are systemically absorbed across the mucosa. Therefore, rapidly disintegrating sublingual tablets are frequently used for prompt relief from an acute angina attack.

The buccal mucosa is considerably less permeable than the sublingual area and is generally not able to provide rapid absorption properties. The buccal mucosa has an expanse of smooth muscle and relatively immobile mucosa, which makes it a more desirable region for retentive systems used for oral transmucosal drug delivery. Thus, the buccal mucosa is suitable for sustained delivery of less permeable molecules, and perhaps peptide drugs. One of the major disadvantages associated with buccal drug delivery is the low flux that results in low drug bioavailability. Therefore, buccal DDSs usually include a penetration (permeability enhancer) to increase the flux of drugs through the mucosa. Another limitation associated with this route of administration is the poor drug retention at the site of absorption. Consequently, bioadhesive polymers have been extensively employed in buccal DDSs. The duration of mucosal adhesion depends on the type and viscosity of the polymer used. Nicotine in a gum vehicle when chewed is absorbed through the buccal mucosa. Glyceryl trinitrate has been found quite effective when administered through this route.

15.2.5 Nasal drug delivery

Although nasal route is traditionally used for locally acting drugs, such as antihistamines and corticosteroids for allergies to reduce mucosal secretion, this route is getting more attention for the systemic delivery of various peptide drugs that are poorly absorbed via the oral route. The major advantages of nasal administration include the fast absorption, rapid onset of action, and avoidance of hepatic and intestinal first-pass effects.

15.2.5.1 Barriers to transnasal drug delivery

There are three major barriers to drug absorption across nasal mucosa. These are

1. *Physical barrier*: A drug or DDS needs to diffuse across the highly viscous mucus and permeate through the epithelial cell lining. Permeation through the epithelial cell lining could utilize either the lipoidal pathway or an aqueous pore pathway. Nasal absorption of weak electrolytes is dependent on the degree of ionization. Systemic bioavailability of nasally administered drugs is generally low.
2. *Temporal barrier*: Dosage forms for nasal absorption must deposit and remain in the nasal cavity long enough to allow effective absorption. The DDS has limited time at the site of administration before it is cleared with the mucus due to the physiological processes of mucociliary clearance and renewal of mucosal secretion.
3. *Enzymatic barrier*: The mucus has proteolytic enzymes. Therefore, protein and peptide drugs that are sensitive to such enzymes may get degraded during the process of drug absorption.

Commonly used dosage forms administered through this route are nasal sprays and drops. The nasal spray deposits drug in the proximal part of the nasal atrium, whereas nasal drops are dispersed throughout the nasal cavity. A nasal spray requires that the particles have a diameter larger than 4 μm to be retained in the nose and to minimize their passage into the lungs. Nasal sprays are commercially available for muserelin (a gonadotropin-releasing hormone agonist), desmopressin, oxytocin, and calcitonin.

15.2.5.2 Drug delivery to the brain through the nasal route

Intranasal administration has also been explored for brain-targeted drug delivery. Treatment of brain disorders presents significant challenges due to the inability of most drugs to cross the tight endothelial blood–brain barrier (BBB). Intranasal drug delivery has been explored for brain targeting because the brain and the nose compartments are connected through the olfactory/trigeminal neural pathway, in addition to the peripheral circulation. The olfactory region of the nasal cavity, however, is a relatively small region and provides a formidable epithelial cell barrier. Nonionic alkyl glycosides such as dodecyl maltoside, decylsucrose, dodecylsucrose, and tetradecylmaltoside, have been used as absorption enhancers to improve drug absorption across the nasal mucosa.

15.2.5.3 Formulation factors affecting transnasal drug delivery

Several formulation factors are important to consider for drug delivery across the nasal barrier. These include the following:

1. *pH*: A pH of the formulation that provides the drug in the nonionic form can enhance drug absorption. Nonetheless, the formulation pH needs to be within the range of 4.5–6.5 to minimize nasal irritation.
2. *Osmolality*: Hypertonic saline solutions inhibit or reduce ciliary activity, thus increasing the residence time of the DDS at the site of absorption.
3. Gelling or mucoadhesive agents can increase the residence time of the DDS at the site of administration.
4. Solubilizers can increase the amount of a drug in the dissolved state within a formulation and increase the diffusible fraction of the drug.
5. Absorption enhancers can open the tight junctions of the endothelial barrier, leading to a higher rate of drug absorption especially for the large molecular weight protein and peptide drugs.
6. Viscosity, volume, and concentration determine the feasibility of a drug for delivery across the nasal mucosa. For example, if the therapeutic dose of a drug is soluble in a volume that is much higher than what can be delivered through the transnasal route, or the viscosity of the solution is too high, a transnasal dosage form may not be feasible.

15.3 NEED FOR ORGAN SPECIFIC AND TARGETED DRUG DELIVERY

Most new chemical entities (NCEs), including molecularly targeted agents, do not present compelling efficacy and safety in terms of the benefit-to-risk ratio in human clinical trials due to either efficacy or toxicity concerns. The toxicity profile of an agent includes general toxicity and effects explained by its mechanism of action. Although the toxicity profile usually remains largely unpredictable and difficult to modify, the safety and efficacy of these agents usually benefits from targeted delivery to either the physiological regions of where their molecular receptors are present in high concentration, or to avoid drug exposure to organs and tissues where significant toxicities exist.

Targeted drug delivery is a method of delivering drug in a manner that selectively increases drug concentration to a biological target. Targeted drug delivery is different from target-based drug development or drug targets that are defined as the molecular targets that the drugs modulate for their pharmacological action. Drug therapy that aims to utilize drug molecules that target a specific protein or receptor for their action is called targeted drug therapy. Targeted drug delivery, on the other hand, refers to the science and technology of presenting a drug to its site of action. The overall goal of all drug-targeting strategies is to improve the efficacy and/or safety profile of a drug substance.

Targeted drug delivery can involve either drug delivery to a specific organ or tissue, or avoiding drug delivery to nontarget organs, tissues, or cells. Targeted drug delivery to a particular physiological location can bring the drug to its primary site of action. Thus, it can help improve the efficacy of a drug and/or prevents its undesired toxicities in other tissues or organs. In addition, sometimes targeted strategies are intended to avoid drug exposure to nontarget organs or tissues.[1] This can help avoid specific drug-related toxicities in particular organs, such as the kidney. For example, intravenously injected liposomal doxorubicin has lower nephrotoxicity and cardiotoxicity than intravenous (IV) injection of doxorubicin solution.

15.3.1 Advantages of targeted drug delivery

The most significant advantages of targeted drug delivery are realized in acute disease states, for example, targeting cytotoxic anticancer drugs to a specific organ (e.g., brain, lungs, liver, kidney, and colon) or the tumor tissue. For example, prodrugs of doxorubicin have been prepared with folate receptor conjugated through bovine serum albumin or polyethylene glycol (PEG), that enable drug targeting to tumors that express folate receptors. Two important design elements of targeted DDS are (a) the selection of the target organ or tissue and (b) the selection of the targeting strategy.

- The selection of the target organ or tissue is governed by the pharmacological need of the disease state and the drug substance. For example, drugs are targeted to the BBB for drug delivery to the brain for neurodegenerative diseases, such as Alzheimer's disease.
- The selection of the targeting strategy for the DDS is governed by the pathophysiology of the target tissue and how it can be utilized to impart stimuli-responsive physicochemical property changes in the DDS. For example, leaky vasculature of the tumor tissue can be utilized for passive drug-targeting by designing a DDS that is smaller in particle size and thus can extravasate to the tumor site after systemic administration. In addition, expression of specific receptors on the cell surface of tumor tissues can be utilized for active targeting of the DDS to tumor cells.

15.3.2 Examples of established drug targeting strategies

Several drug-targeting approaches have successfully transitioned from the proof-of-concept to the clinical application, and have become a state of the art.

Examples of targeted drug delivery platforms that have become well accepted in clinical practice include the following:

- Enteric coating of oral solid dosage forms to overcome chemical instability against acidic pH of the GI tract or adverse effects of the drug in the gastric environment
- Pulmonary drug delivery by dry powder inhalation
- Ocular inserts for drug delivery to the surface of the eye
- Transdermal and implantable DDSs for sustained systemic absorption or local drug delivery.[1]

In addition, several drug delivery strategies being explored are at different preclinical and clinical stages of advancement. Targeted delivery of small and macromolecular drugs has been discussed in-depth in a recent book *Targeted Delivery of Small and Macromolecular Drugs*.[2] In this chapter, we will describe different organ-specific drug targeting strategies.

15.3.2.1 Pulmonary drug delivery

The respiratory tract includes the nasal mucosa, hypopharynx, and large and small airway structures including the trachea, bronchi, bronchioles, and alveoli. This tract provides a larger mucosal surface for drug absorption. Pulmonary drug delivery refers to drug delivery to the local or systemic circulation through the alveoli.

15.3.2.1.1 Advantages of pulmonary drug delivery

Lung epithelium is highly permeable and has low enzymatic/metabolic activity compared to the liver and intestine. With a large surface area (~100 m²) and a highly permeable membrane (~0.2–0.7 mm thickness), alveolar epithelium permits rapid drug absorption into the systemic circulation. There are 200–600 million alveoli in a normal human lung. This route of administration is useful for treating pulmonary conditions and for drug delivery to other organs via the circulatory systems. In general, lipid-soluble molecules are absorbed rapidly from the respiratory tract, and thus, an increasing number of drugs are being administered by this route, including bronchodilators (e.g., beclomethasone dipropionate), corticosteroids, antibiotics, antifungal agents, antiviral agents, and vasoactive drugs.

Lung alveoli can also permit systemic absorption of macromolecules, such as proteins, peptides, oligonucleotides, and genes. For example, DNase alpha (Pulmozyme®, Genentech), an enzyme used to reduce the mucus viscosity in the airways of cystic fibrosis patients, is most effective when administered by inhalation. This protein is delivered directly to its site of action by nebulization. The recent approval of inhaled human insulin by the

FDA for use in diabetes mellitus stands as a major advancement in the field of pulmonary delivery of macromolecules and systemically acting drugs.

15.3.2.1.2 Barriers to pulmonary drug delivery

The lung has evolved to maintain sterility of its pathways and to avoid undesired airborne pathogens and particles through mechanisms such as (a) airway geometry, (b) localized high humidity, (c) mucociliary clearance, and (d) the presence of alveolar macrophages. These mechanisms also present themselves as barriers to pulmonary drug delivery.

15.4 LOCAL DRUG DELIVERY TO THE LUNG

Pulmonary route has long been utilized for localized drug delivery to the lung, which helps overcome several barriers to the treatment of lung disorders such as systemic side effects and metabolism, while allowing high local concentrations and rapid clinical response. In many cases, the dose of the drug required for local treatment can be significantly reduced. For example, orally administered salbutamol at 2–4 mg of dose is therapeutically equivalent to 100–200 µg of salbutamol administered directly into the lung.

15.5 SYSTEMIC DRUG DELIVERY THROUGH THE LUNG

Sufficient drug must be deposited in the alveolar cavity, so that adequate amount of a drug would be absorbed into the systemic circulation to elicit therapeutic efficacy. Aerosolized systems present the drug in a particulate form with an intensity of airflow, which may allow the drug to pass through the lung architecture and reach the alveoli. However, traditional inhalation devices deposited only ~10%–15% of the inhaled drug in the lung.

Certain design elements of the DDS that can permit systemic drug absorption through the lung include the following:

1. *Size and drug loading of the particles*: Fine aerosols deposit better in the lung alveoli and peripheral airways but have a lower amount of a drug per unit surface area. The relatively coarse aerosol particles deposit more drugs per unit surface area, but tend to be deposited in the central, larger pathways that are subject to mucosal clearance, low surface area, and poor drug absorption. For example, particles with a diameter in the range of ~60 µm tend to deposit in the trachea, whereas particles in the size range of ~2 µm tend to deposit in the alveoli. Aerodynamic diameter and size distribution of the particles need to be carefully controlled to maximize particle deposition in the alveoli.

2. *Density of the particles*: Particle density contributes to the inertia of the particle. In addition, being inversely related to particle porosity, particle density also impacts diffusivity of the drug through the particle at the site of absorption.

3. *Particle shape*: Ideal particle shape for pulmonary drug delivery is spherical. However, pharmaceutical formulations tend to have an irregular shape, whereas crystalline drugs may have markedly high aspect ratio (ratio of length to width) of their particles. Deviation from sphericity can reduce the alveolar deposition of a drug.

4. *Aggregation*: Particle surface charge and other surface characteristics that may promote particle adhesion can increase the size of the agglomerates in the respiratory tract, thus compromising the proportion of the formulation from reaching the alveoli.

5. *Hygroscopicity*: Since the respiratory tract presents a high humidity local environment, particles that are hygroscopic and may change their surface adhesion or size characteristics rapidly on exposure to a high humidity environment may not achieve optimal particle deposition in the lung.

15.6 DEVICES FOR PULMONARY DRUG DELIVERY

Pulmonary drug delivery almost invariably requires aerosolization of the drug using a device or mechanism that can achieve controlled drug particle formation and delivery. Examples of devices that have been used for pulmonary delivery include the following:

1. *Metered dose inhaler*: Pressurized spray is a metered dose inhaler (MDI) that incorporates propellant(s), surfactant(s), and the drug in either dissolved or suspended state in its formulation. Pressurized metered dose inhalers provide constant pressure on the liquid formulation and consistent quantity of drug release on actuation of the valve. The propellants used in these formulations include chlorofluorocarbons (CFCs) and hydrofluoroalkanes (HFAs). Effective use of an MDI requires patient coordination of breathing and actuation to provide maximum amount and flow velocity of air going into the lungs. Drug solubility, vapor pressure, surface tension, solubility of oxygen/hygroscopicity, and density affect the effectiveness of drug delivery through the MDIs. MDIs are commonly used for the delivery of drugs for asthma and chronic obstructive pulmonary diseases (COPD).

2. *Nebulizer*: The nebulizer uses an air compressor as a power source instead of a liquid propellant. Compressed air is brought in contact with an aqueous solution of the drug in a device. Liquid shearing leads to the formation of drug droplets that get inhaled by the patient. Droplet size is a key factor in effective pulmonary delivery of the drug through this route. The droplet size is typically controlled through the

orifice diameter of the baffle, pressure of the gas, and the density, concentration, viscosity, surface tension, and flow rate of the drug solution. Nebulizers are frequently used in applications where patient's strong inhalation is not required for effective drug delivery such as in pediatric or hospitalized patients.

3. *Dry powder inhaler*: The dry powder inhaler (DPI) allows the drug to be formulated in dry state, in which it may be more stable, does not require the use of a liquid propellant, and also does not require patient coordination between breathing and actuation. DPIs consist of a suspension of fine particulate drug formulation, which is dispersed by mechanical, pneumatic, or electrical energy, or by the strength of the vacuum generated by the patient's breathing. DPIs have been used to treat diseases such as asthma, bronchitis, emphysema, and COPD.

Optimization of particle properties and device parameters can increase drug penetration to the lungs. For example, simulation results of hydrofluoroalkane (HFA)-propelled metered dose inhaler (pMDI) show increased droplet transport and deposition to the lungs when used with a spacer (Figure 15.1). This study illustrates the impact of a simple spacer in terms of enhanced droplet percentage reaching the tracheobronchial tree.

For further discussion and details on the devices used for pulmonary drug delivery, see Chapter 24.

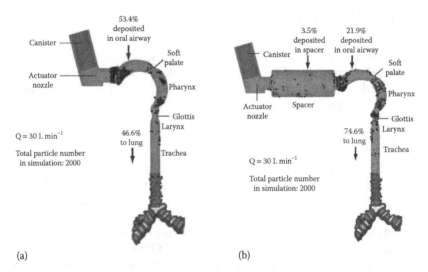

Figure 15.1 Drug delivery to the lung. Role of device in enhanced drug penetration into the lung. Simulation results of hydrofluoroalkane (HFA)-propelled metered dose inhaler (pMDI) droplet transport and deposition: (a) without and (b) with spacer, illustrating the impact of a simple spacer in terms of enhanced droplet percentage reaching the tracheobronchial tree. Q = *Airflow Rate*. (Courtesy of Annual Reviews, Palo Alto, CA.)

15.6.1 Ocular drug delivery

Antibiotics and steroids are the most common classes of drugs typically administered to the eye. These drugs are administered most commonly through the topical route by instilling or application to the surface (cornea) of the eye. Nevertheless, drug delivery is often required for different segments and anatomical regions of the eye that are difficult to access. Treatment of ocular disorders is challenging due to anatomical and physiological constraints of the eye, including its vascular permeation and sequential presence of both lipophilic and hydrophilic barriers to drug penetration upon topical administration. In Section 15.6.1.1, we will discuss the structure of the eye, challenges to drug delivery to the eye, and the approaches that have been taken to overcome these challenges.

15.6.1.1 Structure of the eye

The eye is divided into two chambers—commonly known as the anterior chamber and the posterior chamber (Figure 15.2). The anterior chamber

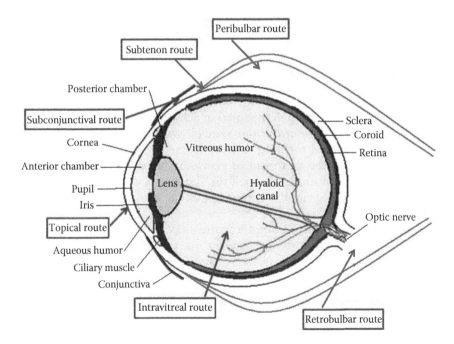

Figure 15.2 Drug delivery to the eye: Structure and schematic representation of various routes of drug delivery to the eye. (Reproduced from Mishra, G.P. et al., Recent advances in ocular drug delivery: Role of transporters, receptors, and nanocarriers, in Narang, A.S., and Mahato, R.I. (Eds.), *Targeted Delivery of Small and Macromolecular Drugs*, Boca Raton, FL: CRC Press, Taylor & Francis Group, pp. 421–453, 2010. With permission.)

is mainly comprised of cornea, conjunctiva, iris, ciliary body, and lens. The posterior chamber includes sclera, choroid, vitreous humor, and retina.[3] Cornea is the outermost, avascular and transparent membrane of the eye. The conjunctiva is a clear mucous membrane that covers the inner part of the eyelid and the visible part of sclera (white part of the eye) and lubricates the eye by producing mucus and some tears. Aqueous humor lies between the lens and the cornea. It circulates from the posterior to the anterior chamber of the eye and through the canal of Schlemm. Impaired outflow of aqueous humor causes elevated intraocular pressure that leads to permanent damage of the optic nerve and consequential visual field loss that can progress to blindness. Retina is a light-sensitive tissue behind the aqueous humor. The vitreous humor is a hydrogel matrix composed of hyaluronic acid and collagen fibrils and is located between the retina and the lens.

15.6.1.2 Routes of drug administration into the eye

Depending on the targeted location of drug action within the various components and compartments of the eye, the selection of the site of administration can play a key role in drug targeting.

1. *Topical route of drug administration*: Treatment of anterior segment diseases usually utilizes topical route of drug administration. This route presents challenges such as precorneal tear clearance, limited conjunctival drug absorption, metabolism by the iris-ciliary body, and elimination through the canal of Schlemm.
2. *Systemic drug administration*: Systemic drug administration is usually not preferred for drug delivery to the eye. However, in certain cases, such as the treatment of glaucoma, administration of drugs, such as acetazolamide, through the systemic route may be preferred to obviate the drug absorption limitation due to high intraocular pressure.
3. *Intravitreal administration*: Intravitreal (IVT) injection is utilized for the treatment of posterior segment diseases, for example, diabetic retinopathy, and viral infections, for example, human cytomegalovirus (HCMV) retinitis and endophthalmitis. Direct administration to the vitreous humor overcomes the blood-retinal barrier (BRB). This route, however, requires injections in the eye and may cause retinal detachment, which could lead to vision loss. Thus, prolonged drug release strategies, including prodrugs, have been utilized to prolong drug residence time in the vitreous humor.
4. *Periocular administration*: The periocular route of administration provides direct access to the sclera, and can result in high drug

concentration both in the anterior and posterior segments of the eye. The periocular drug injections could be retrobulbar, peribulbar, subtenon, and subconjunctival, depending on the site of injection.

5. *Retrobulbar injection*: Direct injection into the retrobulbar space can be useful for drug delivery into the macular region (highly pigmented yellow spot near the center of retina, rich in ganglion cells and responsible for central vision). This injection can, however, result in damage to the blood vessels.

6. *Peribulbar injection*: The peribulbar injection can be circumocular, periocular, periconal, or apical depending on the exact site of injection. This route is generally utilized for the administration of analgesics.

7. *Subtenon injection*: This site of drug injection can be utilized for drug delivery to the posterior segment of the eye. The drug is administered into the tenon space, which is formed by the void between the tenon's capsule and the sclera.

8. *Subconjunctival injection:* This periocular route of drug administration can allow up to 500 µL of drug solution to be injected. This route is utilized for the treatment of both anterior and posterior segment diseases.

15.6.1.3 Challenges to ocular drug delivery

Topically administered drugs can be eliminated via precorneal tear clearance, blinking, and nasolacrimal drainage. This presents challenges to the entry of drug molecules to the anterior segment of the eye (Figure 15.2). Drug delivery to the posterior segment of the eye is challenged by barriers such as inner and outer BRBs and efflux pumps. In addition, the presence of efflux pumps, such as P-glycoprotein (P-gp), multidrug resistance associated proteins (MRPs), and breast cancer resistant protein, also limits the ocular bioavailability of drugs.[3]

For drugs administered through the topical route, the cornea is the main barrier to drug absorption. The cornea and the conjunctiva are covered with a thin film, the tear film, which protects the cornea from dehydration and infection. Following topical administration, a drug is eliminated from the eye by nasolacrimal drainage, tear turnover, productive corneal absorption. and nonproductive conjunctival uptake. The cornea has three anatomical parts: (1) the epithelium, (2) the stroma, and (3) the endothelium. Both the endothelium and the epithelium have high lipid content, and thus are penetrated by drugs in their unionized lipid-soluble forms. The stroma lying between these two structures has high water content. To penetrate the cornea, drugs have to go through both the lipidic and aqueous anatomical components.

For drugs injected into the eye, there are two main barriers to ocular drug adsorption: (a) the blood-aqueous barrier and (b) the blood-retina barrier. The blood-aqueous barrier is composed of the ciliary epithelium, the epithelium of the posterior surface of the iris, and blood vessels within the iris. Drugs enter the aqueous humor at the ciliary epithelium and in the blood vessels. Many substances are transported out of the vitreous humor at the retinal surface.

15.6.1.4 Physicochemical characteristics of the drug for ocular absorption

Drug ionization impacts absorption through the ocular route not only by impacting drug permeability but also by affecting tear turnover. A pH 5 solution induces more tear flow than a pH 8 solution. Greater tear turnover can lead to reduction of concentration gradient in addition to drug loss on blinking. Transport of both ionized and unionized drugs is less at pH 5.

The duration of drug action in the eye can be extended by two approaches:

1. Reducing drainage with viscosity-enhancing agents, suspensions, emulsions, ointments, and polymeric matrices
2. Improving corneal drug penetration with ionophores and liposomes

15.6.1.5 Approaches for enhancing drug delivery to the eye

Drug delivery to the eye can utilize multiple mechanisms to overcome the barriers to drug absorption. These include the modification of physicochemical properties of the drug such as by making prodrugs, targeting natural transporters and receptors for uptake, inhibition of efflux transporters, prolonging the drug residence time at the site of absorption by using nanoparticles, microparticles, micelles, and liposomes; or using mucoadhesive ocular implants and hydrogel-based aqueous formulations to achieve relatively constant drug levels at the target site for a longer duration.

1. *Prodrugs*: Prodrugs are chemical entities that are pharmacologically inactive but can generate an active drug upon absorption through various chemical bond cleavage mechanisms, such as hydrolysis. Prodrugs can be utilized for changing the physiochemical properties of a drug, such as aqueous solubility and lipophilicity, or targeting specific transporters or receptors expressed on cell membranes. For example, lipophilic acyl ester prodrugs of acyclovir (ACV) were investigated to allow high drug permeability through lipophilic membrane. Prodrug hydrolysis to the parent ACV after prodrug permeation and

its transformation to ACV-triphosphate prevent diffusive back transport of the ACV. Amino acid and peptide prodrugs of quinidine and ganciclovir were investigated to bypass efflux pumps and to target peptide transporters for drug absorption.

2. *Permeability and efflux pump modification*: Formulation strategies for topically administered drugs that modify drug permeability and/or inhibit P-gp mediated efflux can be utilized to improve drug permeation into the eye. Several surfactants (such as *d*-alpha-tocopheryl polyethylene glycol 1000 succinate (Vitamin E-TPGS), Cremophor® EL, Polysorbate 80, and Pluronic® F85) and polymers (such as poly-(ethyleneoxide)/poly-(propyleneoxide) block copolymers, and amphiphilic diblock copolymers methoxypolyethylene glycol-block-polycaprolactone) inhibit P-gp efflux pump. Use of these ingredients in the formulation can help delivery of sensitive drugs to the eye.

3. *Nanoparticles and liposomes*: Nanosuspensions of drugs can be utilized to deliver poorly soluble drugs, such as flurbiprofen, methylprednisolone acetate, and glucocorticoids (e.g., hydrocortisone, prednisolone, and dexamethasone) into the eye. The nanosuspensions can enable enhanced retention at the target site and sustained-release (SR) properties to the drug. The retention of nanoparticles in the periocular space versus clearance by blood and lymphatic circulation would depend on the size and surface properties of the nanoparticles. Liposomes, lipid vesicles containing an aqueous core, can protect a drug against enzymatic degradation, increase the capacity to cross the cell membrane, provide SR, and/or prevent drug efflux.

4. *Intraocular implants*: Implants can be utilized for drugs targeted to both the anterior and posterior segments of the eye for diseases such as proliferative vitreoretinopathy, CMV retinitis, and endophthalmitis. The implants can be made with biodegradable or nonbiodegradable polymers such as poly(lactic acid) (PLA), poly(glycolic acid) (PGA), poly(lactide-co-glycolide (PLGA), poly(glycolide-co-lactide-co-caprolactone (PGLC) copolymer, poly(caprolactone) (PCL), polyanhydrides, and polyorthoesters (POE). Drugs that have been investigated for drug delivery by implantable DDS include dexamethasone, cyclosporine, 5-fluorouridine (5-FU), triamcilone acetonide, and recombinant tissue plasminogen activator.

5. *Hydrogels:* Hydrogels are three-dimensional, hydrophilic, polymeric networks capable of absorbing and holding a large amount of water. Thermosensitive hydrogels prepared by cross-linking poly(N-isopropylacrylamide) (PNIPAAm) with PEG have been investigated for drug delivery to the posterior segment of the eye. Drugs such as bevacizumab and ranibizumab have been tested for delivery in hydrogels.

15.6.2 Brain targeted drug delivery

Drug delivery to the brain, or the central nervous system (CNS), in general, is indicated in several clinical situations that include tumors, Alzheimer's disease, epilepsy, migraine, infections, inflammatory diseases, and conditions that are etiologically based on neurotransmitter imbalance.

15.6.2.1 Blood–brain barrier

Drug delivery to the brain and the CNS is particularly challenging because of the BBB. The *BBB* is a term that denotes the special nature of the blood vessels that carry blood to the brain and the spinal cord. The tight junctions of endothelial cells lining these blood vessels are highly impervious to substances in the blood, with electrical resistivity ≥ 0.1 Ωm. The BBB selectively allows the passage of water, certain gases, and lipophilic molecules that can diffuse through the cell membranes (paracellular route), and through active transport (through the transcellular route) of certain molecules that are important to neuronal function such as glucose and amino acids (Figure 15.3). Thus, drugs that are present in the bloodstream are not able to permeate as readily into the CNS tissue as they can into the tissue of other organs.

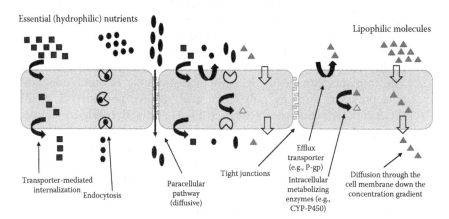

Figure 15.3 Routes of drug transport across the endothelial cell lining of the blood–brain barrier (BBB) that has tight junctions. Hydrophilic molecules, such as essential nutrients, can cross the cellular membranes through endocytosis or transporters. Lipophilic molecules cross the membrane through diffusive transport. Presence of efflux transporters and metabolizing enzymes can reduce the total amount of drug crossing over to the other side.

15.6.2.2 Drug delivery strategies to the brain

Drug delivery strategies to the brain include the following:

1. Drug encapsulation in liposomes or nanoparticles
   - Drug encapsulation in liposomes provides colloidal particles with lipophilic and membranous exterior that can fuse with the endothelial cell membranes and enable transcellular drug transport. Commercially available liposome encapsulated drugs include amphotericin B (AmBisome) and doxorubicin (Caelyx)
   - Commercially available nanoparticulate drugs that target the CNS include colloidal gold nanoparticles (Aurimmune) and gold-coated silica nanoparticles (AuroShell).
2. Drug modification, for example, by preparation of a lipophilic prodrug
   - Small-molecule drugs that are generally low molecular weight (<500 Da), nonionizable at physiologic pH, low hydrogen bond capability, and high lipophilicity can readily cross the BBB through passive diffusion. Examples of such compounds include benzodiazepines, alcohol, and nicotine. The lipophilicity of a drug can be increased through structural modification and/or by preparation of a lipophilic prodrug. Examples of drugs that can diffuse across the BBB include levodopa, γ-Aminobutyric acid (GABA), and valproic acid.
3. Drug conjugation with an active transport substrate to achieve receptor-mediated active transport
   - Drug conjugation with a moiety that has endothelial cell receptors for active transport across the BBB can increase drug uptake in the brain. For example, drug conjugation with the angiopep-2 peptide, which targets the low-density lipoprotein (LDL) receptor, has been utilized to target drugs and nanoparticles to the brain.
4. Transnasal route of administration
   - Nasal mucosa is a highly permeable and vascularized site. The olfactory region in the nasal passage is a small patch of tissue that contains olfactory nerve endings and is considered a port of entry of external chemicals into the brain. In addition, the characteristics of the drug that influence its permeation through the transcellular route, a key challenge in utilizing the nasal route for brain drug delivery is the ability to present the drug to this region of absorption in sufficient concentration and for a long enough duration that can enable the delivery of therapeutic dose. Compounds such as insulin-like growth factor I (IGF-I) have been delivered to the brain through the nasal route.

5. Direct intrathecal or intracerebroventricular injection or placement of DDS during invasive surgery (implant)
  * Intracerebroventricular (ICV) injection of chemotherapy agents, morphine, and antibiotics used in the treatment of meningitis is achieved through a subcutaneously implanted reservoir in the scalp (Ommaya reservoir) that connects to the cerebrospinal fluid (CSF) in a lateral ventricle through an implanted catheter.
  * Biodegradable SR wafers loaded with chemotherapy agents have been used as implants in the brain. These DDSs can be placed directly at the site of tumor following surgery.

15.6.3 Liver-targeted drug delivery

Liver is the major organ responsible for the metabolism, detoxification, and storage of macromolecules; as well as the production and secretion of bile for digestion. It plays an important role in the clearance of pathogens and antigens entering the body via the GI tract. The need and modalities of liver-targeted drug therapy is best understood in the context of cellular components of the liver, the nature of liver diseases, and the cellular receptors on various liver cells that can be utilized for targeted drug therapy.

15.6.3.1 Cellular components of the liver

Liver is designed for the recognition, metabolism, and elimination of foreign material, including bacteria, viruses, and noncellular particulates. This role is served through the anatomical design whereby venous blood is circulated through the liver, including the parts of the GI tract, via the hepatic portal vein, through a sinusoidal system.

The liver consists of four cell types—(1) hepatocyte, (2) endothelial, (3) Kupffer, and (4) stellate cells. The main parenchymal tissue of the liver is composed of *hepatocytes*, which make up 70%–85% of the liver mass and are involved in various liver activities including the formation and secretion of bile. Hepatocytes have metabolic, endocrine, and secretory functions. Liver endothelial cells form the discontinuous lining of the sinusoids and have fenestrations that are ~100 nm in diameter. This relatively large pore size plays an important role in determining the sizes of particles filtering between the blood and the liver parenchymal cells. A space of Disse separates hepatocytes from the sinusoids.

The hepatic sinusoids are lined with the *Kupffer cells*, which are the largest group of tissue macrophages in the liver. Their main function is to phagocytose and destroy foreign material, such as bacteria or colloids.

Hepatic stellate cells (HSCs) localize within the space of Disse in close proximity of both hepatocytes and endothelial cells (Figure 15.4a). HSCs are present in the perisinusoidal space, constituting about 5%–10% of the total number of liver cells. These are the mesenchymal cells that are

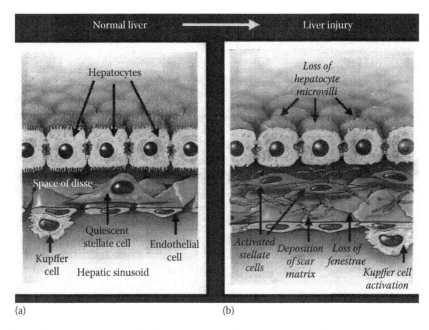

Figure 15.4 Physiology of the (a) normal and (b) diseased liver showing subsinusoidal events during liver injury. In response to liver injury, stellate cells secrete excessive extracellular matrix (ECM), which deposits in the subsinusoidal space of Disse as scar matrix and loss of fenestrae. Liver injury also causes Kupffer cell activation, which contribute to paracrine activation of stellate cells. (Reproduced from Friedman, S.L., *J. Biol. Chem.*, 275(4), 2247–2250, 2000. With permission.)

involved in the liver's response to injury. Stellate cell activation transforms them into myofibroblasts, cells that are phenotypically between a fibroblast and a smooth muscle cell. The myofibroblasts produce fibrinogen, a glycoprotein involved in blood coagulation. Deposition of fibrinogen in the liver can lead to liver fibrosis.

15.6.3.2 Common diseases of the liver

The normal physiology of the liver is affected in the disease state. For example, in response to liver injury, stellate cells secrete excessive extracellular matrix (ECM), which deposits in the subsinusoidal space of Disse as scar matrix and loss of fenestrae. Liver injury also causes Kupffer cell activation, which contribute to paracrine activation of stellate cells (Figure 15.4b). The excessive ECM secretion contributes to the loss of hepatocyte microvilli and sinusoidal epithelial fenestrae, which leads to loss of liver function.

Drug delivery to the liver is indicated in several diseases. For example,

1. *Hepatocellular carcinoma*: Hepatocellular carcinoma (HCC) is the third leading cause of cancer-associated deaths worldwide. HCC has been associated with hepatitis B and C infections, metabolic liver diseases, and nonalcoholic fatty liver diseases.
2. *Cirrhosis*: Activation of HSCs can lead to the deposition of fibrotic tissue. Continuation of the fibrotic process can lead to end-stage liver disease known as cirrhosis. Liver cirrhosis is associated with anatomical alteration of the sinusoidal architecture, reduced liver perfusion, compromised liver function, and increased risk of HCC. Liver cirrhosis is mainly caused by hepatitis B and C infections, alcohol abuse, biliary problems, and fatty liver (steatohepatitis).
3. *Hepatitis*: Hepatitis is a state of inflammation of the liver that is commonly caused by viruses, which are of five main types A, B, C, D, and E. Hepatitis virus types B and C are the most prevalent, lead to chronic diseases, and are the most common cause of liver cirrhosis and cancer. Hepatitis virus types A and E are spread by contaminated food and water. Hepatitis virus types B, C, and D are spread by parenteral contact with infected body fluids by mechanisms such as injection, infusion, sexual contact, and mother-to-baby transmission at the time of birth.

15.6.3.3 Natural mechanisms of hepatic drug uptake

The natural role of the liver in protecting the body from xenobiotics provides mechanisms that allow passive drug targeting to the liver. These include the following:

1. *Hepatic first-pass effect*: Orally absorbed drugs are carried through the hepatic portal vein into the liver before they reach the systemic circulation. The liver metabolizes several drugs (e.g., diazepam and morphine) to a significant extent, leading to reduced oral bioavailability. This phenomenon can also be utilized for liver targeting through
   a. High hepatic exposure of orally administered compound. For example, antiviral drugs targeted for the treatment of hepatitis C, such as ribavirin and telaprevir are administered orally.
   b. Prolonging the circulation time of compounds targeted for the liver provides passive targeting to the hepatocytes through prolonging the duration of time a therapeutic is available for hepatocyte uptake. For example, PEGylated interferons α-2a (PEGASYS®) and α-2b (PegIntron®) have been used effectively in the treatment of hepatitis B and C in combination with ribavirin.

2. *Enhanced permeation and retention effect*: Liver tissue in diseases such as hepatocellular carcinoma (HCC) displays the enhanced permeation and retention (EPR) effect. The EPR effect is attributed to the imperfect endothelium of neovasculature (newly formed blood vessels) of growing tumors that result in larger particulate drug carriers being able to concentrate in the tumor tissue more than the normal tissue with mature vasculature. This mechanism can be utilized for drug delivery to the liver by the utilization of particulate drug carriers such as liposomes and nanoparticles.

15.6.3.4 Cellular targets for disease therapy

Natural xenobiotic scavenging role of the liver cells through endocytotic and specific target/antigen-binding receptors on various cell types affords opportunities for actively targeted drug therapy for various liver cell types. The receptors that can be utilized for drug targeting to specific liver cells include the following:

1. *Hepatocytes*: They are involved in liver diseases such as hepatitis A, B, or C; alcohol-induced or nonalcohol-induced steatohepatitis (NASH); and genetic diseases such as Wilson's disease and hemochromatosis. Hepatocytes can be targeted through the asialoglycoprotein receptors on their cell surface, which bind galactose and lactose.
   a. Asialoglycoprotein receptors on hepatocytes are attractive as a target receptor for drug delivery because of limited distribution of these receptors elsewhere in the body, high binding affinity with the target ligand (e.g., galactose), and rapid ligand internalization. Galactosylated drug carriers (i.e., DDSs that display galactose residues on their surface) are readily delivered to hepatocytes due to the relatively wide sinusoidal gap (~100 nm diameter). Drug delivery carriers that are modified with galactose or lactose have been utilized for drug delivery in HCC.
   b. HCC cells also express several growth factor receptors, such as the epidermal growth factor receptor (EGFR). Antibodies against such growth factor receptors, such as the anti-EGFR antibody cetuximab have shown some activity against HCC.
   c. Coxsackie- and adenoviral-receptor and integrin receptors on their cell surface that help internalize adenoviruses. Adenoviral vectors can be utilized to deliver genes. Hepatocyte selectivity of viral gene delivery can also be achieved from viral vectors that are derived from the human immunodeficiency virus (HIV) and the Sendai virus.
   d. Apolipoprotein E is rapidly cleared from the systemic circulation by hepatocytes. The apolipoprotein E or the high-density lipid (HDL) particles has been utilized for the delivery of short-interfering RNAs (siRNAs) and microRNAs (miRNAs).

2. *Kupffer cells* are highly phagocytic cells that are a part of the reticu-loendothelial system (RES), also called the mononuclear phagocyte system (MPS) or the macrophage system. Kupffer cells can be targeted through a variety of ways. For example,

   a. Sugar (mannose and fucose) receptors that serve to recognize natural foreign particles, such as bacteria and yeast, can be utilized to target proteins and drugs to the RES phagocytic cells. For example, mannose-modified human serum albumin (HSA) selectively accumulates in Kupffer cells.

   b. Kupffer cells phagocytose noncellular particles of 100 nm or higher diameter. Passive targeting to these endocytotic cells can, therefore, be achieved using particulate drug delivery carriers.

   c. Kupffer cells and endothelial cells express scavenger receptors, which predominantly bind negatively charged molecules. Proteins and liposomes with a net negative charge have been utilized for targeting the scavenger receptors. For example, coupling of the electroneutral dexamethasone to HSA through lysine residues increases the net negative charge on HSA, increasing its potential uptake by the scavenger receptors. In addition, incorporation of succinyl-HSA (HSA conjugated with the polyanionic succinic acid) into liposomes has been targeted for drug delivery to the sinusoidal liver endothelial cells.

HSCs are involved in the fibrotic processes that can lead to liver cirrhosis. In the presence of chronic liver injury, HSCs get activated and transform into proliferative myofibroblasts, which are the major source of excessive ECM. The receptors that are highly upregulated on HSCs include the following:

1. Mannose-6-phosphate (M6P)/insulin-like growth factor II receptor
2. Collagen type VI receptor
3. Platelet derived growth factor-β (PDGF-β) receptor.

Conjugation of HSA to M6P or peptides that recognize the collagen type VI or the PDGF-β receptor has been utilized to target HSCs. These carriers have been utilized for the delivery of antifibrotic small-molecule drugs, proteins, siRNAs, and triplex-forming oligonucleotides (TFOs) through direct conjugation with the carrier molecules, complexation/conjugation with carrier molecule-modified HSA, or incorporation in liposomes that have been modified with the carrier molecule.

15.6.4 Colon-targeted drug delivery

Traditionally, colonic drug delivery is focused on the treatment of local conditions such as ulcerative colitis, colorectal cancer, irritable bowel syndrome, amebiasis, and Crohn's disease.[1] However, it has been gaining importance for the systemic delivery of potent compounds such as proteins, peptides, and oligonucleotides that are unstable in the harsh conditions of the upper GI tract. As colon is rich in lymphoid tissues, it offers opportunities for the oral delivery of vaccines targeted for release and absorption in the lower GI tract. In addition, colon delivery can be exploited to improve the bioavailability of drugs that are extensively metabolized by cytochrome P450 enzymes in the upper GI tract, because the activity of these metabolizing enzymes are relatively lower in the colonic mucosa.[5,6] Colon-specific drug delivery may also help overcome GI side effects of drugs. For example, conversion of flurbiprofen to a glycine prodrug, hydrolysable by colonic microfloral enzymes (amidases), reduced its ulcerogenic activity in rats.[7] Targeted drug delivery to the colon has been extensively studied.[1,5–17]

Colon-specific drug delivery is challenged by its distal location in the GI tract. Even localized delivery through the rectum, however, only reaches a small part of the colon and is not a patient-friendly mode of administration. Therefore, oral delivery has been explored, utilizing physiological differences in the colonic microenvironment and physiology. The aspects of colon physiology that have been exploited to develop drug-targeting strategies include the presence of unique colonic microflora, high pH, the relatively predictable transition time in the small intestine, and high intraluminal pressure inside the colon. In addition, osmotically and oxidation potential controlled DDSs, and bioadhesive polymers have been used for colonic drug delivery.

15.6.4.1 Utilization of the unique colonic microflora

Human colonic microflora consists predominantly of bacteria, which also make up to 60% of the dry mass of feces. The metabolic activities of this microflora results in the salvage of absorbable nutrients from diet by fermenting unused energy substrates, trophic effects on the epithelium, and protection of the colonized host against invasion by alien microbes.[18] Colonic bacteria are mostly gram negative and anaerobic, except cecum, which can have high amount of aerobic bacteria. Bacteria in the proximal part of the colon are primarily involved in fermenting carbohydrates, whereas the latter part breaks down proteins and amino acids.

The unique metabolic ability of these microbes has been exploited to develop polymerics and prodrugs that are degraded by the unique enzymatic activities of colonic microflora. In particular, the azo reductase and glycosidase activities of the microflora help degrade the azo bound and glycosidic linkages. Prodrug strategy for colonic drug delivery utilizes drug conjugation with a promoiety through an azo bond, which is degraded by the colonic bacteria. Examples of such prodrugs include sulfasalazine, balsalazide, ipsalazide, olsalazide, and salicylazosulfapyridine for the treatment of inflammatory bowel disease. As shown in Figure 15.5, these

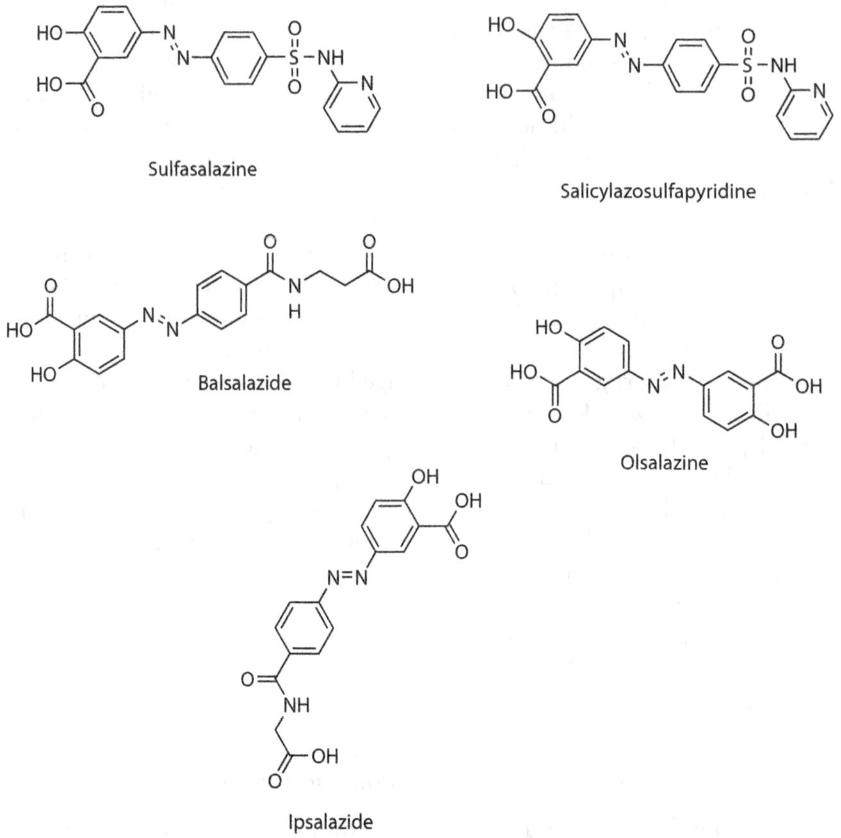

Figure 15.5 Colon-targeted drug delivery by prodrug strategy. Prodrug strategy for colonic drug delivery utilizes drug conjugation with a promoiety through an azo bond, which is reductively cleaved by the colonic anaerobic bacteria to release the parent compound. This figure shows the structure of several prodrugs of 5-amino salicylic acid (5-ASA), an anti-inflammatory compound used for the treatment of inflammatory bowel disease.

prodrugs contain an azo bond, which is reductively cleaved by the colonic anaerobic bacteria to release the anti-inflammatory compound 5-amino salicylic acid (5-ASA). Sulfasalazine was first introduced for the treatment of rheumatoid arthritis and inflammatory bowel disease. In the colon, it degrades into 5-ASA and sulfapyridine, which is responsible for most of the side effects of sulfasalazine. This problem was overcome by the use of other promoieties, such as 4-amino benzoyl glycine in ipsalazide and 4-aminobenzoyl-β-alanine in balsalazide, or azo bond conjugation of sulfasalazine with itself to form olsalazine. In addition, the drug has been covalently conjugated to a polymeric backbone of polysulfonamidoethylene by azo bond (Figure 15.5).[19]

Polymers that degrade specifically in the colon have been used for drug targeting by surface coating to form a barrier to drug release or as matrix systems embedding the drug substance. For example, azo-linked acrylate copolymers and poly(ester-ether) copolymers have been used for the delivery of protein and peptide drugs, and small molecular weight compounds such as ibuprofen, sulfasalazine, and betamethasone.[20-23] For embedding the drug in polymer matrices, natural polysaccharides have been used in oral solid dosage forms to protect the drug during GI transit and release in the colon on polymer degradation by the microflora. They offer advantages such as the presence of derivatizable functional groups and a range of molecular size, in addition to their low toxicity. The hydrogel (hydrophilic and swelling) properties of these polymers, however, can lead to the dosage form swelling and disintegration in the presence of water before reaching the colon. Therefore, these dosage forms require protection from the aqueous environment during upper GI transit. This is usually accomplished by the use of protective surface coating or chemical cross-linking with linkers that are degraded in the colon. Polymers that are stable in the upper GI tract and degraded by colonic microflora include azo cross-linked synthetic polymers and plant polysaccharides, such as amylose, pectin, inulin, and guar gum.[24-32]

A disadvantage of polymeric coating or embedding approaches for colonic drug delivery is their dependence on the bacterial microflora in the large intestine. Although the microflora is fairly constant in the healthy population, it can be affected by the dietary fermentation precursors, type of diet consumed, and coadministration of antibiotics. In addition, the natural polymers are often not available in pure form, which can lead to physicochemical incompatibility with the drug substance and/or inconsistency of product performance.

15.6.4.2 pH-dependent dosage forms

pH-sensitive polymers have been widely used for enteric coating of dosage forms to facilitate pH-dependent drug release. As the pH increases progressively from stomach (pH 1–2) to small intestine (pH 6–7), and the distal ileum (pH 7–8), dosage forms can be coated with polymers that dissolve only the aforementioned specific pH ranges. For colon targeting, the polymeric coating should be able to withstand the acidic pH of the stomach and higher pH of the proximal small intestine, but dissolve in the neutral to slightly basic pH of the terminal ileum. However, most of the commonly used enteric coating polymeric systems have a pH threshold of 6.0 or lower for dissolution. These include the methacrylic acid/methyl methacrylate copolymers, (Eudragits® L100, L-30D, L100-55), polyvinylacetate phthalate (PVAP), hydroxypropyl methylcellulose phthalate (HPMCP), cellulose acetate phthalate (CAP), and cellulose acetate trimelliate (CAT). Only Eudragit® S100 and FS 30D have a higher pH threshold of 6.8 and 7.0, respectively.[12]

Eudragit® S100 coating is used, for example, in the mesalamine (Asacol®, Procter &Gamble)-delayed release tablets for topical anti-inflammatory action in the colon. Eudragit® L100 and S100 are copolymers of methacrylic acid and methyl methacrylate with the ratio of carboxyl to ester groups of 1:1 or 1:2, respectively. The carboxylate groups form salts, leading to polymer dissolution at basic pH. Drug release from these acrylate polymers also depend on the plasticizer, nature of the salt in the dissolution medium, and permeability of the film. Colon-targeted dosage forms utilizing methacrylate resins for coating or matrix formation have been reported in several molecules such as bisacodyl, indomethacin, 5-FU, and budesonide.[33–36]

The use of pH trigger for drug delivery to the colon, however, has the disadvantage of inconsistency in dissolution of the polymer at the desired site due to inter- and intraindividual pH variation, among other factors. For example, Ashford et al. observed significant variability in the disintegration time and location of Eudragit® S coated tablets in human volunteers.[37] In addition, based on GI motility, polymer dissolution can complete toward the end of the ileum or deep in the colon. In addition, factors such as the presence of short-chain fatty acids and residues of bile acids in the luminal contents, and the locally formed fermentation products can reduce the local pH, thus influencing the drug release mechanism.[12]

15.6.4.3 Time-dependent drug release

Human small intestinal transit time for pharmaceutical dosage forms was measured using gamma scintigraphy and found to be about 3–4 h.[38] Although the transit time does vary with the amount of food and the type of dosage form, it is less variable than the gastric emptying time.[39] Timed release of dosage forms to target the colon are, thus, typically formulated to prevent drug release in the acidic gastric environment and to prevent the release of drug until 3–4 s after leaving the acidic gastric environment.

An example of such timed-release dosage form is the Pulsincap® device. In this device, the drug formulation is sealed in an impermeable capsule body with a hydrogel polymer plug. The hard gelatin capsule body may be made insoluble by exposure to formaldehyde vapor, which cross-links gelatin. The plug expands in the aqueous GI tract fluid and exits the body, thus releasing drug, after a time delay determined by the rate of expansion and the length of the plug.[40,41]

Another approach utilized a three-layer coated dosage form with an inner coating of an acid-soluble polymer, Eudragit® E; followed by a water-soluble coat, and the outer enteric coating of Eudragit® L. An organic acid (succinic acid) was used as a part of the formulation. On oral administration, the dosage form is protected in the acidic gastric environment by the enteric coating. In intestinal conditions, water ingress into the formulation lowers the pH inside the dosage form by the dissolution of the organic acid.

This, in turn, causes the inner, acid-labile coat to dissolve, thus releasing the drug. Drug release rate and lag time is controlled by the coating thickness of the acid-soluble layer and the amount of organic acid in the formulation. Using this approach, Fukui et al. prepared timed-release press-coated tablets with the core tablets containing diltiazem hydrochloride (DIL) and the outer, water soluble, layer containing phenylpropanolamine hydrochloride (PPA), as a marker for gastric emptying time.[42] On administration to beagle dogs, the gastric emptying time and lag time after gastric emptying were evaluated by determining the times at which PPA and DIL first appeared in the plasma, which were about 4 and 7 h, respectively. The 3 h lag time between the time of appearance of these drugs in the plasma correlated well with the expected intestinal transit time.

An inherent limitation of the time-dependent drug release systems inter- and intraindividual variability in gastric emptying, and small intestinal and colonic transit time. This can result in variations in the site of drug release in the small intestine or within the colon, which can impact drug absorption as absorption by the transcellular route diminishes in the distal colon.[43]

15.6.4.4 Osmotically controlled drug delivery systems

Osmotic DDSs, such as the OROS-CT® system of Alza Corporation, are based on the incorporation of an osmotic agent, such as a salt, in the dosage form. The dosage form is encapsulated in a semipermeable membrane with an orifice for drug release. On ingestion, osmotic pressure gradient forces the ingress of water, which leads to the formation of flowable gel in the drug compartment and generates pressure to force the drug gel out of the orifice at a controlled rate.[12] Amount of the osmotic agent, rate of water permeation, and size of the laser-drilled orifice primarily determine the drug release rate. The release rate can be extended for 4–24 h in the colon and the each osmotic unit is designed for a 3–4 hpostgastric delay for drug release.

A modification of the osmotic pump suitable for colonic drug delivery involves microbially triggered release mechanism. Liu et al. exploited the gelation of chitosan under acidic conditions and its degradation in the colon to use it as an osmotic agent and as a pore-forming agent in the impermeable cellulose acetate membrane.[44] The authors designed a dosage form containing citric acid and chitosan in the drug containing core, which had a coating of cellulose acetate and chitosan, followed by an enteric coat of methacrylic acid/methyl methacrylate copolymer, Eudragit® L100. As shown in Figure 15.6a, on reaching the small intestine, the enteric coat dissolves followed by water permeation into the core, leading to the formation of a flowable gel through dissolution of citric acid and swelling of chitosan. However, chitosan in the cellulose acetate membrane is completely dissolved only in the colonic microenvironment, thus preventing significant drug release until the dosage form reaches the colon. Figure 15.6b shows drug (budesonide, used as a model drug)-release inhibition at gastric and intestinal pH and controlled

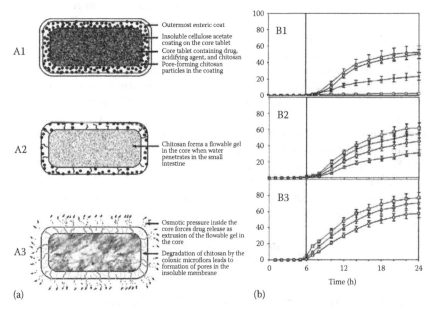

Figure 15.6 Drug delivery to the colon. An osmotic pump colonic drug delivery system that utilizes gelation of chitosan under acidic conditions and its degradation in the colon by the local microflora: (a) core tablets contain both the drug and chitosan. Cores are coated with a semipermeable membrane of cellulose acetate and chitosan, followed by the outermost enteric coating of Eudragit® L 100. The dosage form stays intact in the stomach environment (Figure A1). Dissolution of the enteric coat in the small intestine is followed by water penetration into the core and formation of a flowable gel (Figure A2). When the dosage form arrives in the colon, the colonic microflora degrade chitosan particles in the coating leading to pore formation in the coat (Figure A3). This allows the flowable gel in the core of the tablet to extrude out from the semipermeable cellulose acetate coating in the colon and (b) *in vitro* drug release from this formulation was inhibited in the simulated gastric and intestinal fluids, which represent the first 6 h of dissolution profile. The dosage form was exposed to the simulated colonic fluid (SCF) from 6 h to 24 h. In SCF, drug release was a function of chitosan/citric acid ratio (Sub-Figure B1; where weight ratio of chitosan/citric acid is represented by —□—, 1:1; —✳—, 1:1.6; —◇—, 1:2; and —△—, 1:2.6), amount of chitosan in the core (Figure B2; where the amount of chitosan is represented by —□—, 55 mg; —✳—, 60 mg; —◇—, 50 mg; and —△—, 40 mg), and the thickness of the cellulose acetate coating (Figure B3; where % weight gain of the coating is represented by —□—, 10 %; —△—, 12 %; —◇—, 14 %). (Modified from Liu et al., *Int. J. Pharm.*, 332, 115, 2007. With Permission.)

release in the simulated colonic fluid (SCF), which was a function of the amounts of chitosan and citric acid, and the coating thickness. On similar lines, Kumar et al. designed a metronidazole delivery system using guar gum as a pore-forming agent and showed *in vitro* drug-release characteristics that demonstrated its potential for colon targeting.[15]

15.7 COMBINATION AND OTHER STRATEGIES FOR COLON TARGETING

Physiological differences between the colon and the small intestine, such as intraluminal pressure and the level of hydration, can also be utilized to design a colon-targeted DDS. For example, Takada and colleagues[45] utilized the higher intraluminal pressure in the colon and its low hydration state as a trigger mechanism for drug release. To utilize this as a trigger for drug release, the authors prepared liquid-filled hard gelatin capsules coated with an insoluble ethyl cellulose film. The drug was dissolved in a water soluble or insoluble semisolid base, such as PEG 1000, which liquifies at body temperature. After oral administration of the capsule, it behaves as a flexible membrane balloon with encapsulated drug, thus maintaining integrity during small intestinal transit. On reaching the colon, reabsorption of water leads to increased viscosity of the contents of the ethyl cellulose balloon, leading to its fragility and disintegration under higher pressure. The authors identified the thickness of the water-insoluble ethyl cellulose membrane as the key factor that controls drug release. Using this system, the authors demonstrated targeted delivery to the human colon using caffeine as a model drug[46,47] and glycyrrhizin in dogs.[48]

In addition to targeted drug release in the colon, the dosage form may incorporate a bioadhesive polymer to prolong the duration of time the dosage form stays in the colon. The polymers that can be used for this purpose include polycarbophils, polyurethanes, and poly(ethylene oxide—propylene oxide) copolymers. Utilizing this strategy, Kakoulides et al. synthesized azo cross-linked bioadhesive acrylic polymers. The cross-linking prevents hydration and swelling in the upper intestinal tract.[49] On degradation of azo bonds in the large intestine, hydrogel swelling and bioadhesion was expected to lead to drug release and prolonged residence in the colonic environment.[50,51]

Similarly, Gao et al. synthesized a conjugate of bioadhesive polymer N-(2-hydroxypropyl)methacrylamide (HPMA) and the drug 9-aminocamptothecin (9-AC) via a spacer containing a combination of an aromatic azo bond and a 4-aminobenzylcarbamate group.[52] The spacer was designed to release the drug by azo bond cleavage in the colonic microenvironment. In subsequent studies, the authors observed colon targeting in mouse[53] and rat[54] models for the treatment of colon cancer. After oral administration of equal doses of the polymer conjugate or free 9-AC to mice, colon-specific release of 9-AC produced high local concentrations with the mean peak concentration of 9-AC in cecal contents, feces, cecal tissue, and colon tissue being 3.2, 3.5, 2.2, and 1.6-fold higher, respectively. Therefore, the authors anticipated higher antitumor efficacy of the polymer conjugate due to prolonged colon tumor exposure to higher and more localized drug concentrations.

Combination strategies for colon-specific drug delivery commonly utilize a combination of pH and colonic microflora-based strategies. For example,

Kaur and Kim prepared prednisolone beads with multiple coating layers for colonic delivery of the anti-inflammatory compound.[55] The authors coated prednisolone on nonpareil beads followed by a hydrophobic coat of Eudragit® RL/RS; followed by a layer containing chitosan, succinic acid, and Eudragit® RL/RS; followed by an outermost enteric coat layer (Figure 15.7a). *In vitro* experiments showed absence of drug release in simulated gastric and intestinal fluids, followed by drug release in the

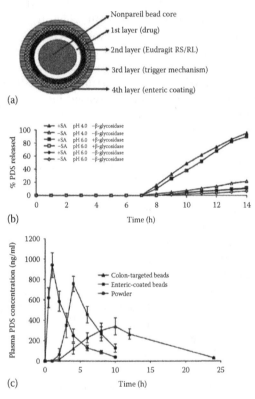

Figure 15.7 Colonic drug targeting. Combination strategy for colonic drug targeting using an oral solid dosage form: (a) design of the targeted drug delivery system. Predniosolone (PDS, drug) was coated on nonpareil beads (1st layer), followed by a hydrophobic coat of Eudragit® RS/RL polymers (2nd layer), which was followed by a layer of Eudragit® RS/RL polymers in combination with chitosan and succinic acid (3rd layer), and the outermost enteric coating layer of Eudragit® L 100 (4th layer), (b) *In vitro* drug release from the system as a function of pH, succinic acid (SA) content in the formulation, and β-glucosidase content in the dissolution medium. The formulation dissolution was carried out in the gastric fluid for the first 2 h, followed by the small intestinal fluid for next 5 h, and the pathological colonic fluid for the last 7 h, and (c) plasma drug concentration after oral administration of powder, enteric coated, or colon targeted drug delivery systems in rats. (Modified from Kaur, K., Kim, K., *Int. J. Pharm.*, 366, 140, 2009. With Permission.)

pathological colonic fluid with rate dependence on the presence of succinic acid in the formulation and the presence of the enzyme β-glucosidase (Figure 15.7b). The authors proposed a combination mechanism of drug release that involved pH-triggered enteric dissolution of the outermost layer, followed by chitosan and Eudragit® swelling in the presence of succinic acid, and biodegradation of chitosan by the colonic bacteria. Organic acid interacts with the amine groups in Eudragit® and chitosan polymers, leading to increased permeability of the coating and facilitated drug release at the colonic site. On oral administration of this formulation to male Sprague-Dawley rats, significant delay in the time to maximum plasma drug concentration (T_{max}) was obtained compared to both unmodified powder and enteric-coated tablet formulations, thus indicating colonic targeting (Figure 15.7c).

15.7.1 Kidney-targeted drug delivery

Kidney-targeted drug delivery is quite promising to improve drug efficacy and safety in the treatment of renal diseases.[1] Renal targeting is valuable to avoid extrarenal side effects of drugs used in the treatment of kidney diseases or to optimize the intrarenal distribution of a drug candidate, thus increasing its therapeutic index. Although renal drug delivery is not well studied, it highlights the challenges and opportunities inherent in developing a targeted DDS. Among the drugs used for the treatment of kidney diseases are anti-inflammatory and antifibrotic compounds. Specific drug delivery to the kidney may also be helpful during shock, renal transplantation, ureteral obstruction, diabetes, renal carcinoma, and other diseases such as Fanconi and Bartter's syndrome.[57] In addition, renal targeting can be helpful for drugs that would otherwise be rapidly metabolized and inactivated before reaching the kidney and to overcome or minimize the effects of pathological conditions, such as proteinuria, on drug distribution to the target site.

15.7.1.1 Cellular drug targets

Three cellular drug targets have been identified within the kidney—(1) proximal tubular cells, (2) mesangial cells, and (3) fibroblasts.[57] Nephron, the functional unit of the kidney, consists of a renal corpuscle and a renal tubule. The renal corpuscle is responsible for blood filtration. It consists of the glomerulus and the Bowman's capsule. The renal tubule consists of proximal and distal convoluted tubules interconnected by the loop of Henle. After blood filtration through the glomerulus, the proximal convoluted tubule is responsible for pH regulation and reabsorption of salts and organic solutes from the filtrate. The luminal surface of the proximal tubular cells has a brush-border epithelium, with densely packed microvilli, which help increase the luminal surface area.

Mesangium, or the mesangial tissue, constitutes the inner layer of glomerulus, within the basement membrane of the renal corpuscle. It surrounds the glomerular arteries and arterioles both within (intraglomerular) or outside (extraglomerular) the glomerulus. The glomerular epithelium is fenestrated and there is no basement membrane between the glomerular capillaries and the mesangial cells. Hence, mesangial cells are separated from the capillary lumen by only a layer of endothelial cells. Mesangial cells are phagocytic in nature and secrete an amorphous, basement membrane-like material, known as the mesangial matrix. These cells generate inflammatory cytokines and are involved in the uptake of macromolecules.

Fibroblasts synthesize ECM and collagen. Excessive production and accumulation of the ECM lead to fibrosis. Renal fibrosis is the underlying process that leads to the progression of chronic kidney disease to end-stage renal disease. It involves changes in the renal vasculature, glomerulosclerosis, and tubulointerstitial fibrosis. Of these, tubulointerstitial fibrosis is considered to be the most consistent predictor of an irreversible loss of renal function and progression to end-stage renal disease.[58] The accumulation of ECM components in fibrotic disease is attributed to the activation of resident interstitial fibroblasts. Therefore, targeted drug delivery to renal fibroblasts has been attempted. For example, Kushibiki et al. used cationized gelatin to complex an enhanced green fluorescent protein (EGFP) expressing plasmid, which was injected into the left kidney of mice through the ureter. The authors observed significant EGFP expression in the fibroblasts residing in the renal interstitial cortex.[59] Similarly, Xia et al. reported the delivery of siRNA targeted against heat shock protein 47 (HSP47) using cationized gelatin microspheres to the mice kidneys with tubulointerstitial fibrosis. The authors observed that the cationized gelatin microspheres enhanced and prolonged the antifibrotic effect of the siRNA.[60]

Of these cell types, the proximal tubular cells have been the target of most drug-delivery strategies. They are metabolically the most active cells in the kidney and are involved with the transport[61] and metabolism[62] of several organic and inorganic substrates. Consequently, they have specific transporter receptors on their luminal and basolateral membranes for substrate exchange between the blood and the urine. These transport and metabolic functions of the proximal tubular epithelial cells are utilized for drug targeting.

15.7.1.2 Particulate systems

The lack of basement membrane in the glomerular capillaries makes mesangial cells in close contact with the bloodstream, being separated from the capillary lumen by only a layer of endothelial cells. The mesangial cells, therefore, can be targeted using particulate carrier systems that may not filter through the glomeruli. Tuffin et al. used OX7-coupled immunoliposomes to target renal mesangial cells.[63] The authors coupled OX7 monoclonal antibody F(ab')$_2$ fragments, directed against the mesangial cell expressing Thy1.1 antigen, on the surface of doxorubicin-loaded immunoliposomes. The authors observed specific targeting to rat mesangial cells *in vitro* and *in vivo* on intravenous administration. Administration of doxorubicin-encapsulated immunoliposomes resulted in significant glomerular damage, with low damage to other parts of the kidney and other organs. The targeted localization was not observed with free drug or liposomes, and immunoliposome localization was blocked by coadministration of free antibody fragments.

In a later study, the authors attempted to correlate the biodistribution of these immunoliposomes with the tissue distribution of the antigen.[64] The Thy1.1 antigen showed high expression in rat glomeruli, brain cortex and striatum, and thymus; and moderate expression in the collecting ducts of the kidney, lung, and spleen. The biodistribution of immunoliposomes did not correlate well with the tissue distribution of Thy1.1 antigen, with the highest levels seen in the spleen, followed by lungs, liver, and kidney. Within the kidney, specific localized delivery to the mesangial cells was observed, which was sensitive to competition with the unbound OX7 monoclonal antibody fragments. The authors concluded that the absence of endothelial barriers and high target antigen density are important factors governing tissue localization of immunoliposomes.

An application of drug targeting to glomerular endothelial cells to reduce systemic side effects of drug therapy was reported by Asgeirsdottir et al., who used immunoliposomes to target glomerular endothelial cells in mice.[65] Glomerulonephritis, a spectrum of inflammatory diseases specifically affecting renal glomeruli, is characterized by the activation of proinflammatory pathways, resulting in glomerular injury and proteinuria. These disorders are frequently treated with glucocorticoids, such as dexamethasone, in combination with cytotoxic agents, such as cyclophosphamide, as anti-inflammatory and immunosuppressive agents. These drugs, however, present serious extrarenal side effects including an increase in blood glucose levels with dexamethasone.[66,67] Asgeirsdottir et al. coupled monoclonal rat anti-mouse E-selectin antibody, MES-1, to the surface of liposomes. The selection of this antibody was designed to target glomerular endothelial cells in glomerulonephritis, wherein endothelial cell expression of inflammation-related cell-adhesion molecules, such as E-selectin and VCAM-1, is upregulated. The authors obtained site-specific delivery of immunoliposome encapsulated anti-inflammatory agent dexamethasone and observed reduction in glomerular proinflammatory gene expression with no effect on blood glucose levels.

In addition to liposomes, nanoparticles have been utilized for drug targeting to the mesangial cells. For example, Manil et al. used isobutylcyanoacrylate nanoparticles for targeting the antibiotic actinomycin D to rat mesangial cells.[68] Compared to the free drug, the uptake of drug-loaded nanoparticles in the whole kidneys was over two fold at both 30 and 120 min after intravenous injection. Similar or higher uptake ratios were obtained for isolated rat glomeruli, but not for tubules. The glomerular uptake of nanoparticles was even higher in rats with experimental glomerulonephritis. Mesangial cell targeting was indicated by *in vitro* experiments, which demonstrated fivefold higher uptake by mesangial cells than the epithelial cells. In a separate study, Guzman et al. also obtained about twofold higher *in vitro* uptake of drug-loaded nanoparticles in rat mesangial cells using polycaprolactone as the polymeric carrier and digitoxin as the drug candidate.[69]

15.7.1.3 The prodrug approach

Prodrugs are drug conjugates designed to modify the physicochemical and/or biopharmaceutical properties of the drug candidate. Their derivatization is bioreversible and is designed to improve drug properties with respect to solubility, stability, permeability, presystemic metabolism, and targeting.[70] Prodrugs retain the advantages of low molecular weight compounds such as low immunogenicity and feasibility of oral administration. Renal specificity of prodrugs would depend on the renal-specific metabolism and/or uptake of the promoiety. For this purpose, amino acid prodrugs, which can be activated by kidney-specific enzymes, have been evaluated for renal targeting.

Amino acid prodrugs have advantages of biodegradability in addition to receptor-mediated uptake, which can help in both oral absorption and organ or tissue-specific targeting. For example, valine prodrugs of acyclovir and ganciclovir showed 3–5 times higher bioavailability than the parent compounds.[71,72] Enhanced oral absorption of amino acid prodrugs is attributed to carrier-mediated intestinal uptake via transporters.[73–76] For organ and tissue-specific drug targeting, the L-glutamate transport system has been commonly utilized.[77]

Prodrug design for renal targeting is aimed at utilizing the kidney-specific enzymes. The proximal tubular cells contain high levels of metabolizing enzymes in the cytosol (such as L-amino acid decarboxylase, β-lyase, and N-acetyl transferase) and at the brush border (such as γ-glutamyl transpeptidase). Examples of renal-targeted prodrugs include the γ-glutamyl prodrugs of L-dopa and sulfamethoxazole.

Gludopa (γ-L-glutamyl-L-dopa) is a kidney-specific dopamine prodrug. Cummings et al. reported its pharmacokinetic and tissue distribution in rats.[78] Gludopa was metabolized primarily in the liver and kidney, with dopamine being the major kidney metabolite. The pharmacokinetics of gludopa in healthy human volunteers indicated urinary dopamine excretion in parallel with urinary levodopa excretion, supporting the view that levodopa was the precursor of urinary dopamine.[79] Based on these results, Boateng et al. indicated that gludopa may be useful in conditions where renal effects of dopamine are indicated. However, Lee noted the limitations in clinical practice posed by its low oral bioavailability in humans.[80]

Kidney-specific delivery of parent compounds after IV administration of γ-L-glutamyl (G) and N-acetyl-γ-L-glutamyl (AG) prodrugs of p-nitroaniline, sulfamethoxazole, and sulphamethizole was investigated by Murakami et al. in rats.[81] The authors observed higher plasma stability of AG over G prodrugs for all compounds. The concentration of parent compounds was higher in the kidney than the pulmonary and hepatic tissue for all compounds, with markedly increased kidney distribution of AG prodrugs of p-nitroaniline and sulfamethoxazole. The activation of AG prodrugs requires the action of two enzymes—deacylation by N-acylamino acid deacylase and hydrolysis by γ-glutamyl transpeptidase, whereas G prodrugs can be activated by the action of γ-glutamyl transpeptidase alone. When biodistribution of G pro-drugs of sulfamethoxazole was studied in mice, relatively high concentrations of sulfamethoxazole were found in nonrenal tissues as well indicating rapid kinetics of enzymatic cleavage of G prodrugs even in tissues with low γ-glutamyl transpeptidase activity.[82] However, kidney selective accumulation was obtained after the administration of AG prodrugs. Drieman et al. hypothesized that the renal selectivity of AG prodrugs of sulfamethoxazole was due to a carrier-mediated transport followed by intracellular conversion of the prodrug to the active compound.[83,84]

Effective utilization of the prodrug strategy requires intensive investigation of the role of variables such as the linker groups and the promoiety modifications.[85,86] This results in an inherent complexity in prodrug design and utilization for organ or tissue-targeted drug delivery.

15.7.1.4 Bioconjugation approaches

Bioconjugation of a drug to a carrier that is significantly larger than the molecular size of the drug allows the biopharmaceutical properties of carrier to dominate the absorption and biodistribution of the conjugate. In the case of renal drug targeting to the proximal tubular cells, the conjugates would need to be filtered through the glomerular capillaries and reabsorbed by the tubular cells. Particles with a hydrodynamic diameter below 5–7 μm are rapidly filtered through the glomerulus.[87]

For this purpose, the carriers that naturally accumulate in the proximal tubular cells can be used as drug carriers. These include the low (less than about 30 kDa) molecular weight proteins (LMWPs), such as lysozyme, aprotinin, and cytochrome C. They are readily filtered through the glomerulus but selectively reabsorbed by the proximal tubular cells (Figure 15.8a).[88] Thus, LMWP–drug conjugates that are stable in the plasma but cleaved within the proximal tubular cells after endosomal/lysosomal uptake can be used as effective vehicles for drug targeting. Drugs may be conjugated to LMWPs directly using the lysine amino groups or with a spacer.[89] Relatively large size of the LMWPs allows the pharmacokinetic properties of the LMWPs to override those of the drug candidate in the LMWP–drug conjugates. This approach, however, is limited by the

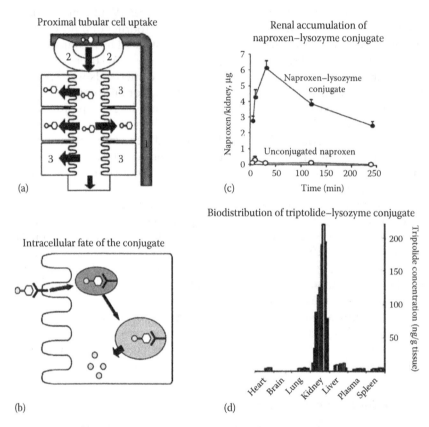

Figure 15.8 Drug delivery to the kidney. Proximal tubular cell targeting pathway and the biodistribution advantage of low molecular weight protein (LMWP) conjugation of small molecule drugs, using lysozyme-drug conjugate as an example: (a) The conjugate (⊶) in the bloodstream (1) is filtered through glomeruli (2) and actively endocytosed by the proximal tubular cells (3) through the megalin receptor (⊶) on its luminal brush border endothelium. (b) The conjugate (⊶) is entrapped within the endosome (⊶), which converts to a lysosome with the lowering of pH and degradation of the protein, thus releasing the drug. (c) Renal accumulation of naproxen after intravenous injection of naproxen (open symbols) or naproxen-lysozyme (closed symbols) conjugate in rats. (d) Biodistribution of triptolide as a function of time (0.08, 0.25, 0.5, 1, 1.5, 2, 4, 8, and 12 h, from right to left in each organ) after intravenous injection of triptolide-lysozyme conjugate in rats. (Modified from Dolman et al., *Int. J. Pharm.*, 364(2), 249–257, 2008; Haas et al., *Kidney Int.*, 52(6), 1693–1699, 1997; Zhang et al., *Biomaterials*, 30(7), 1372–1381, 2009. With Permission.)

requirement of parenteral administration and potential immunogenicity of the conjugates.

The internalization of proteins in the proximal convoluted tubule epithelium cells is mediated via the multiligand megalin and cubilin receptors.[90] These cells have very high endocytic activity. After endocytosis, the protein is degraded in the lysosomes, wherein the attached drug may be released (Figure 15.8b). Lysozyme has been used as a renal carrier for the non-steroidal anti-inflammatory drug (NSAID) naproxen (Figure 15.8c),[91,92] the acetylcholinesterase (ACE) inhibitor compound captoril,[93-95] and the nephroprotective compound triptolide (Figure 15.8d).[96]

Naproxen is a carboxylic acid group-bearing compound that could be conjugated to the amine group in the lysozyme directly by an amide bond or through lactic acid spacer by an ester bond.[92] The biodistribution and degradation of these conjugates was compared with lysozyme and naproxen by themselves in rats. Drug conjugation did not affect the renal uptake or degradation of lysozyme in the rat kidney.[91] The pharmacokinetic profile of the conjugates was similar to that of lysozyme, but markedly different from the drug. The drug was rapidly taken up by and degraded in the kidney with no detectable levels in the plasma (Figure 15.8c). Similar results were obtained when captopril was conjugated with lysozyme through a spacer utilizing disulfide linkage. Targeting this ACE inhibitor to the kidney was hypothesized to prevent attenuation of renoprotective (antiproteinuric) efficacy of captopril under high sodium concentrations. The drug was efficiently targeted to the kidney with the rapid release of the drug.[95]

Triptolide is an immunosuppressive and anti-inflammatory natural compound with low water solubility and significant toxicity. Renal targeting of triptolide–lysozyme conjugate linked through succinyl residue was investigated in rats.[96] The authors obtained significantly higher targeting efficiency of the drug conjugate to the kidney with reversal of disease progression in renal ischemia-reperfusion injury rat model, lower hepatotoxicity, and no effect on immune and genital systems, compared to the free drug (Figure 15.8d). These results demonstrated the potential therapeutic benefits of renal drug targeting.

In addition to the use of LMWPs as drug targeting ligands, their receptor-mediated uptake can also be utilized to mitigate renal toxicity of drugs. For example, endocytosis by proximal tubular cells is responsible for the renal accumulation and toxicity of aminoglycoside antibiotics, such as gentamicin, which is a substrate of the megalin receptors. Watanabe et al. found that coadministration of cytochrome C competes with receptor-mediated renal uptake of gentamicin, thus reducing its renal accumulation in rats.[97] However, the required dose of cytochrome C was quite high; the authors tested the relative efficacy of peptide fragments in reducing the renal accumulation of gentamicin. Three peptide fragments derived from actin-regulating proteins were identified that reduced the renal accumulation of gentamicin without affecting its plasma concentration-time profile.[97]

In addition to the exploitation of LMWPs for modulating the pharmacokinetics and biodistribution of drugs by utilizing their physiological disposition to modify drug biodistribution, drugs and enzymes can also be targeted to the renal proximal tubular epithelial cells by their surface modification. For example, Inoue et al. modified the enzyme superoxide dismutase (SOD), which disproportionates the superoxide free radical into oxygen and hydrogen peroxide—thus reducing free radical and oxidative stress in the cells.[98] Intravenously administered Cu, Zn–SOD is rapidly removed from the circulation with a half-life of about 5 min and appears intact in the urine, thus indicating that it is filtered through the glomerulus. The authors conjugated hexamethylene diamine (AH) to SOD. The conjugate (AH–SOD) was rapidly filtered through the glomeruli but bound apical plasma membranes of proximal tubular cells followed by localized action in these cells. The authors observed more than 80% of the radioactivity derived from AH–SOD localized in the kidney at 30 min after injection, most of which was localized in the proximal tubular cells. *In vitro* kinetic studies revealed that the specific binding of AH–SOD to apical surface of the tubular cells was attributable to AH.[98]

Polymeric carriers have also been described for renal drug targeting. These include the anionized derivatives of polyvinylpyrrolidone (PVP),[99,100] low molecular weight N-(hydroxypropyl) methylacrylamide (HPMA),[101] and low molecular weight chitosan.[102] The use of synthetic polymers requires surface modification and derivatizing groups for optimum renal accumulation. For example, PVP by itself does not accumulate in the tubular epithelial cells, but on copolymerization with maleic acid, it selectively distributed into the kidneys on IV injection in mice.[100] When anionized derivatives of PVP were prepared, the plasma clearance of these derivatives decreased with increasing size of anionic groups. In addition, even though the clearance of carboxylated PVP and sulfonated PVP from the blood was similar, renal accumulation of carboxylated PVP was several fold higher than that of sulfonated PVP.[100] In summary, these studies demonstrate not only the potential for renal targeting of drugs where it may be beneficial but also the potential to prevent accumulation in the kidney for drugs that have renal toxicity.

REVIEW QUESTIONS

15.1 Which of the following routes of drug administration does *not* bypass the hepatic first-pass metabolism?
   A. Oral route of drug delivery
   B. Sublingual absorption of a drug
   C. Transdermal route of drug delivery
   D. Vaginal route of drug absorption
15.2 Which of the following is *not* a barrier to the absorption of lipophilic molecules across the epithelial cell lining of the blood–brain barrier?
   A. Metabolizing enzymes
   B. Efflux transporters
   C. Liophilic cell membrane
   D. Intercellular tight junctions

15.3 Which of the following cell types is primarily responsible for the secretion of fibrinogen in damaged liver, which contributes to hepatic cirrhosis?
A. Hepatocytes
B. Kupffer cells
C. Hepatic stellate cells
D. Epithelial cells

15.4 Which of the following routes of drug administration into the eye is subjected to tear clearance and blinking-associated tear loss as a barrier to drug absorption?
A. Retrobulbar route
B. Intravitreal route
C. Subtenon route
D. Topical route

15.5 Which of the following types of chemical bonds is utilized for colonic drug targeting through enzymatic cleavage by the resident anaerobic bacteria to release the parent drug?
A. Azo bonds
B. Ester bonds
C. Ether bonds
D. Amide bonds

15.6 Which of the following is *not* a strategy to achieve drug release in the latter parts of the intestinal tract after oral drug administration?
A. Enteric coating
B. Osmotic pump-based sustained drug release
C. pH-dependent drug release
D. Time-dependent drug release

15.7 Which of the following is *not* a contributing factor to the low oral bioavailability of protein and peptide drugs?
A. Large hydrodynamic diameter
B. Overall hydrophilic nature
C. Cleavage of peptide bonds by enzymes and the pH environment
D. Presence of side chains on the amino acids constituting the peptide backbone

15.8 Which of the following strategies for drug delivery and targeting necessarily require covalent modification of the drug molecule?
A. Liposomes
B. Micelles
C. Nanoparticles
D. Prodrugs

15.9 Which of the following strategies increases the circulation half-life of a drug by avoidance of the reticuloendothelial cell system (RES) uptake?
A. PEGylation
B. Liposomal incorporation

    C. Prodrug strategy

    D. Incorporation in a micelle

15.10 Which of the following modes of drug absorption does *not* depend on the concentration gradient of the drug?

    A. Passive transport across endothelial cell membranes of the blood–brain barrier

    B. Drug absorption across the sublingual tissue

    C. Transporter-mediated drug uptake

    D. Drug transport across the tight junctions between the endothelial cells

REFERENCES

1. Narang AS and Mahato RI (Eds.) (2010) Targeting colon and kidney: Pathophysiological determinants of design strategy. In *Targeted Delivery of Small and Macromolecular Drugs*, Boca Raton, FL: CRC Press, pp. 351–370.
2. Narang AS and Mahato RI (Eds.) (2010) *Targeted Delivery of Small and Macromolecular Drugs*, Boca Raton, FL: CRC Press, 614 pp.
3. Mishra GP, Gaudana R, Tamboli VM, and Mitra AK (2010) Recent advances in ocular drug delivery: Role of transporters, receptors, and nanocarriers. In Narang AS, and Mahato RI, (Eds.) *Targeted Delivery of Small and Macromolecular Drugs*, Boca Raton, FL: CRC Press, Taylor & Francis Group, pp. 421–453.
4. Friedman SL (2000) Molecular regulation of hepatic fibrosis, an integrated cellular response to tissue injury. *J Biol Chem* 275(4): 2247–2250.
5. Zhang Q-Y, Dunbar D, Ostrowska A, Zeisloft S, Yang J, and Kaminsky LS (1999) Characterization of human small intestinal cytochromes P-450. *Drug Metab Dispos* 27(7): 804–809.
6. Peters WH, Kock L, Nagengast FM, and Kremers PG (1991) Biotransformation enzymes in human intestine: Critical low levels in the colon? *Gut* 32(4): 408–412.
7. Philip AK, Dubey RK, and Pathak K (2008) Optimizing delivery of flurbiprofen to the colon using a targeted prodrug approach. *J Pharm Pharmacol* 60(5): 607–613.
8. Singh BN (2007) Modified-release solid formulations for colonic delivery. *Recent Pat Drug Deliv Formul* 1: 53–63.
9. Gazzaniga A, Maroni A, Sangalli ME, and Zema L (2006) Time-controlled oral delivery systems for colon targeting. *Expert Opin Drug Deliv* 3(5): 583–597.
10. Hovgaard L and Brondsted H (1996) Current applications of polysaccharides in colon targeting. *Crit Rev Ther drug Carrier Syst* 13(3–4): 185–223.
11. Ashford M and Fell JT (1994) Targeting drugs to the colon: Delivery systems for oral administration. *J Drug Target* 2(3): 241–257.
12. Chourasia MK, and Jain SK (2003) Pharmaceutical approaches to colon targeted drug delivery systems. *J Pharm Pharm Sci* 6(1): 33–66.
13. Chourasia MK and Jain SK (2004) Polysaccharides for colon targeted drug delivery. *Drug Deliv* 11(2): 129–148.

14. Kosaraju SL (2005) Colon targeted delivery systems: Review of polysaccharides for encapsulation and delivery. *Crit Rev Food Sci Nutr* 45(4): 251–258.
15. Kumar P and Mishra B (2008) Colon targeted drug delivery systems—An overview. *Curr Drug Deliv* 5(3): 186–198.
16. Haupt S and Rubinstein A (2002) The colon as a possible target for orally administered peptide and protein drugs. *Crit Rev Ther Drug Carrier Syst* 19(6): 499–551.
17. Jain SK and Jain A (2008) Target-specific drug release to the colon. *Expert Opin Drug Deliv* 5(5): 483–498.
18. Guarner F, and Malagelada JR (2003) Gut flora in health and disease. *Lancet* 361(9356): 512–519.
19. Brown JP, McGarraugh GV, Parkinson TM, Wingard RE, Jr., and Onderdonk AB (1983) A polymeric drug for treatment of inflammatory bowel disease. *J Med Chem* 26(9): 1300–1307.
20. Kalala W, Kinget R, Van den Mooter G, and Samyn C (1996) Colonic drug-targeting: In vitro release of ibuprofen from capsules coated with poly(ether-ester) azopolymers. *Int J Pharm* 139(1,2): 187–195.
21. Van den Mooter G, Samyn C, and Kinget R (1995) In vivo evaluation of a colon-specific drug delivery system: An absorption study of theophylline from capsules coated with azo polymers in rats. *Pharm Res* 12(2): 244–247.
22. Saffran M, Kumar GS, Neckers DC, Pena J, Jones RH, and Field JB (1990) Biodegradable azopolymer coating for oral delivery of peptide drugs. *Biochem Soc Trans* 18(5): 752–754.
23. Van den Mooter G, Samyn C, and Kinget R 1992. Azo polymers for colon-specific drug delivery. *Int J Pharm* 87(1–3): 37–46.
24. Vaidya A, Jain A, Khare P, Agrawal RK, and Jain SK (2009) Metronidazole loaded pectin microspheres for colon targeting. *J Pharm Sci* 98(11): 4229–4236.
25. Hodges LA, Connolly SM, Band J, O'Mahony B, Ugurlu T, Turkoglu M, Wilson CG, and Stevens HN (2009) Scintigraphic evaluation of colon targeting pectin-HPMC tablets in healthy volunteers. *Int J Pharm* 370(1–2): 144–150.
26. Freire C, Podczeck F, Veiga F, and Sousa J (2009) Starch-based coatings for colon-specific delivery. Part II: Physicochemical properties and in vitro drug release from high amylose maize starch films. *Eur J Pharm Biopharm* 72(3): 587–594.
27. Calinescu C and Mateescu MA (2008) Carboxymethyl high amylose starch: Chitosan self-stabilized matrix for probiotic colon delivery. *Eur J Pharm Biopharm* 70(2): 582–589.
28. Wilson PJ and Basit AW (2005) Exploiting gastrointestinal bacteria to target drugs to the colon: An in vitro study using amylose coated tablets. *Int J Pharm* 300(1–2): 89–94.
29. Pitarresi G, Tripodo G, Calabrese R, Craparo EF, Licciardi M, and Giammona G (2008) Hydrogels for potential colon drug release by thiol-ene conjugate addition of a new inulin derivative. *Macromol Biosci* 8(10): 891–902.
30. Maris B, Verheyden L, Van Reeth K, Samyn C, Augustijns P, Kinget R, and Van den Mooter G (2001) Synthesis and characterisation of inulin-azo hydrogels designed for colon targeting. *Int J Pharm* 213(1–2): 143–152.
31. Ji CM, Xu HN, and Wu W (2009) Guar gum as potential film coating material for colon-specific delivery of fluorouracil. *J Biomater Appl* 23(4): 311–329.

32. Ji C, Xu H, and Wu W (2007) In vitro evaluation and pharmacokinetics in dogs of guar gum and Eudragit FS30D-coated colon-targeted pellets of indomethacin. *J Drug Target* 15(2): 123–131.

33. Kelm GR, Kondo K, and Nakajima A. (1998) Bisacodyl dosage form with multiple enteric polymer coatings for colonic delivery. Patent: 5843479. The Procter & Gamble Company, Cincinnati, OH.

34. Makhlof A, Tozuka Y, and Takeuchi H (2009) pH-Sensitive nanospheres for colon-specific drug delivery in experimentally induced colitis rat model. *Eur J Pharm Biopharm* 72(1): 1–8.

35. Asghar LF, Chure CB, and Chandran S (2009) Colon specific delivery of indomethacin: Effect of incorporating pH sensitive polymers in xanthan gum matrix bases. *AAPS PharmSciTech* 10(2): 418–429.

36. Zambito Y, Baggiani A, Carelli V, Serafini MF, and Di Colo G (2005) Matrices for site-specific controlled-delivery of 5-fluorouracil to descending colon. *J Control Release* 102(3): 669–677.

37. Ashford M, Fell JT, Attwood D, Sharma H, and Woodhead PJ (1993) An in vivo investigation into the suitability of pH dependent polymers for colonic targeting. *Int J Pharm* 95(1–3): 193–199.

38. Wilson CG, and Washington N (1988) Assessment of disintegration and dissolution of dosage forms in vivo using gamma scintigraphy. *Drug Dev Ind Pharm* 14(2–3): 211–281.

39. Davis SS, Hardy JG, and Fara JW (1986) Transit of pharmaceutical dosage forms through the small intestine. *Gut* 27(8): 886–892.

40. Stevens HNE (2003) Pulsincap and hydrophilic sandwich (HS) capsules: Innovative time-delayed oral drug delivery technologies. *Drugs Pharm Sci* 126(Modified-Release Drug Delivery Technology): 257–262.

41. Stevens HNE, Wilson CG, Welling PG, Bakhshaee M, Binns JS, Perkins AC, Frier M, Blackshaw EP, Frame MW, Nichols DJ, Humphrey MJ, and Wicks SR (2002) Evaluation of Pulsincap to provide regional delivery of dofetilide to the human GI tract. *Int J Pharm* 236(1–2): 27–34.

42. Fukui E, Miyamura N, Uemura K, and Kobayashi M (2000) Preparation of enteric coated timed-release press-coated tablets and evaluation of their function by in vitro and in vivo tests for colon targeting. *Int J Pharm* 204(1–2): 7–15.

43. Hebden JM, Wilson CG, Spiller RC, Gilchrist PJ, Blackshaw E, Frier ME, and Perkins AC (1999) Regional differences in quinine absorption from the undisturbed human colon assessed using a timed release delivery system. *Pharm Res* 16(7): 1087–1092.

44. Liu H, Yang X-G, Nie S-F, Wei L-L, Zhou L-L, Liu H, Tang R, and Pan W-S (2007) Chitosan-based controlled porosity osmotic pump for colon-specific delivery system: Screening of formulation variables and in vitro investigation. *Int J Pharm* 332(1–2): 115–124.

45. Hu Z, Kimura G, Mawatari S, Shimokawa T, Yoshikawa Y, and Takada K (1998) New preparation method of intestinal pressure-controlled colon delivery capsules by coating machine and evaluation in beagle dogs. *J Control Release* 56(1–3): 293–302.

46. Hu Z, Mawatari S, Shimokawa T, Kimura G, Yoshikawa Y, Shibata N, and Takada K (2000) Colon delivery efficiencies of intestinal pressure-controlled colon delivery capsules prepared by a coating machine in human subjects. *J Pharm Pharmacol* 52(10): 1187–1193.

47. Muraoka M, Hu Z, Shimokawa T, Sekino S, Kurogoshi R, Kuboi Y, Yoshikawa Y, and Takada K (1998) Evaluation of intestinal pressure-controlled colon delivery capsule containing caffeine as a model drug in human volunteers. *J Control Release* **52**(1–2): 119–129.

48. Shibata N, Ohno T, Shimokawa T, Hu Z, Yoshikawa Y, Koga K, Murakami M, and Takada K (2001) Application of pressure-controlled colon delivery capsule to oral administration of glycyrrhizin in dogs. *J Pharm Pharmacol* **53**(4): 441–447.

49. Mahkam M, and Doostie L (2005) The relation between swelling properties and cross-linking of hydrogels designed for colon-specific drug delivery. *Drug Deliv* **12**(6): 343–347.

50. Kakoulides EP, Smart JD, and Tsibouklis J (1998) Azocross-linked poly(acrylic acid) for colonic delivery and adhesion specificity: Synthesis and characterisation. *J Control Release* **52**(3): 291–300.

51. Kakoulides EP, Smart JD, and Tsibouklis J (1998) Azocrosslinked poly(acrylic acid) for colonic delivery and adhesion specificity: In vitro degradation and preliminary ex vivo bioadhesion studies. *J Control Release* **54**(1): 95–109.

52. Gao SQ, Lu ZR, Petri B, Kopeckova P, and Kopecek J (2006) Colon-specific 9-aminocamptothecin-HPMA copolymer conjugates containing a 1,6-elimination spacer. *J Control Release* **110**(2): 323–331.

53. Gao SQ, Lu ZR, Kopeckova P, and Kopecek J (2007) Biodistribution and pharmacokinetics of colon-specific HPMA copolymer--9-aminocamptothecin conjugate in mice. *J Control Release* **117**(2): 179–185.

54. Gao SQ, Sun Y, Kopeckova P, Peterson CM, and Kopecek J (2008) Pharmacokinetic modeling of absorption behavior of 9-aminocamptothecin (9-AC) released from colon-specific HPMA copolymer-9-AC conjugate in rats. *Pharm Res* **25**(1): 218–226.

55. Kaur K and Kim K (2009) Studies of chitosan/organic acid/Eudragit RS/RL-coated system for colonic delivery. *Int J Pharm* **366**: 140–148.

56. Dolman ME, Fretz MM, Segers GJ, Lacombe M, Prakash J, Storm G, Hennink WE, and Kok RJ 2008. Renal targeting of kinase inhibitors. *Int J Pharm* **364**(2): 249–257.

57. Haas M, Moolenaar F, Meijer DK, and de Zeeuw D (2002) Specific drug delivery to the kidney. *Cardiovasc Drugs Ther* **16**(6): 489–496.

58. Nangaku M (2004) Mechanisms of tubulointerstitial injury in the kidney: Final common pathways to end-stage renal failure. *Intern Med* **43**(1): 9–17.

59. Kushibiki T, Nagata-Nakajima N, Sugai M, Shimizu A, and Tabata Y (2005) Targeting of plasmid DNA to renal interstitial fibroblasts by cationized gelatin. *Biol Pharm Bull* **28**(10): 2007–2010.

60. Xia Z, Abe K, Furusu A, Miyazaki M, Obata Y, Tabata Y, Koji T, and Kohno S (2008) Suppression of renal tubulointerstitial fibrosis by small interfering RNA targeting heat shock protein 47. *Am J Nephrol* **28**(1): 34–46.

61. Christensen EI, Birn H, Verroust P, and Moestrup SK (1998) Membrane receptors for endocytosis in the renal proximal tubule. *Int Rev Cytol* **180**: 237–284.

62. Pacifici GM, Viani A, Franchi M, Gervasi PG, Longo V, Di Simplicio P, Temellini A et al. (1989) Profile of drug-metabolizing enzymes in the cortex and medulla of the human kidney. *Pharmacology* **39**(5): 299–308.

63. Tuffin G, Waelti E, Huwyler J, Hammer C, and Marti HP (2005) Immunoliposome targeting to mesangial cells: A promising strategy for specific drug delivery to the kidney. *J Am Soc Nephrol* **16**(11): 3295–3305.

64. Tuffin G, Huwyler J, Waelti E, Hammer C, and Marti HP (2008) Drug targeting using OX7-immunoliposomes: Correlation between Thy1.1 antigen expression and tissue distribution in the rat. *J Drug Target* **16**(2): 156–166.

65. Asgeirsdottir SA, Kamps JA, Bakker HI, Zwiers PJ, Heeringa P, van der Weide K, van Goor H et al. (2007) Site-specific inhibition of glomerulonephritis progression by targeted delivery of dexamethasone to glomerular endothelium. *Mol Pharmacol* **72**(1): 121–131.

66. Chadban SJ and Atkins RC (2005) Glomerulonephritis. *Lancet* **365**(9473): 1797–1806.

67. Tam FW (2006) Current pharmacotherapy for the treatment of crescentic glomerulonephritis. *Expert Opin Investig Drugs* **15**(11): 1353–1369.

68. Manil L, Davin JC, Duchenne C, Kubiak C, Foidart J, Couvreur P, and Mahieu P (1994) Uptake of nanoparticles by rat glomerular mesangial cells in vivo and in vitro. *Pharm Res* **11**(8): 1160–1165.

69. Guzman M, Aberturas MR, Rodriguez-Puyol M, and Molpeceres J (2000) Effect of nanoparticles on digitoxin uptake and pharmacologic activity in rat glomerular mesangial cell cultures. *Drug Deliv* **7**(4):215–222.

70. Stella VJ (2004) Prodrugs as therapeutics. *Expert Opin Ther Pat* **14**(3): 277–280.

71. Jung D and Dorr A (1999) Single-dose pharmacokinetics of valganciclovir in HIV- and CMV-seropositive subjects. *J Clin Pharmacol* **39**(8): 800–804.

72. Weller S, Blum MR, Doucette M, Burnette T, Cederberg DM, de Miranda P, and Smiley ML (1993) Pharmacokinetics of the acyclovir pro-drug valaciclovir after escalating single- and multiple-dose administration to normal volunteers. *Clin Pharmacol Ther* **54**(6): 595–605.

73. Amidon GL and Walgreen CR, Jr. (1999) 5'-Amino acid esters of antiviral nucleosides, acyclovir, and AZT are absorbed by the intestinal PEPT1 peptide transporter. *Pharm Res* **16**(2): 175.

74. Borner V, Fei YJ, Hartrodt B, Ganapathy V, Leibach FH, Neubert K, and Brandsch M (1998) Transport of amino acid aryl amides by the intestinal H+/peptide cotransport system, PEPT1. *Eur J Biochem/FEBS* **255**(3): 698–702.

75. Han H, de Vrueh RL, Rhie JK, Covitz KM, Smith PL, Lee CP, Oh DM, Sadee W, and Amidon GL (1998) 5'-Amino acid esters of antiviral nucleosides, acyclovir, and AZT are absorbed by the intestinal PEPT1 peptide transporter. *Pharm Res* **15**(8): 1154–1159.

76. Umapathy NS, Ganapathy V, and Ganapathy ME (2004) Transport of amino acid esters and the amino-acid-based prodrug valganciclovir by the amino acid transporter ATB(0,+). *Pharm Res* **21**(7): 1303–1310.

77. Sakaeda T, Siahaan TJ, Audus KL, and Stella VJ (2000) Enhancement of transport of D-melphalan analogue by conjugation with L-glutamate across bovine brain microvessel endothelial cell monolayers. *J Drug Target* **8**(3): 195–204.

78. Cummings J, Matheson LM, Maurice L, and Smyth JF (1990) Pharmacokinetics, bioavailability, metabolism, tissue distribution and urinary excretion of gamma-L-glutamyl-L-dopa in the rat. *J Pharm Pharmacol* **42**(4): 242–246.

79. Boateng YA, Barber HE, MacDonald TM, Petrie JC, and Lee MR (1991) Disposition of gamma-glutamyl levodopa (gludopa) after intravenous bolus injection in healthy volunteers. *Br J Clin Pharmacol* **31**(4): 419–422.

80. Lee MR (1990) Five years' experience with gamma-L-glutamyl-L-dopa: A relatively renally specific dopaminergic prodrug in man. *J Auton Pharmacol* **10**(1): s103–s108.

81. Murakami T, Kohno K, Yumoto R, Higashi Y, and Yata N (1998) N-acetyl-L-gamma-glutamyl derivatives of p-nitroaniline, sulphamethoxazole and sulphamethizole for kidney-specific drug delivery in rats. *J Pharm Pharmacol* **50**(5): 459–465.

82. Orlowski M, Mizoguchi H, and Wilk S (1980) N-acyl-gamma-glutamyl derivatives of sulfamethoxazole as models of kidney-selective prodrugs. *J Pharmacol Exp Ther* **212**(1): 167–172.

83. Drieman JC, Thijssen HH, and Struyker-Boudier HA (1990) Renal selective N-acetyl-gamma-glutamyl prodrugs. II. Carrier-mediated transport and intracellular conversion as determinants in the renal selectivity of N-acetyl-gamma-glutamyl sulfamethoxazole. *J Pharmacol Exp Ther* **252**(3): 1255–1260.

84. Drieman JC, Thijssen HH, Zeegers HH, Smits JF, Struyker, and Boudier HA (1990) Renal selective N-acetyl-gamma-glutamyl prodrugs: A study on the mechanism of activation of the renal vasodilator prodrug CGP 22979. *Br J Pharmacol* **99**(1): 15–20.

85. Hashida M, Akamatsu K, Nishikawa M, Yamashita F, and Takakura Y (1999) Design of polymeric prodrugs of prostaglandin E(1) having galactose residue for hepatocyte targeting. *J Control Release* **62**(1–2): 253–262.

86. Akamatsu K, Yamasaki Y, Nishikawa M, Takakura Y, and Hashida M (2001) Synthesis and pharmacological activity of a novel water-soluble hepatocyte-specific polymeric prodrug of prostaglandin E(1) using lactosylated poly(L-glutamic hydrazide) as a carrier. *Biochem Pharmacol* **62**(11): 1531–1536.

87. Choi HS, Liu W, Misra P, Tanaka E, Zimmer JP, Itty Ipe B, Bawendi MG, and Frangioni JV (2007) Renal clearance of quantum dots. *Nat Biotechnol* **25**(10): 1165–1170.

88. Haas M, de Zeeuw D, van Zanten A, and Meijer DK (1993) Quantification of renal low-molecular-weight protein handling in the intact rat. *Kidney Int* **43**(4): 949–954.

89. Franssen EJ, Koiter J, Kuipers CA, Bruins AP, Moolenaar F, de Zeeuw D, Kruizinga WH, Kellogg RM, and Meijer DK (1992) Low molecular weight proteins as carriers for renal drug targeting. Preparation of drug-protein conjugates and drug-spacer derivatives and their catabolism in renal cortex homogenates and lysosomal lysates. *J Med Chem* **35**(7): 1246–1259.

90. Christensen EI and Verroust PJ (2002) Megalin and cubilin, role in proximal tubule function and during development. *Pediatr Nephrol* **17**(12): 993–999.

91. Haas M, Kluppel ACA, Wartna ES, Moolenar F, Meijer DKF, de Jong PE, and de Zeeuw D (1997) Drug-targeting to the kidney: Renal delivery and degradation of a naproxen-lysozyme conjugate in vivo. *Kidney Int* **52**(6): 1693–1699.

92. Franssen EJF, Moolenaar F, de Zeeuw D, and Meijer DKF (1993) Low-molecular-weight proteins as carriers for renal drug targeting: Naproxen coupled to lysozyme via the spacer L-lactic acid. *Pharm Res* **10**(7): 963–969.

93. Windt WAKM, Prakash J, Kok RJ, Moolenaar F, Kluppel CA, de Zeeuw D, van Dokkum RPE, and Henning RH (2004) Renal targeting of captopril using captopril-lysozyme conjugate enhances its antiproteinuric effect in adriamycin-induced nephrosis. *JRAAS* 5(4): 197–202.

94. Prakash J, van Loenen-Weemaes AM, Haas M, Proost JH, Meijer DKF, Moolenaar F, Poelstra K, and Kok RJ (2005) Renal-selective delivery and angiotensin-converting enzyme inhibition by subcutaneously administered captopril-lysozyme. *Drug Metab Dispos* 33(5): 683–688.

95. Kok RJ, Grijpstra F, Walthuis RB, Moolenaar F, de Zeeuw D, and Meijer DK (1999) Specific delivery of captopril to the kidney with the prodrug captopril-lysozyme. *J Pharmacol Exp Ther* 288(1): 281–285.

96. Zhang Z, Zheng Q, Han J, Gao G, Liu J, Gong T, Gu Z, Huang Y, Sun X, and He Q (2009) The targeting of 14-succinate triptolide-lysozyme conjugate to proximal renal tubular epithelial cells. *Biomaterials* 30(7): 1372–1381.

97. Watanabe A, Nagai J, Adachi Y, Katsube T, Kitahara Y, Murakami T, and Takano M (2004) Targeted prevention of renal accumulation and toxicity of gentamicin by aminoglycoside binding receptor antagonists. *J Control Release* 95(3): 423–433.

98. Inoue M, Nishikawa M, Sato E, Matsuno K, and Sasaki J (1999) Synthesis of superoxide dismutase derivative that specifically accumulates in renal proximal tubule cells. *Arc Biochem Biophys* 368(2): 354–360.

99. Kamada H, Tsutsumi Y, Sato-Kamada K, Yamamoto Y, Yoshioka Y, Okamoto T, Nakagawa S, Nagata S, and Mayumi T (2003) Synthesis of a poly(vinylpyrrolidone-co-dimethyl maleic anhydride) co-polymer and its application for renal drug targeting. *Nat Biotechnol* 21(4): 399–404.

100. Kodaira H, Tsutsumi Y, Yoshioka Y, Kamada H, Kaneda Y, Yamamoto Y, Tsunoda S et al. (2004). The targeting of anionized polyvinylpyrrolidone to the renal system. *Biomaterials* 25(18): 4309–4315.

101. Kissel M, Peschke P, Subr V, Ulbrich K, Strunz AM, Kuhnlein R, Debus J, and Friedrich E (2002) Detection and cellular localisation of the synthetic soluble macromolecular drug carrier pHPMA. *Eur J Nucl Med Mol Imaging* 29(8): 1055–1062.

102. Yuan ZX, Sun X, Gong T, Ding H, Fu Y, and Zhang ZR (2007) Randomly 50% N-acetylated low molecular weight chitosan as a novel renal targeting carrier. *J Drug Target* 15(4): 269–278.

Part III

Dosage forms

Chapter 16

Suspensions

LEARNING OBJECTIVES

On completion of this chapter, the students should be able to

1. Define suspensions and their applications in clinical practice.
2. Describe factors that affect the physical stability of suspensions.
3. Describe the commonly used approaches for the preparation of stable suspensions.
4. Differentiate between flocculated and deflocculated suspensions.
5. List different types of flocculating agents.
6. Describe Stokes' law and its application in the stabilization of suspensions.

16.1 INTRODUCTION

Suspension is the dispersion of a solid in a liquid or gas. A pharmaceutical suspension is the dispersion of solid particles (usually a drug) in a liquid medium (usually aqueous) in which the drug is not readily soluble. This dosage form is used for providing a liquid dosage form for insoluble drugs. The particle size of the dispersed phase in most of the oral pharmaceutical suspensions is between 1 and 50 μm. The lower the particle size, the larger the surface area of the suspended drug. Lower particle size allows the formation of a uniform, fine particle suspension with low grittiness and high redispersibility. A larger surface area of the dispersed drug promotes rapid drug dissolution on dilution with biological fluids upon administration.

Aerosols, suspensions of a solid or a liquid in a gas (e.g., air), are specialized dosage forms used for pulmonary drug delivery. These dosage forms are discussed in Chapter 15.

In addition to the use of aqueous pharmaceutical suspensions as drug products, suspensions are also used as in-process materials during industrial pharmaceutical manufacturing. For example, tablets are coated with a suspension of insoluble coating materials. Granules manufactured by wet-granulation processes are typically suspended in the air for drying during fluidized bed-drying process. In addition, wet granulation could be carried out on granules suspended in the air in a process called fluid-bed granulation.

16.2 TYPES OF SUSPENSIONS

Suspensions can be classified based on the characteristics of the dispersed phase or the dispersion medium, and also based on their route of administration.

Based on the particle size (diameter) of the dispersed phase, suspensions can be classified as (1) coarse suspension (> 1 μm), (2) colloidal dispersion (< 1 μm), or (3) nanosuspension (10–100 nm). Based on the concentration of the dispersed phase, highly concentrated suspensions are termed as slurries (> 50% w/w), and certain suspensions are considered dilute suspensions (2%–10% w/w). Based on the type of the dispersion medium, suspensions can be aqueous or nonaqueous. Identifying the physical state of the dispersion medium allows the suspensions to be classified as solid-in-liquid or solid-in-gas (aerosols) suspensions.

Based on the route of administration, suspensions can be classified as oral, topical, ophthalmic, otic, or nasal suspensions. Each of these present unique challenges and requirements in terms of desired quality attributes. These are briefly described as follows:

1. *Oral suspensions*: Suspensions meant for peroral route of administration are usually liquid preparations in which solid particles of the active drug are dispersed in a sweetened, flavored, sometimes colored, and usually viscous vehicle. For example, amoxicillin oral suspension contains 125–500 mg dispersed active pharmaceutical ingredient (API) per 5 mL of suspension. When formulated for use as pediatric drops, concentration of suspended API is increased to allow lower volume of administration for pediatric doses. Antacids and radioopaque suspensions generally contain high concentrations of dispersed solids.
2. *Topical suspensions*: Lotions are externally applied suspensions. These are designed for dermatologic, cosmetic, and protective purposes. Topical suspension formulations need to pay particular attention to the lack of grittiness and smooth feel on the skin. These suspensions are typically colored and may have some perfume, but do not need sweeteners and flavors typically used for oral administration.

3. *Injectable suspensions*: Parenteral suspensions may contain from 0.5% to 30% w/w of solid particles. Viscosity and particle size are significant factors because they affect the ease of injection and the availability of the drug in depot therapy. Most parenteral suspensions are designed for intramuscular or subcutaneous administration. For example, procaine penicillin G suspension is intended for intramuscular administration. Sterility is an important consideration for parenteral suspensions. Being a suspension dosage form, they cannot be sterilized by terminal filtration. Thus, the use of sterile API and aseptic processing is required for their manufacturing. In addition, antimicrobial preservatives are not recommended for intravenous (IV) suspensions.

4. *Otic suspensions*: These are intended for administration into the ear. Most otic suspensions are antibiotics, corticosteroids, or analgesics for the treatment of ear infection, inflammation, and pain. For example, cortisporin otic suspension contains polymixin, neomycin, and hydrocortisone for antibiotic and anti-inflammatory effect. Otic suspensions are generally formulated as sterile suspensions since they come in contact with the mucosal surface.

5. *Rectal suspensions*: Local administration through the rectal cavity is used for the treatment or management of local disorders of the colon. For example, 5-acetyl salicylic acid (5-ASA) rectal suspension enema is used as an anti-inflammatory treatment for ulcerative colitis. Formulation and quality considerations for rectal suspensions are similar to the oral suspensions.

6. *Aerosols*: Aerosols are suspensions of drug particles or drug solution in the air and are used for inhalation of drug delivery to the lung. Volatile propellants are frequently used as vehicles for pharmaceutical aerosols.

7. *Liposomes and micro-/nanoparticles*: Suspensions of liposomes, microspheres, microcapsules, nanospheres, or nanocapsules are used for targeted and controlled delivery of drugs. These are usually intended for parenteral administration.

8. *Vaccines*: Vaccines are used for the induction of immunity and are often formulated as suspensions. For example, cholera vaccine and tetanus vaccine are suspensions.

16.3 POWDER FOR SUSPENSION

The inherent physical instability of suspensions and the desirability of a relatively long shelf life have led to the popularity of powder for suspension (PFS) dosage forms. These dosage forms are developed as powder mixtures of typical ingredients required for an aqueous suspension and are marketed in unit dose sachet or multidose bottles. The pharmacist reconstitutes these dosage forms with water before dispensing to the patient. The reconstituted suspension has a limited shelf life under designated storage conditions, such as 14 days under refrigeration.

1. *Unit dose powder for suspension*: A unit dose sachet of powder could be administered to a patient by sprinkling on the top of a semisolid food, such as jelly or ice cream, or by suspending in a suitable vehicle, such as water or juice, immediately before administration. This mode of administration is preferred for pediatric and geriatric populations, who may have swallowing difficulty, and for high dose compounds. A key requirement for this dosage form is the palatability of the drug. Extremely bitter or unpleasant tasting drugs are generally not suitable for formulation as PFS.
2. *Multidose powder for suspension*: The multidose PFS are dispensed as powders in a suitable-sized bottle for reconstitution with water by the pharmacist immediately before dispensing. This allows the advantage of custom flavoring by the pharmacist to increase patient compliance and the reduced requirement for the duration of physical and chemical stability of the formulation. For example, the combination of amoxicillin and potassium clavulanate is dispensed as multidose PFS in a bottle. Superior stability of the powder dosage form allows long shelf life of the commercial product at room temperature of a drug that is very unstable in the presence of water. The pharmacist reconstitutes this PFS immediately before dispensing. The reconstituted suspension is required to be stored by the patient under refrigerated conditions and consumed within 14 days.

16.4 QUALITY ATTRIBUTES

Quality attributes of suspensions include the following:

1. *Uniformity of content* (dose-to-dose within the same bottle and bottle-to-bottle): All the doses dispensed from a given multidose container should have acceptable uniformity of drug content. In addition, the drug content must be uniform between different bottles of a given batch of suspension.
2. *Settling volume*: Once a suspension has been left undisturbed for a sufficient period of time, it is likely to show some degree of separation of the dispersed phase from the dispersion medium. The proportion of the volume occupied by the separated phase, which contains a higher concentration of the dispersed solid, is an indicator of physical stability of the suspension. Higher this volume, more stable is the suspension. Thus, settling volume is measured as a quality attribute indicative of physical stability of the suspension and its changes over storage stability.

3. *Absence of particle size change and active pharmaceutical ingredient crystal growth*: Particle size distribution of the suspension should remain fairly constant over time upon storage. Dissolved drug may crystallize or contribute to the growth of existing drug particles. Crystallization during storage can lead to changes in the particle size distribution of a suspension. Additives in formulation such as hydrophilic polymers can inhibit or minimize crystal growth by adsorption on the surface of dispersed particles. For example, polyvinylpyrrolidone (PVP) can inhibit crystal growth in acetaminophen suspensions.

4. *Palatability*: Palatability of the dosage form is usually enhanced by the use of sweeteners, flavors, and colorants. For especially bitter or otherwise unpleasant tasting drugs, taste-masking approaches such as drug adsorption on an ion exchange resin may be utilized.

5. *Resuspendability*: Suspensions are dispensed with the instruction to the patient to shake gently before administration. Suspended material should settle slowly and should readily redisperse upon gentle shaking of the container.

6. *Physical stability—absence of caking*: Particles that do settle to the bottom of the container should not form a hard cake, but should be readily redispersed into a uniform mixture when shaken. Caking of suspension arises from close packing of sedimented particles, which cannot be eliminated by reduction of particle size or by an increase in the viscosity of the continuous phase. Fine particles have the tendency to cake. Flocculating agents can prevent caking; deflocculating agents increase the tendency to cake.

7. *Deliverability*: The labeled number of doses and the labeled amount of material should be deliverable from a bottle under the normal dispensing conditions by a patient. Deliverability is a function of viscosity of the suspension. Higher viscosity can lead to more of the suspension sticking to the container, reducing deliverable volume.

8. *Flow*: Suspensions must not be too viscous to pour freely from a bottle or to flow through a needle syringe (for injectable suspensions). Suspensions are non-Newtonian flowing liquids. Suspensions should be designed as thixotropic or shear-thinning systems rather than shear-thickening systems.

9. *Lack of microbial growth*: Use of antimicrobial preservatives is deemed sufficient for oral and topical suspensions, whereas parenteral, nasal, and ophthalmic suspensions must be sterile.

10. *Physical integrity*: The suspension should not show any unexpected change in color, or any other change in physical appearance or perception of the dosage form, such as odor, during storage.

11. *Particle adhesion to the package*: When the walls of a container are wetted, an adhering layer of suspension particles may build up, and this may subsequently dry to a hard and thick layer. Adhesion often increases with increase in suspension concentration. Surfactants can

modify the adhesion of suspension particles by decreasing surface tension and adsorption on the particle surface, leading to modification forces of interaction between the suspended particles and the container.

12. *Polymorphic integrity*: Crystallization of the drug could lead to a change in its polymorphic form. A change in the polymorphic form of the drug could lead to changes in its biopharmaceutical properties, such as dissolution rate and absorption. Therefore, the drug must not recrystallize and/or change its polymorphic form during the storage of the formulation.

13. *Chemical stability*: Refers to a lack of unacceptable chemical degradation of the drug during the shelf life of the product under the recommended packaging and storage conditions. The drug product must meet the predetermined requirements of minimum potency of the API and maximum levels of known and unknown impurities.

14. *Drug release*: The drug in a suspension must dissolve in the biological fluids at the site of absorption on administration. Since suspension contains the drug in a dispersed, particulate form, the release of the drug into solution in an appropriate dissolution vessel is used as a quality control tool. The rate and extent of drug dissolution must remain consistent throughout the shelf life of a suspension.

In addition, there are special requirements for suspensions depending on their specific usage. For example, suspensions for external use, such as *lotions* should be fluid enough to spread easily but not so fluid that it runs off the surface too quickly. They must dry quickly and provide an elastic film that will not rub off easily. They must also have pleasant color and odor, although sweetener is not needed.

The quality attributes of a suspension reconstituted from a *PFS* are same as those of a suspension that is marketed in a ready-to-use form. In addition, there are quality requirements for the unit dose PFS powder sachets or the multidose PFS powder in a bottle. For example:

1. *Fill amount*: The amount of powder per container must be tightly controlled to be as close as possible to the amount listed on the label. For a unit dose container, the dispensable or deliverable amount, in addition to the label amount, is measured.

2. *Reconstitution time*: As PFS are meant for reconstitution by the patient or the pharmacist, the suspension should be readily formed on addition of water and reasonable manual agitation.

3. *Uniformity of content*: Container-to-container uniformity of content of the PFS is important to assure uniformity of the drug amount dispensed across different containers.

4. *Physical and chemical stability*: The PFS must maintain physical and chemical stability throughout the labeled shelf life under the labeled storage conditions.

16.5 FORMULATION CONSIDERATIONS

Suspensions are formulated to meet key quality requirements as outlined earlier. Additional formulation considerations for suspensions include managing the bitterness and grittiness of the API, and dose volume. For example, a highly bitter API is likely to impart an unpleasant taste to the suspension due to its solubility in the suspension vehicle, even though this solubility may be extremely low. A gritty particle shape of an API, such as needle-shaped crystals, is likely to have poor mouthfeel unless the particle size of the suspension is reduced significantly. In addition, reasonable dose volume for a patient is one teaspoon (5 mL) or one tablespoon (15 mL) or other nondecimal multiples of these measures. Total dose volume that may be administered per day is also limited by the maximum allowed daily dose of other ingredients, such as the artificial sweetener and the preservative.

Typically, the following ingredients are used in suspensions.

1. *Drug*: A water-insoluble drug is usually the dispersed phase in an aqueous suspension. Drugs should be of a narrow particle size distribution within the range of 1–50 μm.
2. *Wetting agents*: The surface of dispersed drug particles can be either hydrophilic or hydrophobic. Drugs with hydrophobic surfaces are usually difficult to disperse in an aqueous medium. Wetting agents are surfactants that reduce the surface tension of an aqueous medium and facilitate the wetting of hydrophobic particles. Wetting agents adsorb onto the hydrophobic particle surface to either partially coat the surface or form a complete monolayer. Examples of typical wetting agents are sodium lauryl sulfate and polysorbate 80.
3. *Suspending agents*: Suspending agents are hydrophilic colloids, such as cellulose derivatives, acacia, and xanthan gum that are added to a suspension to increase viscosity inhibit agglomeration, and decrease sedimentation. Suspending agents may also interact with the suspended particle's surface to facilitate wetting and reduce the tendency to agglomerate upon interparticle collisions. Typical suspending agents are listed in Table 16.1. Although increasing the viscosity of a suspension improves its physical stability, on oral administration highly viscous suspensions may prolong gastric emptying time, slow drug dissolution, and decrease the absorption rate. Thus, the dose volume of a suspension and *in situ* fluid viscosity on dilution of

Table 16.1 Commonly used suspending agents

Name	Ionic charge	Typical concentration range (% w/w)
Cellulose derivatives		
Methylcellulose	Neutral	1–5
Hydroxypropyl methylcellulose	Neutral	0.3–2
Sodium carbodymethylcellulose	Anionic	0.5–2
Polymers		
Carbomer	Anionic	0.1–0.4
Povidone	Neutral	5–10
Gums		
Xanthan gum	Anionic	0.3–3
Carrageenan	Anionic	1–2

Source: Ofner, C.M., Schnaare, R.I., *Suspensions*, http://www.fmcbiopolymer.com/ Portals/Pharm/Content/Docs/PS/09_Suspensions.pdf, last accessed December 2016.

the suspension in the gastric fluid must be considered to understand potential impact on oral drug absorption.

4. *Flocculating agents*: Suspended particles that have high charge density display deflocculation and caking upon sedimentation. Such suspensions have high particle–particle electrostatic repulsion and do not settle rapidly. Such suspensions are called deflocculated because they display uniform distribution of particles without any settling/separation or flocculation for extended periods of time. However, once they settle, they form a strong cake that is difficult to redisperse. Addition of oppositely charged formulation ingredient(s) to such a suspension results in partial neutralization of effective charge (i.e., zeta potential) on the particles through the formation of the electrical double layer. This results in suspended particles being weakly linked together in loose aggregates or flocs. These flocs settle rapidly but form large fluffy sediment, which is easily redispersed. Such suspensions are called flocculated suspensions—and the formulation ingredients that promote flocculation are called flocculating agents. Flocculated suspensions are preferred over deflocculated systems to enable rapid redispersibility upon shaking.

5. *Preservatives*: Preservatives are often added in aqueous suspensions because suspending agents and sweeteners are good media for microbial growth. Some preservatives are ionic, such as sodium benzoate, and

Table 16.2 Commonly used antimicrobial preservatives

Name	Typical concentration range (% w/w)
Alcohols	
Ethanol	>20
Propylene glycol	15–30
Benzyl alcohol	0.5–3
Surfactants	
Benzalkonium chloride	0.004–0.02
Acids	
Sorbic acid	0.05–0.2
Benzoic acid	0.1–0.5
Parabens	
Methylparaben	0.2
Propylparaben	0.05

Source: Ofner, C.M., Schnaare, R.I., *Suspensions*, http://www.fmcbiopolymer.com/Portals/Pharm/Content/Docs/PS/09_Suspensions.pdf, last accessed December 2016.

may interact or form complexes with other suspending ingredients—thus reducing their preservative efficacy. Thus, effective aqueous concentration of the preservative must be monitored and controlled. Solvents, such as alcohols, glycerin, and propylene glycol, may also have some preservative effect depending on their concentration. Typical microbial preservatives are listed in Table 16.2.

6. *Sweeteners, flavors, and colorants*: Sweeteners are often added to suspensions to reduce any unpleasant taste of the partially dissolved drug and to improve palatability in general. Examples include sorbitol, corn syrup, sucrose, saccharin, acesulfame, and aspartame. Flavors are added to enhance patient's acceptance of the product. Colorants are added to provide a more esthetic appearance to the final product. Choice of colorant is usually tied to the choice of flavor, and their choices are also linked to the patient population, such as age group and geographic region, and the therapeutic need. For example, red colorant is usually used with strawberry flavor for pediatric formulations.

Table 16.3 shows two examples of suspension formulations. One is benzoyl peroxide topical suspension, which is used for treating mild to moderate acne. The other is triamcinolone diacetate parenteral suspension, which is used for treating allergic disorders.

Table 16.3 Examples of suspension formulations

Ingredients	% w/w	Use
A. Benzoyl peroxide topical suspension		
Benzoyl peroxide	5.0	Drug
Hydroxypropyl methylcellulose	1.5	Suspending agent
Xanthan gum	1.5	Suspending agent
Polysorbate 20	5.0	Wetting agent
Isopropyl alcohol	10	Solvent
Phosphoric acid	0.03	pH adjustment
Purified water q.s.	100	Solvent
B. Triamcinolone diacetate parenteral suspension		
Triamcinolone	4.0	Drug
Polyethylene glycol (3400 Da)	3.0	Suspending agent
Polysorbate 80	0.2	Wetting agent
Sodium chloride	0.85	To adjust tonicity
Benzyl alcohol	0.9	Microbial preservative
Water for injection q.s.	100	Solvent

16.5.1 Flocculation

The large surface area of the suspended fine particles is associated with high surface-free energy that makes the system thermodynamically unstable. Generation of fine particles by milling, which is commonly used for pharmaceutical suspensions, generates particles with higher energy per unit surface area than the parent, unmilled API crystals. This is attributed to the preferred orientation of functional groups of a molecule during crystallization. During crystallization, hydrophobic regions of an API, which interact less favorably with the solvent of crystallization, get embedded on the inside of the crystal structure, whereas hydrophilic regions are exposed to the surface. When such a material is milled, the inner hydrophobic surface gets exposed. This is a high-energy surface in an aqueous environment because it resists interaction with water and has the propensity to self-aggregate to decrease the total surface area and surface-free energy.

Stabilization of high-energy fine particles can be accomplished through the use of suspending agents that increase solution viscosity, reduce surface tension, and/or coat the surface of the dispersed particles. In addition, the use of formulation ingredients, such as hydrophilic polymers, that facilitate the formation of lose associations of dispersed particles through the formation of relatively weak bonds with each other, can contribute to a phenomenon called flocculation.

Flocculation is the formation of loose, light, and fluffy flocs (associations of particles) held together by weak van der Waals forces. In contrast, particles in deflocculated suspensions tend to exhibit strong interparticle

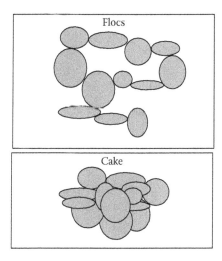

Figure 16.1 Formation of flocs and cake in pharmaceutical suspensions. Suspensions often form loose networks of flocs that settle rapidly, do not form cakes, and are easy to suspend. However, settling and aggregation may result in the formation of cakes that is difficult to resuspend.

attraction forces, leading to aggregation. *Aggregation* occurs in a compact cake situation, that is, growth and fusing together of crystals in the precipitates to form a solid cake. Figure 16.1 illustrates the difference between flocs and cake in pharmaceutical suspensions.

Forces at the surface of the dispersed particles affect the degree of flocculation and agglomeration in a suspension. Generally, the forces of attraction are of the van der Waals type, whereas the repulsive forces arise from the interaction of the electric double layers surrounding each particle (Figure 16.2). When the repulsion energy is high, collision of the particles is opposed. The system remains deflocculated. However, when sedimentation is complete, the particles form a close-packed and strongly bound structure. Those particles lowest in the sediment are gradually pressed together by the weight of the ones above. The repulsive energy barrier is thus overcome, allowing the particles to come into close contact with each other. The reduced bonding potential energy at a critical interparticulate distance allows the forces of attraction to dominate and caking to occur (Figure 16.3).

To resuspend and redisperse caked particles in a suspension, it is necessary to overcome the high-energy barrier. Since this is not easily achieved by agitation, the particles tend to remain strongly attracted to each other and form a hard cake. When the particles are flocculated, the particles equilibrate in the second energy minimum, which is at a distance of separation of ~1000–2000 Å—sufficient to form the loosely structural flocs.

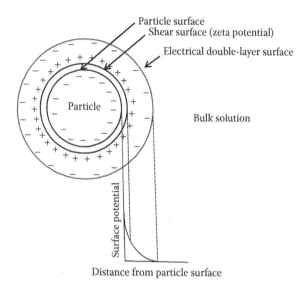

Figure 16.2 Illustration of particle surface charge and zeta potential on the surface of a particle. Interparticle interactions in a suspension are determined by the zeta potential, the net charge at the end of an electrical double layer on the particle surface. This electrical double layer is formed by the selective adsorption of oppositely charged ionic species in solution to an electrostatically charged particle surface.

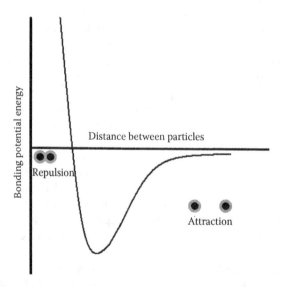

Figure 16.3 Bonding potential energy between particles as a function of distance, in the absence of surface charge. The forces of attraction between particles are dependent on the distance between the particles and are maximized at an optimum distance. Caking in a suspension is facilitated if interparticulate distance allows the forces of attraction to dominate and form strong bonds.

Caking is undesirable, since a caked dispersed phase is difficult to redisperse. Flocculating agents can prevent caking, whereas deflocculating agents increase the tendency to cake. To convert a suspension from a deflocculated to a flocculated state, the following flocculating agents are often used:

1. *Electrolytes*: Electrolytes act as flocculating agents by reducing the electric barrier between the particles. The addition of an inorganic electrolyte to an aqueous suspension alters the zeta potential of the dispersed particles. Flocculation occurs when the zeta potential is lowered sufficiently. The most widely used electrolytes include sodium salts of acetate, phosphate, and citrate.
2. *Surfactants*: Ionic surfactants may also cause flocculation by neutralization of the charge on each particle. An example of ionic surfactant is sodium lauryl sulfate.
3. *Hydrophilic polymers*: Particles coated with hydrophilic polymers are less prone to caking than uncoated particles. Especially for particles that lack strong electrostatic surface charge, using nonionic hydrophilic polymers, which act as protective colloids, can contribute to flocculation. These polymers exhibit pseudoplastic (i.e., shear-thinning, viscosity reduces upon exposure to shear) flow in solution. This property serves to promote physical stability within the suspension by maintaining high viscosity when the suspension is stagnant, whereas allowing easy pourability by reduction of viscosity when the suspension is mixed.

Starch, alginates, cellulose polymers (sodium carboxymethylcellulose), gum (tragacanth), carbomers, and silicates are examples of polymeric flocculating agents. Their linear branched chain molecules form a gel-like network within the system, thus increasing the viscosity of the aqueous vehicle, and become adsorbed on the surfaces of the dispersed particles, thus acting as protective colloids.

Whether a suspension is flocculated or deflocculated depends on the relative magnitudes of the electrostatic forces of repulsion and the forces of attraction between the particles. When *zeta potential* is relatively high, the repulsive forces usually exceed the attractive forces. Consequently, dispersed particles remain as discrete units and settle slowly. The suspension is deflocculated. The slow rate of settling prevents the entrapment of liquid within the sediment, which thus becomes compacted and can be very difficult to redisperse. This phenomenon is called caking.

Flocculated systems form lose sediments, which are easily redispersible. However, the sedimentation rate is faster than deflocculated suspensions. Association of particles with each other in a flocculated system leads to a rapid rate of sedimentation because each unit is composed of many individual particles and is, therefore, larger. Supernatant of a deflocculated

Table 16.4 Properties of flocculated and deflocculated suspensions

Property	Flocculated suspension	Deflocculated suspension
Sedimentation rate	Faster	Slower
Clarity of supernatant	Higher	Lower
Ease of redispersibility	Higher	Lower
Zeta potential on particles	Lower	Higher
Forces between particles	Forces of attraction predominate over the repulsive forces	Repulsive forces exceed attractive forces
Tendency to cake	Lower	Higher

system remains cloudy for an appreciable time after shaking due to the very slow settling rate of the smallest particles in the suspension. In contrast, the supernatant of a flocculated system quickly becomes clear as the flocs, composed of lose agglomerates of particles of all sizes, settle rapidly. If the sedimentation rate is too fast, the dispensed dose may not be accurate. Therefore, an optimum suspension formulation should only be partially flocculated. In addition, viscosity is controlled so that the sedimentation rate is minimized. Controlled flocculation is usually achieved by a combination of particle size control, the use of electrolytes to control zeta potential, and the addition of polymers to enable the formation of weak networks in solution that entangle and form weak bonds between the dispersed particles. Differences between flocculated and deflocculated suspensions are summarized in Table 16.4.

16.5.2 Quantitating the degree of flocculation

Sedimentation studies can quantitatively define the sedimentation volume and degree of flocculation/deflocculation of a system. As illustrated in Figure 16.4, the sedimentation volume, F, is defined as the ratio of the final volume of the sediment, V_u, to the original volume of the suspension, V_0 before settling. Thus,

$$F = \frac{V_u}{V_0}$$

The sedimentation volume can have values from less than 1 (particle settling) to greater than 1 (particle swelling). It is usually less than 1. That is, the final volume of sediment is smaller than the original volume of suspension. Particle swelling can occur for a freshly prepared suspension that has not been allowed enough time to equilibrate to fully hydrate all solid components. If the volume of sediment in a flocculated suspension is equal to the original volume of suspension, then $F = 1$. Such a product is believed to be in flocculation equilibrium and shows no clear supernatant on standing.

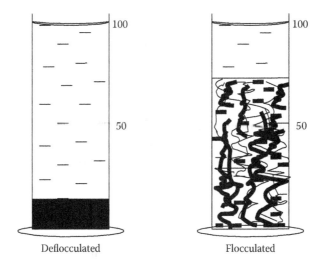

Deflocculated   Flocculated

Figure 16.4 A schematic illustrating the differences in the sediment quality and volume between a deflocculated and a flocculated suspension. On placing undisturbed in a measuring cylinder, an equal volume of a flocculated suspension forms a larger mass of a more porous sediment than a deflocculated suspension.

16.5.3 Physics of particle sedimentation: Stokes' law

The control of sedimentation rate of dispersed particles is required to ensure uniform dosing of a pharmaceutical system. Sedimentation of a disperse system depends on the motion of the particles, which may be thermally or gravitationally induced. If a suspended particle is sufficiently small in size, random Brownian motion dominates over unidirectional gravitational pull. When the radius of the suspended particles is increased, Brownian motion becomes less important and sedimentation becomes dominant. These larger particles, therefore, settle gradually under gravitational forces.

Stokes' law describes the sedimentation of suspended particles in suspensions:

$$V = \frac{2gr^2(\rho_1 - \rho_2)}{9\eta} = \frac{gd^2(\rho_1 - \rho_2)}{18\eta}$$

where:

V is the velocity of sedimentation
r is the particle radius
d is the particle diameter
ρ_1 and ρ_2 are the densities of the particles and dispersion medium, respectively
g is the acceleration of gravity
η is the viscosity of the medium

As the diameter is squared in Stokes' law, a reduction in particle size by ½ will reduce the sedimentation rate by $(½)^2$ or a factor of 4. Thus, particle size control is an important element in the formulation of stable suspensions.

In addition, doubling the viscosity of a suspension will decrease the sedimentation rate by a factor of 2. Increasing the viscosity of a suspension reduces the rate of settling of dispersed particles, changes the flow properties of a suspension, and affects the spreading qualities of a lotion. Viscosity can be increased by the addition of hydrophilic polymers or gums that act as suspending agents. An ideal suspending agent should have a high viscosity at negligible shear and should be free flowing during agitation, pouring, and spreading. A suspending agent that is thixotropic as well as pseudoplastic should prove to be useful since it forms a gel on standing and becomes fluid when distributed.

Stokes' law further indicates that if the difference in density between the suspended particle and the suspension medium can be matched, the sedimentation rate would be reduced to zero. Density of a drug particle is an inherent property of the crystal structure and packing, and may not be altered readily. However, the density of the suspension medium can be increased by increasing the solids content of the liquid or solution. This, however, needs to be balanced with an increase in solution viscosity and reduction of pourability.

16.6 MANUFACTURING PROCESS

Suspensions are typically manufactured using a high-energy mill to disperse the insoluble powder ingredients in the suspension vehicle. A high-energy mill is required to ensure thorough mixing because the vehicle is usually viscous. In addition, the high-energy process can lead to the desired reduction or narrowing of the particle size distribution. A high-shear hand mixer is frequently used in laboratory scale for suspension manufacture. A colloid mill is usually used for the manufacture of pilot and production-scale manufacture of suspensions on a commercial scale.

The powder properties of incoming raw materials are critical and closely controlled to assure the quality attributes of powder blend. These include particle size, shape, charge, size distribution, residual moisture content, flowability, compatibility, and any aggregation tendency. Each ingredient is screened to ensure it is homogeneous and free of agglomerates, followed by mixing with other ingredients in an order that ensures uniform mixing. Preparation of a suspension involves the mixing of water-soluble components with water to form an aqueous solution. The solid ingredients are then added to this solution under high shear-mixing process in a sequential manner to form a suspension. The suspension is dispensed into bottles using automated liquid dispensing machines.

PFSs are manufactured as dry powders. These formulations are designed to be rapidly redispersible by gentle mixing in the presence of water. The manufacturing process for PFSs involves mixing the bulk powders of the formulation components followed by dispensing into commercial containers using an automated bottle or sachet-filling machine. Mixing of low quantity ingredients, such as colorants, can be challenging. Usually, such ingredients are premixed and/or adsorbed on the surface of another higher-quantity ingredient before being mixed with the rest of the material. In addition, ingredients that may be liquid at room temperature, such as liquid flavors, are adsorbed onto another material before mixing with the bulk of the ingredients. Ingredients can also be coscreened or comilled to ensure their thorough mixing.

REVIEW QUESTIONS

16.1 Indicate which statements are TRUE and which are FALSE?
   A. Flocculation is desirable for pharmaceutical suspensions.
   B. Deflocculation is not desirable for pharmaceutical suspensions.
   C. Motion of dispersed particles in a suspension is induced by thermal and gravitational forces.
   D. Viscosity of the suspension affects the settling of particles
   E. Crystal growth of particles in a suspension is due to temperature fluctuation on storage and due to wider particle size distribution
16.2 How does the increase in viscosity of the suspending medium affect the rate of sedimentation when assuming the density of the particles is larger than that of the suspending medium?
   A. Sedimentation rate will not change
   B. Sedimentation rate will be slower
   C. Sedimentation rate will be faster
   D. No particle sedimentation will take place
16.3 Which one of the following phenomena is undesirable in pharmaceutical suspensions?
   A. Slow settling of particles
   B. Particles agglomerate to dense cake
   C. Particles readily redisperse upon agitation
   D. Suspension pours readily
16.4 Define and differentiate flocculated and deflocculated suspensions. Why is deflocculation not desirable, whereas flocculation is an acceptable characteristic for pharmaceutical suspension dosage forms?
16.5 A course powder with a true density of 2.44 g/cm³ and a mean diameter, d, of 100 µm was dispersed in a 2% carboxymethylcellulose dispersion having a density, ρ_0, of 1.010 g/cm³. The viscosity of the medium at low shear rate was 27 poise. Using Stokes' law, calculate the average velocity of sedimentation of the powder in cm/s.

16.6 Using Stokes' law, compute the velocity of sedimentation in cm/s of a sample of zinc oxide having an average diameter of 1 μm (radius of 5×10^{-5} cm), a true density, ρ, of 2.5 g/cm³ in a suspending medium having a density, $ρ_0$, of 1.1 g/cm³, and a Newtonian viscosity of 5 poise.

16.7 Which of the following parameters control the rate of sedimentation of particles in a suspension?
A. Particle diameter
B. Viscosity of the suspending vehicle
C. Surface charge on the particles
D. Density of the particles

16.8 Under what circumstances would a powder for suspension preferred dosage form for commercialization compared to a ready-to-use suspension dosage form?
A. High settling volume of suspension
B. Low viscosity of vehicle
C. Poor stability of API in suspension
D. High viscosity of vehicle

16.9 Which of the following ingredients in a suspension could help in flocculating the dispersed particles?
A. Surfactant
B. Hydrophilic polymer
C. Electrolyte
D. Cosolvent

16.10 Which of the following is not a typical requirement for lotions?
A. Must dry quickly
B. Must be fluid, not highly viscous
C. Must have a smooth feel to the skin
D. Must be sweet

FURTHER READING

Allen LV, Popovich NG, and Ansel HC (2005) *Ansel's Pharmaceutical Dosage Forms and Drug Delivery Systems*, 8th ed., New York: Lippincott Williams & Wilkins.

Im-Emsap W, Siepmann J, and Paeratakul O (2002) In Banker GS and Rhodes CT (Eds.) *Modern Pharmaceutics*, 4th ed.: New York: Marcel Dekker, pp. 237–285.

Kolling WM and Ghosh TK (2005) Oral liquid dosage forms: Solutions, elixirs, syrups, suspensions, and emulsions. In Ghosh TK and Jasti BR (Eds.) *Theory and Practice of Contemporary Pharmaceutics*, Boca Raton, FL: CRC Press, pp. 367–385.

Ofner CM and Schnaare RI (2016) Suspensions. http://www.fmcbiopolymer.com/Portals/Pharm/Content/Docs/PS/09_Suspensions.pdf, last accessed December 2016.

Sinko PJ (2005) *Martin's Physical Pharmacy and Pharmaceutical Sciences* 5th ed., Philadelphia, PA: Lippincott Williams & Wilkins, pp. 561–583.

Subramanyan CVS (2000) *Textbook of Physical Pharmaceutics*, 2nd ed., Delhi: Vallabh Prakashn, pp. 366–394.

Young SA and Buckton G (1990) Particle growth in aqueous suspensions: The influence of surface energy and polarity. *Int J Pharm* **60**: 235–241.

Chapter 17

Emulsions

17.1 INTRODUCTION

An emulsion consists of at least two immiscible liquid phases, one of which is dispersed as globules (dispersed phase) and into the other liquid phase (continuous phase). Emulsions are thermodynamically unstable and are usually stabilized by the presence of an emulsifying agent. The process of formation of an emulsion is termed emulsification. The diameter of the dispersed phase globules is generally in the range of about 0.1 to 10 μm, although it can be as small as 0.01 μm or as large as 100 μm. Emulsions in which the size of the dispersed phase is so small, in the nanometer range, that all of the dispersed phase is subvisible appear as a homogeneous, isotropic liquid. Such systems are called microemulsions or nanoemulsion. Emulsified systems range from lotions of relatively low viscosity to ointments and creams, which are semisolid in nature. Pharmaceutical emulsions are used for the administration of nutrients, drugs, and diagnostic agents. Topical creams and lotions are popular forms of emulsions for external use.

The main advantages of emulsions as drug delivery systems include the following:

1. *Increased drug bioavailability*: Many drugs are highly hydrophobic, with high log *P* values (partition coefficient between oil and water). These drugs are usually poorly soluble in water but readily soluble in oils. Formulation of drug-dosage form as an emulsion allows the administration of a hydrophobic drug in a soluble/dissolved state. This can improve the oral bioavailability of a biopharmaceutics classification system (BCS) class II (low solubility, high permeability) and class IV (low solubility, low permeability) drugs because absorption from an emulsion does not require the dissolution step. In addition, oral administration of a drug as an emulsion involves coadministration of lipid, which increases bile secretion. Higher concentration of bile in the gastrointestinal (GI) fluids further enhance oral absorption of hydrophobic drugs.
2. *Increased drug stability*: Drugs that are more stable in an oily phase compared to an aqueous medium can show improved stability in an emulsion dosage form.
3. *Prolonged drug action*: The oily phase can serve as a reservoir of the drug, which slowly partitions into the aqueous phase for absorption. This phenomenon, especially with semisolid emulsions, can help prolong drug action. For example, intramuscular injection of an emulsion can prolong drug absorption time.

A key disadvantage of using emulsions as a delivery system is their physical instability and complex, energy-intensive manufacturing process. In addition, solid dosage forms are generally preferred over liquid dosage forms for oral administration.

17.2 TYPES OF EMULSIONS

Emulsions typically consist of a polar (e.g., aqueous) and a relatively nonpolar (e.g., an oil) liquid phase. Based on the nature of the internal and/external phase, emulsions can be classified into different types (Figure 17.1).

17.2.1 Oil-in-water emulsion

When the oil phase is dispersed as globules throughout an aqueous continuous phase, the system is referred to as an oil-in-water (o/w) emulsion. An o/w emulsion is generally formed if the aqueous phase constitutes more than 45% of the total weight and a hydrophilic emulsifier, such as sodium lauryl sulfate, triethanolamine stearate, sodium oleate, and glyceryl monostearate is used. The emulsifier is present in the external,

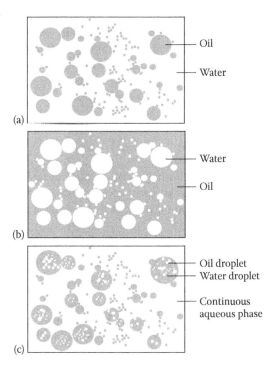

Figure 17.1 Types of emulsions: (a) o/w emulsions, (b) w/o emulsion, and (c) w/o/w multiple emulsion.

continuous phase and helps stabilize the interface with the dispersed phase globules.

17.2.2 Water-in-oil emulsion

When the aqueous phase is dispersed, and the oil phase is the continuous phase, the emulsion is termed as water-in-oil (w/o) emulsion. A lipophilic emulsifier is used for preparing w/o emulsions. The w/o emulsions are used mainly for external applications and may contain one or several of the following emulsifiers: calcium palmitate, sorbitan esters (Spans), cholesterol, and wool fats. Thus, the use of a lipophilic emulsifier enables the formation of w/o emulsions with the oil phase as the external, continuous phase.

17.2.3 Multiple emulsions

Multiple emulsions are emulsions whose dispersed phase contains droplets of another emulsion. Both water-in-oil-in-water (w/o/w) and oil-in-water-in-oil (o/w/o) multiple emulsions are of interest as delayed- and/or sustained-action drug delivery systems. They also have applications in cosmetics. Emulsifying a w/o emulsion using water-soluble surfactants (which stabilize an oily dispersed phase) can produce w/o/w emulsions with an external aqueous phase, which generally has a lower viscosity than the primary w/o emulsion. Multiple emulsions can also be used for the encapsulation of peptides/proteins and hydrophilic drugs.

17.2.4 Microemulsions

Microemulsions are visually homogeneous, transparent/isotropic systems of low viscosity. In their simplest form, microemulsions are small droplets (diameter 5–140 nm) of one liquid dispersed throughout another by virtue of the presence of a fairly large amount of surfactant(s) and cosolvent(s). Microemulsions have a very finely subdivided dispersed phase, and often contain a high concentration of the emulsifier(s) and a cosolvent (such as ethanol).

Microemulsions are thermodynamically stable for prolonged periods of time. They can be dispersions of o/w or w/o. The type of microemulsion (w/o or o/w) formed is determined largely by the nature of the surfactants. Microemulsions can be used to increase the bioavailability of poorly water-soluble drugs by incorporating them into the oily phase. Incorporation of etoposide and methotrexate diester derivative into w/o microemulsion has been suggested as a potential carrier for cancer therapy.

17.2.4.1 Self-emulsifying drug delivery systems and self-microemulsifying drug delivery systems

A solution of drug in the oil–surfactant–cosolvent mixture can spontaneously form an emulsion or microemulsion with minimal agitation at room temperature. Whether this mixture forms an emulsion or a microemulsion depends on the composition of this mixture and the amount of water added. A higher proportion of oil and a lower proportion of cosolvent lead to the formation of an emulsion. Self-microemulsifying mixtures typically contain a higher proportion of the cosolvent and the surfactant, whereas the proportion of oil is lower. These mixtures are termed as self-emulsifying drug delivery system (SEDDS) or self-microemulsifying drug delivery systems (SMEDDS). The SEDDS and SMEDDS can be administered orally for *in vivo* emulsion or microemulsion formation in the patient's GI tract. For example, cyclosporine is available as a self-microemulsifying preconcentrate (Neoral®), which is more rapidly and consistently absorbed than the original self-emulsifying formulation of cyclosporine (Sandimmune®). Both these show greater and more consistent bioavailability than unformulated cyclosporine.

17.3 QUALITY ATTRIBUTES

Emulsion dosage forms are designed to meet the following quality attributes:

1. *Uniformity of content* (dose-to-dose within the same bottle and bottle-to-bottle): All the doses dispensed from a given multidose container should have acceptable uniformity of drug content. In addition, the drug content must be uniform between different bottles of a given batch of emulsion.

2. *Separation volume or creaming*: Once an emulsion has been left undisturbed for some time, it may show some degree of separation of the dispersed phase from the dispersion medium. For example, in the case of an o/w emulsion, *creaming* of an emulsion is sometimes observed, which indicates a higher concentration of the dispersed oil phase in the top layer of the emulsion. This top phase is visually distinguishable from the bottom layer due to greater light obscuration and diffraction by a higher concentration of the dispersed phase globules. More the concentration of the dispersed phase, more the light obscuration in a smaller thickness of the *creamed* layer and greater the instability. Thus, the proportion of the volume occupied by the separated phase is an indicator of physical instability of the emulsion. Higher this volume, more stable is the emulsion.

3. *Dispersed phase size distribution*: Size distribution of dispersed phase should remain fairly constant during the shelf life of the emulsion. Brownian motion, agitation during handling, and gravitational motion of the dispersed phase lead to collisions of globules with each other, which can cause coalescence or agglomeration resulting in an increase in the size of the dispersed phase. A change in the dispersed phase globule size on storage is indicative of inherently low physical stability of the emulsion.

4. *Drug concentration*: In cases where drug concentration in the emulsion is close to the drug solubility, crystallization can sometimes occur due to temperature fluctuations during storage, preferential evaporative loss of one phase, incompatibility with packaging components, or unintended nucleation. Crystal growth can be inhibited by the use of appropriate solubilizers and surfactants, and by formulating an emulsion at a lower concentration than the drug's thermodynamic solubility. Changes such as drug crystallization or evaporative loss of the continuous phase can reflect changes in the drug concentration during storage stability or shelf life.

5. *Palatability*: Use of an emulsion dosage form can improve the palatability of particularly bitter drugs by dissolving them in the dispersed phase. However, incorporation of the drug in the dispersed phase may not be adequate because some drug would inadvertently partition into the continuous phase depending on the partition coefficient (logP) of the compound. Palatability of the emulsion can be increased

by the use of sweeteners, flavors, and colorants. In certain cases, specialized taste-masking approaches, such as complexation, may be needed. These considerations, of course, are not pertinent for parenteral emulsions. In the case of parenteral emulsions, tissue irritability and osmotic pressure are important considerations.

6. *Redispersability*: A separated or creamed emulsion should readily redisperse upon gentle shaking of the container.

7. *Absence of phase separation*: Coalescence leading to phase separation is irreversible. Although creaming of an emulsion is, to some extent, unavoidable, the dispersed phase should not coalesce and separate from the dispersion medium. This needs to be designed into the formulation by the use of right surfactants in an appropriate concentration.

8. *Deliverability*: The labeled number of doses and the labeled amount of emulsion should be deliverable from a bottle under the normal dispensing conditions by a patient. This is usually ensured by pouring out the labeled number of doses from the container and ensuring that the remaining dose can be poured completely within a reasonable period of time.

9. *Flow*: The emulsion must not be too viscous to pour freely from a bottle or to flow through a needle syringe or an IV infusion set (for parenteral emulsions).

10. *Lack of microbial growth*: Use of antimicrobial preservatives could be sufficient for oral and topical emulsions, whereas parenteral, nasal, and ophthalmic suspensions must be sterile.

11. *Physical integrity*: The dosage form should not show any unexpected change in color, or any other change in physical appearance or perception of the dosage form, such as odor, that may alarm the patient and/or the health-care provider with respect to the physical integrity of the emulsion.

12. *Adhesion to the package*: Preferential adsorption or adhesion of one phase or component of the emulsion, such as the drug, the chelating agent, or the emulsifier, can adversely affect the uniformity and stability of an emulsion.

13. *Leachables and extractables*: Primary packaging components of the emulsion can leach out small amounts of chemical components used in the manufacturing of those components. This behavior can be exacerbated at certain pH values of the formulation. The packaging components must be selected appropriately, and their compatibility with the emulsion determined to make sure no chemical compounds leach into or are extracted by the emulsion from the container on storage.

14. *Chemical stability*: There should not be any unacceptable chemical degradation of the drug during the shelf life of the product under recommended packaging and storage conditions. The drug product must meet the predetermined requirements of maximum levels of known and unknown impurities.

15. *Drug release*: Since an emulsion contains the drug in the dispersed phase, the release of drug from the dispersed phase into an aqueous solution in an appropriate dissolution vessel is quantified and controlled as an indicator of its bioavailability. This could be particularly important for semisolid emulsions.

In addition, topical emulsions should be fluid enough to spread easily but not so fluid that the emulsion runs off the surface too quickly. The emulsion must dry quickly and provide an elastic film that should not be too oily. In addition, the dosage form must have pleasant color and odor, although a sweetener is not needed.

17.4 FORMULATION CONSIDERATIONS

Emulsions are inherently *thermodynamically unstable* due to the differences in the molecular forces of interaction between the molecules of the two liquid phases. The oxygen and hydrogen atoms in the water molecules in the aqueous phase bond with surrounding water molecules through dipolar and hydrogen-bonding interactions, whereas the carbon atoms in the oil phase bond with the surrounding molecules predominantly through weak hydrophobic and Van der Waals interactions. Creation of surface of interaction between the two phases is thermodynamically unfavorable. Therefore, production of emulsions requires the introduction of energy into the system. This is accomplished by trituration on the small scale and homogenization on the pilot and large scale. In addition, interfacial molecules of both phases must be stabilized against the tendency for the self-interaction of phases, which can lead to the coalescence and collapse of the dispersed phase.

17.4.1 Minimization of interfacial free energy

Dispersion of insoluble phases results in thermodynamic instability and high total free energy of the system. Since every system tends to spontaneously reduce its energy to a minimum, all emulsions tend to separate into the two insoluble phases with time. When one liquid is broken into small globules, the interfacial area of the globules is much greater than the minimum surface area of that liquid in a phase-separated system. Thus, the dispersed phases tend to coalesce or phase separate to minimize the surface of interaction between the two phases. This phase separation is driven by greater forces of interaction between molecules of the same phase than between the molecules of different phases. A phase-separated system represents the state of minimum surface-free energy.

The surface-free energy of an emulsion is evident as the *interfacial tension* between the two phases. Addition of a surfactant to an emulsion leads to preferential translocation of the surfactant molecules to the interface of the two liquid phases since a surfactant is amphiphilic and molecules in both phases form attractive interactions with the surfactant molecule. The presence of a surfactant at the interface of dispersed-phase globules lowers the interfacial tension. Reduction of the interfacial tension delays the kinetics of coalescence of the two phases. Frequently, combinations of two or more emulsifying agents are used to adequately reduce the interfacial tension by forming a rigid interfacial film.

17.4.2 Optimum-phase ratio

The ratio of volume of the disperse phase to the volume of the dispersion medium (*phase ratio*) greatly influences the characteristics of an emulsion. The optimum-phase volume ratio is generally obtained when the internal or dispersed phase is about 40%–60% of the total quantity of the product. A conventional emulsion containing less than 25% of the dispersed phase has a high propensity toward creaming or sedimentation. Nevertheless, a combination of proper emulsifiers and suitable processing technology makes it possible to prepare emulsions with only 10% disperse phase without stability problems. Such a combination of emulsifiers includes the use of a hydrophilic emulsifier in the aqueous phase and a hydrophobic emulsifier in the oil phase. Such a combination leads to the formation of a closely packed surfactant film at the interface. For example, a combination of sodium cetyl sulfate and cholesterol leads to a closely packed film at the interface that improves emulsion stability. On the other hand, sodium cetyl sulfate and oleyl alcohol do not form a closely packed or condensed film. Consequently, this combination results in a poor emulsion.

17.4.3 Size, viscosity, and density: Stoke's law

Creaming or sedimentation of the dispersed phase in an emulsion is mathematically modeled by Stoke's law (see Chapter 16), which indicates that the physical stability of an emulsion can be enhanced by

1. Decreasing the globule size of the internal phase. The dispersed globule size less than 5 μm in diameter contributes to good physical stability and dispersion of the emulsion.
2. Increasing the viscosity of the system. Gums and hydrophilic polymers are frequently added to the external phase of an o/w emulsion to increase viscosity, in addition to reducing the interfacial tension and forming a thin film at the interface. Higher the viscosity of the continuous phase, lower the Brownian motion, collision frequency, and energy of collisions of the dispersed-phase globules. Increasing the viscosity can have an unwanted effect of reduction of deliverable

volume from the container because highly viscous liquids tend to adhere to the container. In addition, the non-Newtonian viscosity behavior of the emulsion is an important consideration. For example, shear-thinning system may be preferred, whereas a shear-thickening system would be undesirable.

3. Reducing the density difference between the dispersed phase and the dispersion medium. This reduces the creaming tendency by minimizing the driver for separation and preferential accumulation of the dispersed phase in a particular direction.

17.4.4 Zeta potential

Emulsions can be stabilized by electrostatic repulsion between the droplets. High *zeta potential* (see Chapter 16) on the surface of the droplets causes the dispersed-phase droplets to repel each other and thereby resist collisions due to Brownian motion, mixing, and gravitational forces. Thus, the droplets remain suspended for a prolonged period of time. For example, if negatively charged lecithin is adsorbed at the droplet surface it creates a net negative charge and zeta potential on the dispersed-phase droplets. Highly charged dispersed-phase droplets, however, can coalesce irreversibly when colliding with high enough energy to overcome the repulsive forces. Optimizing the zeta potential of the dispersed phase to facilitate flocculation can minimize a system's propensity toward undesirable coalescence. The zeta potential can be optimized by the addition of positively charged electrolytes to the outer, continuous phase. At appropriate zeta potential, the system achieves a state where *flocculation* is facilitated over coalescence.

Table 17.1 lists the composition of two typical o/w pharmaceutical emulsions.

Table 17.1 Examples of emulsion formulations

A. Protective w/o emulsion of calamine		
Ingredients	Amount	Role
Calamine	1 g	Protective
Zinc oxide	1 g	Protective
Olive oil	15 mL	External phase
Lime water	15 mL	Internal phase
B. Benzoyl benzoate o/w emulsion		
Benzoyl benzoate	25 mL	Drug and internal phase
Emulsifying wax	2 g	Emulsifier
Water q.s.	100 mL	External phase

17.5 EMULSIFICATION

17.5.1 Mechanisms of emulsification

Emulsification can be facilitated by the following three mechanisms:

1. Reduction of interfacial tension.
2. Formation of a monomolecular film at the interface that physically inhibits coalescence of dispersed-phase granules.
3. Changing the zeta potential of the dispersed phase.

Emulsifying agents can be surfactants, hydrophilic colloids, or finely divided solid particles. Table 17.2 lists some of the commonly used emulsifying agents.

17.5.2 Surfactants

Surfactants are amphiphilic molecules, which contain both a polar hydrophilic region and a nonpolar hydrophobic region. Depending on the functional groups and relative surface areas of the two regions, surfactants could have a range of hydrophilic and hydrophobic properties. The use of predominantly hydrophilic emulsifying agent leads to the formation of an o/w emulsion since it has stronger and/or greater area of interaction with the aqueous than the oily phase. Conversely, the use of a predominantly hydrophobic emulsifying agent tends to form a w/o emulsion

Table 17.2 Typical emulsifying agents

Type	Examples
Surfactants	
Nonionic	Sorbitan oleate (Span 80)
	Polyoxyethylene sorbitan oleate (Tween 80)
Anionic	Potassium laurate
	Triethanolamine stearate
	Sodium lauryl sulfate
Cationic	Quaternary ammonium compounds
	Benzalkonium chloride
Hydrophilic colloids	
Polysaccharides	Acacia
Phospholipids	Lecithin
Sterols	Cholesterol
Finely divided solid particles	
Colloidal clays	Bentonite
Metallic hydroxides	Magnesium hydroxide

because it has stronger and/or greater area of interaction with the oily than the aqueous phase.

Since the same surfactant molecule has attractive interactions with both the oily and aqueous phase, surfactants get adsorbed at the oil–water interfaces to form *monomolecular films*. This results in a decrease in interfacial tension and physical barrier to collision of dispersed-phase globules. Often, simultaneous use of a predominantly hydrophilic with a predominantly hydrophobic surfactant is used to form more stable emulsions, due to the close packing, strength, and flexibility of the interfacial layer. In addition, the use of ionized surfactants can impact the *zeta potential* of the dispersed phase. This mechanism can further improve emulsion stability by increasing or decreasing electrostatic repulsive forces and facilitating flocculation.

17.5.2.1 Ionic and nonionic surfactants

Surfactants could be anionic (containing anionic, or acidic, functional groups, such as sulfates and carboxylates that become negatively charged at solution pH higher than their pK_a), cationic (containing cationic, or basic, functional groups, such as amines that become positively charged at solution pH higher than their pK_a), amphoteric (containing both anionic and cationic functional groups, with a propensity to become either or both positively and negatively charged, depending on the pH), and nonionic (without any ionizable functional groups, such as alcohols). In addition, surfactants that bear a quaternary ammonium ion bear a permanent positive charge.

Ionized surfactants tend to have strong and specific interactions with a variety of molecules and are, consequently, more toxic than nonionic surfactants. Nonionic surfactants are less sensitive to variations in the electrolyte content and pH of the formulation. Nonionic surfactants, such as the alkyl or aryl polyoxyethylene ethers, sorbitan polyoxyethylene derivatives, and sorbitan are widely used for producing stable emulsions.

17.5.2.2 Hydrophile-lipophile balance value

The relative hydrophobicity and hydrophilicity of a surfactant is indicated by its hydrophile–lipophile balance (HLB) value. A typical HLB value scale ranges from 0 to 20. An emulsifying agent with high HLB (~9–12) is preferentially soluble in water and favors the formation of an o/w emulsion. Conversely, surfactants with low HLB value (~3–6) are preferentially oil soluble and tend to form w/o emulsions.

The HLB system assumes the hydrophilic contribution of the surfactant from ionizable and water-miscible functional groups, such as polyhydric alcohols, ethylene oxide group, fatty acid, or fatty alcohol groups. The hydrophilic portion of a molecule is calculated on a molecular weight basis and divided by 5 to arrive at the HLB value. In general, surfactants with an HLB value of 1–3 can be used for mixing oils, 4–6 for making w/o emulsions, 7–9 for wetting powders into oils, 7–10 for making self-emulsifying systems, 8–16 for making o/w emulsions, 13–15 for making detergents, and 13–18 for making self-microemulsifying systems.

17.5.3 Hydrophilic colloids

Hydrophilic colloids are polymeric materials that bear several electronegative atoms, such as oxygen and nitrogen, thus having strong hydrophilicity through dipole–dipole interactions and hydrogen bond formation. Several hydrophilic colloids, such as gelatin, casein, acacia, cellulose derivatives, and alginates, are used as emulsifying agents. Hydrophilic colloids are used for formation of o/w emulsions since the films are hydrophilic. These materials adsorb at the oil–water interface and form *multilayer films* around the dispersed droplets of oil in an o/w emulsion. Most cellulose derivatives are not charged, but can sterically stabilize the systems.

Hydrated hydrophilic colloids differ from surfactants because they do not cause an appreciable lowering in interfacial tension. They stabilize emulsions by the formation of multilayer films that are strong and resist coalescence. In addition, they increase the viscosity of the dispersion medium.

17.5.4 Finely divided solid particles

Finely divided solid particles that are wetted to some degree by both oil and water can act as emulsifying agents by concentrating at the interface, where they produce a film of particles around the dispersed droplets and act as a physical barrier to coalescence. Finely divided solid particles that are predominantly wetted by water form o/w emulsions, whereas those that are predominantly wetted by oil form w/o emulsions. Examples include bentonite, magnesium hydroxide, and aluminum hydroxide.

17.6 MANUFACTURING PROCESS

Emulsions are manufactured by a high shear-mixing process, such as homogenization. The two phases of the emulsion are assembled separately, by dissolving and mixing of the ingredients to form appropriate solutions. Then, phases are combined by slow addition of the dispersed phase into the continuous phase with continuous mixing. An optimum amount of mixing shear and time are determined based on the rate of change of the size distribution of the dispersed phase with mixing. The resulting emulsion can then be packaged and/or dispensed.

The sequence of addition of formulation ingredients to the emulsion can be critical for the stability of the emulsion. For example, if an o/w emulsion is desired and the system contains two surfactants with different HLB values, the surfactant with the higher HLB value should be added first.

In addition, volatile ingredients, such as flavors, and thermosensitive ingredients should be added last, after the emulsion has been formed, to minimize loss during processing. The API may be predissolved in one of the phases or added last depending on drug's solubility, stability, and partitioning properties.

Self-emulsifying and self-microemulsifying drug delivery systems are manufactured as nonaqueous preconcentrates by simple mixing to dissolve all ingredients. The resulting formulations can then be packaged in single or multidose containers for distribution. In cases where the SEDDS or the SMEDDS is to be administered as a unit dose without dilution prior to administration, the dosage form can be packaged in a soft gelatin capsule.

17.7 STABILITY

Emulsions must demonstrate physical, chemical, and microbial stability throughout their shelf life under recommended packaging and storage conditions.

17.7.1 Physical instability

Physical stability of an emulsion is characterized by the maintenance of elegance with respect to appearance, odor, color, taste, opacity, and viscosity. Four major phenomena are associated with the physical instability of emulsions: (1) flocculation, (2) creaming, (3) coalescence, and (4) breaking. These phenomena are schematically illustrated in Figure 17.2. Flocculation is discussed under the chapter on suspensions.

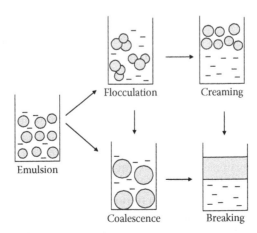

Figure 17.2 Schematic illustrations of different types of instability of emulsions.

17.7.1.1 Creaming and sedimentation

Creaming is the upward movement of dispersed oil droplets in an o/w emulsion, whereas *sedimentation,* the reverse process, is the downward movement of dispersed-phase droplets. Creaming involves visually evident separation of two layers that differ primarily in the number density of the dispersed phase, and, thus, show optical differences. These processes take place due to the density differences in the two phases and can be reversed by shaking. Creaming is undesirable because a creamed emulsion increases the likelihood of coalescence due to the closer proximity of the globules in the cream and because of the nonuniformity of the creamed emulsion. The propensity and rate of creaming is influenced by factors similar to those involved in the sedimentation of suspensions and are indicated by the Stoke's Law, as discussed in a preceding Section 17.4.3.

17.7.1.2 Aggregation, coalescence, creaming, and breaking

Aggregation involves close packaging/contact of the dispersed-phase droplets, but the droplets do not fuse. Aggregation is, to some extent, reversible. Coalescence is the process by which emulsified globules merge with each other to form larger globules. Coalescence is an irreversible process because the film that surrounds the individual globules is destroyed. It leads to progressive increase in the size of the dispersed phase, ultimately leading to breaking of the emulsion. Breaking of an emulsion refers to complete separation of the two liquid phases. Creaming is a reversible process, whereas breaking is irreversible. When breaking occurs, simple mixing fails to resuspend the globules in a stable emulsified form, since the film surrounding the particles has been destroyed and the oil tends to coalesce. The proportion of the volume of emulsion occupied by creamed layer is an indicative of the stability of an emulsion. Greater the proportion of the creamed layer, more stable the emulsion. Flocculation, which leads to weak interactions between dispersed-phase droplets, can stabilize an emulsion by increasing the duration of time it takes for creaming and the proportion of the creamed phase.

Formation of a thick interfacial film is essential to minimize coalescence. In addition, increasing the mechanical strength of the interfacial barrier, such as by closer packing of the interfacial surfactant monolayer, reduces the propensity toward coalescence. Increasing the viscosity of the continuous phase helps to stabilize the dispersed phase and minimizes coalescence.

Particle size does not correlate well with increased/decreased breaking, nor does viscosity. Phase volume is an important consideration in the stability of an emulsion. For example, at greater than ~74% of oil in an o/w emulsion, the oil globules often coalesce and breaking occurs. Thus, a critical concentration is defined in terms of the concentration of the internal phase above which the emulsifying agent cannot produce a stable emulsion of the desired type. Generally, a phase-volume ratio of 50:50 results in the most stable emulsion.

17.7.1.3 Phase inversion

An emulsion is said to invert when it changes from an o/w to a w/o emulsion, or vice versa. Phase inversion can occur by the addition of an electrolyte or by changing the phase volume ratio. Addition of monovalent cations promotes the formation of o/w emulsions, whereas the addition of divalent cations increases the propensity toward the formation of w/o emulsions. For example, an o/w emulsion stabilized with sodium stearate can be inverted to a w/o emulsion by adding calcium chloride to form calcium stearate.

17.7.2 Chemical instability

The API must be chemically stable in the dosage form throughout the shelf life of the product under recommended packaging and storage conditions in terms of both potency and impurities. The drug product must meet predetermined requirements of minimum potency of the API and maximum levels of known and unknown impurities. Factors governing the reaction kinetics of the API, such as the reactivity of functional groups and the kinetics of reactions are no different for emulsion dosage forms than other solution-based dosage forms. Nevertheless, separation of the reacting species in the oily and aqueous phases can minimize reactivity and improve stability of a drug in an emulsion.

17.7.3 Microbial growth

Microbial load of a dosage form must be controlled within the compendial and the regulatory levels. In addition to the health risks of microbial growth, microorganisms in an emulsion can cause physical separation of the phases. Preservatives must be added in adequate concentrations in the formulations to resist microbial growth. The preservative should be concentrated in the aqueous phase because bacterial growth will normally occur there. The oil and water partition coefficient of the preservatives should be considered to calculate the concentration of the surfactant in the aqueous phase, which needs to be above the antimicrobial concentration. The parabens (methylparaben, propylparaben, and butylparaben) are the commonly used preservatives in emulsions.

REVIEW QUESTIONS

17.1 Coalescence can be reduced by
   A. Decreasing the difference between the density of the dispersed phase and the density of the medium
   B. Adding an agent that reduces the viscosity of the medium

C. Increasing the droplet size of the dispersed phase

D. All of the above

17.2 When compounding an emulsion that contains a flavoring agent, the flavoring agent should be in the

A. Continuous phase

B. Discontinuous phase

C. Aqueous phase

D. Oil phase

E. Emulsifier

17.3 Define and differentiate between the following:

A. Creaming and breaking

B. Creaming and sedimentation

C. Coalescence and aggregation

D. Phase inversion and self-emulsification

E. Multiple emulsions and microemulsions

F. SEDDS and SMEDDS

17.4 Explain how sedimentation and creaming in emulsions can be minimized.

17.5 Why is a surfactant needed to make stable emulsions? Explain which properties of a surfactant are important in formulating emulsions. Enlist two factors that determine whether an emulsion is o/w or w/o.

17.6 List the three mechanisms of emulsification.

17.7 What are emulsifying agents? List the three types of emulsifying agents and differences in their mechanism of stabilization of an emulsion, for example, in terms of the type of film formed around the dispersed phase and the zeta potential on the dispersed phase.

17.8 Which surfactants will you select for o/w and w/o emulsification.

17.9 Identify the type of self-emulsifying system most appropriate for the following statements (SEDDS or SMEDDS):

A. Has lower dispersed-phase globule size after emulsification

B. Has higher content of oil

C. Has higher content of cosolvent

D. Is transparent in appearance after emulsification

E. Is likely to have higher oral bioavailability

17.10 Which of the following surfactants is suitable for the formulation of a o/w emulsion?

A. Surfactant with an HLB value of 1–3

B. Surfactant with an HLB value of 3–6

C. Surfactant with an HLB value of 6–9

D. Surfactant with an HLB value of 9–12

E. Surfactant with an HLB value of 12–15

F. Surfactant with an HLB value of 15–18

17.11 Which of the following surfactants is suitable for the formulation of a w/o emulsion?

A. Surfactant with an HLB value of 1–3

B. Surfactant with an HLB value of 3–6

C. Surfactant with an HLB value of 6–9
D. Surfactant with an HLB value of 9–12
E. Surfactant with an HLB value of 12–15
F. Surfactant with an HLB value of 15–18

FURTHER READING

Allen LV Jr. (2002) *The Art, Science, and Technology of Pharmaceutical Compounding*, 2nd ed., Washington, DC: American Pharmaceutical Association, pp. 263–276.

Florence AT and Attwood D (2006) *Physicochemical Principles of Pharmacy* 4th ed., London: Pharmaceutical Press.

Im-Emsap W, Siepmann J, and Paeratakul O (2002) Disperse systems. In: Banker GS and Rhodes CT (Eds.) *Modern Pharmaceutics*, 4th ed., New York: Marcel Dekker, pp. 237–285.

Lieberman HA, Rieger MA, and Banker GS (Eds.) (1996) *Pharmaceutical Dosage Forms: Disperse Systems*, Vols. 1, 2 and 3, New York: Marcel Dekker.

Narang AS, Delmarre D, and Gao D (2007) Stable drug encapsulation in micelles and microemulsions. *Int J Pharm* 345: 9–25.

Chapter 18

Pharmaceutical solutions

LEARNING OBJECTIVES

On completion of this chapter, the students should be able to

1. Describe different types of solutions.
2. Identify quality attributes of solution dosage forms.
3. Describe three common approaches for improving drug solubility.
4. Describe buffer and buffer capacity.
5. Describe physical, chemical and microbial stability of solution dosage forms.

18.1 INTRODUCTION

Solutions are homogeneous mixtures of one or more solutes molecularly dispersed in a suitable solvent or a mixture of mutually miscible solvents. A solution composed of only two substances is a binary solution. The components making up a binary solution are termed the solute and the solvent depending on their relative proportions (component in lower proportion is termed solute).

Pharmaceutical solutions are used for many routes of administration, including oral, rectal, vaginal, ophthalmic, parenteral, and otic. The most common solution dosage form is the oral liquid, which includes aqueous solutions, syrups, and elixirs. The physicochemical (e.g., solubility) and stability characteristics of the active drug determine whether an oral solution dosage form can be prepared. The required solubility of a drug and its solubility in water and biocompatible water-miscible solvents help decide the dosage form composition. For example, if the drug is water soluble, a simple aqueous solution can be prepared. However, if it is soluble in a water–alcohol–glycerin cosolvent system, an elixir is appropriate.

Drugs are commonly given in solution in cough/cold remedies and in medications for the young (pediatric) and elderly (geriatric).

Saturated solutions are solutions which, at a given temperature and pressure, contain the maximum amount of solute that can be dissolved in the solvent. *Buffer solutions* contain a combination of weak acid and its salt with a strong base *or* a weak base and its salt with a strong acid. These solutions resist changes in pH upon the addition of small quantities of acid or base. Solubility and stability of most ionic drugs change with pH. Therefore, most pharmaceutical solutions are pH controlled using an appropriate buffer. *Isotonic solutions* have similar tonicity as biological fluids. These solutions cause no swelling or contraction of the tissues with which they come in contact and produce no discomfort when instilled in the eye, nasal tract, blood, or other body tissues. Parenteral solutions or solutions for direct administration to mucosal tissues should be isotonic or hypotonic to avoid local tissue stress and pain upon administration. Tonicity is usually adjusted using dextrose or sodium chloride. Isotonic sodium chloride is a 0.9% w/v concentration of NaCl in water and is also called normal saline. A 5% w/w dextrose, also known as glucose, solution in water is also isotonic. These are commonly used infusion fluids for intravenous administration.

Solutions intended for oral administration usually contain sweeteners, flavors, and colors to make the medication more attractive and palatable to the patient. They may contain stabilizers to maintain the physicochemical stability of the drug and preservatives to prevent the growth of microorganisms in the solution.

A drug dissolved in an aqueous solution is generally in the most bioavailable form. As the drug is already in solution, no dissolution step is necessary before systemic absorption occurs.

18.2 TYPES OF SOLUTIONS

18.2.1 Syrup

Syrup is a saturated sugar solution. Thus, aqueous solutions containing sugar at or close to its saturation concentration of 67% w/w are called syrups. In special circumstances, it may be replaced in whole or in part by other sugars (e.g., glucose/dextrose) or nonsugars (e.g., sorbitol, glycerin, and propylene glycol). Syrups containing flavoring agents but no drugs are called nonmedicated syrups. Syrups provide a pleasant means of administering a liquid form of a disagreeable tasting drug. Syrups are appropriate for water-soluble drugs. Cold and cough syrups are the most common examples of medicated syrups.

Most syrups contain the following components in addition to the purified water and drug(s): (a) sugar, usually sucrose or other sugar substitutes are used to provide sweetness and viscosity, (b) antimicrobial preservatives, (c) flavorants, and (d) colorants. Syrups may also contain solubilizing agents, thickeners, or stabilizers.

Sucrose is the sugar that is most frequently employed in syrups. Sucrose not only provides sweetness and viscosity to the solution but it also renders the solution inherently antimicrobial. Although dilute sucrose solutions can provide an efficient nutrient medium for the growth of microorganisms, concentrated sugar solutions are hypertonic and resist microbial growth because of the unavailability of the water required for the growth of microorganisms. Glycine, benzoic acid (0.1%–0.2%), sodium benzoate (0.1%–0.2%), and various combinations of methylparabens, propylparabens, and butylparabens or alcohol are commonly used as antimicrobial preservatives.

Most syrups are flavored with synthetic flavorants or with naturally occurring materials, such as orange oil and vanillin, to render the syrup pleasant tasting. To enhance the appeal of the syrup, a coloring agent that correlates with the flavorant employed (i.e., green with mint, brown with chocolate) is used.

18.2.2 Elixir

Sweetened hydroalcoholic (combinations of water and ethanol) solutions are termed elixirs. Compared to syrup, elixirs are usually less sweet and less viscous, because they contain a lower proportion of sugar and are consequently less effective than syrups in masking the taste of drugs. In contrast to aqueous syrups, elixirs are better able to maintain both water-soluble and alcohol-soluble components in solution due to their hydroalcoholic properties. These solubility characteristics often make elixirs preferable to syrups.

All elixirs contain flavoring and coloring agents to enhance their palatability and appearance. Each elixir requires a specific proportion of alcohol and water to maintain all of the components in solution. Elixirs containing over 10%–12% alcohol are usually self-preserving and do not require the addition of antimicrobial agents for preservation. Alcohols precipitate tragacanth, acacia, agar, and inorganic salts from aqueous solutions. Therefore, such substances should either be absent from the aqueous phase or present in such low concentrations so as not to promote precipitation on standing. Examples of some commonly used elixirs include dexamethasone elixir USP, phenobarbital elixir, pentobarbital elixir USP, diphenhydramine HCl elixir, and digoxin elixir.

18.2.3 Tincture

Tinctures are alcoholic or hydroalcoholic solutions of chemical or soluble constituents of vegetable drugs. Most tinctures are prepared by the extraction process. Depending on the preparation, tinctures contain alcohol in amounts ranging from approximately 15% to 80%. The alcohol content protects against microbial growth and keeps the alcohol-soluble extractives in solution. Because of the alcoholic content, tinctures must be tightly stoppered and not exposed to excessive temperatures to avoid or minimize the evaporation of alcohol.

18.2.4 Oil-based solutions

Although most solution dosage forms are aqueous based, certain solutions are oil based. For example, progesterone injection is a solution of the hormone in a suitable vegetable oil for intramuscular use. In addition, solution of cyclosporine A in olive oil is available for ophthalmic and oral use. Oily solutions are generally not preferred as oral dosage forms due to palatability concerns. When a drug needs to be administered as a solution in oil, dosage forms such as emulsions and self-emulsifying or self-microemulsifying drug delivery systems are preferred. These systems contain high concentrations of surfactant and sometimes also ethanol so that they spontaneously make an emulsion or a microemulsion upon gentle mixing with water.

18.2.5 Miscellaneous solutions

Hydroalcoholic solutions of aromatic materials are termed *spirits*. *Mouthwashes* are solutions used to cleanse the mouth or treat diseases of the oral membrane. *Antibacterial topical solutions* (e.g., benzalkonium chloride and strong iodine) kill bacteria when applied to the skin or mucous membrane.

18.2.6 Dry or lyophilized mixtures for solution

Some drugs, particularly certain antibiotics, have insufficient stability in aqueous solution to withstand long shelf lives. Thus, these drugs are formulated as dry powders or granule dosage forms for reconstitution with purified water immediately before dispensing to the patient. The dry powder mixture contains all of the formulation components including drug, flavorant, colorant, buffers, and others, except for the solvent. Once reconstituted, the solution remains stable for the in-use period. Examples of dry powder mixtures intended for reconstitution to make oral solutions include cloxacillin sodium, nafcillin sodium, oxacillin sodium, and penicillin V potassium. Several proteins and antibody–drug conjugates are commercially available as lyophilized solids in vials for reconstitution into solutions immediately before parenteral administration.

18.3 QUALITY ATTRIBUTES

Important quality attributes of a solution dosage form include potency; physical, chemical, and biological stability; palatability; deliverability of the dose; and dosage uniformity. Physical stability refers to the lack of precipitation, change in color, or any other change in physical appearance or perception (e.g., viscosity or pain upon injection) of the dosage form. Chemical stability refers to the lack of unacceptable chemical degradation of the drug during the shelf life of the product under recommended packaging and storage conditions. The drug product must meet the predetermined requirements of minimum potency of the active pharmaceutical ingredient (API) and maximum concentration of known and unknown impurities. In addition, since most of the solutions are formulated in aqueous vehicle, special measures must be taken that they remain free of any microbial growth.

Palatability of the dosage form is usually enhanced by the use of sweeteners, flavors, and colorants. Additional taste masking approaches such as drug adsorption on an ion-exchange resin may be utilized for especially bitter or otherwise unpleasant tasting drugs. Deliverability of the dose refers to the ability to retrieve the labeled amount of liquid from the dispensed container under normal usage conditions. The uniformity of content of API in each dispensed unit dose is demonstrated for interdose variability within the doses dispensed from a given multidose container (bottle) and also for bottle-to-bottle uniformity in the concentration of the API. Viscosity of the formulation is an important determinant of its deliverability and uniformity of content.

18.4 FORMULATION COMPONENTS AND MANUFACTURING PROCESS

Typical formulation components of oral solution dosage forms include

- API.
- Vehicle, which is usually aqueous but could also be vegetable oil.
- Buffer for maintaining desired solution pH.
- Sweetener, flavor, and color for improving palatability.
- Taste masking agent, if required.
- Antimicrobial preservative(s).
- Antioxidant(s) or other stabilizer(s) (such as chelating agent), if and when needed.
- Cosolvent(s) and/or surfactant(s), if and when needed.

Typical manufacturing process for solution dosage forms involves simple mixing of all ingredients to make a solution. However, several process variables need to be carefully controlled to ensure a reproducible and high-quality manufacturing process, such as sequence of addition of ingredients, process equipment and parameters to control foaming and mixing dynamics, and temperature control.

Vehicle considerations include the selection of appropriate type and concentrations of pH-controlling buffer, flavor(s), sweetener(s), color(s), preservative(s), viscosity control agents in terms of their functionality, compatibility with each other and the API, and stability in solution. Vehicles used in oral solutions primarily include water, ethanol, glycerin, syrups, and various blends of these ingredients. Aqueous-miscible cosolvents used in smaller concentrations include propylene glycol and polyethylene glycol.

Most of the vehicles used for oral solutions can be used in topical solutions. In addition, topical solutions may also contain some amount of acetone, isopropanol, propylene glycol, polyethylene glycols, many oils, and numerous polymers.

18.5 SOLUBILITY

The required concentration of API in an aqueous solution is determined by the drug's dose and reasonable amount of solution that can be administered. In addition, factors such as drug's solubility and taste play a role in determining drug concentration. For example, the taste of bitter or unpleasant drugs tends to be concentration dependent. In addition, taste-masking strategies, such as drug adsorption to ion-exchange resin, limit the maximum drug concentration in solution depending on the maximum amount of drug that can be adsorbed on the resin and resin concentration in solution.

Solubilization of the API, that is, increasing the soluble concentration of the API in the vehicle, is frequently required to prepare aqueous solutions. The most commonly used approaches for solubilizing API are the use of one or more of pH control, surfactant(s), and/or cosolvent(s). Drugs that are poorly soluble in water may be dissolved in a mixture of water and a water-miscible solvent such as alcohol, glycerol, polyethylene glycol, or propylene glycol. The proper selection of a solvent depends on the physicochemical characteristics of the solute and the solvent.

Temperature is an important factor in determining the solubility of a drug and in preparing its solution. Sometimes the manufacturing process requires the use of elevated temperature to prepare a solution. After manufacturing, the solution can be brought back to room temperature without drug precipitation or crystallization.

18.5.1 pH and buffer capacity

The pH of the vehicle is an important determinant of solubility of an ionizable drug. Most drugs are weak acids (e.g., having a carboxylate group) or weak bases (e.g., having an amine group). Weak acids are ionized at basic pH. Weak bases are ionized at acidic pH. Ionized forms of the drugs are more soluble than unionized forms. Thus, pH affects the solubility of the drug. Depending on the slope of the pH-solubility profile of a drug, a slight increase or decrease in pH can cause some drugs to precipitate from a solution. Therefore, an adjustment of pH can aid in solubilizing ionizable drugs, and use of buffer to prevent pH shift on storage can minimize the risk of precipitation or crystallization.

Buffers are binary mixtures of compounds in solution that resist changes in solution pH upon the addition of small quantities of acid or base. These binary mixtures could be (a) a combination of a weak acid and its conjugate base (i.e., its salt) *or* (b) a combination of a weak base and its conjugate acid (i.e., its salt). A weak acid buffer is exemplified by the combination of acetic acid and sodium hydroxide, which forms the salt sodium acetate. A weak base buffer is exemplified by histamine and hydrochloric acid, which forms protonated histamine chloride salt. Buffer solutions are generally not prepared from weak bases and their salts because bases are usually highly volatile and unstable.

18.5.1.1 pH and buffering capacity

The most important characteristics of a buffer solution are its pH, which can be calculated using the Henderson–Hasselbach equation, and its buffer capacity, which is defined as the magnitude of the resistance of a buffer to pH changes. The stable pH of the solution generated by a buffer depends on the concentration of the two species and the pK_a of the weak acid or the weak base. It is determined by the Henderson–Hasselbalch equation.

$$pH = pK_a + \log\frac{[Salt]}{[Acid]}$$

$$pH = pK_a + \log\frac{[Base]}{[Salt]}$$

The extent to which a buffer resists change in solution pH is known as the buffering capacity. Buffering capacity of a buffer is related to the concentration of the acid and the base, that is higher the concentration, greater the buffering capacity. Buffering capacity is generally expressed as the concentration of the buffer. Thus, a 2 M acetate buffer has 10× more buffering capacity than a 0.2 M acetate buffer. The ratio of the acetate salt to the acetic acid may be the same in both buffers.

If strong acid, such as 0.1N HCl, is added to a 0.02 M solution containing equal amounts of acetic acid and sodium acetate, the pH is changed only 0.09 pH units because the base acetate (Ac⁻) ties up the hydrogen ions according to the reaction:

$$Ac^- + H_3O^+ \leftrightarrow HAc + H_2O$$

If strong base, such as 0.1 N NaOH, is added to the buffer mixture, acetic acid neutralizes the hydroxyl ions as follows:

$$HAc + OH^- \leftrightarrow H_2O + Ac^-$$

Example of How to Make a Buffer

BUFFERS OF WEAK ACIDS

An acetate buffer is created by the addition of sodium acetate to acetic acid. Alternatively, sodium hydroxide can be added to a solution of acetic acid. In the presence of the strong base, sodium hydroxide, an equimolar amount of acetic acid, converts to the sodium acetate salt or the acetate ion *in situ*.

When sodium acetate is added to acetic acid, the dissociation constant, K_a, for the weak acid is expressed by the equation:

$$K_a = \frac{[H_3O^+][Ac^-]}{[HAc]}$$

The dissociation constant is a known constant for each acid. The pK_a of acetic acid is 4.75. This means that an equal concentration of acetic acid and sodium acetate in solution will result in a solution pH of 4.75.

The pH of the final solution is obtained by rearranging the equilibrium expression for acetic acid:

$$[H_3O^+] = K_a \frac{[HAc]}{[Ac^-]} = K_a \frac{[Acid]}{[Salt]}$$

The aforementioned equation can be expressed in logarithmic form, with the sign reversed as follows:

$$-\log\left[H_3O^+\right] = -\log K_a - \log\left[Acid\right] + \log\left[Salt\right]$$

This is the Henderson–Hasselbalch equation for a weak acid:

$$pH = pK_a + \log\frac{[Salt]}{[Acid]}$$

The term, pK_a, is the negative logarithm of K_a, which is called the dissociation constant.

BUFFERS OF WEAK BASES

The buffer equation for solutions of weak base and their salts can be derived in a manner similar to that for the weak acid buffers. Accordingly,

$$[OH^-] = pK_a \frac{[Base]}{[Salt]}$$

Using the relationship $[OH^-] = K_w/[H_3O^+]$, we can obtain the following buffer equation:

$$pH = pK_w - pK_b + \log \frac{[Base]}{[Salt]}$$

18.5.2 Surfactants and cosolvents

Surfactants are commonly used in the dosage form to impart an amphiphilic character to the aqueous vehicle and/or associate with the hydrophobic drug to increase its solubility. When low concentrations of surfactants are added to the aqueous solution, they associate with the hydrophobic parts of a solute and increase the solubility of the solute in a concentration-dependent manner. At a certain concentration, known as the critical micelle concentration (CMC), there are enough surfactant molecules in solution that several surfactant molecules self-associate, with hydrophobic parts of the molecule buried inside and the hydrophilic part on the outside, facing the aqueous environment, to make structures known as micelles. Typical micelles contain 6–12 molecules of the surfactant. Micelles are subvisible soluble colloidal structures with a hydrophobic core. This allows the partition and retention of hydrophobic drug in the core of the micelle, thus dramatically increasing total drug solubility. The slope of concentration dependence of solubilization of a solute by a surfactant is significantly higher above the CMC than below.

Cosolvents increase drug solubility by altering the dielectric constant and hydrogen bonding capability of the vehicle and by providing a hydrophobic microenvironment. Commonly used cosolvents include ethanol, polyethylene glycol, and propylene glycol. In addition, cyclic polysaccharides, such as cyclodextrins, that have a hydrophobic cavity and a hydrophilic exterior are often used for drug solubilization.

18.6 STABILITY

18.6.1 Physical stability

A solution consists of drug substance solubilized in a vehicle commonly with the aid of pH control, surfactant(s), or cosolvent(s). Physical or chemical changes during storage, such as decrease in the storage temperature, microbial growth resulting in pH change and cosolvent evaporation or loss by selective adsorption, can lead to supersaturation of the drug in the vehicle. Supersaturated solutions can form crystal nuclei of the drug when the supersaturated drug concentration reaches above the threshold for nucleation (Figure 18.1). The crystal nuclei tend to grow slowly (crystallization)

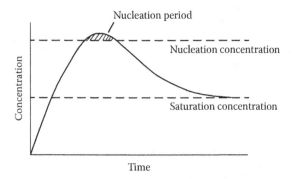

Figure 18.1 Time dependence of concentration required for monodispersity. This figure represents supersaturation region of drug solubility between saturation and the concentration that leads to nucleation. (From Narang, A.S. et al., *Int. J. Pharm.* 345, 9–25, 2007. With Permission.)

resulting in reduced solution concentration of drug and formation of particulates, which, when sufficiently large, can become visible to the naked eye. In addition, sudden changes in temperature, such as freezing, can result in instantaneous *precipitation* of the drug in the form of small, amorphous particles. Formulating a drug solution much below its saturation concentration is preferred to avoid physical instability by precipitation or crystallization.

18.6.2 Chemical stability

Solution dosage form presents an environment with high molecular mobility of reacting species, resulting in higher degradation liability than other dosage forms, such as tablets. Common modes of drug degradation in solution include hydrolysis and oxidation. Drug degradation pathways and stabilization strategies are discussed in Chapter 7.

Degradation of drug in the dosage form leads to decrease in drug potency and formation of impurities. Depending on the therapeutic window and dose of the drug, and the toxicological nature and quantity of impurities formed, the national compendia such as the United States Pharmacopeia (USP) and the international bodies such as the International Council on Harmonization (ICH) recommend maximum limits on the permissible impurities. These limits are identified in terms of reporting, identification, or qualification thresholds—requiring the sponsor of the new drug application (NDA) to report, identify, or provide toxicological safety data on the given impurity to identify and justify a maximum permissible concentration. In addition, impurities that are suspected to be genotoxic are rigorously controlled.

In addition to chemical stability of the drug, adequate potency of other additives critical to the stability and performance of the dosage form, such as antimicrobial agents and antioxidants, must be demonstrated throughout a product's shelf life.

18.6.3 Microbial stability

Pharmaceutical aqueous solutions generally contain organic compounds, including carbohydrates, thus providing a suitable growth environment for bacteria and other microbes. Except in the case of broad-spectrum antibiotics or self-preserving solutions, such as syrups, antimicrobial preservatives are frequently required in solution formulations. Methylparaben, propylparaben, and sodium benzoate are the commonly used antimicrobial agents. Methylparaben and propylparaben are commonly used in 9:1 w/w ratio at combination at 0.2% w/v total concentration.

REVIEW QUESTIONS

18.1 Indicate which statements are TRUE and which are FALSE.
  A. Buffers are used to avoid fluctuations in the pH of a solution
  B. Tinctures and elixirs contain alcohol, whereas syrups contain sucrose
  C. Concentrated sucrose solutions are good for microbial growth, but not the diluted sucrose solution.
  D. Pharmacologically active agents should be in solution before they can exert their effect.
  E. In general, solution dosage forms have a longer shelf life than the same drug formulated as a tablet.
18.2 Which of the following describes the solution dosage form?
  A. A homogeneous system
  B. The product contains at least two components
  C. The solute is in a monomolecular dispersion
  D. All of the above
  E. None of the above
18.3 Define the following terminologies: pharmaceutical solutions, elixirs, spirits, and tinctures.
18.4 Which of the following formulation components are antimicrobial preservatives?
  A. Sodium benzoate
  B. Methylparaben
  C. Propylparaben
  D. All of the above
  E. B and C of the above

18.5 Which of the following characteristics will increase drug solubility in an aqueous solution?
   A. Presence of a polar group
   B. Low melting point
   C. High boiling point
   D. Presence of an ionized group

FURTHER READING

Allen LV Jr. (2002) *The Art, Science, and Technology of Pharmaceutical Compounding*, 2nd ed., Washington, DC: American Pharmaceutical Association, pp. 231–248.

Allen LV Jr., Popovich NC, and Ansel HC (2005) *Ansel's Pharmaceutical Dosage Forms and Drug Delivery Systems,* 8th ed., Philadelphia, PA: Lippincott Williams & Wilkins.

Billany MR (1988) Solutions. In Aulton ME (Ed.) *Pharmaceutics: The Science of Dosage Form Design,* New York: Churchill Livingstone.

Block LH and Yu ABC. (2001) Pharmaceutical principles and drug dosage forms. In Shargel L, Mutnick AH, Souney PH, Swanson LN (Eds.) *Comprehensive Pharmacy Review,* New York: Lippincott Williams & Wilkins, pp. 28–77.

Narang AS, Delmarre D, and Gao D (2007) Stable drug encapsulation in micelles and microemulsions. *Int J Pharm* 345: 9–25.

Rolling WM and Ghosh TK (2004) Oral liquid dosage forms: Solutions, elixirs, syrups, suspensions, and emulsions. In Ghosh TK and Jasti BR (Eds.) *Theory and Practice of Contemporary Pharmaceutics*, Boca Raton, FL: CRC Press, pp. 367–385.

Chapter 19

Powders and granules

LEARNING OBJECTIVES

On completion of this chapter, the students should be able to

1. Differentiate between powders and granules.
2. Describe methods of production of granules.
3. Differentiate between amorphous and crystalline powders.
4. Describe methods of production of amorphous powders.
5. Define polymorphism.
6. Discuss desired quality attributes of powders and granules and techniques for their quantitation.

19.1 INTRODUCTION

Almost all the pharmaceutical dosage forms involve the handling of powders at one or more stages of their preparation. For example, the manufacture of tablets and capsules requires the compression or filling of powders in a tableting or a capsule filling machine, respectively. Most of the drugs are in solid state at room temperature. Therefore, even the processing of liquid dosage forms, such as oral, ophthalmic, or parenteral solutions, requires the powder drug to be dissolved in a solvent during their manufacture.

In many cases, pharmaceutical powders are dispensed as a dosage form. For example, the lyophilized powders of protein and peptide drugs are provided in a vial for reconstitution before parenteral administration. Similarly, powders for oral solution or suspension are provided in a bottle for reconstitution with water before oral administration. These powder dosage forms offer the advantage of better physicochemical stability and longer shelf life over the corresponding liquid dosage forms.

The widespread usage of powders in the pharmaceutical industry and practice, therefore, requires a thorough understanding of their properties and behavior to ensure their appropriate and efficient utilization. Powders—both excipients and active pharmaceutical ingredients (APIs)—can be amorphous or crystalline and possess different density, size, surface area, and flow properties. Mixtures of powders need to be uniform in the distribution of different powder components.

A key quality attribute of powders is the size and size distribution of their constituent particles. Often, efficient pharmaceutical processing or dosage form bioavailability requires increase or reduction of the particle size distribution (PSD) of the powder particles. PSD can be increased by powder agglomeration into larger, relatively stable aggregates in which the original particles maintain structural integrity and can still be identified. These aggregates are called granules. The process of preparing granules is commonly called granulation. Granulation of powders is frequently carried out during pharmaceutical manufacturing to improve the bulk properties of the starting materials. Granulation can be carried out with or without the use of water—and is accordingly called dry or wet granulation. Reduction in the PSD of powders can be achieved by milling or comminution.

This chapter will introduce basic concepts of powder properties, their importance to pharmaceutical processing and dosage forms, and common techniques for modification of powder properties.

19.2 PRODUCTION OF POWDERS AND GRANULES

Most of the materials used in pharmacy and pharmaceutical industry occur as finely divided solid materials, known as powders. Understanding the origin and nature of these powders is important for their effective usage.

19.2.1 Origin of powdered excipients

Powdered raw materials for pharmaceutical applications can be of natural, synthetic, or semisynthetic origin. This is exemplified by the common excipients used in pharmaceutical manufacturing. For example[1]:

- *Natural origin: animal products*
  - Lactose is produced from the whey of cows' milk, whey being the residual liquid of the milk following cheese and casein production. Lactose is a commonly used fragile and water-soluble filler.
- *Natural origin: plant products*
  - Microcrystalline cellulose is manufactured by the controlled hydrolysis, with dilute mineral acid solutions, of α-cellulose, which is obtained as a pulp from fibrous plant materials. Following

hydrolysis, the hydrocellulose is purified by filtration, and the aqueous slurry is spray dried to form dry, porous particles of a broad-size distribution. Microcrystalline cellulose is commonly used as plastic deformable filler.

- Starch is extracted from plant sources through a sequence of processing steps involving coarse milling, repeated water washing, wet sieving, and centrifugal separation. The wet starch obtained from these processes is dried and milled before use in pharmaceutical formulations. Starch is used as a filler, binder, and disintegrant in tablet formulations.
- Pregelatinized starch is chemically and/or mechanically processed starch that possesses better flow, compressibility, and binding properties.
- *Semisynthetic product*
  - Sodium starch glycolate is a substituted and cross-linked derivative of potato starch. Starch is carboxymethylated by reacting with sodium chloroacetate in an alkaline medium followed by neutralization with citric, or some other acid. Cross-linking may be achieved by either physical methods or chemical methods by using reagents such as phosphorus oxytrichloride or sodium trimetaphosphate. Cross-linking hydrophilic polymer chains creates an excipient that swells but does not dissolve in the presence of water and can serve as a tablet disintegrant.
- Hydroxypropyl cellulose (HPC) is water-soluble cellulose ether produced by the reaction of cellulose with propylene oxide. This is a long chain hydrophilic polymer that serves as a binder in wet granulation.
- *Synthetic product*
  - Pyrrolidone is produced by reacting butyrolactone with ammonia. This is followed by a vinylation reaction in which pyrrolidone and acetylene are reacted under pressure. The monomer, vinylpyrrolidone, is then polymerized in the presence of a combination of catalysts to produce polyvinylpyrrolidone, also known as povidone. Povidone, a long chain hydrophilic polymer, is also a commonly used binder in wet granulation.
- Water-insoluble cross-linked polyvinyl pyrrolidone (PVP, crospovidone) is manufactured by a polymerization process, where the cross-linking agent is generated *in situ*. Cross-linking of the hydrophilic polymer generates a swellable but insoluble polymer that serves as a disintegrant.
- Magnesium stearate is prepared either by chemical reaction of aqueous solution of magnesium chloride with sodium stearate or by the interaction of magnesium oxide, hydroxide, or carbonate with stearic acid at elevated temperatures. Magnesium stearate is a fine hydrophobic powder that serves as a lubricant in the manufacture of solid dosage forms.

19.2.2 Origin of powdered active pharmaceutical ingredients

Most of the drug substances are of synthetic origin. A chemical synthesis process is preferred over natural raw materials to assure adequate material purity, availability, and consistency of the process. Few drug compounds, such as taxol, are of semisynthetic origin. Most drugs are purified as pure powder crystals at the end of their synthesis. In the case of amorphous drugs, processes such as lyophilization and spray drying are used to convert the drugs into solid forms for purification, long shelf life, and ease of handling. Crystalline APIs are preferred because they are more stable, and the process of crystallization removes non-API molecules, such as intermediates used during synthesis, from the crystal lattice—thus ensuring purity.

19.2.3 Amorphous and crystalline powders

The powders used in pharmaceutical industry and pharmacy practice could be either crystalline or amorphous in nature.

- Crystalline powders have a well-defined and repeating, long-range order of the arrangement of molecules due to intermolecular interactions in the solid state. The term *long range* indicates that many molecules may be involved in the intermolecular interactions that define the fixed arrangement of molecules in a crystalline structure. This smallest fixed arrangement of molecules that repeats throughout the crystal is known as a unit cell. In a crystalline material, the arrangement of molecules with respect to each other is well defined and not random.
- Amorphous powders do not have a well-defined and repeating, long-range order of the arrangement of molecules. The molecules of an amorphous solid may show intermolecular interactions, but these interactions may not repeat consistently over several molecules. Therefore, in an amorphous material, the orientation of molecules with respect to each other is largely random.

Most of the APIs are crystalline in nature and are produced by a process known as crystallization.

19.2.4 Production of crystalline powders

Crystallization is the production of solid crystals in a solution of the solute being crystallized, which is followed by the subsequent separation of those crystals from the solution. Crystallization is accomplished by creating a state of supersaturation of the solute in a solution. A supersaturated solution has a solute concentration greater than the thermodynamic equilibrium solubility of the solute in the solvent.

Supersaturation can lead to crystallization through the spontaneous formation or extraneous addition (seeding) of nuclei. Nuclei are the associations of few (10s to 100s) molecules with the same intermolecular spatial arrangements that characterize the crystal form. Supersaturation can be achieved in one of several ways:

- Evaporation of solvent from a solution.
- Changing the temperature of the solution. For example, cooling the solution could lead to supersaturation if the solute has a positive heat of solution (increase in solubility with increase in temperature).
- Production of additional solute in the solution by chemical reaction.
- Change in solution by the addition of other soluble solute(s).
- Change in solution by the addition of other solvent(s). For example, addition of a miscible solvent that has lower solubility for the solute could lead to the formation of a cosolvent system with lower overall solute solubility than the solute concentration.

19.2.4.1 Crystallization versus dissolution

The crystals of a solute in its solution can undergo either of the two processes—growth of the crystals involving transfer of solute from the solution to the crystal state (crystallization) or loss of solute molecules from the crystals into the solution (dissolution). Crystallization would be expected in the case of supersaturated solutions, and dissolution of the crystals is expected when the solution concentration is lower than the saturation concentration.

Although the driving forces for these processes (relative strength of solute–solute, solute–solvent, and solvent–solvent interactions) are the same, they can have very different rates for the same concentration gradient. Generally, the rate of dissolution is greater than the rate of crystallization.

19.2.4.2 Polymorphism

Polymorphism refers to the ability of a solid to exist in more than one crystal structure or form. Intrinsic properties of a molecule together with crystallization conditions determine the possibility of existence of different crystalline (polymorphic) or amorphous forms of a molecule. For example, certain molecules may only exist in one form in the solid state. Some other molecules can have several crystalline forms and may also exist in an amorphous state. For example, flufenamic acid exists in nine different polymorphic forms.

Polymorphism can be of different types depending on the driving forces or the underlying molecular reason for polymorphism:

- The existence of polymorphism due to differences only in the spatial arrangement of molecules in a crystal, or crystal packing, is termed packing polymorphism.
- When a solute can exist in different crystal types depending on its state of solvation or hydration, the polymorphism is termed pseudopolymorphism.
- Polymorphism attributable to different conformers of a molecule, formed by rotation along single bond(s), is known as conformational polymorphism.

At a molecular level, polymorphs differ in the strength and nature of intermolecular interactions. Polymorphs differ in the surface exposure of functional groups of the molecule on different faces of a crystal. Accordingly, different polymorphic forms of a molecule usually differ in their surface properties such as wettability and interparticle interactions leading to differences in dissolution rate, bioavailability, and/or chemical stability.

Changing the conditions of crystallization can generate spatial polymorphs. For example, type of solvent, degree of supersaturation, pH of solution, rate of cooling, extent of mixing, or the presence of impurities in solution can lead to the formation of different crystalline forms of a molecule. Seeding the solution with a small quantity of the desired crystal form generates the same crystal form.

Crystalline forms have lower free energy and are thermodynamically more stable than amorphous forms of a molecule. Thus, an amorphous form tends to convert to a crystalline form on storage.

Different crystalline polymorphic forms usually differ in their thermodynamic stability. When a drug substance exists in different polymorphic forms, the greater thermodynamic stability of one crystalline form over another is often attributable to the higher strength of intermolecular interactions and/or closer or dense crystal packing. These differences often reflect in the melting point of various crystalline forms. Thus, the higher melting crystal form is usually also the more stable form.

A metastable (less stable) polymorphic form tends to transform into a more stable polymorphic form on storage. A change in the polymorph of an API during pharmaceutical manufacturing or in a finished drug product can lead to unintended consequences with respect to drug stability or bioavailability. Therefore, identification and characterization of polymorphic forms of a drug substance are carried out during new product development. In addition, the thermodynamically most stable polymorphic form is usually preferred for use in a drug product.

19.2.5 Production of amorphous powders

Amorphous forms of a solute can be produced by several means. For example, a high rate of solvent evaporation from a solution of the solute can result in the precipitation of solute in an amorphous form. High rate of solvent evaporation can be achieved, for example, by spray drying. Spray drying involves atomization of a solution followed by solvent evaporation in a continuous flow gaseous phase at a temperature higher than the boiling point of the solvent. Large evaporating surface area of small droplets of solution facilitates the rapid rate of solvent evaporation. Changes in process parameters for spray drying, for example, droplet size, solute concentration, and rate of solvent evaporation can lead to significantly different powder properties, such as size, of the precipitated material.

Solvent removal from a solution is also utilized to generate powders that contain two or more solid substances intimately mixed together in a fixed composition. This process generates powder particles that have one solid dispersed or dissolved in another solid of higher quantity. These systems are termed solid dispersions or solid solutions, respectively. These systems can be utilized to generate and stabilize amorphous forms of a drug substance. The choice of the other component (non-API or excipient) in these systems can determine the stability and dissolution rate of a drug from its solid dispersion or solid solution. Commonly used hydrophilic excipients that are used to prepare stable amorphous solid dispersions of APIs include povidone and hydroxypropyl methylcellulose acetate succinate (HPMC-AS). Use of hydrophobic excipients, such as Eudragits®, can produce solid dispersions with slow or sustained drug release properties.

19.3 ANALYSES OF POWDERS

Characterization of powders and granules typically involves analysis and quantification of physicochemical properties that are of significance to their proposed use. These properties can be identified as particle or bulk properties.

- Particle properties refer to the properties or characteristics of individual particles, such as their size, hardness, and porosity.
- Bulk properties refer to the properties or behavior of a collection of particles, such as flow and bulk density (BD).

Characterization of pharmaceutical powders involves analysis and quantification of both particle and bulk properties.

19.3.1 Particle shape and size

19.3.1.1 Defining particle shape and size

The size of a sphere can be defined in terms of its radius, or more commonly, diameter. The size of a cube can be described in terms of the length of its side or diagonal. However, as shown in Figure 19.1, particles can have a diverse range of shapes from needle shape to irregular polygonal. Quantitatively measuring and defining the size of these particles can be a challenge. Nevertheless, the use of finely divided powders in pharmaceutical unit operations requires a numerical description of particle size, preferably as a single number, to enable comparison of different powder types and also of different batches of the same material. Using a one-dimensional property of a particle (such as its surface area or volume) and describing it in terms of an equivalent sphere allow the description of a three-dimensional object by a single number with respect to the property of interest. The criterion of equivalency of particle size to the size of a sphere is based on the powder's intended use or application. For example, use of a powder for surface catalysis or comparison of dissolution rate of different batches would require surface area-based equivalency.

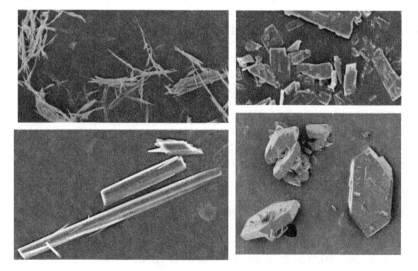

Figure 19.1 Examples of particle shapes commonly encountered for active pharmaceutical ingredients.

Irregular-shaped particles can be defined in terms of two parameters:

1. Diameter of an equivalent sphere. Powder processing technologies, such as milling and granulation, tend to change the shape of particles toward or closer to a spherical shape.
2. Aspect ratio, which is the ratio of longest to the smallest axis of a particle. It would be one for a sphere and the largest for a needle-shaped particle. Aspect ratio helps define the deviation of a shape from a perfect sphere.

Many commonly used particle size measurement methods define the size of a particle in terms of the diameter of an equivalent sphere. There are several assumptions and/or limitations associated with this description. For example, defining particle size in terms of the diameter of an equivalent sphere requires a consideration of the criterion used to define *equivalency*. For example, two particles can be described as equivalent in terms of volume or surface area. Thus, size of a particle can be expressed as the diameter of a sphere of equivalent volume or surface area of the particle being analyzed.

19.3.1.2 Defining particle size distribution

Powders are a collection of particles of different sizes. Therefore, powders have a PSD rather than a single particle size. A single numeric descriptor of the PSD of a powder can be the mean particle size. The mean diameter of a set of particles in a powder sample can be described using either arithmetic mean or geometric mean. When using arithmetic mean diameter, the presence of fewer, larger diameter particles can skew the calculated average result toward the large particle size, which may not be truly representative of the batch. The distribution of particles of a powder often follows a unimodal (one peak) lognormal distribution (i.e., when log of particle size is plotted against the frequency of occurrence of the particles of each size—a Gaussian or normal distribution is obtained). Therefore, geometric mean diameter (GMD) is generally preferred to define the particle size of a powder.

The method for defining PSD needs to have the following properties:

- Be independent of the statistical type of distribution in the sample, for example, normal or lognormal.
- Be descriptive of the particle characteristics of interest to the intended application that is, expressing particle size as spheres of equivalent surface area or volume depending on the application.

The statistical measures listed in Table 19.1 are frequently used to characterize the PSD of a powder sample.

Table 19.1 Statistical measures used to define a particle size distribution

Statistical parameter	Nomenclature or meaning of the parameter	Calculation of the parameter	Equation	Comments on its usage
$d(90)$, $d(50)$, and $d(10)$	90%, 50%, or 10%, respectively, of particles by number are below this diameter (of an equivalent sphere)	Percentile calculation	None	Most commonly utilized parameters for quality control in pharmaceutical manufacturing
$d[1,0]$	Number length mean diameter, number mean diameter	Mean diameter calculated by dividing the sum of diameter terms in the numerator with the number of particles in the denominator	For three particles of diameter d_1, d_2, and d_3, $$d[1,0] = \frac{d_1 + d_2 + d_3}{3}$$	Represents the mean diameter of a sphere of equivalent diameter to the particles of the powder
$d[2,0]$	Number surface mean diameter	Mean diameter calculated by dividing the sum of squares of diameter terms in the numerator with the number of particles in the denominator, followed by taking a square root	For three particles of diameter d_1, d_2, and d_3, $$d[2,0] = \sqrt{\frac{d_1^2 + d_2^2 + d_3^2}{3}}$$	Represents the mean diameter of a sphere of equivalent surface area to the particles of the powder
$d[3,0]$	Volume mean diameter, number volume mean diameter, number weight mean diameter	Mean diameter calculated by dividing the sum of cubes of diameter terms in the numerator with the number of particles in the denominator, followed by taking a cube root	For three particles of diameter d_1, d_2, and d_3, $$d[3,0] = \sqrt[3]{\frac{d_1^3 + d_2^3 + d_3^3}{3}}$$	Represents the mean diameter of a sphere of equivalent volume or weight to the particles of the powder

(Continued)

Table 19.1 (Continued) Statistical measures used to define a particle size distribution

Statistical parameter	Nomenclature or meaning of the parameter	Calculation of the parameter	Equation	Comments on its usage
d[4,3]	Volume moment mean diameter; weight moment mean diameter (if density is known)	Sum of fourth power of diameters divided by the sum of cubes of diameters of particles in the sample	For three particles of diameter d_1, d_2 and d_3, $$d[4,3] = \frac{d_1^4 + d_2^4 + d_3^4}{d_1^3 + d_2^3 + d_3^3}$$	Represents the mean diameter of a sphere of equivalent volume or weight to the particles of the powder. It is preferred over d[3,0] because the calculation of d[4,3] does not require the number of particles
d[3,2]	Sauter mean diameter; surface area moment mean diameter	Sum of cubes of diameters divided by the sum of squares of diameters of particles in the sample	For three particles of diameter d_1, d_2 and d_3, $$d[3,2] = \frac{d_1^3 + d_2^3 + d_3^3}{d_1^2 + d_2^2 + d_3^2}$$	Represents the mean diameter of a sphere of equivalent surface area to the particles of the powder. It is preferred over d[2,0] because the calculation of d[3,2] does not require the number of particles

19.3.1.3 Desired particle shape and size

The desired particle size and shape of a powder is determined by its usage in the downstream unit operations. For example,

- Uniform mixing of powders is greatly facilitated if they are of equivalent size by volume. Therefore, the mixing of two or more powders with similar particle size and shape is the most likely to produce uniform distribution of each material in the mix.
- Use of particles in inhalation devices requires particles to be of similar sedimentation rate in the air. Particle size expressed as diameter of spheres with equivalent sedimentation rate in the air is called aerodynamic diameter.
- The surface area per unit weight or volume (specific surface area) of the powder determines the extent of physicochemical properties of a material that are of surface origin. For example, for crystal packing structures that lead to the exposure of functional groups on the surface, a polymorphic form with greater specific surface area is more likely to show greater intensity of such surface phenomenon than another polymorphic form with lower specific surface area. Examples of such crystal surface-dependent physical properties include chemical reactivity or surface adsorption in the solid state and the sticking tendency of a material to the stainless steel processing equipment during pharmaceutical manufacturing.

Notably, the spherical shape offers the least surface area per unit volume or weight of the material.

19.3.1.4 Factors determining particle shape

Particle shape is primarily determined by the intrinsic properties of the material and its manufacturing process. For example, crystal habits of a compound determine the crystal faces exposed to the surface of the solid. Solute–solvent interactions during crystallization determine which faces of a crystal grow faster than others. In general, faces of the crystal that interact more with the solvent grow at a slower pace than the faces that have less interaction with the solvent. Thus, crystal shape is a function of the solvent used during crystallization, and one can produce crystals of different shape having the same crystalline or polymorphic form.

Particle shape can be altered after crystallization. For example, milling of a drug substance results in smaller, irregular-shaped crystals that are closer to the spherical geometry. Also, pharmaceutical processes such as granulation, spheronization, and spray drying can produce larger particles that are closer to the spherical shape.

19.3.1.5 Techniques for quantifying particle shape and size

Particle size is commonly measured using one or more of the following techniques:

- *Sieve analysis*: This is a conventional technique that involves mass fractionation of a powder sample on a set of sieves, or wire meshes, of defined size openings using mechanical vibration. Several sieves of increasing size openings are placed one over another. A powder sample is loaded on the top sieve. Mechanical vibration is applied to allow the powder to sift through as many sieves as it would until it reaches a state where powder particles do not move through the sieve openings any more. The amount of powder on each sieve is weighed and expressed as the size fraction is lower than sieve opening diameter above and is higher than the one below (on which the powder was retained). Thus, sieve analysis produces a weight distribution of particles in different sieve fractions. The sieve analysis data can be used to compare the PSD of two or more samples graphically, or by using the calculated mean particle diameter and/or the proportion of fines.
- *Laser diffraction analysis*: Laser diffraction analysis is based on the size dependence of scattering of incident laser light by particulates in the sample. A powder sample is dispersed in an insoluble liquid or air and is passed through a beam of laser light. The scattered laser light intensity is recorded using a detector. The angle of light scattering decreases and the intensity of scattered light increases with the increasing particle size. Measuring the intensity of scattered light at a particular angle allows the estimation of size of the particle scattering the light. Cumulative plotting of size of all the particles in a powder sample produces a PSD of the sample.
- *Focused beam reflectance measurement*: *In situ* measurement of particle or droplet size and size distribution in dispersed systems is often carried out using focused beam reflectance measurement (FBRM). This is an inline technique used to generate real-time data during chemical synthesis, such as crystallization, and pharmaceutical processing, such as granulation. A fast spinning laser beam is focused on the sample through a quartz lens in a conical pattern. The laser light that encounters a particle is reflected back to the lens, where a fiber optic collects the light and passes to a detector that quantifies the intensity. The time period between the incident and the reflected light, the speed of the rotating lens, and the speed of laser light are used to calculate the length of a particle passing through the focus of the laser light. This is called the chord length. Collective plot of chord length of several particles

produces a chord length distribution. Changes in the chord length distribution during processing are used as a fingerprint of the process dynamics.

- *Microscopy*: Microscopy allows direct visual examination of powder particles. However, it provides only a two-dimensional image of a three-dimensional particle. Although this technique allows versatility with respect to sample types that can be examined, the sample preparation process can introduce bias into the sample. It is a qualitative tool for most of the applications. Automated image analysis software is frequently used when quantitation is desired. Several commercial instruments are available that automate the process of image collection and analysis, allowing the examination of several hundreds or thousands of particles in a sample.

- *Sedimentation*: Sedimentation involves gravity or centrifugal force-assisted separation of the dispersed phase from the dispersion medium over time. The density difference between the dispersed phase and the dispersion medium leads to particle separation. Sedimentation is not a preferred method for the assessment of particle size and size distribution. It is more commonly used for the quality assessment of colloidal systems, such as suspensions and emulsions, functionality assessment of superdisintegrants, such as croscarmellose sodium, and separation of particles of extremely small size from the dispersion medium.

- *Electrozone sensing*: Changes in the electrical conductance through a small aperture with the flow of a fluid containing suspended particles are used to estimate the size and number of particles in the dispersion medium. The electrical conductance changes when a particle flows through the aperture, with the change in conductance being proportional to the size of the particle. It is commonly used for counting biological cells and bacteria, using a coulter counter.

A comparison of these techniques with respect to their merits, demerits, range of particle size measured, and principle of operation is provided in Table 19.2.

19.3.1.6 Changing particle shape and size

Reduction in particle size of the API is frequently desired to improve the biopharmaceutical properties of the dosage form, such as its dissolution and absorption. Increase in particle size of the bulk powder is generally desired to improve its processability, such as flow properties. Particle size of the powders can be decreased by controlled crystallization or milling (also called comminution) of preformed particles. Particle size can be increased by controlled agglomeration through granulation.

Table 19.2 Techniques for measurement of particle size distribution

Technique	Principle	Approximate particle size range	Disadvantages and limitations	Advantages and applications
Sieve analysis	Weight fractionation based on particle diameter	40–40,000 µm	• Only for relatively large particles, such as pharmaceutical granules • Not suitable for dry powders under 38 µm diameter, emulsions, and sprays, and for cohesive and agglomerated materials • Low resolution • Does not produce true weight distribution as particle orientation and sieving time determine weight fraction	• Intuitive, simple, inexpensive, and reliable method • Commonly utilized for characterizing granulations
Laser diffraction	Angle of scatter of incident laser light depends on particle size	0.05–500 µm	• Relatively expensive and involved technique requiring operator training, careful selection of dispersion medium, and method development for each sample type	• Can run diverse sample types, wide dynamic measuring range (0.02 µm to few mm diameter), rapid procedure, and good repeatability • Generates volume-based particle size distribution, can be converted to weight-based distribution if true density of particles is known • Commonly utilized for characterizing and quality control of raw materials, such as drug substances and excipients
Focused beam reflectance measurement	Time taken for the backscatter of incident laser light, detected by a rotating lens, depends on the chord length of the particle on probe's surface	0.25–1,000 µm	• Expensive and involved technique requiring operator training and method development for each sample type • Application specific probe design frequently necessary • Absence of direct correlation with conventional methods for particle size distribution analysis	• Inline monitoring capability of chemical and pharmaceutical manufacturing processes • Commonly used for crystallization and particle growth monitoring

(Continued)

Table 19.2 (Continued) Techniques for measurement of particle size distribution

Technique	Principle	Approximate particle size range	Disadvantages and limitations	Advantages and applications
Microscopy	Direct, two-dimensional examination of particles under magnification	0.01–10,000 μm	• Low sample size and possible effects of sample preparation on unrepresentative observation • Subjectivity and individual judgment involved • Not suitable as a quality control or routine monitoring tool	• Direct visual examination of particles • Inexpensive, versatility with respect to sample types • Can be used to observe crystallinity of particles by the observation of birefringence • Commonly used in research investigations
Sedimentation	Gravity or centrifugal force-assisted separation of the dispersed phase from the dispersion medium	0.01–10,000 μm	• Affected by several factors such as particle shape, temperature, dispersed phase viscosity, and interparticle interactions • Not commonly used for particle size distribution measurement. More commonly used for assessing stability of dispersed systems	• Simple, intuitive, and inexpensive • Commonly used in research investigations
Electrozone sensing	Changes in electrical conductance of fluid flowing through an aperture as nonconducting dispersed phase passes through	0.4–1,200 μm	• Requires suspension of particles in a conducting (weak electrolyte containing) liquid medium. Not suitable for dry powders, sprays, and emulsions • Nature of particles, such as density and porosity affects method capability	• Rapid and reliable method for total particle counting • Commonly used for cell count determination in blood samples • Pharmaceutical applications limited to research investigations

Processing steps to change the size of the particles invariably also results in changes in particle shape. Milling of odd-shaped particles, such as needles, tends to reduce their aspect ratio and to change the shape toward spherical dimensions. Granulation is often accompanied by shear force and consolidation of particles into larger particles, which tend to have an irregular shape with low aspect ratios. Both milling and granulation tend to increase the sphericity of particles.

19.3.2 Surface area

19.3.2.1 Significance of surface area

Surface-dependent physicochemical phenomena are of pharmaceutical significance. For example,

- Absorption of a drug from a dosage form involves dissolution of the drug substance into the absorption medium. The rate of dissolution is proportional to the surface area of the drug substance.
- Lubricants, such as magnesium stearate, used during pharmaceutical processing are intended to cover the surface of the granules to provide adequate lubricity during unit operations such as tableting. Changes in the surface area of the granules or the lubricant can directly impact the surface coverage and effectiveness of the lubricant.
- Wet granulation is a surface phenomenon, involving wetting and agglomeration of particles. Changes in the surface area of the raw materials can significantly influence the reproducibility of granulation.

19.3.2.2 Defining surface area

Total surface area available in a powder sample is a function of both its particle size and porosity. Particle size is relatively easier to measure and compare among different powders. Porosity of the particles refers to air-filled solvent accessible channels inside particles. Thus, porosity contributes to the surface area of the particles without impacting particle size or shape. A higher porosity particle of the same size and shape as a lower porosity particle will have greater surface area. The rate of disintegration and drug dissolution from granules depends on the penetration of the dissolution medium inside the granules, which is determined by the porosity of the granules.

In determination of total surface area of a powder sample, it is difficult to distinguish the area contributed by the surface of the granules from the area contribution attributable to the porosity. For all practical purposes, this distinction is ignored. It is assumed that the surface area accessible to the penetrating medium is representative of the surface area relevant to the pharmaceutical applications of the powder.

19.3.2.3 Quantitation of surface area by gas adsorption

Surface area is commonly measured by the adsorption of an inert gas on a solid surface. It is commonly expressed as specific surface area, which is the surface area per unit weight of the powder.

Adsorption of an inert gas (the adsorbate) on a solid surface (the adsorbent) is driven by the weak van der Waals forces of attraction. The rate and extent of adsorption of the gas is primarily driven by the partial pressure of the gas (P). At isothermal (constant temperature) conditions, Freundlich proposed that the mass of gas adsorbed (x) per unit mass of adsorbent (m) is given by

$$\frac{x}{m} = k * P^{1/n}$$

where k and n are constants.

Freundlich isotherm considers multiple layers of the adsorbate on the adsorbent. Langmuir proposed an alternative equation to describe gas adsorption on the solid surface that relies on the assumption of monolayer adsorption. The number of sites occupied on the surface of a solid (θ) is given by

$$\theta = \frac{k * P}{1 + k * P}$$

where, $k = k_a / k_d$, k_a and k_d representing the rate constants of adsorption and desorption processes, respectively.

Both Langmuir and Freundlich adsorption isotherms explain gas adsorption at low pressures, but not at high pressures. Multilayer formation during gas adsorption was explained by the Brunauer–Emmett–Teller (BET) equation:

$$W_{\text{total}} = \frac{W_m * C * \left(\dfrac{P}{P_0} \right)}{\left(1 - \dfrac{P}{P_0} \right) * \left(1 + C * \dfrac{P}{P_0} - \dfrac{P}{P_0} \right)}$$

where:

P and P_0 are the equilibrium and saturated vapor pressure of the adsorbate

W_{total} is the total amount of gas adsorbed

W_m is the amount of gas adsorbed to form a monolayer

C is the BET constant that depends on the heat of adsorption for the first layer (E_1), the heat of adsorption for the second and subsequent layers or the heat of liquefaction of the adsorbate (E_L), gas constant (R), and absolute temperature (T) as

$$C = e^{\frac{E_1 - E_L}{RT}}$$

BET adsorption isotherm adequately describes physical gas adsorption for $\theta = 0.8$ to 2.0. This range covers the formation of the monolayer. The BET equation can also be expressed as a linear equation:

$$\frac{1}{W_{total}\left(\frac{P_0}{P} - 1\right)} = \frac{C-1}{W_m * C} * \frac{P}{P_0} + \frac{1}{W_m * C}$$

The determination of surface area of pharmaceutical powders is most frequently carried out using this equation. Assessment of binding interactions of a dissolved drug with solid particles in solution is carried out using Langmuir adsorption isotherm. Freundlich isotherm is used to characterize the types of adsorption profiles of different solids.

For the determination of powder surface area using BET method, adsorption of an inert gas, such as nitrogen, is carried out at isothermal conditions. The number of moles of the gas adsorbed (W_{total}) as a function of the equilibrium pressure (P) is recorded. The use of BET equation allows the calculation of the amount of gas that would form a monolayer (W_m), which allows the calculation of total surface area using the molecular area of the gas (nitrogen, 15.8 Å^2) and the Avogadro's number of molecules per mole of substance.

19.3.2.4 Altering powder surface area

Specific surface area of excipients and drug substances is primarily determined by their manufacturing process, which affects their PSD and porosity. Therefore, making changes to their manufacturing process can change surface area of raw materials. For example, the use of spray drying instead of slow solvent evaporation techniques, such as drum drying, results in the production of higher porosity particles. Changes in crystalline polymorphic form produced as result of crystallization process, such as the solvent used for crystallization, can also result in changes to the specific surface area of the material.

High-specific surface area of APIs is often desired to increase their dissolution rate from the dosage forms. This is commonly achieved by communition or particle size reduction. In addition, certain excipients, such as magnesium stearate, have a unique *plate*-type structural organization of the molecules, such that the application of shear and mixing results in the separation of plates leading to increase in surface area.

Reduction of particle surface area is desired for applications where, for example, reduction of undesired, surface-induced phenomena is needed. For example, sticking of the powder material to the stainless steel processing equipment during pharmaceutical manufacture is a function of the surface characteristics of the APIs. Therefore, reduction in the surface of the APIs per unit powder weight can minimize or mitigate this processing risk. This is commonly achieved by decreasing drug load in the formulation and granulation of the APIs with low proportion of fine particles.

19.3.3 Density and porosity

19.3.3.1 Significance of density determination

Density of powders and granules plays an important role in pharmaceutical processing. For example, handling and processing of pharmaceutical powders often require mixing of the drug with excipients. The drug must be uniformly mixed for the dosage form to have a uniform amount of drug between different dosage units. Adequate flow of a powder and the uniformity of mixing of two or more powders are significantly affected by powder density. Uniform mixing generally requires the powders to be of similar density. Mixing of powders whose particles have significantly different density may not achieve uniform mixing.

19.3.3.2 Defining powder density

Density of powders and granules is defined by their measurement technique and application to processing as follows:

- *Bulk density*: BD represents the combined mass of many *loosely* packed particles of a powder sample divided by the total volume they occupy. This total volume reflects the interparticulate (void volume) and intraparticulate (porosity of the particle) volume occupied by air, in addition to the volume occupied by the solid component(s) of the particle. BD is important for material handling considerations because it directly measures the volume that a given mass of powder would occupy under undisturbed conditions.
- *Tapped density*: Tapped density (TD) represents the *settled* or packed volume of a given mass of particles under well-defined rate and extent of agitation. For example, a measuring cylinder containing a given mass of powder can be manually or instrumentally tapped on a solid surface at a fixed rate and distance from surface, for a fixed number of taps, to cause the consolidation of the sample. TD is then determined by dividing the combined mass of the consolidated sample by the total volume it occupies.

  TD enables the determination of the extent of powder consolidation that may be expected under routine handling and equipment vibration conditions during pharmaceutical manufacturing. This is referred to as the compressibility index or Carr's index (CI) of the powder. It is defined in terms of a powder BD and TD as

$$CI = \frac{TD - BD}{BD}$$

Thus, CI represents the proportion of the bulk volume that gets consolidated under vibrational and routine handling stress. In addition, a parameter that defines the ratio of consolidated to bulk volume, Hausner ratio (IIR), is defined as:

$$HR = \frac{TD}{BD}$$

These ratios help compare the relative degree of consolidation and estimated flow characteristics of different powders.

- *True density*: True density refers to the density of the *solid* phase of the particles. It excludes the volume contribution of both inter- and intraparticulate spaces. Therefore, true density of a powder is independent of powder porosity, compaction, and pretreatment of the sample. True density is important for understanding, for example, the solid fraction of a tablet—which represents the proportion of total volume that is occupied by the solid mass.

19.3.3.3 Methods for quantifying powder density and porosity

The bulk and tapped powder densities are estimated using a simple volumetric cylinder. The compendia, such as the United States Pharmacopeia (USP), have standardized the equipment and process for the measurement of bulk and tapped densities, and also for the pretreatment of the sample before loading in the measuring cylinder. This harmonization of testing procedure helps reduce variability due to material handling and other subjective parameters that may differ between personnel and laboratories.

True density of a powder can be determined by the following:

- Volumetric measurement using Archimedes' principle and Boyle's law using an instrument called helium pycnometry.

  This method is based on the penetration of an inert gas inside a chamber of known volume that contains the powder sample under constant temperature and pressure. Estimation of the actual amount of gas penetrated against that expected based on the ideal gas law, as mentioned in the following equation, allows the calculation of the volume occupied by the solid mass and, thus, the determination of total porosity of the sample.

  $$PV = nRT$$

where:

　　P is the pressure
　　V is the volume
　　n is the mole of gas
　　R is the gas constant
　　T is the temperature

The powder sample is placed inside a chamber of defined volume, which is then filled and emptied with a defined volume of an inert gas, such as helium. The pressures observed during the filling and emptying of the sample chamber with the inert gas allow the computation of solid phase volume of the sample.

These calculations are based on the Archimedes' principle that fluid displacement by the solid phase of the particles is proportional to the volume of the solid phase. Boyle's law describing the inverse proportionality of pressure and volume of a gas at a constant temperature allows the determination of volume occupied by the gas in the sample chamber as a function of its pressure.

This method is commonly used for true density determination of powders and granules.

- Mass measurement using Washburn equation (mercury intrusion porosimetry).

  Total pore volume in a defined mass of powder can be estimated by the penetration of mercury, a nonwetting (high contact angle) liquid, inside the sample under externally applied pressure.

  In this technique, the sample is placed in a sealed chamber of known volume. Mercury is filled in the chamber under vacuum to occupy all interparticulate spaces (easily accessible, around the sample). This is followed by forced ingress of mercury inside the pores of the particles by application of external pressure. Total amount of mercury penetrated inside the pores is determined as a function of pressure. Washburn equation, describing the capillary penetration of a liquid as a function of its viscosity and surface tension, is used to estimate pore diameter at the pressures used. A plot of pressure applied against volume of mercury penetrated into the sample allows the calculation of pore volume or size as a function of the penetrated volume—which allows the calculation of not only total penetrable porosity but also the porosity as a function of pore diameter.

  Mercury intrusion porosimetry is commonly used for the comparison of granule porosity among different samples because it represents the fluid-penetrable portion of the total porosity.

19.3.3.4 Changing powder density and porosity

Control of particle density is important to ensure uniformity of mixing of two or more powders. Powders with significant differences in particle density tend to segregate during processing. In addition, particle density influences powder flow. Changing powder's bulk or TD is frequently required to achieve desired flow properties. For example, a very low-density powder may not flow well. Particle density can be changed during several pharmaceutical unit operations.

The true density and porosity of particles are determined by the intrinsic or inherent characteristics of a material, which can be influenced by the material's manufacturing process. For example, the crystalline polymorphic form of the API determines the closeness of molecular packing and the size of the cell in the crystal lattice, which impacts particle density. In the case of excipients, different density grades are sometimes available commercially. For example, Avicel PH 101 (FMC Corp.) and Avicel PH 301 have the same PSD but significantly different particle density. Particle density can be changed by changes in the manufacturing process, such as spray drying versus drum drying for the preparation of raw materials, the amount of water and shear used during wet granulation, or the pressure applied on the rolls during roller compaction. Granulation techniques such as roller compaction or wet granulation lead to shear-induced consolidation of particles, in addition to the binding and agglomeration of fine particulates.

High compressibility index (CI) or HR of a powder can also lead to flow problems, attributable to the consolidation and densification of powder bed in a localized region of the processing equipment such as the outlet nozzle of a hopper. This issue can be addressed by reducing the TD of the powder. The consolidation characteristics, and hence the TD, of a powder bed mainly depend on the PSD of the powder. Therefore, reducing the spread of the PSD by granulation and reduction of fines can reduce the CI or HR of a powder.

19.3.4 Flow

Flowability of a powder refers to its rate of passage, mass per unit time, through an aperture of given dimensions. Flowability can be assessed and expressed as either the lowest pore opening of a funnel through which a powder can flow well or the rate of flow of a powder through a defined pore opening of a funnel.

19.3.4.1 Importance of flowability of powders

Flowability of a powder is critical to most pharmaceutical unit operations. For example, adequate flow is important for ensuring

- Mixing and blend homogeneity during blending of two or more powders.

- Adequate control of dosage form weight variation during tablet and capsule filling unit operations.
- Uniformity of roller compaction of the powder.
- Transfer of powders between different unit operations through bins.

19.3.4.2 Factors influencing flow of powders

Powder flow is mainly influenced by particle shape, size, density, and size distribution. For example,

- High aspect ratio and irregularity of particle shape can hinder smooth flow of particles. Needle-shaped crystals have poorer flow compared to sphere-shaped particles.
- Powder blend with a large proportion of fines can lead to flow issues arising due to higher tendency for consolidation of powder blend. High proportion of fines can lead to the localized consolidation of powder bed, leading to stagnation, in a system requiring mass flow of the powder, such as a hopper.
- For a given particle density, particle size is the primary determinant of gravitational and inertial force on the particles. Therefore, a powder bed consisting of very fine particles, even though they may possess a narrow size distribution, tends to have flow problems compared to a similar powder bed of coarse particles.

In addition, surface characteristics of powders such as electrostatic charge and surface roughness can increase interparticle cohesiveness, resulting in flow problems.

19.3.4.3 Quantitation of powder flow

Methods for the quantitation of powder flow are designed to simulate the large-scale manufacturing conditions. A typical flow test consists of passing a predetermined mass of powder through a small hopper, or funnel, with an aperture of known diameter and quantifying the time it takes for the powder to pass through the aperture with or without any agitation of the powder bed in the hopper. Powder flow is typically expressed in weight/time units, for example, g/s. Several commercially available equipments, for example, Erweka powder flow tester (Erweka GmbH, Germany), use this principle. A limitation of these techniques is the need for strict adherence to the experimental protocol for all the powder samples whose flow needs to be compared.

A more reliable, although indirect, technique that enables powder flow comparison irrespective of the sample size or testing equipment is the measurement of angle of repose. The angle of repose is the angle of the slope of a cone of powder, from the horizontal base, when the powder is made to fall on a horizontal surface in a uniform stream and allowed to settle undisturbed. The ease of particles sliding over each other is a complex cumulative function of particle size, density, particle shape, and PSD. Lower angle of repose is indicative of ease of particle sliding across each other and interpreted to indicate better flow characteristics of the powder.

19.3.4.4 Manipulation of flow properties of powders

Changing the particle size and shape can change flow properties of a powder or powder mixture. A coarse powder with low PSD and aspect ratio tends to flow better. Flow problems attributable to the consolidation characteristics of the powder, for example, high TD, can be altered by changing powder density.

In addition, flow problems that arise from electrostatic charging or cohesive nature of the particles often require surface modification of the particles. For example, the use of lubricant, such as magnesium stearate, can alter the surface characteristics of the powder by forming a hydrophobic layer on particle surface.

19.3.5 Compactibility

In the manufacture of the most common pharmaceutical dosage forms, tablets, and capsules, powders and granules are compacted into solid masses of a given dosage unit. This process of compaction involves application of pressure on a fixed quantity of the powder within a die using stainless steel punches. The ability of a powder to form a compact on application of pressure is important for the ability to manufacture tablets.

19.3.5.1 Compressibility, compactibility, and tabletability

As illustrated in Figure 19.2, the ability of a powder to form a compact on application of pressure can be defined in terms of three parameters:

- Compressibility is defined as the ability of material to reduce in volume under applied pressure and can be measured by plotting tablet porosity as a function of compression pressure. Tablet porosity is measured by the solid fraction of the compact.
- Compactibility represents the ability of material to produce tablets with sufficient strength under the effect of densification. Compactability can be measured by plotting the mechanical strength of the compact (tensile strength) as a function of its porosity.
- Tabletability is the ability of powder material to be transformed into tablets of a specified strength under compaction pressure. Tabletability is measured by plotting the mechanical strength of the compact (tensile strength) as a function of the compaction pressure used.

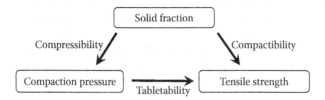

Figure 19.2 An illustration of the interrelationship between the concepts of compactibility, compressibility, and tabletability.

19.3.5.2 Importance of compactibility

The ability of a powder blend to form a strong and physically stable compact depends on its interparticle adhesion characteristics, and the balance of plastic deformation and elastic recovery under mechanical stress. Plastic deformation refers to the ability of a powder blend to permanently deform under pressure. Elastic recovery, on the other hand, represents the percent expansion of the compact from its most consolidated state under pressure.

Under compressive stress, powder particles may maintain their size but only deform in shape (plastic deformation, e.g., microcrystalline cellulose) or may break into several smaller particles (brittle fracture, e.g., dibasic calcium phosphate). Such material behavior can affect bonding between different components of the powder and affect adhesion of the powder blend. For example, lubricated powder particles—where the surface has been coated with hydrophobic magnesium stearate—show better interparticle bonding for materials that exhibit brittle fracture due to particle breakage and exposure of new uncoated surfaces upon compression compared to the materials that undergo plastic deformation.

Powder blends that show high elastic recovery or lack of adhesive bonding with other components of the powder tend to form physically unstable compacts. Such compacts tend to show problems such as capping and lamination of the tablets. Capping refers to breakage and separation of one layer of tablet close to the edge, whereas lamination refers to breakage in the middle.

19.3.5.3 Determination of compaction characteristics

During pharmaceutical development, compactibility of a powder blend is estimated using simulated tableting equipment, such as Presster® tablet press simulator or a compaction simulator. These equipments apply well-defined and controlled compression pressure within a defined period of time on the powder blends. They allow the study of a powder blend's compaction characteristics under a range of compression pressures, dwell times (duration of time for which the blend is subjected to the compression pressure), and compaction speeds (speed of compaction of powder as determined by the speed of punch movement during compression).

19.3.5.4 Factors affecting compactibility

Compactibility of a powder is a function of its intrinsic mechanical behavior, such as plastic deformation or brittle fracture, and surface interactions with other powder particles. Compactibility of a powder can also be affected by its particle size and moisture content, which act by impacting particle packing and interparticle interactions, respectively.

In pharmaceutical operations, usually powder blends are used for tableting. The pharmaceutical unit operations, such as granulation, and the use of excipients in the dosage form can adjust the compaction characteristics of a powder blend. The compactibility of a blend is a result of the compaction behavior of its individual components, which can be influenced by changing the composition of the blend.

19.3.6 Content uniformity

The powder or granulation used in pharmacy or pharmaceutical industry is commonly a mixture of two or more distinct components. Adequate performance of the powder blend at different stages of manufacturing or use depends on the uniformity of distribution of the different component materials throughout the powder.

19.3.6.1 Importance of uniform mixing

Uniform distribution of each component in a powder mixture is desired to assure uniform subdivision of the individual components when the powder mixture is subdivided. For example, compression of granules of a combination drug product, containing two different drugs, requires good content uniformity of both drugs in the granulation so that each tablet would have both drugs at the desired dose level.

Uniform distribution of components is also critical for the excipients used in the drug product manufacture. For example, magnesium stearate as a lubricant can function effectively only when it is uniformly distributed throughout the granulation. Any lack of uniformity distribution of magnesium stearate can lead to overlubrication and underlubrication of portions of the granulation, which can lead to potential drug dissolution and processability issues, respectively.

19.3.6.2 Factors affecting mixing uniformity

Uniformity of mixing of two or more components is affected by the similarity of particle characteristics of the components. Components having similar particle size, shape, density, and size distribution tend to produce uniform powder mixtures. Uniformity of content of a drug in a dosage form is usually good if the drug loading in the dosage form is high (e.g., 50% w/w or more of the dosage form weight is attributable to the drug weight) and the drug particles exhibit good flow, have a shape that is close to spherical, and possess density that is comparable to other ingredients used in the dosage form.

In addition, the choice of mixing equipment and blending protocol can affect the uniformity of content. For example,

- A V-shaped blender tends to produce better mixing than a bin blender.
- In terms of the blending protocol, minor (lower quantity) components of the powder mixture are often *sandwiched* between the major components by controlling the sequence of addition of the components to the blender. This is particularly important for critical excipients that have a tendency to segregate, such as magnesium stearate.
- Components that have atypical particle characteristics, such as the very low BD of colloidal silicon dioxide, are often premixed with a small quantity of another component before addition to the blender.
- Mixing time plays a key role. Although a minimum amount of time is required to achieve desired content uniformity, prolonged mixing does not necessarily result in better uniformity of content. In fact, prolonged mixing can compromise the uniformity. Therefore, optimum time of mixing is carefully determined and controlled.

Uniformity of a powder mixture can get compromised after mixing, such as during the storage and handling of powders. For example, vibration in the storage bins due to the operation of large-scale equipment can lead to segregation of a uniform mixture of components especially if they differ in particle size and/or density. Segregation can also happen during material transfer. For example, flow of a powder blend through the hopper from a closed chamber can result in a counter-current flow of air, which can partially fluidize the powder leading to segregation based on differences in the fluidization potential of particles of different components.

19.3.6.3 Assessment of content uniformity

Uniformity of content of the APIs in the finished drug product is an important criterion to ensure consistency of the dose delivered to the patient. The USP and other compendia define the acceptance criterion for determining the uniformity of content. This criterion is based on statistical probability considerations and is based on the requirement that the potency of each individual dosage unit must be within a given range, and no more than a given number of dosage units may exceed a narrower range.

To ensure the uniformity of content of the API in the finished drug product, pharmaceutical manufacturing also typically tests the content uniformity of the powder blend at the end of certain unit operations, such as blending and granulation. These may also provide a prospective guidance to adjust the operating parameters of such unit operations. The testing of content uniformity in powders and granules typically involves sampling a fixed quantity of the powder from several different, predefined locations in the storage container or process equipment and testing them for the content of the APIs. The acceptance criteria for the uniformity of content on these powder samples are typically same as the compendial criteria for finished drug products.

19.3.6.4 Addressing content nonuniformity issues

Selection of appropriate manufacturing process and its parameters plays a key role in ensuring good content uniformity of the drug in the final dosage form. For example, wet granulation or roller compaction-based dry granulation processes can improve the uniformity of distribution of segregation prone drugs, such as due to low drug loading or atypical particle shape or density. Granulation adds an additional mixing step and leads to the aggregation of drug particles with those of excipients, thus changing both particle size and shape. The selection of drug loading in the dosage form also plays a key role. Higher the drug loading, lower the chances of segregation of the drug.

Content uniformity issues arising from segregation in powder blends can also be addressed by engineering considerations in the design and operation of large-scale equipment. These include the handling operations that minimize vibration on the equipment and material transfers. For example, conventional tablet manufacturing processes involved preparation of the powder blends for compression and their storage in drums, which were then transferred to bins for loading on the tablet press for compression. In the redesigned process, the powder blend is prepared in a modified bin that can be used on the tablet press, thus minimizing two transfer operations. Another example of equipment redesign is designing a vent for air inlet in closed powder transfer processes to minimize fluidization of powder.

19.4 POWDER PROCESSING

19.4.1 Increasing particle size: Granulation

The size of individual particles in a powder determines bulk properties of the powder such as its flow, density, and compactibility. In addition, the surface characteristics of these particles, such as electrostatic charge and cohesivity, determine interparticle interactions that further influence bulk properties of the powder. Often, the bulk properties of a powder need to be changed for facilitating the processing and use of powders. For example, a cohesive, finely powdered API may not mix well with the inactive ingredients of a formulation (excipients) and may not flow rapidly and uniformly through the equipment used in pharmaceutical manufacturing. These problems can compromise the dose uniformity of a drug between different dosage units. Therefore, size and surface characteristics of powders are often modified in pharmaceutical processing by granulation of powders.

Granulation is the process of preparing granules, or physical aggregates of powders, in which the original particles can still be identified. Granulation commonly involves adhesion of multiple particles of more than one type of powders. This may be achieved with or without the use of water or naturally adhesive hydrophilic polymers, known as binders. Accordingly, granulation is classified based on the means of achieving the adhesion of its powder components into dry granulation or wet granulation.

- Dry granulation involves compaction of a powder under compressive force of stainless steel rolls, followed by breaking of the compacts into granules of a desired size range. It does not involve any addition of water. The characteristics of the powder particles, such as adhesion, cohesion, fragility, and plasticity, determine the compactibility of a powder.
- Wet granulation involves the addition of water, and a binder, to a powder, followed by mixing and removal of water. The binder that becomes well mixed and forms interparticle bonds during granulation maintains granules as loosely adhered powder masses even after drying and removal of water.

19.4.1.1 Dry granulation

Dry granulation involves compaction of a powder mixture. Compaction is usually carried out by roller compaction. As shown in Figure 19.3, roller compaction involves a continuous flow of powder through two rolls concurrently counter rotating in the direction of the powder flow. The rolls are hydraulically pressurized to press on the powder as it passes through the rolls. This causes the powder particles to be deformed and/or fragmented, resulting in the formation of a compact ribbon of material. This ribbon of compacted material is then force passed through an appropriate-sized screen, using equipment such as a comil. This results in the production of granules.

The important quality attributes of the granules produced by roller compaction include the percentage of fines, or the proportion of powder that did not get compacted when the compacts were passed through the comil, and the density of the granules. These can be modified using process parameters such as the distance between the rolls, pressure applied to the rolls, and the feeding rate of the powder.

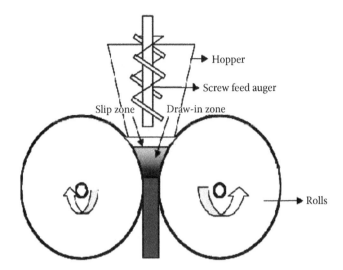

Figure 19.3 Roller compaction process. (From He, X., *Am. Pharm. Rev.*, 6, 26–33, 2003. With Permission.)

19.4.1.2 Wet granulation

Wet granulation involves the use of a binder and water to aid the agglomeration of particles. A binder is a substance with intrinsic cohesive and adhesive properties that can help form particle agglomerates. Typically, the binders used in pharmaceutical processing are hydrophilic polymers, such as PVP (also known as povidone), HPC, and starch. The binder can be added to the powder in either a dry or a solution form.

- A dry binder addition process of wet granulation involves addition and mixing of the binder as a dry powder to the powder mixture to be granulated. Granulation is carried out by the addition of water, whereas mixing is carried out in a granulator mixer. After the addition of water and mixing are complete, the granulation is force passed through an appropriate screen, using equipment such as a comil, followed by drying to obtain granules of desired size. The dried granules are passed again through the screen using the comil to obtain the final granules.
- A wet binder addition process of wet granulation involves dissolving the binder in water prior to granulation. The powder mixture to be granulated is loaded in a granulator mixer. Granulation is carried out by the addition of the binder solution followed by force passed through an appropriate screen using a comil, drying, and passing the dried granules again through the comil to obtain the final granules.

The wet granulation process is further classified as a high-shear or a low-shear process depending on the equipment used for granulation.

- A high-shear granulation process is carried out in a granulator that imparts high shearing and compacting force on the powder mixture. As shown in Figure 19.4, a typical high-shear granulator involves the movement of horizontally placed impellers at the bottom of the powder bed. The weight of powder bed increases the shear in this granulator design.
- A low-shear granulation process is carried out in a granulator that imparts relatively less shearing and compacting force on the powder mixture. As shown in Figure 19.5, a typical low-shear granulator involves the movement of vertically placed impellers around the powder bed.

The drying process involves exposure of the wet granules to a dry and hot air, which leads to the drying of granules. It is typically carried out in a tray drier or a fluid bed dryer.

- A tray drier represents a static drying process whereby the granules are spread on flat metallic trays and exposed to dry and hot air in a convection oven. This process is less efficient, time consuming, and may lead to uneven drying of granule surfaces.
- A fluid bed drying process involves suspending the granules in a current of dry and hot air that flows vertically upward through the powder bed. This process is usually more efficient but can lead to greater attrition of the granules during drying due to interparticle collisions.

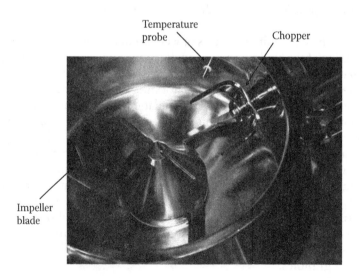

Figure 19.4 A high-shear granulator.

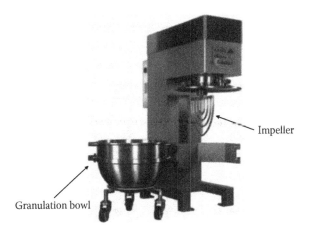

Figure 19.5 **A low-shear granulator.**

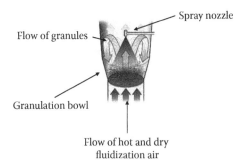

Figure 19.6 **Fluid bed process, showing the granulation chamber with the flow dynamics of granules, granulating fluid spray, and the fluidization air.**

An alternative process for low-shear granulation involves fluid bed granulation (Figure 19.6). This process involves spray of water or binder solution on the powder suspended in a vertical current of dry and hot air, leading to simultaneous equilibrium processes of wetting, granulation, and drying of the particles.

The binder fluid used in wet granulation can be other than water, or it can be a mixture of water with another fluid. For example, ethanol or hydroethanolic solutions have been used as binder fluids. The use of non-aqueous fluids places stringent requirements on the processing plant to control potential explosive potential and environmental egress of the solvents. Therefore, most, if not all, modern wet granulation processes use water as a granulation fluid.

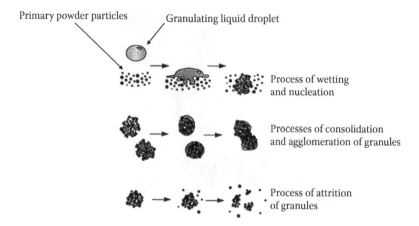

Figure 19.7 An illustration of the mechanisms involved in wet granulation. (From Iveson, S.M. et al., *Powder Technol.,* 117, 3–39, 2001. With Permission.)

The mechanism of wet granulation involves four processes (Figure 19.7)[2]:

1. Agglomeration of primary powder particles into coarse aggregates or granules
2. Breakage of large aggregates into two or smaller aggregates due to the shear or impact of collision
3. Consolidation, involving the densification of granules by shear and compressive forces leading to reduced porosity of granules
4. Attrition due to shear forces and interparticle collisions leading to breakage of particles from the surface of granules

The important quality attributes of the granules produced by wet granulation include the PSD and density of the granules. These can be modified using process parameters such as the amount of water and binder, duration and speed of mixing during granulation, use of a high- or low-shear granulator, and the size of the screen used for sizing the granulation.

19.4.2 Decreasing particle size: Comminution or milling

Comminution, or milling, is the mechanical process of size reduction of powder particles or aggregates. Particle size reduction is often also called micronization, which indicates reducing the size of powder particles to micrometer level in diameter. A finely divided particulate nature of powders is frequently needed for their efficient use. In addition to the reduction of size, milling also changes the shape of the particles toward a spherical shape. This can improve the cohesivity and flow of powders with needle- or irregular-shaped particles. Powders of similar particle size flow better and are more likely to show good uniformity of content when mixed together. Also, dispensing of powders can be more precise if the powders are of finely divided and of uniform nature.

19.4.2.1 Techniques for particle size reduction

Based on the type of equipment employed, comminution may be termed as follows:

- *Cutting*: For example, extrusion spheronization and hot melt granulation involves cutting a uniform stream of granulation mix into smaller particles that are then rounded off into uniform granules. This may also be necessary for the production of fibrous materials, such as cellulosic excipients, used in pharmaceutical manufacturing.
- *Grinding*: For example, colloid mill operates on the principle of grinding a coarse suspension between static and rotating stones, leading to the reduction of particle size of suspended particles.
- *Trituration*: For example, the manual process of using a pestle and a mortar to crush and/or mix fine powders together leads to some reduction of particle size.
- *Milling*: Several mills are utilized in the pharmaceutical industry. Depending on their principal of operation, they may be subclassified as follows:
  - *Ball mill*, which utilized steel balls to impact powders in a close container. The size of balls and duration and intensity of impact are the process parameters that determine the extent of particle size reduction.
  - *Air jet mill*, which utilizes a high-speed stream of air impacting the powder flowing through a closed loop. Air pressure and material flow rate are the key process parameters in this case.
  - *Fitzmill*, which impacts the powder with a high-speed rotating blade or hammer configuration of steel rods. Process parameters that determine the extent of particle size reduction, in this case are the material flow rate and the speed of the mill.
- *Comil*: This mill operates on the principle of forcing by scrubbing the granules through a screen of defined pore size and shape. It is commonly used for the sizing of granules.

19.4.2.2 Selection of size reduction technique

Selection of appropriate techniques for particle size reduction depends on the characteristics of powders as starting materials, desired particle size of the milled powder, and their use. Examples of material characteristics that influence the selection of particle size reduction method include the following:

- *Strength and plasticity*: Size reduction of high melting point (which indicates high strength of their crystalline lattice) crystalline solids can be carried out using high-impact processing equipment. However, low melting point solids, such as polyethylene glycols, may not be efficiently processed using high-impact equipment. The heat generated during processing can lead to plastic deformation or melting of these solids and compromise the unit operation. This is also true for materials that are inherently soft or pliable. In addition, the presence of moisture can frequently increase the plasticity of materials, leading to difficulty in processing.
- *Brittleness*: Powders that contain highly brittle particles can be easily processed using, for example, a fitzmill or air jet mill. However, strong particles that are not brittle may require relatively low efficiency and high-impact processing equipment such as a ball mill.
- *Chemical stability*: Particle size reduction is an inherently high-energy process that frequently also involves generation of heat. Therefore, powders that are chemically unstable may not be suitable for one or more of the size reduction techniques. For example, colloid milling may be preferred over ball mill for powders that show thermal degradation because the presence of the aqueous suspending medium in the colloid mill helps dissipate the heat generated during the process.

19.5 POWDERS AS DOSAGE FORMS

Although the use of powders as a dosage form has been replaced largely by the use of tablets and capsules in modern medicine, they represent one of the oldest dosage forms and present certain advantages that have led to their continued use as pharmaceutical dosage forms.

19.5.1 Types of powder dosage forms

Powders as dosage forms can be classified based on their usage and/or physical characteristics as detailed in Sections 19.5.1.1 through 19.5.1.6.

19.5.1.1 Oral powders in unit dose sachets

Powders containing drugs intended for children, such as antibiotics, are commonly made available in powder-filled unit-dose sachets. These powders are intended for administration after premixing with a food product, such a yogurt or juice. For example, the antibiotic Augmentin (amoxicillin in combination with clavulanic acid) and probiotics are available as a sachet. The powder blend is required to have a sweet taste, pleasant flavor, appealing color, and an acceptable mouthfeel. Uniform filling of the powder blend in sachets is the only major concern in the dispensing of this dosage form.

19.5.1.2 Powders for oral solution or suspension

Powders for reconstitution into an oral solution or suspension are commonly dispensed to the patient in multidose bottles. The pharmacist reconstitutes the powder using water, and the patient is instructed to consume a defined dose, by volume, of the resulting suspension. This mode of drug dispensing is intended to minimize the effects of physical instability of the suspension and/or the chemical instability of the drug compound on storage. This dosage form is exemplified by amoxicillin powder for oral suspension.

The powder blend is required to have a sweet taste, pleasant flavor, appealing color, and an acceptable mouthfeel after reconstitution. Stability of both the dry powder and the reconstituted suspension are important considerations. Also, in addition to the uniform filling of the powder blend in bottles, dose-to-dose uniformity of dispensed solution or suspension after reconstitution of a bottle of powder needs to be established.

19.5.1.3 Bulk powders for oral administration

Herbal medicines, such as laxatives, are commonly dispensed in bulk powder containers for dose dispensing and administration by the patient. The husk of the plant ispaghula as a laxative exemplifies these. These powders must be relatively nontoxic with a wide range of well-tolerated doses. These are generally over the counter products that are meant for self-medication by the patient.

19.5.1.4 Effervescent granules

Effervescent granules are sold as bulk powders intended for dispensing of a unit dose and reconstitution with water to form a solution by the patient immediately before administration. Upon contact with water, effervescence is produced by the reaction between an acidic component, such as succinic acid or tartaric acid, and a carbon dioxide-releasing basic component, such as sodium carbonate or bicarbonate. Effervescent granules must be kept in dry state to prevent this reaction before reconstitution by the patient.

19.5.1.5 Dusting powders

Dusting powders are intended for external, local application. The antibiotics in powder form for application to open skin wounds exemplify these.

Characteristics of powder blends for their use as dusting powders include low and flexible dose, low and relatively uniform particle size, high density and low aerosolization, and nongrittiness.

19.5.1.6 Dry powder inhalers

Dry powder inhalers (DPIs) are devices that deliver medication to the lungs using an inhalation device in the form of a dry powder. These devices are commonly used for drug delivery for local action, for example, for asthma, bronchitis, and emphysema.

Powder characteristics required for their use in DPIs include good flow, lack of adhesion to the material of package, low and uniform particle size for deposition in the appropriate region of the lung, and an adequate low drug dose.

19.5.2 Advantages of extemporaneous compounding of powders

Compounding of powders for dispensing in pharmacy presents the advantages of flexibility in dosing and a relatively good chemical stability. There are, however, disadvantages to extemporaneous compounding of powders as a dosage form. The preparation methods are time consuming and are generally not suitable for drugs that are highly potent, unpleasant tasting, or hygroscopic.

The compounded powders can either be dispensed in unit doses or as bulk powders in a multidose container. The dispensing of bulk powders has a further disadvantage of dosage inaccuracy resulting from several factors such as the BD of powder, consolidation during handling, and the method of measuring the dose by the patient. For these reasons, the dispensing of bulk powders is restricted to drugs with some dosage flexibility. These include, for example, herbal and other natural products such as laxatives and nutraceuticals and dusting powders intended for external, local application.

19.5.3 Extemporaneous compounding techniques

Extemporaneous compounding of powders as dosage forms in the pharmacy utilizes the same basic pharmaceutical processes, such as weighing, mixing, and sifting—with differences in the equipment used and scale of compounding. For example:

- Efficient mixing by geometric dilution of the component in the least quantity, such as the potent drug, is carried out by mixing it with an equal quantity of the larger component, such as a diluent, followed by repeated mixing with double the quantity of the larger component.

- A pestle-and-mortar is typically used for mixing powders in a circular motion with the application of shearing force (trituration). In addition to mixing, trituration helps reduce the bulkiness, and tends to reduce and normalize the particle size of powder components.
- Powdered solids can be incorporated into ointments and suspensions by forming a paste using water as an insoluble liquid medium. The paste is triturated using a pestle-and-mortar or a spatula on an ointment slab for uniform mixing and particle size reduction (levigation).

REVIEW QUESTIONS

19.1 Identify which of the following represent an intrinsic (inherent) characteristic of powder particles or a bulk property of powders: crystallinity, porosity, density, flow, content uniformity, compactibility, size, and shape.

19.2 Which of the following unit operations are likely to affect particle size and shape: mixing, compaction, granulation, and milling?

19.3 Density.
   A. Rank the three kinds of density of a powder in the expected increasing order of magnitude: true density, BD, and TD.
   B. Which of these densities is related to the porosity of the powder particles?
   C. Which density classification is expected to have the highest interparticulate spaces?
   D. Which density is most relevant to the equipment capacity determination during pharmaceutical manufacturing?
   E. Which densities are the most involved in determining the flow characteristics of the powder?

19.4 Identify which of the following represent potentially surface-mediated powder properties: sticking to the tablet tooling, true density, electrostatic charge, plastic deformation during compaction, flow, and/or crystallinity.

19.5 Particle size.
   A. Identify which of the following particle diameters represent the sphere of equivalent surface area: $d[1,0]$, $d[2,0]$, $d[3,0]$, $d[4,3]$, $d[3,2]$, $d(90)$, $d(50)$, $d(10)$.
   B. Identify which of the following particle diameters represent the sphere of equivalent volume: $d[1,0]$, $d[2,0]$, $d[3,0]$, $d[4,3]$, $d[3,2]$, $d(90)$, $d(50)$, $d(10)$.
   C. Identify which of the following particle diameters represent a percentile of particles: $d[1,0]$, $d[2,0]$, $d[3,0]$, $d[4,3]$, $d[3,2]$, $d(90)$, $d(50)$, $d(10)$.

D. Identify which of the following particle size determination techniques involve the use of a beam of laser light: sieve analysis, laser diffraction, microscopy, focused beam reflectance measurement, electrozone sensing, and sedimentation.

E. Identify which of the following particle size determination techniques can also provide information regarding the crystallinity of the particles: sieve analysis, laser diffraction, microscopy, focused beam reflectance measurement, electrozone sensing, and sedimentation.

19.6 Which of the following processes are NOT used for increasing the average size of powders?

A. Crystallization
B. Wet granulation
C. Dry granulation
D. Direct compression

19.7 Which of the following techniques are NOT used for the generation of an amorphous form of API?

A. Solid dispersion
B. Spray drying
C. Slow solvent evaporation
D. Extrusion spheronization

19.8 Polymorphism refers to

A. Two forms of a crystalline solid that differ in unit cell structure
B. Two forms of a crystalline solid that differ in the number of unit cells assembled in each dimension
C. Two forms of a solid such that one is crystalline and the other is amorphous
D. Two forms of a crystalline solid that differ in the solvent molecule entrapped in the crystal lattice

19.9 Which of the following methods can be used for characterizing particle shape?

A. Microscopy
B. Laser diffraction
C. Sieve analysis
D. Sedimentation
E. Electrozone sensing

19.10 Which of the following powder characteristics affects flow?

A. Aspect ratio
B. BD
C. Electrostatic charge
D. Surface cohesiveness
E. All of the above
F. None of the above

REFERENCES

1. Narang AS, Rao VM, and Raghavan K (2009) Excipient compatibility. In Qiu Y, Chen Y, Zhang GGZ et al. (Eds.) *Developing Solid Oral Dosage Forms: Pharmaceutical Theory and Practice,* Burlington, MA: Elsevier, pp. 125–46.
2. Iveson SM, Litster JD, and Hapgood K et al. (2001) Nucleation, growth and breakage phenomena in agitated wet granulation processes: A review. *Powder Technol* 117: 3–39.

FURTHER READING

He X (2003) Application of roller compaction in solid formulation development. *Am Pharm Rev* 6(3): 26–33.

Chapter 20

Tablets

LEARNING OBJECTIVES

On completion of this chapter, the students should be able to

1. Identify and describe different types of tablets.
2. Describe roles of formulation components of tablets.
3. Identify the three general processes of preparation of powder blends for compression.
4. Describe the tableting process.
5. Describe the requirements of powder blends for successful tableting.
6. Identify and describe quality attributes of tablets.
7. Describe the relationship between tablet disintegration, dissolution, and absorption.

20.1 INTRODUCTION

The development of a new chemical entity (NCE) requires the testing of its biological activity at various stages of development. For systemically acting drugs, animal studies are carried out at early stages of development using parenteral administration of a solubilized form of the drug. As drug development proceeds to later stages, human clinical studies are preferred with an orally administered dosage form that is both simple to formulate and provides adequate bioavailability. Preferred drug product (DP) dosage form choices are determined based on the drug substance's (DS's) physico-chemical properties, patient and disease state constraints and preferences, dose, manufacturability, and commercial factors such as other therapeutic options available to the patient.

An oral tablet dosage form is usually the most preferred dosage form because of patient convenience and acceptance. Most drugs are formulated in tablet dosage forms. Tablets are available in a wide variety of shapes, sizes, colors, and surface markings. This chapter would describe the types of tablets and discuss its formulation components, manufacturing processes, quality attributes, and some key considerations in the design and development of an oral tablet dosage form.

20.2 TYPES OF TABLETS

Depending on the physicochemical properties of the drug, site and extent of drug absorption in the gastrointestinal (GI) tract, stability to heat, light, or moisture, biocompatibility with other ingredients, solubility, and dose, the following types of tablets are commonly formulated (Table 20.1):

20.2.1 Swallowable tablets

The most common types of tablets are swallowed whole. These tablets disintegrate and release their contents in the GI tract.

20.2.2 Effervescent tablets

These tablets are formulated to allow dissolution or dispersion in water prior to administration and should not be swallowed whole. In addition to the DS, these tablets contain sodium carbonate or bicarbonate and an organic acid such as tartaric acid. In the presence of water, these additives react, liberating carbon dioxide, which acts as a disintegrator and produces effervescence. The drug is released into the aqueous medium as a solution, if it is highly soluble, or suspension. Ingestion of a dissolved or finely dispersed drug provides a rapid rate of drug absorption. Therefore, effervescent tablets can be suitable for acute conditions that require immediate relief, such as pain and gastric acidity. For example, cephalon's fentanyl effervescent tablet can be used to reduce the intensity of breakthrough pain in cancer patients.

20.2.3 Chewable tablets

Chewable tablets are used when a faster rate of dissolution and/or buccal absorption is desired. Chewable tablets consist of the drug dispersed throughout a saccharide base that provides mild sweetness. Flavors, sweeteners, and colors are also added to chewable tablets to improve palatability and organoleptic appeal. The drug is released from the dosage form by physical disruption associated with chewing and dissolution in the fluids of the oral cavity, and the presence of a effervescent material. For example, some antacid tablets can be chewed to obtain quick indigestion relief. Chewable tablets are typically prepared by compression and usually contain mannitol or sorbitol as saccharide, mildly sweet, fillers. Mannitol is sometimes preferred as a chewable base diluent, because it has a pleasant cooling

Table 20.1 Types of tablets

Basis of classification	Type of tablets	Description and special advantages	Drug substance requirements and other considerations	Physicochemical principle of formulation	Example
Route of administration	Oral	• Most of the tablets fall in this category. These tablets are designed for peroral administration by swallowing.	• Oral absorption and stability in the dosage form.	• Drug substance is mixed with excipients to aid manufacturability and drug release on administration.	• Tylenol tablets
	Buccal	• Buccal tablets are designed for placement under the cheek mucosa or between the lip and the gum. • They are typically designed for slow drug release and absorption through the oral cavity and/or the upper gastrointestinal tract. • Buccal administration is used for local drug action or avoiding extensive degradation in the gut or metabolism in the liver. • Allow drug administration without water.	• Drug should not be bitter or have other sharp taste. • Drug should be soluble.	• Buccal tablets typically have a mucoadhesive component. • These tablets do not contain a disintegrant and are fairly soft. • These tablets are usually small and flat. • The drug is released by dissolution from the surface.	• Testosterone • Fentanyl (analgesic) • Nitroglycerin • Miconazole (antifungal)

(Continued)

Table 20.1 (Continued) Types of tablets

Basis of classification	Type of tablets	Description and special advantages	Drug substance requirements and other considerations	Physicochemical principle of formulation	Example
	Sublingual	• Sublingual tablets are designed for placement under the tongue. • They allow rapid drug release and absorption through the blood vessels under the tongue, avoiding gut and hepatic exposure. • Sublingual administration is used for rapid, systemic drug action and/or avoiding extensive degradation in the gut or metabolism in the liver. • Allow drug administration without water.	• Drug should not be bitter or have other sharp taste. • Drug should be soluble. • Typically, low dose drugs are formulated as sublingual tablets.	• Typically have soluble components in a small-sized tablet formulation. • These tablets do not contain a disintegrant and are relatively soft. • The drug is released by dissolution from the surface.	• Vitamin B12 • Isoprenaline sulfate (bronchodilator) • Nitroglycerin (vasodilator)
	Orally disintegrating	• Orally disintegrating tablets (ODTs) are designed to be dissolved on the tongue rather than swallowed whole. • Useful for patients who suffer from dysphagia (difficulty in swallowing). • Allow drug administration without water.	• Drug should not be bitter or have other sharp taste. • Drug should be soluble. • Typically, low dose drugs are formulated as sublingual tablets.	• Designed to disintegrate and dissolve in the mouth within 60 s or less.	• Clonazepam

(Continued)

Table 20.1 (Continued) Types of tablets

Basis of classification	Type of tablets	Description and special advantages	Drug substance requirements and other considerations	Physicochemical principle of formulation	Example
Release characteristics of the drug substance	Immediate release (IR)	• Most of the conventional tablets fall in this category. • IR tablets are designed to start releasing the drug as soon as they come in contact with the tablet disintegrating/dissolving fluids.	• Most drugs are formulated as IR tablets. • The drug should not show significant instability in the gastric environment.	• Tablets are typically designed for manufacturability and rapid drug release on administration.	• Tylenol tablets
	Controlled release (CR)	• CR tablets are designed to release the drug at a predetermined and controlled rate.	• Usually used for drugs for chronic ailments that requires repeated administration. • Used for drugs that would benefit the patient from consistent maintenance of drug's plasma levels and/or reduced dosing frequency.	• The release rate is controlled by the use of slow-release matrix (e.g., carnauba wax) or insoluble coating. • Insoluble coating-based controlled release typically provides osmotic or diffusion-limited drug release. • Insoluble matrix-based controlled release system typically provides dissolution or diffusion-limited drug release.	• Oxycodone HCl CR tablet (analgesic)

(Continued)

Table 20.1 (Continued) Types of tablets

Basis of classification	Type of tablets	Description and special advantages	Drug substance requirements and other considerations	Physicochemical principle of formulation	Example
	Extended release (XR) or sustained release (SR)	• XR tablets are designed to release drug over an extended period of time, but not necessarily at a predetermined and/or controlled rate.	• Same as CR tablets.	• Same as CR tablets. • An XR/SR tablet can also be a CR tablet, but is not necessarily so.	• Methylin ER (methylphenidate for narcolepsy and attention deficit disorder) • Ritalin CR
	Delayed release (DR), for example, enteric coated	• DR tablets are designed to delay the release of the drug from the time it first comes in contact with the tablet disintegrating/dissolving fluid. • Useful for drugs that degrade in the acidic gastric environment (e.g., peptides) or are irritating to the gastric mucosa (e.g., aspirin)	• Enteric-coated tablets are coated with a coating material that does not dissolve under acidic conditions. • Time-controlled colon release tablets are coated with a coating material that has a slow rate of dissolution.	• Drug release is delayed by a physiologically controlled mechanism such as gastric acidity or a defined period of time. • Commonly used polymers for enteric coating are cellulose acetate phthalate (CAP), hydroxypropyl methylcellulose phthalate (HPMCP), polyvinyl acetate phthalate (PVAP), cellulose acetate trimellitate (CAT), and Eudragits®, which are copolymers of methacrylic acid and methylmethacrylate.	• Enteric coated tablets • Time-controlled colon-targeted drug release • Aspirin

(Continued)

Table 20.1 (Continued) Types of tablets

Basis of classification	Type of tablets	Description and special advantages	Drug substance requirements and other considerations	Physicochemical principle of formulation	Example
Mode of administration	Effervescent	• Effervescent tablets are designed for dispersion and/or dissolution in water prior to administration. • Effervescent tablets provide rapid drug dissolution into solution and immediate availability for absorption upon administration. • These tablets are usually large in size and should not be swallowed whole.	• Drug should be soluble or easily dispersible in water. • Drug should be absorbed through the stomach. • Drug should be compatible with the acidic and basic components of the dosage form.	• In addition to the drug substance and necessary functional excipients, these tablets contain an acidic and a carbon dioxide-generating basic component. The acidic component can be tartaric acid or succinic acid and the basic component is usually sodium carbonate or bicarbonate. In the presence of water, these additives react, liberating carbon dioxide, which rapidly disintegrates the tablets, and produces effervescence. • Tablets typically also contain sweetening, flavoring, and/or coloring agents to improve palatability.	• Fentanyl-effervescent buccal tablet (to reduce the intensity of breakthrough pain in cancer patients) • Zantac effervescent tablet (for relief of gastric acidity)

(Continued)

Table 20.1 (Continued) Types of tablets

Basis of classification	Type of tablets	Description and special advantages	Drug substance requirements and other considerations	Physicochemical principle of formulation	Example
	Chewable	• Chewable tablets provide faster rate of drug dissolution, buccal absorption, and/or smooth mouthfeel. • They are used to improve palatability, especially for the pediatric population.	• The drug should not have a bitter or other sharp taste.	• Tablets contain a significant quantity of a soluble, mildly sweet, smooth tasting base such as sorbitol or mannitol. • Mannitol is sometimes preferred as a chewable base diluent, because it provides a cooling sensation due to its negative heat of solution (−28.9 cal/g). • Tablets typically also contain sweetening, flavoring, and/or coloring agents to improve palatability. • The dosage form is physically disrupted by chewing and the drug is released by dissolution in the saliva.	• Amoxicillin chewable tablets

(Continued)

Table 20.1 (Continued) Types of tablets

Basis of classification	Type of tablets	Description and special advantages	Drug substance requirements and other considerations	Physicochemical principle of formulation	Example
	Dispersible	• Dispersible tablets are designed for rapid dispersion in a spoonful of water immediately before administration.	• The drug should not have a bitter or other sharp taste.	• Tablets contain rapid disintegrating agents. • Tablets typically also contain sweetening, flavoring, and/or coloring agents to improve palatability.	• Amoxicillin
	Lozenges	• Lozenges are slow releasing tablets designed for drug release in the saliva by surface dissolution from a candy-sucking action.	• Drug should not be bitter or have other sharp taste. • Drug should be soluble.	• Lozenges do not contain a disintegrant. • Lozenges contain soluble ingredients and are designed for drug release by slow dissolution from surface.	• Cough drops • Vitamin supplements

sensation in the mouth due to negative heat of dissolution and can mask the taste of some objectionable medicaments.

20.2.4 Buccal and sublingual tablets

Buccal and sublingual tablets dissolve in the cheek pouch (buccal) or under the tongue (sublingual). Buccal or sublingual route of drug absorption bypasses hepatic metabolism, often referred to as the first-pass effect on oral administration, and is preferred for low dose drugs that have extensive hepatic metabolism. Sublingual administration also allows rapid drug absorption, which may be critical in cases such as nitroglycerin for chronic heart failure. Other examples include isoprenaline sulfate (bronchodilator), glyceryl trinitrate (vasodilator), and testosterone tablets. These tablets are usually small and flat, do not contain a disintegrant, and are intended for dissolution in the local fluids.

20.2.5 Lozenges

Lozenges are slow dissolving compressed tablets that do not contain a disintegrant. Some lozenges contain antiseptics (e.g., benzalkonium) or antibiotics for local effects in the mouth. Lozenges are also used for systemic effect, such as those containing vitamin supplements. Lozenges are palatable and organoleptically appealing by the addition of flavors, sweeteners, and colors.

20.2.6 Coated tablets

Most tablets are coated for one or more of the following reasons:

- To prevent decomposition of drugs sensitive to air (oxygen), light, or humidity
- To minimize the unpleasant taste of certain drugs that may come during partial dissolution of the drug in buccal fluids during absorption
- To improve swallowability and palatability by increasing surface smoothness in the mouth
- To provide visual appeal and consistency, smooth surface texture, and uniform distribution of color
- To serve as anticounterfeiting medium by incorporating tracer compounds in the coating material
- To allow containment of highly potent compounds in the core of the tablet and, thus, avoid exposure to personnel handling of the tablets

Coating is not used on buccal, sublingual, chewable, effervescent, or dispersible tablets to avoid any delay in drug release due to the time required for the rupture or dissolution of the coating material (Table 20.2).

Coating of core (compressed, uncoated) tablets is carried out by loading the tablets in a moving, perforated pan supplied with dry hot air and spraying the coating dispersion onto the tablet bed at a rate matched with the rate of evaporation of the solvent. This leads to the deposition of a film

Table 20.2 Characteristics of types of tablet coatings

Characteristics	Sugarcoating	Film coating	
		Using organic solvent(s)	Water-based
Coating thickness	20%–50% w/w of total tablet weight	2%–5% w/w of total tablet weight	2%–5% w/w of total tablet weight
Coating material composition	Sugar, plasticizer (e.g., polyethylene glycol), opacifier (e.g., titanium dioxide), and colorant (e.g., iron oxide red and/or yellow)	Polymer (e.g., ethyl cellulose), plasticizer (e.g., polyethylene glycol), opacifier (e.g., titanium dioxide), glidant (e.g., talc), and colorant (e.g., iron oxide red and/or yellow)	Polymer (e.g., hydroxypropyl methyl cellulose or polyvinyl alcohol), plasticizer (e.g., polyethylene glycol), opacifier (e.g., titanium dioxide), glidant (e.g., talc), and colorant (e.g., iron oxide red and/or yellow)
Usage	Historically predominant, it is now relatively uncommon for prescription pharmaceuticals. More commonly used for consumer products, such as candies.	Less commonly used due to environmental and safety concerns associated with the use of organic solvents during the coating process.	Most commonly used.

of the coating material on the surface of the coated tablets. This process is carried out for a sufficient duration of time to allow uniform and elegant coverage of the entire surface of the tablet by the coating material. The coating material consists of an opacifier (such as fine particle size titanium dioxide), color, plasticizer (such as polyethylene glycol), and a polymer (such as hydroxypropyl methylcellulose [HPMC] or polyvinyl alcohol [PVA]). Typically, about 3% w/w application of the coating material provides complete coverage by the formation of a thin film around the tablet. The coated tablets are called *film-coated tablets*.

Other types of coated tablets are sugarcoated, gelatin-coated (gel caps), and enteric-coated tablets. Sugarcoated tablets are produced by the application of sucrose solution, containing preservatives, colorants, sweeteners, and flavors, to the core with a relatively high (~30% w/w) weight buildup. Film coating has almost completely taken over sugar coating in the pharmaceutical industry because of shorter processing time and the smaller size of the coated tablet. A common example of sugarcoated tablet is the M&Ms, which are consist of a solidified liquid-chocolate center and a hard-candy shell which is a combination of sugar and corn syrup. Characteristics of sugarcoated and film-coated tablets are summarized in Table 20.3.

Table 20.3 Examples of sustained release tablets

Brand names	Manufacturer	Active ingredients	Indications
Theo-Dur™	ALZA Corp.	Theophylline	Asthma
Abacavir™ (Ziagen)	Glaxo Wellcome Inc	Nucleoside reverse transcriptase inhibitor	HIV-1 infection
Sinemet®	Bristol Myers Squibb	Carbidopa+Levodopa	Parkinson's disease
Volmax®	ALZA Corp.	Salbutamol	Bronchospasm, asthma
Voltaren®	Novartis	Diclofenac sodium	Osteoarthritis and rheumatoid arthritis
Efidac®-24	ALZA Corp.	Chlorpheniramine	Allergy, nasal congestion
DynaCirc®	ALZA Corp.	Isradipine	Hypertension

Gelatin-coated tablets, commonly known as gelcaps, are capsule-shaped compressed tablets coated with a gelatin layer. These tablets are produced by dipping the core tablets in a solution of gelatin with colors and preservatives, followed by drying. This allows the product to be smaller than an equivalent capsule filled with an equivalent amount of powder, and provides an elegant visual appeal. Many over-the-counter (OTC) medications are marketed as gelcaps.

20.2.7 Enteric-coated tablets

GI fluid pH increases progressively from acidic to basic from the stomach through the intestines to the colon. The changes in GI pH can be utilized to release a drug at a particular physiological location in the GI tract. In particular, oral solid dosage forms can be coated with a polymer that is insoluble at the acidic stomach pH and soluble at basic intestinal pH. Such polymers are known as enteric polymers and such coatings are termed enteric coatings. Enteric-coated tablets are the tablets coated with enteric polymers. Complete coating of the tablet with these polymers allows the polymer to form a barrier between the core of the tablet and the surrounding aqueous medium. Thus, the enteric-coated tablet remains insoluble in the low pH environment of the stomach, but dissolves readily on passage into the small intestine with its elevated pH.

Enteric coating is used to minimize irritation of the gastric mucosa by certain drugs and/or protect sensitive drugs against decomposition in the acidic environment of the stomach. For example, aspirin produces less gastric bleeding when formulated as enteric-coated sustained-release (SR) tablets than conventional immediate release dosage forms.

Commonly used polymers for enteric coating are acid-impermeable polymers, such as cellulose acetate phthalate (CAP), HPMC phthalate (HPMCP), polyvinyl acetate phthalate (PVAP), and Eudragits®, which are copolymers of methacrylic acid and methylmethacrylate. Eudragits are available in different grades. The drug release behavior of a Eudragit-coated tablet is controlled through the ratio of methacrylic acid copolymers. For example, the ratio of carboxyl to ester groups is approximately 1:1 in fast dissolving Eudragit L100 and 1:2 in slow dissolving Eudragit S100.

20.2.8 Immediate release tablets

Most tablets (discussed earlier) are immediate release (IR) tablets, that is, they make all the drugs available to the dissolution medium immediately on coming in contact with the aqueous medium. The drug dissolves at a rate determined by the composition of the dissolution medium (such as pH) and physicochemical properties of the drug (such as solubility and particle size).

20.2.9 Controlled release tablets

In contrast to the IR tablets, certain dosage forms, such as controlled-release (CR) or extended-release (XR) tablets, are designed to control or extend, respectively, the rate at which drug dissolves in the aqueous medium. Thus, CR tablets reduce the rate of drug release to a slow, controlled rate, which is typically zero order. XR tablets, on the other hand, extend the duration of drug release by slowing down the rate of drug release but may not have control on the rate (i.e., may not provide zero-order kinetics of drug release). CR or XR tablets are sometimes also called SR tablets.

CR tablets can reduce dosing frequency, increase patient compliance, and may reduce side effects of certain drugs. The rate-controlling feature of the CR tablets could be either the matrix or the film coating. Coformulating or mixing drugs with water-insoluble polymers prepare matrix tablets. A slow dissolving polymer matrix, such as high molecular weight HPMC, can be used to prepare SR tablets of highly water-soluble drugs. On coming in contact with the aqueous dissolution medium, the core tablet dissolves by surface erosion, and the rate of surface erosion is controlled by the rate of dissolution of the polymer matrix. For example, metformin hydrochloride XR tablets (Glucophage XR tablets®) are matrix tablets.

Membrane-coated SR tablets utilize an insoluble membrane to reduce or control the rate of drug release. The insoluble membrane can be impregnated with a soluble polymer to provide pores from which the drug can diffuse out or have a hole for osmotically CR tablets. Often, a combination of IR and SR components is included in a dosage form to provide a loading dose (by the IR component) followed by a slow releasing maintenance dose (by the SR component). The IR and the SR component can form two different layers of a bilayer tablet, for example. These components can also be mixed together and packaged in a capsule. The SR and IR components can also be compressed together. For example, Theo-Dur® CR tablet of theophylline consists of two components: (1) a matrix of compressed theophylline crystals and (2) coated theophylline granules embedded in the matrix. In contact with fluids in GI tract, theophylline diffuses slowly through the wall of the free granules, which dissolves with time. After oral administration of Theo-Dur® 300 mg tablets to human subjects, serum theophylline concentrations over 1 mg/ml were maintained over 24 hours.

Examples of commonly used SR drug delivery products are listed in Table 20.3.

20.3 TABLET FORMULATION

In addition to the active drug, called DS or active pharmaceutical ingredient (API), tablets may contain one or more of functional ingredients such as diluents (also known as fillers), binders, disintegrants, glidants, lubricants, coating materials, coloring agents, stabilizer(s), sweeteners, and flavoring agents. These ingredients are called *excipients*. Excipients are added to improve one or more of the three key functional properties of a dosage form: (1) bioavailability, (2) manufacturability, and (3) stability.

- Excipients facilitate optimum bioavailability, that is, rate and extent of drug absorption from the dosage form, by providing reproducible and optimum rate, extent, and the site of drug release.
- Excipients facilitate manufacturability by converting the API powder into a powder blend that flows, compresses, and can be manufactured on high-speed equipment.
- Excipients facilitate stability of the API for the duration of time a product may be stored between the manufacture and the consumption (called shelf life).

Functionality and examples of these excipient types are listed in Table 20.4. Two examples of tablet formulations are listed in Table 20.5.

20.3.1 Diluents

A tablet should weigh at least about 50 mg for ease of handling by the patient. Therefore very low-dose drugs invariably require a diluent (also known as filler) or bulking agent to bring overall tablet weight to at least 50 mg. In addition, enabling manufacturability of powder blends on high-speed equipment requires adequate properties such as flow and compressibility, which are improved by the addition of diluents.

Commonly used diluents are microcrystalline cellulose (MCC), lactose, mannitol, sorbitol, and dicalcium phosphate. Each excipient presents characteristics that define its preferred application. For example, lactose generally is not preferred for use with drugs that have primary amine groups due to the propensity for Maillard reaction (Figure 20.1).

Table 20.4 Functional excipients used in tablets

Functional role	Examples	Description and functionality
Filler	• Microcrystalline cellulose (MCC) • Lactose monohydrate or anhydrous • Mannitol • Sorbitol	• Add bulk to the dosage form • May contribute to dissolution and disintegration characteristics
Binder	• Polyvinylpyrrolidone (PVP) • Hydroxypropyl cellulose (HPC) • Starch	• Bind the powder ingredients to form granules for processing
Disintegrant	• Croscarmellose sodium (CCS) • Crospovidone (xPVP) • Sodium starch glycolate (SSG) • Starch	• Disintegration of the tablet to granules and powders on coming in contact with water
Glidant	• Colloidal silicon dioxide	• Aid the flow of granules/blend
Lubricant	• Magnesium stearate • Stearic acid • Sodium stearyl fumarate	• Aid the flow of granules/blend and ejection of tablets in the tablet press
Coating material	• Polymers such as hydroxypropyl methyl cellulose (HPMC), ethyl cellulose (EC), polyvinyl alcohol (PVA) • plasticizer (e.g., polyethylene glycol) • opacifier (e.g., titanium dioxide) • glidant (e.g., talc) • colorant (e.g., iron oxide red and/or yellow)	• Provide a physical barrier coating on the surface of the compressed core tablets
Coloring agent	• Iron oxide red and/or yellow • FD&C Blue #6	• Visual appeal of color
Stabilizer	• Antioxidants such as ascorbic acid, butylated hydroxyanisole (BHA), butylated hydroxytoluene (BHT), α-tocopherol	• Stabilization of the drug in the dosage form from stresses such as oxidation
Sweetener	• Aspartame, saccharin sodium, sucralose, acesulfame potassium	• Sweetening to overcome drug taste and/or improve palatability for some types of tablets
Flavoring agent	• Proprietary flavors (orange, pineapple, etc.)	• Flavoring to overcome drug taste and/or improve palatability for some types of tablets

Mannitol has a negative heat of solution and, thus, provides a cooling sensation in the mouth. It is typically used for chewable and orally dissolving tablets. MCC can absorb water and improves compressibility by undergoing plastic deformation on compression. Lactose is a fragile excipient that fragments to undergo brittle fracture during compression. Addition of lactose to powder blends can improve interparticle bonding during compression.

Table 20.5 Examples of immediate release tablet compositions

A. Acetaminophen tablets

Ingredient	Quantity per tablet	Use
Acetaminophen	325 mg	Drug
Sucrose	60 mg	Filler
PVP 10% in alcohol	q.s.	Binder
Stearic acid	6 mg	Lubricant
Talc	15 mg	Lubricant, glidant
Corn starch	30 mg	Disintegrant
Alginic acid	20 mg	Disintegrant

B. Acetaminophen Tablets USP (direct compression)

Acetaminophen USP (granular or large crystal)	70.00	325.00	Drug
Avicel PH 101	29.65	138.35	Filler
Stearic acid (fine powder)	0.36	1.65	Lubricant
	100.00	465.00	

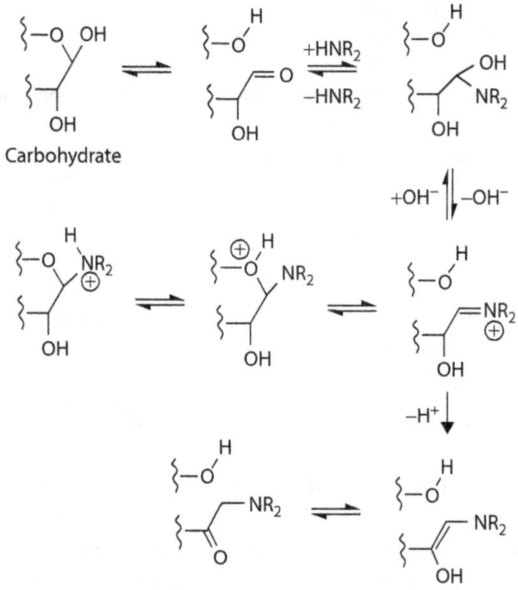

Figure 20.1 Example of Maillard reaction, followed by Amadori rearrangement, for a secondary amine compound. (From Wirth, D.D. et al., *J. Pharm. Sci.*, 87(1), 31–39, 1998. With Permission.)

20.3.2 Adsorbents

Adsorbents are substances capable of holding fluids in an apparently dry state. These are the powder particles that can adsorb the liquid while maintaining the ability to be handled as powders. Oil-soluble drugs or fluid extracts can be mixed with adsorbents to bring them to a solid form for compression into tablets. For example, fumed silica, microcrystalline cellulose, magnesium carbonate, kaolin, and bentonite.

20.3.3 Granulating fluid

Granulating fluids are liquids that are used for wet granulation. Typically, these are water, ethanol, or isopropanol—or the solution of a hydrophilic polymer (binder) in one of these liquids. Addition of granulating fluid while mixing a powder blend of the API with excipients promotes surface adhesion of particles. After granulation, when the granulating fluid is dried off, the particles tend to continue to adhere in relatively strong agglomerates through multiple, weak noncovalent forces of adhesion among primary particles promoted by the hydrophilic polymer (binder). These particles tend to have higher sphericity than primary particles and can be milled or sifted to the desired particle size distribution. Thus, granulation enables flow and compressibility while modifying the surface properties of primary particles such as sticking to stainless steel equipment. The granulating fluid is completely removed during drying to ensure that the residual fluid does not act as a plasticizer and contribute to chemical instability in the finished DP.

An alternate process, known as moisture activated dry granulation (MADG), involves the spray addition of just enough amounts (e.g., 1%–3% w/w) of granulating fluid, typically water, to dry powder blend to promote surface adhesion during compression. These processes are typically direct compression processes that do not require separate granulation and drying unit operations.

20.3.4 Binders

Hydrophilic polymers are added in either dry or liquid (solution in water) form to promote the transformation of primary powder particles into cohesive agglomerates (granules) during wet granulation or to promote cohesive compacts during direct compression. These hydrophilic polymers are called binders, when used for such purposes in these formulations. Common binders used in wet granulation are cellulose derivatives such as hydroxypropyl cellulose (HPC), polyvinylpyrrolidone (PVP), and starch.

In wet granulation, the binder can either be dissolved in the granulating fluid or added dry to the powder mixture followed by the addition of the granulating fluid for wet granulation. The type and concentration of binder affect the granule strength, friability, and the granule growth rate during the wet-granulation process. Size and size distribution of granules can affect the flow and compressibility. Densification of granules can affect the rate of drug release. For example, higher molecular weight HPC or PVP binders produce larger and denser granules.

20.3.5 Glidants

Glidants are added to tablet formulations to improve the flow properties of the granulations during transfer operations, such as from the hopper to the roller compactor or tablet press. They improve flow by reducing interparticulate friction. The commonly used glidants are fumed (colloidal) silica, starch, and talc. These glidants are very small size powder particles that occupy surface ridges and irregularities in coarse powder particles, thus increasing the sphericity and reducing the tendency to adhere to surfaces.

20.3.6 Lubricants

Lubricants help prevent adherence of the tablet material to the stainless steel processing equipment surfaces under compression forces. Thus, lubricants reduce or prevent adhesion of powder particles to tablet compression punch faces and dies or to the rolls of a roller compactor. They promote flow, reduce interparticle friction, and facilitate the smooth ejection of compressed tablets from the die cavity. Commonly used lubricants are magnesium stearate, stearic acid, and sodium stearyl fumarate. These lubricants are small hydrophobic particles that tend to coat the surface of larger powder particles by spreading out under the mild shear during mixing. Hydrophobic surface coating reduces noncovalent hydrophilic interparticle and particle-equipment forces that are generally responsible for adhesion and sticking.

Among these, magnesium stearate is the most commonly used lubricant. The lubricity of magnesium stearate in a formulation can be increased with either the concentration or the duration of mixing. Most lubricants are used in concentration $\leq 1\%$ w/w. Overlubrication, due to the use of high concentration or excessive mixing can result in reduced compactibility of the blend and/or rate of drug release from the tablets. Sodium stearyl fumarate is the only water soluble or hydrophilic lubricant and is used in formulations that are highly sensitive to hydrophobic lubricants.

20.3.7 Disintegrants

Disintegrants are added to the tablets to facilitate the breakup or disintegration when tablets come in contact with aqueous fluids in the GI tract or dissolution medium during *in vitro* testing. Disintegrants help break a compressed tablet into constituent granules and primary powder particles. The dispersed primary API particles then provide a high surface area for the dissolution of the API in the aqueous fluid. Thus, breaking of the tablets by disintegrants increases the effective surface area and promotes rapid release and dissolution of the drug.

Disintegrants act by either or both (a) swelling in the presence of water and bursting tablet and granule open and/or (b) capillary action to promote rapid ingress of water into the center of the tablet or capsule. Examples of commonly used disintegrants are starch, cross-linked PVP (xPVP), cross-linked sodium carboxymethylcellulose—also known as croscarmellose sodium (CCS), and sodium starch glycollate (SSG). Among them, xPVP, CCS, and SSG are called superdisintegrants because these particles swell significantly in contact with water. Starch swells moderately in contact with water. Mild disintegrants, such as microcrystalline cellulose (MCC), can also act by capillary action through their pores to promote the rapid ingress of water into the center of the tablet.

Both disintegrants and binders are hydrophilic polymers. For example, PVP is a binder and xPVP is a disintegrant. Long open chains of hydrophilic polymers can serve as a binder by promoting hydrogen bonding and hydrophilic bonding forces on the surface of powder particles. When these polymer chains are cross-linked, they can imbibe water and swell due to their high hydrogen bonding capacity—thus acting as a disintegrant. Thus, PVP and xPVP have the same polymer backbone—while the former is a binder, the latter is a superdisintegrant.

20.3.8 Miscellaneous

Organoleptic ingredients, such as colors, sweeteners, and flavors, are used for taste masking and improving palatability especially in products such as chewable and dispersible tablets. Artificial sweeteners, such as acesulfame potassium and saccharin sodium, are preferred because smaller quantities produce similar or higher sweetness than sucrose.

Chemical stabilizers commonly used in tablet formulations include antioxidants, such as ascorbic acid and vitamin E, and heavy metal chelators, such as ethylenediaminetetraacetic acid (EDTA) and ethylene glycol-bis (β-aminoethyl ether)-N,N,N',N'-tetraacetic acid (EGTA).

20.4 MANUFACTURING OF TABLETS

20.4.1 Requirements for tableting

As shown in Figure 20.2, tableting involves compression of a powder blend in a die cavity between the upper and the lower punches. Several punches and dies are arranged on three rotary turrets on a high-speed rotary tablet press that move concurrently in a circular motion as the tablets are made. As the turret moves, the powder is fed into the dies at one port through a

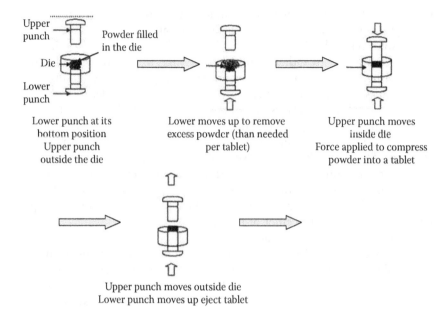

Figure 20.2 Tabletting process.

hopper and feed frame, dose adjusted, compressed, ejected, and the tablets are collected at another port. This process requires

- Uniform flow of blend into the die cavity through a hopper and feedframe.
- Nonsegregation of powder blends in the hopper and during loading in the die cavity.
- Compactibility (ability to reduce in volume and compress on application of force by the punches) of the powder in the die cavity during compression.
- Nonsticking of the powder blend to walls of dies and surfaces of punches.
- Adequate cohesion of the powder blend to form a strong tablet.

20.4.2 Powder flow and compressibility

Powder flow is required for transporting the materials through the hopper of a tableting machine and roller compactor (for dry granulation or roller compaction-based processes). Inadequate powder flow can lead to variable die filling, which produces tablets that vary in weight, drug content, and strength (hardness). Therefore, steps must be taken to ensure that the proper powder flow is maintained. Incorporation of a glidant and/or a lubricant into the formulation enhances the powder flow. Increasing the sphericity of particles also improves the flow. Processes such as spray drying, fluid-bed granulation or extrusion spheronization increase the sphericity of granules. In addition, increasing the density of granules, such as by granulation, also improves the powder flow. The most popular method of increasing the flow properties of powder is by granulation. As discussed earlier, granulation could be either dry granulation, which does not use a granulating fluid, or wet granulation, which involves wetting with a fluid followed by drying.

Compressibility is the property of forming a stable, intact compact mass when pressure is applied. Some materials compress better than others do. Compressibility is an outcome of the extent of plastic deformation that a material can undergo combined with cohesive forces among the powder blend that will keep the material in the compressed state. Most materials exhibit different degrees of elastic recovery, that is, expansion toward original higher volume on removal of stress. Low elastic recoveries coupled with high plastic deformability and high interparticle adhesion promote the formation of strong compacts at low compression forces.

Granulation generally improves compressibility. Materials that do not compress well produce soft tablets.

20.4.3 Types of manufacturing processes

Based on the characteristics of the starting materials that influence the properties of the powder blend, three general processes are used for preparing granulation blends for compression:

- *Direct compression*
- *Dry granulation or roller compaction*
- *Wet granulation*

The purpose of both wet and dry granulation is to improve the flow of the mixture and to enhance its compression properties by increasing particle size, density, and sphericity. The selection among these processes is based on the physicomechanical properties of the API and the raw material blend (API with excipients). For example, stability of the API to other ingredients used for preparing granulation blends and processing conditions (e.g., use of water during wet granulation) is critical. For example, dry granulation may be preferred for moisture and/or heat sensitive APIs.

1. *Direct compression*: Direct compression is the preferred method if powder blend has adequate flow, compactibility, and cohesion with low segregation potential. This is the simplest process that involves the least extent of material handling. Direct compression involves simply mixing the required ingredients and compressing them into tablets on the press.

2. *Dry granulation*: Dry granulation is preferred in circumstances where powder flow, cohesion, and/or segregation potential need to be improved, but compactibility is adequate. This process involves compacting a powder blend. It can be carried out by either of two processes: (a) slugging, which involves compression using large punches and dies in a tablet press; or (b) roller compaction, which involves forcing the powder blend between two counterrotating rolls that are pressed together under hydraulic pressure. This squeezes the powder blend into a solid cake between rollers. As shown in Figure 20.3, the compacted material is milled to form granules, which are generally larger in particle size than starting powder blend. These granules are then mixed with extragranular excipients and compressed on the tablet press.

3. *Wet granulation*: Wet granulation is preferred when compactibility of the powder is not very high and there is a need to improve the flow, cohesion, and/or segregation potential of the powder blend. The powder blend is loaded in a granulator (vessel with a rotating blade to mix the powder, Figure 20.4) and granulated with a solution of the binder or water (if a dry binder is added to the powder mixture). Water is the most widely used blender vehicle. The use of nonaqueous granulation liquids, such as ethanol, is no longer preferred for safety and environmental reasons. The formed granules are dried in a tray or fluid bed dryer at moderately elevated temperatures. Dried granules are then mixed with extragranular excipients and compressed on the tablet press.

(a)   (b)

Figure 20.3 (a) A high shear granulator. (From Vector Corporation, http://www. vectorcorporation.com.) and (b) a low shear granulator. (From Hobart Corporation, http://www.hobartcorp.com. With Permission.)

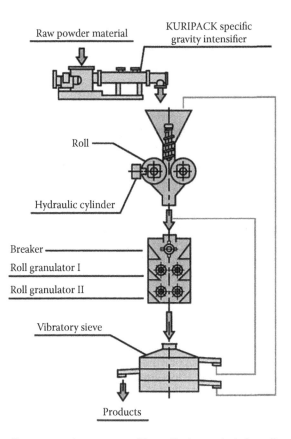

Raw powder material

KURIPACK specific
gravity intensifier

Roll

Hydraulic cylinder

Breaker

Roll granulator I

Roll granulator II

Vibratory sieve

Products

Figure 20.4 A roller compaction process. (From Kurimoto Ltd., http://www.kurimoto. co.jp. With Permission.)

a. *Low or high-shear wet granulation*: Depending on the design of the granulator, wet granulation could impart low or high levels of shear to the powder blend and are termed accordingly. For example, a flat-bottom bowl granulator with horizontal blades that move in a circular motion at the bottom of the powder bed (Figure 20.4a) leads to high shear, whereas the use of vertical blades in an oval bowl (Figure 20.4b) lead to low shear. The extent of shear can affect porosity, compactibility, and density of granules. Low-shear granulation generally yields higher porosity, higher compactibility, and lower density of the formed granules. A choice between low- and high-shear granulation is based on the sensitivity of the desired product quality attributes to process conditions.

b. *Fluid-bed granulation*: Fluid-bed granulation involves the spray of the granulating liquid on the fluidized powder bed. This process combines the drying step with the granulation step. In this process, the evaporation of the granulating liquid is concurrent with the granulation of the powder blend. It is a relatively slow, but a well-controlled process that leads to the generation of granules, which are more porous, less dense, and more uniform in shape and size.

4. *Moisture-activated dry granulation*: Other processes commonly employed for preparing powder blend for compression involve a combination of the three basic processes. For example, MADG involves spraying of a minimum amount of water on the powder blend before compression to improve powder adhesion.

5. Continuous granulation: Use a continuous granulation process minimizes the material transfers and enables flexibility of batch size. Continuous processes are based on a tunnel or channel of powder flow with sequential positions where different steps of a process—such as water addition, drying, and milling in the case of wet granulation—are carried out in tandem.

20.4.4 Packaging and handling considerations

Following compression and coating, tablets are stored in tight containers and protected from high temperature and humidity places. Products that are prone to decomposition by moisture generally are copackaged with desiccants, such as silicon dioxide. Drugs that are adversely affected by light are packaged in light-resistant containers.

20.5 EVALUATION OF TABLETS

The compendia, such as the United States Pharmacopeia (USP), and the regulatory bodies, such as the United States Food and Drug Administration (FDA), in addition to historic product-development experience, inform the desired quality attributes of the tablets. Tablets are usually tested for the following characteristics:

20.5.1 Appearance

All tablets should have identical size, shape, thickness, color, and surface markings. The general appearance of the tablet allows monitoring a lot-to-lot and tablet-to-tablet uniformity. Tight control of tablet thickness is required to ensure automated machine operations during its packaging and handling. Tablet-to-tablet thickness within a batch and average thickness of tablets across all batches are defined and controlled.

20.5.2 Uniformity of content

All tablets must be demonstrated to contain the labeled active ingredient and there should be tablet-to-tablet uniformity in drug content. This is usually tested by an analytical method for drug potency (such as high-performance liquid chromatography) in a several individual tablets.

20.5.3 Hardness

Tablet hardness refers to the amount of force required to diametrically crush a tablet. It is representative of the tensile strength of a tablet and is determined by the cohesion characteristics of the powder blend. Tablet hardness impacts tablet disintegration, dissolution, and friability. If tablets are too hard, they may not disintegrate within a reasonable period of time. This can lead to reduced bioavailability and failure to meet the dissolution specification. If they are too soft, then they may not withstand the handling and shipping operations, leading to tablet breakage or chipping (breaking away from edges) and failure during friability testing. Friability is the tendency of the tablets to chip or break by tumbling motion.

20.5.4 Friability

Tablet friability represents the tendency of a tablet to shed powder or break into smaller pieces under mechanical stress, such as falling from a fixed distance. It is a function of the fragility of the compressed powder blend, tablet shape, cohesion, and hardness. Low tablet friability is desired to ensure its physical integrity during packaging, shipment, and handling.

20.5.5 Weight uniformity

Tablets are compressed at a predefined weight. Under the assumption of normality of statistical distribution of tablet weight, all tablets are required to be within a certain range of the predefined tablet weight. Several tablets are weighed individually, and both the average weight and variation of individual tablet weight from the average are calculated and controlled during the manufacturing to ensure that the tablets contain the desired amounts of drug substances, with no more than acceptable variation among tablets within a batch.

20.5.6 Disintegration

Disintegration of tablets is evaluated to ensure that the tablet dissolves or breaks apart into smaller particles or granules on contact with water under agitation. This allows the DS to dissolve from its primary particles, being fully available for dissolution and absorption from the GI tract. Tablet disintegration is evaluated in a standardized apparatus that subjects six tablets to a defined mechanical stress in individual reciprocating cylinders in a suitable aqueous medium at 37°C, to reflect the conditions on oral ingestion. The time it takes for the last of six tablets to disintegrate into smaller particles and disappear from the reciprocating cylinders is called *disintegration time*. The disintegration media required varies depending on the type of tablets to be tested. The disintegration time is generally not more than 15 min for IR tablets.

The disintegration test is used as a control for tablets intended to be administered by mouth, but not for the tablets intended to be chewable and SR.

20.5.7 Dissolution

As drug absorption and physiological availability depend on having the DS in the dissolved state at the site of absorption, dissolution, also termed drug release, is an important property of tablets. The rate and extent of dissolution of a drug are tested *in vitro* by a suitable dissolution test. Dissolution is used as both a quality control tool to ensure batch-to-batch and tablet-to-tablet uniformity in drug-release characteristics of the tablets and sometimes also as a tool for *in vitro–in vivo* correlation (IVIVC) of drug release (*in vitro*) and drug absorption (*in vivo*). Dissolution test provides a means of control in ensuring that a given tablet formulation is similar with respect to the rate and extent of drug release as the batch of tablets were shown initially to be clinically effective.

20.6 RELATIONSHIP AMONG DISINTEGRATION, DISSOLUTION, AND ABSORPTION

Oral absorption from the GI tract takes place from the drug in aqueous solution at the site of absorption. Therefore, a tablet must disintegrate into primary drug particles and release the drug, that is, the drug should dissolve in the fluids at the site of absorption, before absorption can take place (Figure 20.5). However, tablets that are intended for chewing or SR do not have to undergo disintegration. Excipients for tablet formulation affect the rates of disintegration, dissolution, and absorption. Systemic absorption of most products consists of a succession of rate processes, such as

- Disintegration of the DP into granules and primary drug particles
- Dissolution of the drug from the granules and primary drug particles in an aqueous environment
- Absorption of drug solution across cell membranes into the systemic circulation

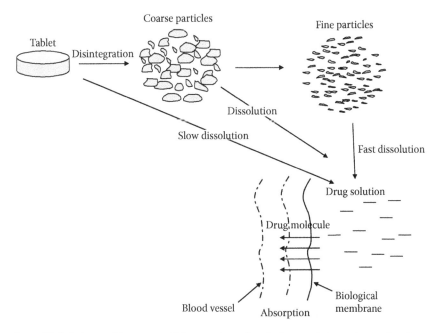

Figure 20.5 Relationship among disintegration, dissolution, and drug absorption from an intact tablet.

Tablet disintegration, dissolution, and drug absorption are influenced by physicochemical properties (e.g., solubility, compactibility, density, and flow) and stability (e.g., to heat, moisture, and light) of the DS; its compatibility with the excipients in the dosage form; site and extent of drug absorption in the GI tract; and dose. In these processes, the rate at which drug reaches the circulatory system is determined by the slowest step in the sequence. Disintegration of a tablet is usually more rapid than drug dissolution and absorption. For the drug that has poor aqueous solubility, the rate at which the drug dissolves (dissolution) is often the slowest step, and therefore exerts a rate-limiting effect on drug bioavailability. In contrast, for the drug that has a high aqueous solubility, the dissolution rate is rapid and the rate at which the drug crosses or permeates cell membranes is the slowest or rate-limiting step.

Aqueous solubility and permeability across the intestinal mucosa form the basis of identifying the relative difficulty in formulating drugs for oral delivery through what is known as the biopharmaceutics classification system (BCS). BCS classifies both the rate of drug permeation and drug solubility relative to the highest dose in two categories: high and low. Thus, all drugs can be divided into four distinct categories: I—high solubility, high permeability; II—low solubility, high permeability; III—high solubility, low permeability; and IV—low solubility, low permeability. This classification system has been used to understand and influence (by formulation design) the oral pharmacokinetics of drugs.

REVIEW QUESTIONS

20.1 Which condition usually increases the rate of drug dissolution from a tablet?
   A. Increase in the particle size of the drug
   B. Decrease in the surface area of the drug
   C. Use of the ionized or salt form of the drug
   D. Use of sugarcoating around the tablet
20.2 Which of the following is NOT true for tablet formulations?
   A. A disintegrating agent promotes granule flow
   B. Lubricants prevent adherence of granules to the punch faces of the tableting machine
   C. Glidants promote flow of the granules
   D. Binding agents are used for adhesion of powder into granules
   E. All of the above
   F. None of the above
20.3 Agents that may be used in the enteric coating of tablets include
   A. Hydroxypropyl methylcellulose
   B. Carboxymethylcellulose
   C. Cellulose acetate phthalate
   D. All of the above
   E. None of the above
20.4 Patients who cannot swallow enteric-coated tables should
   A. Dissolve the tablet before taking
   B. Crush before taking it
   C. Swallow tablet without water
   D. Consult a pharmacist for an alternative
20.5 Adequate powder flow ensures that after tableting
   A. Tablets of constant weight are produced
   B. Rapid drug release
   C. Drug molecules are crushed
   D. Smooth tablets are produced
20.6 To provide enough bulk for compression, which of the following excipients is often added to tablet formulation?
   A. Glidants
   B. Diluents
   C. Lubricants
   D. Disintegrants

20.7 Mixing of magnesium stearate with tablet granules will
  A. Decrease the crushing strength of tablets
  B. Increase tablet dissolution
  C. Increase tablet hardness
  D. Increase tablet disintegration

20.8 Which of the following excipients can be used as a binder in granulation?
  A. Magnesium stearate
  B. Starch mucilage
  C. Fumed silica
  D. Isopropanol

20.9 Your lab is designing a tablet dosage form of a highly insoluble compound, Lisinopril. You have recently faced the problem of tablet sticking to the punches during the tablet compression operation.
  A. Explain what modification in the formulation would be the easiest way to solve the problem
  B. What problem do you anticipate this step to result in and why? How do you correlate this with Fick's law?

20.10 Explain how the role of a glidant in a tablet formulation is different from the role of a lubricant, during the process of tablet compression.
  A. Define briefly disintegration, dissolution, and absorption.
  B. You desire to formulate a highly insoluble compound into an oral pharmaceutical formulation. The formulation you prepared has excellent disintegration characteristics but the dissolution profile in water or acid media is very low (less than 10% dissolved in 60 minutes). You desire to redesign the dissolution conditions so as to achieve higher dissolution rates. Suggest what all experiments would you conduct for this purpose.

20.11 What are the two main properties of drugs that are used to categorize drugs in the biopharmaceutics classification system (BCS)?
  A. Solubility
  B. Stability
  C. Permeability
  D. Bioavailability
  E. Manufacturability

FURTHER READING

Allen LV Jr., Popovich NC, and Ansel HC (2005) *Ansel's Pharmaceutical Dosage Forms and Drug Delivery Systems,* 8th ed., Lippincott Williams & Wilkins, Philadelphia, PA.

Allen LV Jr. (2002) *The Art, Science, and Technology of Pharmaceutical Compounding,* 2nd ed., Washington, DC: American Pharmaceutical Association, pp. 231–248.

Kottke MK and Rudnic EM (2002) Tablet dosage forms. In Banker GS and Rhodes CT (Eds.) *Modern Pharmaceutics* 4th ed., New York: Marcel Dekker, pp. 287–333

Qiu Y, Chen Y, and Zhang GGZ (Eds.) (2009) *Developing Solid Oral Dosage Forms: Pharmaceutical Theory and Practice.* 1st ed., Burlington, MA: Academic Press.

Rudnic EM and Schwartz JD (2000) Oral dosage forms. In Gennaro AR (Eds.) *Remington: The Science and Practice of Pharmacy,* 20th ed., Easton, PA: Mack Publishing Company, pp. 858–893.

Shukla AJ and Chang RK (1998) Introduction to coatings. In Avis KE, Shukla AJ and Chang RK (eds.) *Pharmaceutical Unit Operations—Coating,* Chicago, IL: Interpharm Press.

Wirth DD, Baertschi SW, Johnson RA et al., (1998) Maillard reaction of lactose and fluoxetine hydrochloride, a secondary amine. *J Pharm Sci* 87(1): 31–39.

Capsules

21.1 INTRODUCTION

Capsules are the dosage forms in which the drug formulation in a powder, semisolid, or liquid form is enclosed in a shell. This shell is generally made from gelatin, but can be made from other polymers such as hydroxypropyl methylcellulose (HPMC), polyvinyl alcohol (PVA), seaweed, or starch. Depending on the composition of the gelatin shell, the capsules can be hard or soft gelatin capsules. Soft gelatin capsules (also known as *softgels*) are made from a relatively more flexible, plasticized gelatin film than hard gelatin capsules. Hard capsules, such as hard gelatin or HPMC capsules, are typically used for powder or solid fills, whereas soft gelatin capsules are used for semisolid or liquid fills. Lately, hard capsules have also been used for liquid or semisolid fills.

Most soft and hard capsules are intended to be swallowed as a whole. Some soft gelatin capsules are intended for rectal or vaginal insertion as suppositories. Some soft gelatin capsules are intended to be cut open by the patient to remove and externally apply the contained medicament, for

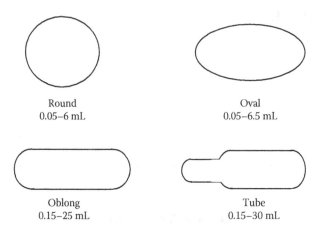

Figure 21.1 Schematic diagrams illustrating different shapes of soft gelatin capsules. The range of fill volumes is also indicated.

Table 21.1 Typical sizes of hard gelatin capsules

Size designation	Fill volume (mL)	Fill volume (oz)	Height of locked capsule (mm)	Outer diameter (mm)
000	1.37	1/20	26.1	10.0
00	0.95	1/30	23.3	8.5
0	0.68	1/40	21.7	7.7
1	0.50	1/55	19.4	6.9
2	0.37	1/75	18.0	6.4
3	0.30	1/100	15.9	5.8
4	0.21	1/135	14.3	5.3
5	0.13	1/220	11.1	4.9

example, ophthalmically. Figure 21.1 shows the common shapes of soft gelatin capsules. Table 21.1 shows the examples of commonly used capsule dosage forms.

The capsule shell dissolves rapidly on contact with gastrointestinal (GI) fluids, thus releasing the capsule's contents. Drug's bioavailability from capsules is usually high and similar to those of immediate-release (IR) tablets. Coating of capsule shell or drug particles (within the capsule) with sustained-release (SR) polymers can prolong drug release and affect bioavailability.

Hard gelatin capsules have a significant amount of bound water. These capsules are generally not physically stable in low humidity conditions, such as in the presence of desiccant in the packaged drug product. They tend to become fragile and crack at low humidity. On the other hand, HPMC capsules have lower equilibrium moisture contents than gelatin capsules and have better physical stability (i.e., do not become fragile and crack) on exposure to low humidity. The majority of capsule products manufactured today are hard gelatin capsules.

21.2 HARD GELATIN CAPSULES

Gelatin is a colorless, almost tasteless, translucent proteinaceous substance that is brittle when dry and elastic when mixed with controlled amount of moisture. It is produced by irreversible, partial hydrolysis of collagen, which is obtained from animal skin and bones. It forms a semisolid colloid gel in the presence of water, which displays a temperature-dependent gel–sol transformation and viscoelastic flow. It has crystallites (microscopic crystals formed during the cooling phase of manufacture of capsule shells) that stabilize the three-dimensional gel network structure and are responsible for streaming birefringence in gelatin solutions.

A hard gelatin capsule shell consists of two pieces: a cap and a body. The body has slightly lower diameter than the cap and fits inside the cap. They are produced empty and are then filled in a separate operation. During the capsule filling unit operation, the body is filled with the medicament, followed by the insertion of the cap over the body.

The shapes and interlocking arrangement of the body and the cap have evolved to meet the manufacturing and use requirements of hard gelatin capsules as shown in Figure 21.2.

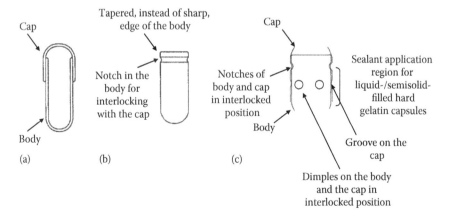

Figure 21.2 Schematic diagrams (a–c) of hard gelatin capsules illustrating their design features. The larger, narrower part of the capsules is the body and the smaller, wider part is the cap.

- Conventionally, the body and the cap had smooth edges with a diameter of the cap being slightly higher than that of the body. The two components could slide over each other (Figure 21.2a).
- To minimize defects during the production process, the design of the edge of the body was tapered to allow smooth penetration into the cap with minimum defects during high-speed production operation (Figure 21.2b).
- The capsules were modified to have an encircling groove each on the cap and the body (Figure 21.2c) and/or a notch to allow firm locking of the cap on the body (Figure 21.2b and c).
- To accommodate the need for a firm seal in the case of liquid and semisolid-filled hard gelatin capsules, raised circular bands (*dimples*) were introduced on the body and the cap along the sealing zone (Figure 21.2c).
- For the use of hard gelatin capsules in double-blind clinical trials, it was necessary to have hard gelatin capsules that could not be reopened after closing. To meet this objective, capsules with the cap that covers most of the body were developed.

For human use, empty gelatin capsules are manufactured in eight sizes, ranging from 000 (the largest, fill volume 1.37ml) to 5 (the smallest, fill volume 0.13ml), as shown in Table 21.1. The powder-filling capacity of these capsules varies depending on the packed density of the formulation. Modern high-speed capsule-filling machines are capable of filling up to 200,000 capsules per hour, matching the production capacity of tablets. The formulation filled weight in the capsules can range from 30 to 1400 mg, depending on the powder's bulk and compact densities.

Hard gelatin capsules can be filled with powders, granules, pellets, microtablets, tablets, capsules, liquids, or semisolids. Most of the marketed products contain powders or granules. Recently, the liquid- or semisolid-filled hard gelatin capsules have gained popularity.

After ingestion, the gelatin shell imbibes water, softens, swells, and dissolves in the GI tract. Encapsulated drugs are released rapidly and dispersed easily, leading to rapid absorption.

21.2.1 Advantages and disadvantages of hard gelatin capsules

21.2.1.1 Comparison with tablets

Hard gelatin capsules often provide formulation capability for uniquely challenging drug molecules. For example, a drug candidate with a low melting point or that is liquid at room temperature usually has poor manufacturability as a tablet, especially if it requires a high dose. Such a compound can be encapsulated in a liquid- or semisolid-filled hard gelatin capsule.

In addition, very low-dose drugs (in μg) can have content uniformity challenges when formulated as a tablet. The distribution of these drugs can be significantly better when encapsulated as a solution in a liquid or semisolid matrix in a hard gelatin capsule.

Hard gelatin capsules generally require less formulation components and place less stringent requirement on the powder properties of the formulation. They can also allow flexibility in formulation with the possibility of filling one or more of diverse systems including powders, granules, pellets, and small tablets. In addition, hard gelatin capsules can be used for blinding in clinical studies.

The disadvantages of hard gelatin capsules are owed to the inherent high moisture content requirement of gelatin. For example, highly soluble salts, such as iodides, bromides, and chlorides, of drugs are generally not formulated in hard gelatin capsules because these can draw moisture from the shell, thus making the shell brittle. Storage under low humidity conditions, such as with the use of desiccant in packaging, can also make the shell brittle. In addition, gelatin is prone to cross-linking in the presence of very low (in parts per million range) concentrations of formaldehyde, which may be present in certain pharmaceutical excipients such as polyethylene glycol (PEG).

21.2.1.2 Comparison with soft gelatin capsules

In comparison to soft gelatin capsules, the manufacturing process of hard gelatin capsules is less demanding, tedious, and costly. This is because the soft gelatin capsule manufacture requires the formation of gelatin ribbons during the encapsulation process itself, whereas the hard gelatin capsules use premanufactured capsule shells. The hard gelatin capsule manufacture also does not require a curing or moisture-loss step after encapsulation of the drug formulation.

The residual water in the capsule shells is lower (~10–16% w/w) for hard gelatin capsules than for soft gelatin capsules (~30% w/w). This can affect the stability of the encapsulated formulation directly by chemical degradation (e.g., hydrolysis) of water-sensitive compounds or plasticization of the reaction medium with water, thus increasing the rate of degradation. In addition, soft gelatin capsule shells have a high oxygen permeation rate, which can contribute to the oxidation of sensitive drugs.

21.2.2 Solid-filled hard gelatin capsules

21.2.2.1 Main applications

Hard gelatin capsules are often preferred over tablets as the dosage form for initial (Phase I and Phase IIA) clinical studies of new molecular entities (NMEs). This is because the effect of limited availability of the active pharmaceutical ingredient (API) to conduct necessary screening for the development of tablets. Many initial clinical studies simply use a drug-in-capsule (DIC) product, which is the only drug manually encapsulated in the hard gelatin or HPMC capsules.

Hard capsules are also preferred for the comparator and blinded clinical studies. These clinical studies require that the patient and/or the doctor should not be able to identify the actual drug product being administered to the patient. In these studies, two or more drug products are administered after encapsulating them in hard gelatin capsules of the same specifications and such that the capsules cannot be opened.

21.2.2.2 Formulation considerations

Hard gelatin capsule-manufacturing process places a relatively less stringent requirement on the powder properties of the fill formulation than tablets. The important formulation considerations include the following:

1. *Flow*: Adequate flow through the hopper and into the dosing device (dosator) for reproducible filling of the capsules.
2. *Density*: Reproducible density of the powder is important for fill weight uniformity of capsules because the dosing devices in high-speed capsule-filling machines are filled based on the volume of the powder for a target weight.
3. *Lubricity*: Magnesium stearate is typically added to most powder formulations. When mixed with other particles, magnesium stearate coats their surface and acts as a lubricant. Lubricants facilitate the lack of adhesion to metallic machine parts, especially the dosing device used to form a plug in high-speed machines, and adequate flow of the formulation. Other lubricants commonly used are stearic acid and sodium stearyl fumarate.
4. *Compactibility*: Some high-speed capsule-filling machines form a plug of the powder before filling into the capsule. In cases where plug formation is required for encapsulation, some level of compactibility of the powder is needed.
5. *Noninteraction with capsule shell*: Lack of interaction between the drug substance and/or formulation components with the capsule shell, either gelatin or HPMC. This interaction could be in the form of solubilization or changing the water content of the shell. Hygroscopic and volatile components are usually unsuitable. The fill should not contain more than 5% w/w of water. In addition, chemical interactions between the components can lead to bioavailability or stability problems. For example, the use of PEG in drug formulation can lead to cross-linking of gelatin on storage due to the unintended presence of formaldehyde in PEG, which can diffuse into the shell and react with gelatin. Similar problems have been observed due to the presence of residual peroxides in excipients.

6. *Dose*: Dose and drug loading (i.e., %w/w of the formulation, that is the API) influences drug content uniformity between the capsules, the extent to which the powder properties of the formulation are affected by the physicochemical characteristics of the drug substance, and manufacturability of the capsule dosage form. For example, it may be difficult to assure adequate uniformity of the content of the API for drugs with extremely low doses (e.g., in μg), and it may not be possible to fill a capsule of acceptable size for extremely high-dose drugs (e.g., more than 600 mg). For intermediate doses, the percent drug loading in the formulation can range widely. Drug properties predominantly govern the powder properties of the formulation for high drug-loading formulations (e.g., more than 60% w/w).

7. *Particle size, shape, and density*: Particle size and shape influence the flow, uniformity, and thus content of the active in a formulation. A drug substance with irregular or spherical-like crystals is more likely to flow well than the needle-shaped crystals. Drug content uniformity is also affected by particle density, if it is significantly different than the density of the excipients.

8. *Moisture sorption–desorption isotherm*: Moisture sorption and retention properties of the drug and excipients, indicated by a hysteresis in the sorption–desorption isotherm, can affect the physical stability of gelatin during storage and the chemical stability.

9. *Solubility and wettability*: Solubility and wettability of the drug substance affect its dissolution characteristics. A low-solubility drug substance might require the addition of a wetting agent (e.g., surfactant such as polysorbate 80) in the formulation.

21.2.2.3 Formulation components

The powder formulations for encapsulation into hard gelatin capsules require a careful consideration of the filling process requirements, such as lubricity, compactibility, and flow. Additives present in capsule formulations, such as the amount and choice of fillers, lubricant, disintegrant, and surfactant, and the degree of plug compaction, can influence drug release from the capsule. The functional categories of formulation components are as follows:

1. *Fillers (or diluents)*: Active ingredient is mixed with a sufficient volume of a diluent, usually microcrystalline cellulose, lactose, mannitol, starch, or dicalcium phosphate, to increase the bulk of the formulation.

2. *Glidants*: Glidants are finely divided dry powders added to the formulation in small quantities to improve their flow rate from the hopper and into the body of the capsule during the filling process. Glidants,

such as colloidal silicon dioxide, powdered silica gel, starch, talc, and magnesium stearate, improve flow by

a. Reducing the roughness by filling surface irregularities.
b. Reducing attractive forces.
c. Reducing electrostatic repulsion.

The optimal concentration of the glidant used to improve the flow of a powder mixture is generally less than 1% w/w.

3. *Lubricants*: Capsule formulations usually require a lubricant just as the tablet formulations to reduce powder adhesion to the machine parts, especially during plug formation. Lubricants ease the ejection of plugs by reducing the adhesion of powder to metal surfaces and friction between the sliding surfaces in contact with the powder. The most common lubricants for capsule formulations are hydrophobic stearates, such as magnesium stearate, calcium stearate, and stearic acid.

4. *Surfactants and wetting agents*: Surfactants may be included in capsule formulations of poorly water-soluble drugs to reduce the contact angle, increase the wettability of drug particles, and enhance drug dissolution. The most commonly used surfactants in capsule formulations are sodium lauryl sulfate and sodium docusate (sodium dioctyl sulfosuccinate).

   In addition, a hydrophilic polymer, such as HPMC, is sometimes used as a wetting agent in the formulations of poorly soluble drugs. Powder wettability and dissolution rate of several drugs, such as hexobarbital and phenytoin, were enhanced with the inclusion of methylcellulose or hydroxyethylcellulose in their capsule formulations.

5. *Disintegrants*: A disintegrant is frequently included to aid rapid disintegration and dissolution of the contents. Common disintegrants used in hard gelatin capsule formulations include croscarmellose sodium, crospovidone, and sodium starch glycolate.

   Controlled-release beads and minitablets are often filled into gelatin capsules for convenient administration of an oral controlled-release (CR) dosage form. For example, SR antihistamines, antitussives, and analgesics are first manufactured into extended-release (XR) microcapsules or microspheres, and then placed inside a gelatin capsule. Another example is enteric-coated lipase minitablets that are placed in a gelatin capsule for more effective protection of these enzymes from the acidic environment.

21.2.2.4 Manufacturing process

Very small-scale and experimental filling of the hard gelatin capsules can simply be carried out manually, that is, by removing the cap from the body of an empty capsule shell, filling the body with a preweighed amount of API or formulation, and attaching the cap. This can be carried out in early clinical studies by the sponsor or by the pharmacist. Compounding by

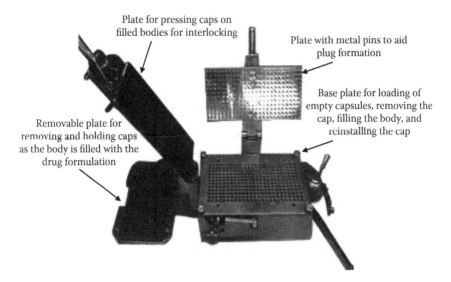

Plate for pressing caps on filled bodies for interlocking

Plate with metal pins to aid plug formation

Removable plate for removing and holding caps as the body is filled with the drug formulation

Base plate for loading of empty capsules, removing the cap, filling the body, and reinstalling the cap

Figure 21.3 Hand-filling machine used to fill hard gelatin capsules.

the pharmacist is preferred when the stability of the drug in the capsule is unknown and is called on-site compounding.

Small-scale manufacture (several hundred capsules) can be done by using a manual capsule-filling machine. As illustrated in Figure 21.3, the manual-filling operation involves the following steps:

1. Placing empty gelatin capsules on the removable plate with bodies facing downward. This removable plate is then placed on the base plate and the bodies of the capsules are locked in position with the base plate using a lever.
2. The removable plate is removed with the caps on it. The body is filled with the formulation manually using a plastic spetula, and the excess powder is removed.
3. The removable plate is placed back on the base plate and pressing the flat plate seals the capsule caps. The sealed capsules are removed from the base plate by opening the lock on the body using the lever and inverting the base plate.

Large-scale filling of hard gelatin capsules follows the same principles using a high-speed capsule-filling machine, with two significant improvements:

• Capsule alignment and separation are driven by vacuum, instead of mechanical interlocking.
• Powder filling may require a soft compact (plug) formation depending on the formulation weight and capsule fill volume. This compact is usually much softer than a typical tablet. The compaction force

used for plug formation is typically 20–30 N, compared to 10–30 kN typically used for tableting.

- The high-speed powder filling is accomplished by either of the two dosing devices: (a) dosator device or (b) dosing disk/tamping device.

  1. The *dosator device* uses an empty tube that dips into powder bed, which is maintained at a height approximately two-fold greater than the desired length of the plug. The dosator piston's forward movement helps form the plug, which is then transferred to the body of the capsule, and released.

  2. The *tamping device* operates by filling the cavities bored into the dosing disk, similar to the die-filling operation during tableting. A tamping punch slightly compresses the filled powder by repeated action, which is followed by the ejection of the plug into the capsule body.

21.2.3 Liquid- and semisolid-filled hard gelatin capsules

21.2.3.1 Main applications

Liquid- and semisolid-filled hard gelatin capsules are sometimes used to improve the bioavailability of drug substances with low solubility and wettability. Lipids in the formulation tend to increase the bile flow *in vivo* and promote drug absorption. For example, mixtures of mono-, di-, and triglycerides of mono- or dicarboxylate esters of PEGs, commercially available as Gelucire®, are available in various melting point and hydrophilic–lipophilic balance (HLB) ranges. Oral availability of drug solution in Gelucire® or in PEG is frequently higher than that of powder drug formulation. In addition, self-emulsifying and self-microemulsifying drug delivery systems (SEDDS and SMEDDS, respectively) can significantly improve drug's bioavailability, for example, in the case of cyclosporine A and fenofibrate.

Liquid filling of hard gelatin capsules may also be indicated in the case of drugs with extremely low dose (e.g., in μg) and drug loading (e.g., less than 5% w/w) in the formulation to assure uniformity of content. Uniformity of drug distribution between different dosage units can be higher with a drug solution in a liquid or semisolid base than a blended powder.

Drugs with manufacturability issues in a tablet dosage form may also be formulated as a liquid-filled hard gelatin capsules. For example, drugs with low melting points can show significant sticking issues in both tablet- and powder-filled capsule dosage forms. Certain drugs with significant instability to light or moisture can show better stability in liquid or semisolid filled, compared to a powder-filled, hard gelatin capsule. The presence of an opaque waxy base and a molecular mixture of the antioxidant with the drug can increase the effectiveness of environmental protection in the capsule dosage form.

Examples of drug substances formulated as liquid-filled hard gelatin capsules are listed in Table 21.2.

Table 21.2 Examples of commonly used capsule dosage forms

Formulation type	Active ingredient(s)	Brand names	Manufacturer	Indications
Solid-filled hard gelatin capsules	Cinoxacin	Cinobac	Eli Lilly and Co.	Urinary tract infection
	Amphetamine and dextroamphetamine	Adderal XL	Shire Pharmaceuticals	Attention deficit disorder
	Methylphenidate hydrochloride	Ritalin LA	Novartis	Attention deficit disorder
	Didanosine	Videx EC	Bristol Myers	HIV-1 infection
Liquid- or semi-solid-filled hard gelatin capsules	Vancomycin	Vancodin	Lilly	Colitis
	Captopril	Captopril-R	Sankyo	Hypertension
	Ibuprofen	Solufen	SMB Ivax	Pain
	Piroxicam	Solicam	SMB	Arthritis
Soft gelatin capsules	Saquinavir	Fortovase	Roche	HIV
	Dutasteride	Avodart	GSK	Benign prostate hyperplasia
	Cyclosporine A	Neoral	Novartis	Immunosuppressant
	Progesterone	Prometrium	Abbott	Hormone replacement therapy

21.2.3.2 Formulation considerations

The main formulation considerations for liquid-filled hard gelatin capsule are similar to those for soft gelatin capsules:

1. *Noninteraction with capsule shell*: Physicochemical compatibility between the drug/formulation excipients and the capsule shell are required for any capsule formulation. As described earlier, known drug–gelatin interactions include pH effect on gelatin hydrolysis or tanning, hygroscopicity or water effect on shell integrity, and the role of diffusible aldehydes in cross-linking gelatin shell.
2. *Dose*: The capsule size imposes a limit on the maximum amount of formulation that can be filled into a hard gelatin capsule.
3. *Hygroscopicity*: The formulation components should not significantly affect the moisture level of the shell. For example, highly hygroscopic excipients such as glycerol, sorbitol, and propylene glycol are not suitable for liquid-filled hard gelatin capsules in high concentrations, although they may be used for soft gelatin capsules. This is because of the lower inherent moisture content of the hard gelatin shell.

21.2.3.3 Formulation components

Drug solution in an appropriate base formulation can be filled into hard gelatin capsules at room or slightly higher temperature. The functional categories of formulation components are as follows:

1. *Triglycerides* for solubilization of the drug substance. These include either the medium chain triglycerides, such as Miglyol® 810 and 812, or the long chain triglycerides, such as soybean oil, olive oil, and corn oil.
2. *Surfactants* can be included in the formulation as solubility, dissolution, and/or absorption enhancers, such as Cremophor®, Gelucire®, Labrafil®, and Tween®.
3. *Cosolvents* can be used in low concentrations, especially for SEDDS and SMEDDS, such as ethanol, propylene glycol, and PEG.

21.2.3.4 Manufacturing process

The main consideration and process risk in the manufacture of liquid-filled hard gelatin capsules is their tendency to leak at the joint between the body and the cap. This concern has been addressed in one of the two ways:

1. Applying a zone of gelatin film on the joining region of the body and the cap. This is known as *banding*, because a band of gelatin is formed on the outside of the capsule.

2. Spraying a solution of ethanol and water on the overlapping areas of the body and the cap along with the application of heat (e.g., 40°C–60°C for several seconds). This process is known as *sealing*. The low surface tension of the solvent mixture allows it to diffuse into and dissolve gelatin, which also melts during heating, to allow the fusion of gelatin from the cap with that from the body.

21.3 SOFT GELATIN CAPSULES

Soft gelatin capsules consist of a hermetically sealed outer shell of gelatin that encloses a liquid or semisolid medicament in the unit dosage. Soft gelatin capsules are a completely sealed dosage form and cannot be opened without destroying the capsules. Drugs that are commercially prepared in soft capsules include cyclosporine, declomycin, chlorotrianisene, digoxin, vitamin A, vitamin E, and chloral hydrate.

Figure 21.1 shows different shapes of soft gelatin capsules.

21.3.1 Advantages and disadvantages of soft gelatin capsules

Soft gelatin capsules provide a patient-friendly dosage form for peroral administration of nonpalatable and/or oily liquids. Solutions or suspensions with an unpleasant odor or taste can be easily ingested in a soft gelatin capsule dosage form, which offers tidy appearance and convenient ingestion.

This dosage form can be particularly advantageous for low dose drugs that are lipid soluble because it can allow greater uniformity of content between dosage units than the conventional tablet dosage form. It can also be more suitable than a tablet dosage form for the encapsulation of liquid, water-insoluble drugs. The capsules can be formulated to be immediate release (IR), slow or sustained release (SR), or enteric coated.

The use of soft gelatin capsule shell imposes significant limitations on the drug formulations that can be encapsulated in this dosage form, that is, restricted to liquids and semisolids. The manufacturing process is relatively tedious and difficult to optimize (e.g., ribbon thickness, fill weight, and weight variation). In addition, the breakage of even one capsule during the manufacturing can lead to the coating of drug formulation on the outer surface of several other capsules. This can also happen during storage in multiple use containers, such as high-density polyethylene (HDPE) bottles.

Soft gelatin capsules have certain disadvantages compared to liquid-filled hard gelatin capsules. Due to the relatively higher water content in soft gelatin shell (20–30% w/w) compared to hard gelatin capsules (13–16% w/w) moisture-sensitive drugs may not be stable in soft gelatin capsules. In addition, the maximum temperature of the formulation that can be filled into soft gelatin capsule without deformation of the shell and other production issues is about 35°C, whereas a formulation can be filled at up to 70°C in hard gelatin capsules without shell deformation. Extreme acidic and basic pH must also be avoided because a pH below 2.5 hydrolyzes gelatin, whereas a pH above 9 has a tanning effect on the gelatin.

21.3.2 Drivers for development of soft gelatin capsules

Soft gelatin capsules are often developed for one or more of the following reasons:

1. *Line extension* products for strategic marketing advantage in a therapeutic area with intense competition. For example, cough and cold medicines available as a soft gelatin capsule can offer patient benefit, such as ingestion without water and portability.
2. *Technological advantage* such as good content uniformity of a low dose drug or formulation of a water-insoluble drug that is liquid at room temperature.
3. *Safety* reasons during product manufacturing, dispensing, and usage. For example, most of the product manufacturing unit operations of tablets and hard gelatin capsules involve handling of fine powders. In the case of soft gelatin capsules, the powder handling is restricted to drug dissolution or dispersion in a liquid medium. Powders inherently have greater exposure hazards than liquids. Therefore, soft gelatin capsules provide greater operator safety during manufacturing. In addition, as the drug formulation is hermetically sealed in a shell, the exposure to the medication is minimized during dispensing as well as use.
4. *Improved oral bioavailability*: The use of certain lipids can be associated with increased oral bioavailability and reduced intra- and interpatient variability by modification of GI digestive processes. In addition, presentation of the drug in a predissolved state can lead to shorter duration to the onset of action. By formulating nifedipine or ibuprofen into soft gelatin capsules after being dissolved in PEG, the bioavailability of these drugs can be improved.

21.3.3 Formulation of soft gelatin capsule shell

The composition of the soft capsule shell consists of three main ingredients: (1) gelatin, (2) plasticizer, and (3) water. In contrast to hard gelatin capsules, a relatively large amount (~30 % w/w) of plasticizers is added in soft gelatin capsule shell formulation to ensure adequate flexibility. Water is used to form the capsule, and other additives are often added as needed.

Table 21.3 Typical composition of a soft gelatin capsule shell

Component	Function	Typical content (% w/w)
Gelatin	Polymeric base	66.3
Glycerin	Plasticizer	33.0
Methylparaben + propylparaben (80/20 ratio)	Preservative	0.1
Color	Colorant	0.1
Titanium dioxide	Opacifier	0.5
Water	Solvent/process aid	q.s. (0.7–1.3 × of gelatin)

A typical composition of the soft gelatin capsule shell is listed in Table 21.3 and the functional components are described as follows:

1. *Gelatin*: Similar to hard gelatin shells, the basic component of soft gelatin shell is gelatin. The properties of gelatin shells are controlled by the choice of gelatin grade and by adjusting the concentration of plasticizer in the shell. The physicochemical properties of gelatin are controlled to allow
   - Adequate flow at desired temperatures to form ribbons of defined thickness, texture, mechanical strength, and elasticity.
   - Ribbons to be easily removed from the drums, stretch during filling, seal the temperature below the melting point of the film, and dry quickly under ambient conditions to an adequate and a reproducible strength.
   Physicochemical properties of gelatin important to capsule formation include gel strength, viscosity, change in viscosity with temperature and shear, melting point, settling point (temperature), settling time, particle size (affects time to dissolve), and molecular weight distribution (affects viscosity and strength).
2. *Plasticizer*: A plasticizer interacts with gelatin chains to reduce the glass transition temperature (T_g) of the gelatin shell and/or promotes the retention of moisture (hygroscopicity). The most common plasticizer used for soft gelatin capsules is glycerol. Sorbitol, maltitol, and polypropylene glycol can also be used in combination with glycerol. Glycerol derives its plasticizing ability primarily from its direct interactions with gelatin. In contrast, sorbitol is an indirect plasticizer because it primarily acts as a moisture retentive agent. Compared to hard gelatin capsules and tablet film coatings, a relatively large amount (~30% w/w) of plasticizers are added in a soft gelatin capsule formulation to ensure adequate flexibility.
3. *Water*: The desirable water content of the gelatin solution used to produce a soft gelatin capsule shell depends on the viscosity of the specific grade of gelatin used. It usually ranges between 0.7 and 1.3 parts of

water to each part of dry gelatin. After the capsule is formed, most of the water is removed by drying. The finished soft gelatin capsules contain 13–16 % w/w water.

4. *Preservative*: Preservatives are often added to prevent the growth of bacteria and mold in the gelatin solution during storage. Potassium sorbate, and methyl, ethyl, and propyl hydroxybenzoate are commonly used as preservatives.

5. *Colorant and/or opacifier*: A colorant and/or opacifier (e.g., titanium dioxide) may be added to the shell for visual appeal and/or reducing the penetration of light for the encapsulation of a photosensitive drug. The color of the capsule shell is generally chosen to be darker than that of its contents.

6. *Other excipients*: Other, infrequently, used excipients can include flavors and sweeteners to improve palatability and acid-resistant polymers to impart enteric release characteristics. They can also be used to formulate chewable soft gelatin capsules, for example, ChildLife's Pure DHA chewable 250 mg soft gel caps. A chelating agent, such as ethylene diamine tetracetic acid (EDTA), can be added to prevent chemical degradation of oxidation sensitive drugs catalyzed by free metals in gelatin, such as iron.

21.3.4 Drug formulation for encapsulation in soft gelatin capsules

Soft gelatin capsules may contain a liquid or semisolid solution, suspension, or preconcentrate of a self-emulsifying or self-microemulsifying system. For example, Accutane® is a suspension of isotretinoin in oil, Sandimmune® is a self-emulsifying preconcentrate, and Neoral® is a self-microemulsifying preconcentrate.

Formulation considerations for the contents of the soft gelatin capsules include the following:

- *Noninteraction with gelatin*: The contents of the soft gelatin capsule should not interact with the gelatin shell.
- *Nonmoisture sensitivity*: The moisture content of soft gelatin capsules plasticized with glycerol is considerably higher than that of hard gelatin capsules. Therefore, to ensure chemical stability of the drug, moisture-sensitive drugs should not be formulated in soft gelatin capsules.
- *Nontemperature sensitivity*: The molten gelatin mass usually has a pourable viscosity at 60°C–70°C. Therefore, the sealing operation is usually carried out at a higher than ambient temperature. Hence, highly thermolabile drugs may not be encapsulated in soft gelatin capsules.
- *pH*: Extreme acidic and basic pH should be avoided because a pH below 2.5 hydrolyzes gelatin (leading to leakage), whereas a pH above

9 has a tanning effect on the gelatin. Tanning process involves cross-linking of gelatin, which results in hardening of the shell. The shell becomes insoluble in water and resistant to digestion by GI enzymes: trypsin and chymotrypsin.

Drugs for encapsulation in a soft gelatin capsule are usually dissolved or suspended in a suitable carrier. Insoluble drugs are often dispersed or suspended in an agent such as beeswax, soybean oil, or paraffin. Surfactants are often added to promote wetting of the ingredients. The use of water or ethanol in the fill composition is only possible with special modifications of the capsule shell. Drugs can be dispersed in ethylcellulose for an SR effect.

21.3.5 Manufacturing process

Soft gelatin capsules are filled with solutions or suspensions of drugs in liquids, and sealed in a single operation. They are prepared from a more flexible plasticized gelatin by a rotary-die process. As shown in Figure 21.4, this process involves the following sequential operations:

1. Two heated sheets of gelatin of similar thickness are produced by the controlled flow of the fluid gelatin from its heated storage container (gelatin tank) by using a controlled pore opening and fill in a *spreader box*.
2. The gelatin film flows through a series of *oil rolls* that stretch the sheets and direct them appropriately toward *die rollers*.
3. The two sheets of gelatin merge on the metallic rollers that contain dies of appropriate shape and size and move in the opposite direction toward each other. The application of vacuum inside the rollers combined with pressurized filling of the components enables the formation of a cavity. The application of heat and mechanical pressure enables sealing of the shells as they pass through the rollers.
4. As the gelatin sheets are being annealed, a calibrated amount of the drug formulation is pumped into each cavity by the *product pump* through an *injection wedge*.
5. The concurrent process of drug product injection into the die cavity and sealing of the cavity is either accompanied by the cutting and release of individual soft gelatin capsules (if the rollers are suitably designed) or, as shown in Figure 21.4, the capsules may be cut from the sheets in a separate, subsequent operation.
6. The filled capsules are dried at ambient conditions to remove moisture from the outer surface and may be tray dried for an extended period of time (e.g., up to 48 hours).
7. Finished capsules are passed on a conveyor belt for the next unit operations of packaging and labeling.

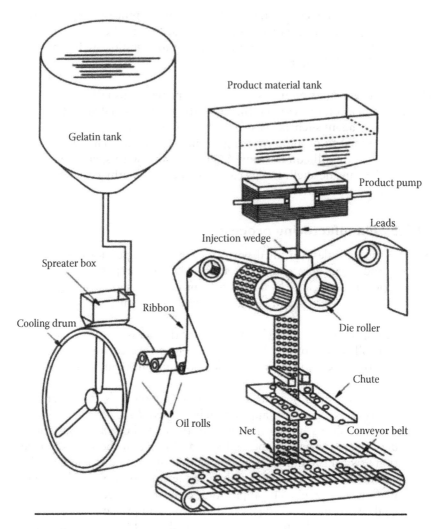

Figure 21.4 Manufacturing process of soft gelatin capsules. (Adapted from http://www.sunkingpm.com/htm/PM/SCP/5.html)

21.3.6 Nongelatin soft capsules

The use of alternate polymers for the formation of soft capsules is driven by marketing or formulation requirements. For example, Vegicaps® are animal-free. Their shell is made from seaweed extract and gluten-free starch. For moisture sensitive drugs, HPMC capsules may be preferred, which generally have lower equilibrium moisture content than gelatin capsules. HPMC capsules also have better physical stability on exposure to low humidity.

21.4 EVALUATION OF CAPSULE DRUG PRODUCTS

Drug product testing is generally divided into three stages:

1. In-process testing, during the manufacture of the drug product. These batteries of tests are carried out at predefined intervals during the product manufacturing, by the manufacturing personnel, and their results recorded on the batch record. Adverse findings in these tests can be used as a guide to alter the manufacturing-process parameters.
2. Finished product testing, after the whole batch has been manufactured. These tests help identify whether the batch is acceptable for marketing or its intended usage.
3. Shelf-life testing, after the whole batch has been packaged. These tests are frequently carried out after defined periods of storage at predetermined conditions. They help to assign and verify the shelf life and usability of the drug product.

21.4.1   In-process tests

Visual inspection of soft gelatin capsules is done to ensure absence of clearly malformed, damaged, or improperly filled capsules. During the encapsulation of soft gelatin capsules, the following parameters are usually closely monitored and controlled:

- Gel ribbon thickness and uniformity across the ribbon
- Seal thickness
- Weight of the capsule fill and its variation from capsule-to-capsule
- Weight of the capsule shell and its variation from capsule-to-capsule
- Moisture level of the capsule shell before and after drying

Visual inspection, fill weight, and fill-weight uniformity are the key in-process tests used for hard gelatin capsules.

21.4.2 Finished product quality control tests

21.4.2.1   Permeability and sealing

Soft gelatin capsules are tested for physical integrity (absence of leakage) by visual inspection. Similarly, hard gelatin capsules are tested for any breach of physical integrity (breakage or opened cap and body).

21.4.2.2   Potency and impurity content

All capsules are tested for drug content (potency, as a percent of label claim). In addition, most drug products are tested for the related substances or impurities. These must meet predefined specifications for a batch to be acceptable.

21.4.2.3　Average weight and weight variation

Ten hard gelatin capsules are usually weighed individually and the contents are removed. The emptied shells are individually weighed and the net weight of the contents is calculated by subtraction. The content of active ingredient in each capsule may be determined by calculation based on the percent drug content in the formulation for high drug load formulations.

For soft gelatin capsules, the gross weight of 10 gelatin capsules is determined individually. Then each capsule is cut open, and the contents are removed by washing with a suitable solvent (that dissolves the fill but not the shell). The solvent is allowed to evaporate at room temperature, followed by weighing of the individual washed shells. The net contents are calculated by subtraction and the content of active ingredient in each of the capsules can be determined by calculation based on the percent drug content in the formulation.

Fill-weight variation of capsules is often a function of equipment setup and filling operation. An automated capsule sizing machine and/or weight checker is frequently used to discard over- or underfilled capsules.

21.4.2.4　Uniformity of content

Uniformity of content of the active ingredient can be determined by weight variation of the fill of hard or soft gelatin capsules for high drug load (API $\geq 25\%$ w/w of the total fill weight), high fill-weight (250 mg/capsule) formulations. For low drug load and low fill-weight formulations, each capsule must be analyzed individually by the potency method for the content of the active ingredient. The uniformity of content is assured if predetermined criteria for the range and variation in the content of the active ingredient are met.

21.4.2.5　Disintegration

Disintegration of hard and soft gelatin capsules is evaluated to ensure that the drug substance is fully available for dissolution and absorption from the GI tract. The disintegration media varies depending on the type of capsules to be tested.

21.4.2.6　Dissolution

Drug absorption and physiological availability depend on the drug substance being in the dissolved state at the site of drug absorption, viz. the GI fluids. The rate and extent of dissolution of the drug from the capsule dosage form is tested by a dissolution test. Dissolution test provides means of quality control in ensuring that (a) different batches of the drug product have similar drug release characteristics and (b) that a given batch has similar dissolution as the batch of capsules that was shown initially to be clinically effective.

21.4.2.7 Moisture content

Water content of the entire capsule or the capsule contents are determined by Karl Fisher titrimetry to enable the correlation of water content with the degradation profile or drug-release characteristics of capsules.

21.4.2.8 Microbial content

The capsules are tested to ensure lack of growth of bacteria and mold by microbiological tests. These tests are usually carried out by incubation of the capsule contents in a growth medium and counting the colonies formed after a predefined period of time. Selection of the growth medium and duration of the test, as well as maintenance of aseptic conditions during the testing, are critical to successful assessment of microbial contamination by this method.

21.5 SHELF-LIFE TESTS

Stability testing of capsules is performed to determine the physicochemical stability of the active drug molecule in the finished drug product under specified package and recommended storage conditions. Shelf-life tests are usually same as the finished product tests. Since the shelf life of the product at recommended storage conditions can be long, the product is often subjected to accelerated (higher than normal levels of environmental conditions) storage for predicting shelf life under recommended storage conditions. These storage conditions that are accelerated for stability testing include temperature, humidity, and light.

For example, for a product intended for sale in the United States with recommended storage at room temperature and ambient humidity, real-time stability is carried out at 25°C and 60% relative humidity (RH) with periodic testing up to the recommended shelf life, for example, 2 years. Accelerated stability testing on such a product is usually carried out at 40°C and 75% RH for a limited duration of time, for example, 3 months. The exact conditions for real-time and accelerated storage testing depend on the geographic and climatic region where the drug product is intended to be manufactured and marketed.

REVIEW QUESTIONS

21.1 Why should highly soluble chloride salts not be dispensed in hard gelatin capsules?
  A. Capsules will dissolve slowly
  B. Salts will decompose
  C. Rapid release may cause gastric irritation
  D. The capsule shell will disintegrate

21.2 The main difference between soft and hard gelatin capsules is
  A. The level of plasticizer
  B. Hard gelatin shells are not plasticized
  C. Hard gelatin shells are plasticized
  D. The basic composition of soft shells is not gelatin
  E. Dyes are added to the capsule shell

21.3 Leakage from soft gelatin capsules can be caused by
  A. Hydrolysis of gelatin at low pH
  B. Addition of surfactants
  C. Addition of polyethylene glycol
  D. All of the above

21.4 The ideal powder characteristics for successful filling of hard gelatin capsules include
  A. Poor compatibility
  B. Poor lubrication
  C. Have adequate flow properties
  D. Have low bulk density

21.5 The decrease in solubility of gelatin capsules has been attributed to
  A. Acid hydrolysis
  B. Gelatin cross-linking
  C. Trace amount of glycine
  D. None of the above

21.6 Following is a commonly used plasticizer in soft gelatin capsules
  A. Polyethylene glycol
  B. Polypropylene glycol
  C. Glycerol
  D. Sorbitol

FURTHER READING

Allen LV Jr., Popovich NC, and Ansel HC (2005) *Ansel's Pharmaceutical Dosage Forms and Drug Delivery Systems*, 8th ed., Lippincott Williams & Wilkins, Philadelphia, PA.

Augsburger LL (2002) Hard and soft shell capsules. In Banker GS and Rhodes CT (Eds.) *Modern Pharmaceutics*, 4th ed., New York: Marcel Dekker, pp. 335–380.

De Villiers MM (2004) Oral conventional solid dosage forms: Powders and granules, tablets, lozenges, and capsules. In Ghosh TK and Jasti BR (Eds.) *Theory and Practice of Contemporary Pharmaceutics*, Boca Raton, FL: CRC Press, pp. 279–331.

Jones BE, Seager H, Aulton ME, and Morton SS (1988) Capsules. In Aulton ME (Ed.) *Pharmaceutics: The Science of Dosage Form Design*, New York: Churchill Livingstone, pp. 322–340.

Chapter 22

Parenteral drug products

LEARNING OBJECTIVES

On completion of this chapter, the students should be able to

1. Enlist the common parenteral routes of drug administration and discuss circumstances where one route may be preferred over another.
2. Identify different types of parenteral dosage forms.
3. Identify key quality attributes of parenteral drug products.
4. Define sterilization and describe the methods of sterilization of injectable products.

22.1 INTRODUCTION

Parenteral drug products are the dosage forms intended for administration by a route that does not involve the gastrointestinal (GI) tract (thus, parenteral). Most of the parenteral drug products are injectable dosage forms that are intended for administration by injection using a syringe and a needle.

Parenteral dosage forms are preferred for one or more of the following reasons:

- Low oral bioavailability and/or high variability in oral drug absorption.
- Instability of the drug in the GI tract. For example, most protein drugs are highly unstable.
- Rapid onset of drug action is desired.
- Ability to immediately stop drug administration is important. For example, most emergency room medications and anesthetics.
- High degree of flexibility in dosage adjustment with or without real-time patient physiological response is needed. For example, emergency medications such as analgesics, anticancer drugs, and fertility medications.

Many drugs are available only as parenteral dosage forms. These include most protein and peptide drugs, some antibiotics, heparin, lidocaine, protamine, glucagon, and many anticancer compounds. Certain drugs, on the other hand, are available both as parenteral and oral dosage forms for different clinical settings. For example, analgesics and antihistamine drugs for patient self-administration may be available as oral tablets, whereas they are also available as infusions and injections for use in an emergency room or hospital settings where rapid onset of drug action may be desired. Similarly, hormonal drugs, such as progestins and antiprogestins, are available as tablets for use in contraception, and are also available as injectable dosage forms for use in fertility therapy.

22.2 PARENTERAL ROUTES OF ADMINISTRATION

Most injections are designed for administration into a vein (*intravenous*, IV), into a muscle (*intramuscular*, IM), into the skin (*intradermal*, ID), or under the skin (*subcutaneous*, SC). Nevertheless, drugs may be administered into almost any organs or area in the body, including the joints (*intraarticular*), joint fluid area (*intrasynovial*), spinal column (*intraspinal*), spinal fluid (*intrathecal*), arteries (*intraarterial*), and in the heart (*intracardiac*). In addition, parenteral routes of administration include dosage forms such as sublingual tablets, transdermal patches, and inhalers—which will not be discussed in this chapter.

22.2.1 Intravenous route

The IV administration provides immediate access of the drug to the systemic circulation, resulting in the rapid onset of drug action. Depending on the rate of drug administration, IV injections could be a bolus or an infusion. A *bolus* means the drug is injected into the vein over a short period of time. A bolus is used to administer a relatively small volume and is often written as *IV push* (IVP). An *infusion* refers to the introduction of larger volumes (100–1000 mL) of the drug over a longer period of time. A *continuous infusion* is used to administer a large volume of drug at a constant rate. *Intermittent infusions* are used to administer a relatively small volume of drug over a specified amount of time at specified intervals.

IV infusion can be administered through peripheral veins, typically in the forearm or the peripherally inserted central catheter. The commonly administered IV infusion products include Lactated Ringers Injection USP; Sodium Chloride Injection USP (0.9%), which replenish fluids and electrolytes; and Dextrose Injection USP (5%), which provides fluid plus nutrition; and various combinations of dextrose and saline. Other solutions of essential amino acids or lipid emulsions are also used as infusions.

22.2.2 Intramuscular route

IM injections of drugs into the striated muscle fibers that lie beneath the SC layer provide effects that are less rapid but generally longer lasting than those obtained from IV administration. Aqueous or oleaginous solutions or suspensions of drugs may be administered intramuscularly. Drugs in aqueous solution are absorbed more rapidly than those in oleaginous preparations or in suspensions. An IM medication is injected deep into a large muscle mass, such as the upper arm, thigh, or buttocks. Up to 2 mL of the drug may be injected into the upper arm and 5 mL in the gluteal medial muscle of each buttock.

Numerous dosage forms are administered through this route of administration, including solutions (aqueous- or oil-based), emulsions (o/w or w/o), suspensions (aqueous- or oil-based), colloidal suspensions, and reconstitutable powders. Slow drug absorption leading to a sustained-release (SR) effect can be achieved with highly insoluble drugs or formulations that are oleaginous or particulate. IM injections are often painful and nonreversible, that is, the administered drug cannot be withdrawn if needed. Antibiotics are often administered by this route.

22.2.3 Subcutaneous route

The SC route is used for small volume injections, typically 1 mL or less. SC injections are administered beneath the surface of the skin, between the dermis and muscle. Medications administered by this route are slowly absorbed and consequently have a slower onset of action than medications given by IV or IM routes. Drugs often given by this route include epinephrine, insulin, heparin, scopolamine, and vaccines. Small injection volume often puts limitations on the drugs that can be administered by this route. For example, high dose drugs that tend to become highly viscous at high concentrations, such as most globular proteins, are usually difficult to formulate as subcutaneous injectable dosage forms.

22.2.4 Other routes

Certain types of injections are intended for specific purposes. For example,

- *Intradermal* administration involves injection just beneath the epidermis, within the dermal or skin layers. The usual site for intradermal injection is the anterior forearm. The volume of solution that can be administered intradermally is limited to 0.1 mL. The onset of action and the rate of absorption of medication from this route are slow. This route is used for diagnostic agents, desensitization, testing for potential allergies, or immunization.
- *Intrathecal* route involves drug administration into the cerebrospinal fluid (CSF). This route is needed if CSF is the desired site of drug

action because most drugs do not reach the CSF from the systemic circulation. Drugs administered intrathecally include antineoplastics, antibiotics, anti-inflammatory, and diagnostic agents.

- An *intraarticular* injection is made into the synovial cavity of a joint, usually to obtain a local therapeutic effect. For example, an intraarticular injection of a corticosteroid provides an anti-inflammatory action in an arthritic joint.
- An *intraarterial* injection is made directly into an artery that has been surgically isolated if it is necessary to deliver a high concentration of drug to a diseased organ, such as kidney, with minimal distribution to other systemic locations.
- An *intraocular* injection is made directly into the eye. For example, an injection into the vitreous humor provides access of drug to the rear regions of the eye, such as the retina, which does not receive high drug concentration on topical administration.

22.2.5 Rate and extent of absorption

The route of administration has a significant impact on the rate and extent of systemic absorption of a drug. Drugs injected intravenously are immediately available in the systemic circulation. Systemic availability of the drug from other sites of injection, such as SC, IM, and intraperitoneal (IP), requires drug absorption. The rate of drug absorption from the site of administration to the systemic circulation depends on the blood flow to the site and drug diffusivity in the tissue. Thus, increase in local blood flow increases the rate of drug absorption. Increasing tissue diffusivity in the extracellular matrix (ECM) of the injection site also increases the rate of drug absorption. Thus, hyaluronidase, which breaks down the ECM, increases drug diffusion and absorption. The extent of drug absorption from a parenteral route could be lower if the drug is metabolized in the tissues.

22.2.6 Factors affecting selection of route

Selection of a parenteral route of administration for a new therapeutic moiety depends on several considerations, such as

- *Desired rate of onset of action*: IV route provides the most rapid onset of action, whereas the SC, IM, and IP routes have slower rate of drug absorption into the systemic circulation. SC route is often preferred for SR dosage forms when slow drug absorption over a prolonged period is desired.
- *Location of drug action*: Intraarterial injections are preferred for localized drug action in an organ, whereas IP route is preferred if drug action is desired in the lymphatic system.

- *Tissue irritability*: Injection of an irritant drug is likely to be more painful by the IM than the SC route due to higher blood flow and sensory innervations in the muscles.
- *Injection volume*: The volume of drug injected is lower for SC than for IM or IV routes. In certain cases, formulation of low volume injections is not feasible, especially for protein drugs with high doses.

22.3 TYPES OF PARENTERAL DOSAGE FORMS

22.3.1 Small-volume parenterals versus large-volume parenterals

Injectable parenteral drug products are available as single or multiuse containers in different container–closure systems and volumes. Small-volume parenterals (SVPs) are available in volumes of less than 1 ml, and up to 50 ml. Large-volume parenterals (LVPs) are usually packaged in volumes up to 1000 mL.

SVPs include both unit-dose and single-dose and multidose containers. Unit dose containers are usually hermetically sealed ampoules that are intended to be discarded after a single injection. Multidose containers, on the other hand, are usually rubber-stoppered and sealed glass vials that are intended for multiple injections. The drug for each injection is withdrawn by inserting the needle through the rubber stopper, which self-seals after the needle is withdrawn.

SVPs for IV injection may not be isotonic because the large volume of blood rapidly dilutes them. However, hypertonic solutions tend to be tissue irritants. The pH of SVPs can also vary from the physiological pH because the blood buffering system rapidly readjusts the pH after a small volume injection. SVPs for single-dose administration may be free of antimicrobial preservatives, but multidose vials usually have the preservatives to ensure sterility over multiple uses over a certain period of time.

22.3.2 Injections versus infusions

Injection and infusion are the predominant methods of parenteral administration. Injection via different routes of administration usually utilizes a SVP. An infusion involves the IV administration of a LVP over a prolonged period of time. Infusions are commonly used for fluid replacement, administration of drugs with a short plasma half-life, and/or dilution of a drug immediately before administration.

22.3.3 Types of formulations

Parenteral products can be formulated as solutions, suspensions, emulsions, or lyophilized products (solid) for reconstitution immediately before use.

22.3.3.1 Solutions

Most injectable products are solutions. Although usually aqueous, they may also contain cosolvent(s), such as glycols (e.g., polyethylene glycol [PEG] or propylene glycol), alcohols (e.g., ethanol), or other nonaqueous solvents (e.g., glycerin). These solutions are usually filtered through a 0.22 µm membrane to achieve sterility. Solutions that do not contain any antimicrobial agents should be terminally sterilized. Autoclaving is the preferred method for terminal sterilization whenever drug solutions can withstand heat. An antimicrobial agent is often added to SVPs that cannot be terminally sterilized.

22.3.3.2 Suspensions

Parenteral suspensions should be easily resuspended and passed through an 18 to 21-guage needle throughout their shelf lives. To achieve these properties, it is necessary to select and carefully maintain particle size distribution, zeta potential, rheological properties, and wettability. Injectable suspensions often consist of the active ingredient suspended in an aqueous vehicle containing an antimicrobial preservative, a surfactant, a suspending agent, a buffer, and/or a salt.

Due to the inherent long-term physical instability of suspensions, parenteral suspension dosage forms are formulated as dry powders for reconstitution immediately before administration. The sterile dry powder could be produced by freeze-drying, sterile crystallization, or by spray-drying. Parenteral suspensions are prepared by mixing dry powders in sterile vehicles immediately before administration. Examples of parenteral suspensions include penicillin G procaine injectable suspension USP and testosterone injectable suspension USP.

Lyophilization or freeze-drying is used to prepare powder cakes for reconstitution immediately before administration. It has inherent advantages over other methods of preparation of dry powders, such as

- Water is removed at low temperatures, avoiding damage to heat-sensitive drugs.
- Freeze-dried product usually has high-specific surface area, facilitating rapid reconstitution.
- Freeze-dried dosage form allows drugs to be filled into vials as a solution, which can then be freeze dried into the final, marketed dosage form. Thus, it does not require powder filling, which is technologically more challenging than filling solutions.

Despite the advantages of freeze-drying, cautions must be taken for lyophilizing proteins, liposomal systems, and vaccines, because they tend to get damaged by freezing, freeze-drying, or both. These damages can often be minimized by using protective agents, such as polyols, polysaccharides, disaccharides and monosaccharide.

22.3.3.3 Emulsions

Because emulsions can cause pyrogenic reactions and hemolysis, and require autoclave sterilization in addition to their inherent physical instability, their use as IV dosage forms has been limited. Total body nutrition is often administered as an IV emulsion to enable coadministration of both water-soluble and water-insoluble nutrients. IV fat emulsion usually contains 10% oil. Fat emulsions yield triglycerides that provide essential fatty acids and calories during total parenteral nutrition of patients who are unable to absorb nutrients through the GI tract. IV lipid emulsions are usually administered in combination with dextrose and amino acids in the aqueous phase.

22.4 QUALITY ATTRIBUTES AND EVALUATION

In addition to meeting the physical and chemical stability attributes of the dosage form being formulated, all parenteral products must be sterile, nonpyrogenic, and free from extraneous insoluble materials. Injectable products are usually required to be tested for the following characteristics:

22.4.1 Sterility

Sterility testing is carried out by incubating the drug product in a conducive environment for microbial growth. Such conducive environment includes appropriate temperature, humidity, and nutrient media. Microbial growth is monitored after a given period of time, determined by standard protocols for each type of microbes.

There are two methods of sterility testing:

- *Direct inoculation*: The drug product is added to the nutrient media and incubated, followed by observation for any microbial growth.
- *Membrane filtration*: Whenever the nature of the drug product is likely to hinder the detectability of microbial growth, the product is filtered through a membrane and the membrane is incubated in nutrient media for observation of microbial growth.

Typically, two culture media are used: (1) trypticase soy broth and (2) fluid thioglycollate medium. The sterility of each sterilized batch of medium is confirmed by incubating a portion of the batch at 20°C–25°C when trypticase soy broth is used as a culture medium, but at 30°C–35°C when fluid thioglycollate medium is used.

22.4.2 Pyrogens

22.4.2.1 Endotoxins, exotoxins, and pyrogens

Bacterial toxins could be endotoxins or exotoxins. Endotoxins are the structural molecules of certain gram-negative bacteria that are recognized by the human immune system, resulting in fever and immune reaction. Exotoxins, on the other hand, are the toxins secreted by microorganisms, such as bacteria, fungi, and algae. When injected, both endotoxins and exotoxins can induce fever, that is, pyrogenic. Such substances are termed as pyrogens. Some of the effects caused by pyrogens in the body are an increase in body temperature, chills, cutaneous vasoconstriction, a decrease in respiration, an increase in arterial blood pressure, nausea and malaise, and severe diarrhea. When compounding a sterile injectable product from nonsterile components, there is always a concern about endotoxin contamination.

22.4.2.2 Endotoxin components and tolerance limits

The endotoxins might originate from the microbes that get destroyed during sterilization. The lipopolysaccharide (LPS) portion of the cell wall that gets released during cell lysis, is the principal constituent of the endogens that cause the pyrogenic response. The LPS can be sloughed off the bacteria, which do not have to be living for the LPS to be pyrogenic. Gram-negative bacteria produce more potent endotoxins than gram-positive bacteria and fungi.

Endotoxin levels higher than 5 endotoxin units (EU)/kg/h can elicit pyrogenic response on IV injection. The maximum permissible levels in the United States (mandated by the U.S. FDA) are 0.2 EU/kg products for intrathecal, 5 EU/kg products for nonintrathecal injectable, and 0.25–0.5 EU/mL for sterile water. Intrathecal injections of 0.2 EU/kg can cause pyrogenic response. One EU is approximately 100 pg (picogram, i.e., 10^{-12} g) of *Escherichia coli* LPS, present in approximately 105 bacteria.

22.4.2.3 Sources

Water is the main source of pyrogens. This is because *Pseudomonas*, a gram-negative bacterium, grows readily in water. Other sources of endotoxins or pyrogens are raw material, processing equipment, and human contamination.

22.4.2.4 Depyrogenation

Endotoxins are not completely removed by filtration and steam sterilization. Endotoxins can be destroyed by dry heat. Thus, when compounding a sterile product from nonsterile starting material that can withstand the heat of 200°C, it should be depyrogenated. If a particle is depyrogenated, it is also sterile.

22.4.2.5 Detection

A preferred method for the detection of pyrogens is the limulus amebocyte lysate (LAL) test. A test sample is incubated with amebocyte lysate from the blood of the horseshoe crab, *Limulus polyphemus*. A pyrogenic substance causes gelling.

22.4.3 Particulate matters

Parenteral solutions are carefully inspected for the presence of any foreign particles, such as glass, fibers, precipitates, and any floating material by microscopy, video imaging, visual inspection, and/or particle counters. Sources of particulate matter include the raw materials, processing and filling equipment, the container, and environmental contamination. Any parenteral product samples found containing particulate matter are discarded. If the quantity and the type of discard exceed a predetermined quality threshold, an investigation is initiated to determine and remediate the cause of the particulate.

22.5 FORMULATION COMPONENTS

Injectable products contain active drugs and inactive ingredients, also called excipients or adjuvants. Adjuvants are excipients that are added to vaccines to help boost body's immune response. The excipients could be vehicles, cosolvents, buffers, preservatives, antioxidants, inert gases, surfactants, complexing agents, and chelating agents.

- Vehicle is the larger continuous phase or the medium in which the formulation is prepared. Water is the most common vehicle, although oil-based injections are formulated in a vegetable oil, such as corn oil, sesame oil, and cottonseed oil or peanut oil.
- Sesame oil is the preferred oil for most of the official injections in oil. Sesame oil has also been used to obtain slow release of fluphenazine esters given intramuscularly. Examples of injectable products formulated with nonaqueous solvents are Diazepam injection USP and Phenytoin Sodium USP.
- Water for injection is prepared by distillation of deionized water or reverse osmosis, and stored in a manner to ensure that it is pure and free from pyrogens.
- Sodium chloride injection USP is a sterile solution of 0.9% w/v sodium chloride in water for injection. It is often used as a vehicle in preparing parenteral solutions and suspensions.
- Cosolvents, such as ethyl alcohol, glycerin, propylene glycol, or PEG, may be used to increase drug solubility in the medium. When cosolvents

are used as vehicles, the preparations should not be diluted with water or precipitation may occur.

- Buffer systems are used to maintain a desired pH of optimum drug solubility and stability.
- Preservative is used in drug products packaged in multiple-dose vials to prevent the growth of microorganisms that may be introduced when the container is pierced for dosing. When preservatives are used, their compatibility with drugs should be carefully examined. For example, benzyl alcohol is incompatible with chloramphenicol sodium succinate, and the parabens and phenol preservatives are incompatible with nitrofurantoin, amphotericin B, and erythromycin.
- Antioxidants are used to prevent oxidative degradation of sensitive drugs. Salts of sulfur dioxide, including bisulfite, metasulfite, and sulfite, are the most common antioxidants used in aqueous parenterals.
- Chelating agents are added to inactivate metals, such as copper, iron, and zinc that generally catalyze oxidative degradation of drug molecules. Ethylenediaminetetraacetic acid (EDTA) in 0.01–0.05 % w/v concentration is a commonly used chelating agent.
- Tonicity modifiers, such as dextrose, sodium chloride, or potassium chloride, are commonly used to achieve isotonicity in a parenteral formulation.
- An isotonic solution has an equal amount of dissolved solute in it compared to the solution it is being introduced into, such as blood for IV injection.
- Typically in humans and most other mammals, the isotonic solution corresponds to 0.9% w/v sodium chloride or 5% w/v dextrose. An isotonic solution has an osmotic pressure close to that of the body fluids. This minimizes patient discomfort and damage to red blood cells.
- A *hypertonic* solution contains a higher concentration of dissolved substances than the red blood cells, which cause the red blood cells to shrink. In contrast, a *hypotonic* solution contains a lower concentration of dissolved substances than the red blood cells, causing the red blood cells to swell and possibly burst.

22.6 STERILIZATION

All parenteral products must be sterile. Sterility is assured by a three-step process: (1) use of sterile starting materials and process equipment; (2) use of special technique in drug product manufacture that minimizes the possibility of contamination from human or extraneous material during manufacture; and (3) sterilization postmanufacture, preferably in final marketed sealed containers. The are several methods of sterilization for parenteral products, including dry heat, steam, filtration, gas, and radiation.

22.6.1 Filtration

Sterilization by filtration is a process that removes, but does not destroy, microorganisms. Filtration is the method of choice for solutions that are unstable to other types of sterilizing processes, for example, thermolabile products. Membrane filters of 0.22 μm pore size are commonly used as sterilizing filters. However, macromolecules, such as proteins and peptides, may be damaged by filtration due to shear stress, leading to alteration in their three-dimensional structure. In certain cases, formulation might affect filter integrity and clogging. In addition, some filters adsorb drug. Therefore, drug interactions with filter materials are carefully investigated before implementing this method of sterilization. Common filter materials include nylon and teflon.

22.6.2 Dry heat sterilization

Dry heat sterilization is the simplest and most economical method of sterilization. However, this method requires higher temperature ($\sim$160°C–250°C) and longer exposure ($\sim$30–180 min) to achieve sterility. A major problem associated with dry heat sterilization is nonuniform distribution of temperature. Furthermore, dry heat sterilization cannot be used with materials that are heat sensitive. It is mainly used for sterilization of glass- and metal-processing equipment.

22.6.3 Steam sterilization (autoclaving)

Steam sterilization is carried out in an autoclave, which is an airtight jacketed chamber designed to maintain a high pressure of saturated hot steam, with the typical temperature of 121°C. Steam sterilization is the method of choice for sterilization of aqueous solutions, glassware, and rubber articles. However, steam sterilization cannot be used with materials that are heat sensitive or nonaqueous formulations.

22.6.4 Radiation sterilization

Radiation sterilization is accomplished by exposure to ultraviolet (UV) light or high-energy ionizing radiation. UV radiation is useful in reducing the number of airborne microorganisms. Microorganisms are often killed by using β-rays, γ-rays, X-rays, and accelerated electron beams. Thermolabile drugs, such as penicillin, streptomycin, thiamine, and riboflavin have been effectively sterilized by ionizing radiation. However, the retail and hospital pharmacists have little opportunity to use radiation sterilization.

REVIEW QUESTIONS

22.1 All parenteral products must be
- A. Sterile
- B. Pyrogen free
- C. Isotonic
- D. Sterile and pyrogen free
- E. All of the above

22.2 An intravenous injection is desirable when
- A. A rapid action is required
- B. An oral administration is ineffective
- C. A prolonged action is required
- D. A and B

22.3 Which of the following statements is TRUE and which one is FALSE?
- a. Systemic drug absorption occurs more rapidly than from oral administration compared to intravenous administration.
- b. All parenteral products must be isotonic.
- c. Filtration cannot be used to sterilize parenteral suspensions.
- d. Buffers are used in parenteral products to stabilize the solution against pH changes.
- e. Sterilization by filtration prevents thermal stress on the product.
- f. During aseptic filtration, the solution is passed through a sterile filter of 2 µm pore size.
- g. Filtration cannot be used to sterilize parenteral suspensions.
- h. Heat and radiation sterilization methods are intended to eliminate viable microorganism from the final products.

FURTHER READING

Allen LV Jr., Popovich NC, and Ansel HC (2005) *Ansel's Pharmaceutical Dosage Forms and Drug Delivery Systems,* 8th ed., Philadelphia, PA: Lippincott Williams & Wilkins.

Borchert SJ, Abe A, Aldrich DS, Fox LE, and White RD (1986) Particulate matter in parenteral products: A review. *J Parenter Sci Technol* 40: 212–241.

Boylan JC and Nail SL (2202) Parenteral products. In Banker GS and Rhodes CT (Eds.) *Modern Pharmaceutics*, 4th ed., New York: Marcel & Dekker, pp. 381–414.

Ford JL (1988) Parenteral products. In Aulton ME (Ed.) *Pharmaceutics: The Science of Dosage Form Design*, New York: Churchill Livingstone, pp. 359–380.

Rojanasakul R and Malanga CJ (2004) Parenteral routes of delivery. In Ghosh TK and Jasti BR (Eds.) *Theory and Practice of Contemporary Pharmaceutics*, Boca Raton, FL: CRC Press, pp. 387–419.

Thoma LA (2005) Sterile products. In Gourley DR (Ed.) *APhA's Complete Review for Pharmacy*, 3rd ed., New York: Castle Connolly Graduate Medical Publishing, pp. 83–106.

Turco SJ (1994). *Sterile Dosage Forms: Their Preparation and Clinical Application*, 4th ed., Philadelphia, PA: Lippincott Williams & Wilkins.

Chapter 23

Semisolid dosage forms

LEARNING OBJECTIVES

On completion of this chapter, the students should be able to

1. Define and differentiate ointments, creams, gels, lotions, pastes, and jellies.
2. Describe different types of ointment bases.
3. Differentiate between hydrogels and organogels.

23.1 INTRODUCTION

Dosage forms that are in a plastic (i.e., change shape upon application of force), malleable semisolid state at room temperature include ointments, creams, gels, pastes, lotions, jellies, and foams. These semisolid preparations may contain dissolved and/or suspended drugs. These preparations are designed to stay in physical contact with the surface of application for a reasonable duration of time, before they are inadvertently or intentionally removed or washed off. Their semisolid state and plastic rheological behavior is designed to aid their application to the target surface as a film. Many cosmetics would be considered semisolid.

Most of the semisolid formulations are used topically to deliver drugs to/through the skin. They can also be used for topical or systemic drug action in/through the eye, nose, ear, vagina, rectum, buccal tissue, or the urethral membrane. In addition, unmedicated semisolid formulations are frequently used as protectants or lubricants. Topical applications can be designed for either local effects or systemic absorption. For example, a topical dermatological product is designed to deliver a drug into the skin for treating dermal disorders. A transdermal product is designed to deliver drugs through the skin (percutaneous absorption) to the underlying tissue or the systemic circulation.

The major classes of agents that are used topically include corticosteroids, antifungals, acne products, antibiotics, emollients, antiseptics, and local anesthetics. Topical agents are used as protectives, adsorbents, emollients, and cleansing agents.

23.2 OINTMENTS

Ointments are semisolid preparations that incorporate a lipid or hydrophobic excipient and are intended for external application to the skin or other mucosal membranes. An ointment usually contains <20% water and other volatile ingredients, such as ethanol, and >50% hydrocarbons, waxes, or polyols. Ointments are designed to soften or melt at body temperature, spread easily, and have a smooth, nongritty feel. Ointments are typically used as (1) emollients to make the skin more pliable, (2) protective barriers to prevent harmful substances from coming in contact with the skin, and (3) vehicles for hydrophobic drugs.

23.2.1 Types of ointment bases

An ointment base forms the body of any ointment. Ointment bases are classified into four general groups: (1) hydrocarbon bases, (2) absorption bases, (3) emulsion or water-removable bases, and (4) water-soluble bases (Table 23.1).

23.2.1.1 Hydrocarbon bases

Oily or oleaginous bases include hydrocarbons derived from petroleum, which are called hydrocarbon bases. These bases are anhydrous and insoluble in water. These bases are used for their emollient effect (to hydrate the skin) and as an occlusive dressing. They cannot absorb or contain water. Thus, they can be protective to water labile drugs, such as bacitracin and tetracycline. However, they are greasy and not water washable. Thus, they can stain clothing and are generally not preferred. Oily- or fatty-base ointments may have hard, soft, or liquid paraffin bases, or mixtures of these, in such proportions as will render an ointment to be of suitable consistency.

Common hydrocarbon bases include the following:

- *Petrolatum*: It is used as a base for water-insoluble ingredients. Yellow petrolatum or petrolatum jelly, for example, Vaseline®, melts at 38°C–60°C. Decolored petrolatum is known as white petrolatum. Petrolatum forms an occlusive film on the skin and absorbs less than 5% water under normal conditions. Wax can be incorporated to stiffen the base. For example, yellow ointment contains 5% w/w yellow wax and 95% w/w petrolatum.

Table 23.1 Various types of ointment bases

Types of ointment bases	Characteristics	Applications	Examples
Hydrocarbon/ oleaginous	• Anhydrous • Water insoluble • Not water washable • Form occlusive film on skin	• Incorporation of hydrophobic drugs	• Petrolatum • Wax • Synthetic esters, for example, glycerol monostearate
Absorption	• w/o emulsions or oleaginous bases that allow incorporation of aqueous solution to form w/o emulsions • Not easily water washable	• Emollients	• Anhydrous: hydrophilic petrolatum and anhydrous lanolin • w/o emulsion: lanolin and cold cream
Emulsion	• o/w emulsions • Leave a hydrophobic film on the surface of the skin when water evaporates	• Drug carriers • Foundation for makeup	• Hydrophilic ointment • Vanishing cream
Water soluble	• Hydrophilic polymer (e.g., PEG) mixture	• Drug carriers	• PEG 400 + PEG 4000 in 40:60 ratio • Propylene glycol + ethanol with 2% w/w HPC

- *Liquid petrolatum*, also known as mineral oil, is a mixture of refined saturated hydrocarbons obtained from petroleum that are liquid at room temperature. It is used as a levigating agent to incorporate lipophilic solids into ointments.
- Synthetic esters are used as constituents of oleaginous bases. These esters include glycerol monostearate, isopropyl myristate, isopropyl palmitate, butyl stearate, and butyl palmitate.
- Long-chain alcohols, such as cetyl alcohol and stearyl alcohol, are sometimes also incorporated in oleaginous bases. In addition, lanolin derivatives, such as lanolin oil and hydrogenated lanolin, are sometimes used.
- Plastibase® (ER Squibb & Co., Princeton, NJ) is a commercially available polyethylene-base gelled mineral oil. It is useful for the extemporaneous preparation of ointments by cold incorporation of drugs, thus being suitable for heat-labile compounds.

23.2.1.2 Absorption bases

Absorption bases contain an oleaginous material and a water-in-oil (w/o) emulsifier so that they can absorb water to form or expand w/o emulsions. Absorption bases are useful as emollients, although they do not provide the degree of occlusion afforded by the oleaginous bases. Emollients are preparations that soften and soothe the skin. These preparations may be used to reduce the dryness and scaling of skin. However, they are greasy because the external phase of the emulsion is oily. Absorption bases are not easily removed from the skin with water.

Absorption bases are of two types:

1. Anhydrous bases that permit the incorporation of aqueous solutions, resulting in the formation of w/o emulsions. These absorption bases are anhydrous vehicles composed of a hydrocarbon base and an additive. The hydrocarbon base could be, for example, hydrophilic petrolatum and anhydrous lanolin. The additive is a miscible substance with polar groups (a surfactant), which functions as a w/o emulsifier. For example, cholesterol, lanosterol and other sterols, acetylated sterols, or the partial esters of polyhydric alcohols, such as monostearate or monooleate, can serve as additives.

2. Bases that are already w/o emulsions (emulsion bases) and permit the incorporation of small additional quantities of aqueous solutions. For example, lanolin and cold cream.
   a. *Lanolin* is a w/o emulsion that can form an occlusive film on the skin and serve as an emollient, effectively preventing epidermal water loss. It retards but does not completely inhibit, transepidermal water loss. It can restore the water in the skin to a normal level of 10%–30%. Lanolin is a pale yellow substance obtained from sheep wool. It is chemically a wax, consisting of high molecular weight alcohols (e.g., sterols) and fatty acids. Lanolin can absorb twice its own weight of water. It is self-emulsifying and produces stable w/o emulsions. Lanolin is used to help prevent drying and chapping of the skin.
   b. *Cold cream* is a semisolid white w/o emulsion prepared with cetyl ester wax, white wax, mineral oil, sodium borate, and purified water. Sodium borate combines with free fatty acids present in the waxes to form sodium salts of fatty acids (soaps) that act as emulsifiers. Cold cream is employed as an emollient and ointment base. For example, Eucerin cream is a w/o emulsion of petrolatum, mineral oil, mineral wax, wool wax, alcohol, and bronopol. It contains urea as the active ingredient and is used to help rehydrate dry, scaly skin.

23.2.1.3 Emulsion or water-removable bases and creams

Emulsion or water-removable bases are oil-in-water (o/w) emulsions. As these emulsion bases have an aqueous external phase, they are water washable or water removable. They are non/less greasy and occlusive than oleaginous bases. They can be diluted with water and have a better cosmetic appearance. *Highly viscous emulsion bases are commonly referred to as creams.* These represent the most commonly used type of ointment base. The majority of dermatologic drug products are formulated in an emulsion or cream base.

An emulsion base has three component parts: (a) an internal oil phase, which is typically made of petrolatum and/or liquid petrolatum together with cetyl or stearyl alcohol; (b) an emulsifier; and (c) an aqueous phase. Drugs can be included in one of these phases before forming the emulsion or can be added to the formed emulsion.

Emulsion bases are of the following types:

- *Hydrophilic ointment* is an o/w emulsion that uses sodium lauryl sulfate as an emulsifying agent. It is readily miscible with water and is easily removed from the skin. A typical composition of hydrophilic ointment is listed in Table 23.2. In addition to these basic components, this base may also contain preservatives to control microbial growth. The preservative(s) could be methylparaben, propylparaben, benzyl alcohol, sorbic acid, or quaternary ammonium compounds. The aqueous phase contains the water-soluble components of the emulsion system, together with any additional stabilizers, antioxidants, and buffers that may be necessary for drug stability and pH control.
- *Vanishing cream* is an o/w emulsion that contains a large percentage of water as well as a humectant (e.g., sorbitol, glycerin, or propylene glycol)

Table 23.2 A typical composition of hydrophilic ointment

S. No.	Component	Function	Content (% w/w)
1.	White petrolatum	Oil base of o/w emulsion	25
2.	Stearyl alcohol	Hydrophobic, oil soluble component, used as an emollient, emulsifier, and thickener	25
3.	Propylene glycol	Hydrophilic viscous liquid used in the aqueous phase to increase viscosity	12
4.	Sodium lauryl sulfate	Surfactant/emulsifier	1
5.	Water	Aqueous base of o/w emulsion	37

Table 23.3 A typical composition of vanishing cream

S. No.	Component	Function	Content (% w/w)
I.	Stearyl alcohol	Oil base of o/w emulsion	14
2.	Other hydrophobic ingredients, for example, cetyl esters wax, glyceryl monostearate, and polyoxyethylene stearyl ether	Emollient, emulsifier, and/ or thickener	10
3.	Surfactant	Emulsifier	1
4.	Water	Aqueous base of o/w emulsion	65
5.	Sorbitol	Water-soluble component, used as a humectant and thickener	10

that retards surface evaporation of water. A typical composition of vanishing cream is listed in Table 23.3. It is a cosmetic product that is colorless when applied and is used as a foundation for powder or as a cleansing or moisturizing cream. The hydrophobic stearyl alcohol component in the formula helps to form a thin film when the water evaporates.

23.2.1.4 Water-soluble bases

Water-soluble bases absorb water to the point of solubility. They are water washable and may be anhydrous, or contain some water. Water-soluble bases are made of carbowax or polyethylene glycol (PEG) as the base. They are oil/lipid free and non/less occlusive. However, they may absorb water from the skin, thus dehydrating the skin, and may hinder percutaneous absorption.

PEGs are water soluble, nonvolatile, stable, and do not support the growth of mold. PEGs are polymers of oxyethylene units with different molecular weights. The number at the end of PEGs indicates their average molecular weight. Their melting point increases with increasing molecular weight. Thus, PEGs with a molecular weight $\leq 400-600$ are liquid at room temperature; PEGs with a molecular weight 800–2,000 are waxy or semisolid; and PEGs with a molecular weight >2,000 are solid at room temperature.

A typical composition of water-soluble base is listed in Table 23.4. The ointment is a blend of water-soluble PEG that forms a semisolid base. The base of PEGs alone is highly water soluble and does not allow incorporation of more than 5% w/w water or aqueous solution to make an ointment. If greater quantities of water or aqueous component need to be added, a

Table 23.4 A typical composition of water-soluble base

S. No.	Component	Function	Content (% w/w)
Base with low (< 5% w/w water incorporation capacity)			
1.	PEG 400	Nonaqueous, hydrophilic base that is liquid at room temperature	40
2.	PEG 4,000	Nonaqueous, hydrophilic base that is solid at room temperature	60
Base with higher (> 5% w/w water incorporation capacity)			
1.	PEG 400	Nonaqueous, hydrophilic base that is liquid at room temperature	47.5
2.	PEG 4,000	Nonaqueous, hydrophilic base that is solid at room temperature	47.5
3.	Cetyl alcohol	Hydrophobic component	5.0

modified composition, such as with the addition of 5% w/w hydrophobic component may be used. A water-soluble base can solubilize water-soluble drugs and some water-insoluble drugs. The water-insoluble drugs are solubilized by the cosolvent action of the nonaqueous hydrophilic polymers present in the base. These bases are compatible with a wide variety of drugs.

Another water-soluble base is the ointment prepared with propylene glycol and ethanol, which form a clear gel when mixed with 2% w/w hydroxypropyl cellulose (HPC). This base is commonly used as a dermatologic vehicle.

23.2.2 Selection of ointment bases

An ointment base is chosen depending on

- The solubility characteristics of the drug and the desired rate of drug release. For example, hydrophilic drug incorporated in an o/w base would be released immediately, whereas incorporation in a w/o emulsion would lead to slower drug release.
- Whether the final product is intended for drug absorption by the skin (percutaneous drug absorption) or not (topical application).
- Typical properties of various ointment bases, such as water washability and tendency for skin occlusion.
- Intended usage of the ointment, for example, a cosmetic use would require due attention to customer convenience factors such as water washability and nonstaining on the clothing. On the other hand, usage in a clinical setting, such as occlusive barrier on wounds that would be bandaged, might not require such considerations.

23.2.3 Methods of incorporation of drugs into ointment bases

In addition to the active drug, ingredients in ointment preparations can include oleaginous components, aqueous components, emulsifying agents, stiffeners, penetration enhancers, preservatives, and antioxidants. Oleaginous ointments may be prepared by levigation and fusion.

- *Levigation* involves dispersing and/or grinding an insoluble drug into small particles while wet. Mixing of a base and other components over an ointment slab using a spatula can carry it out. Components such as liquid petrolatum serve as levigating agents by promoting the wetting of powders for incorporation into bases. Hydrophobic ointments and w/o emulsions and suspensions are typically prepared by levigation process to incorporate a powder and/or a small quantity of water or hydrophilic component into an oil base.
- *Fusion* process involves melting components (such as paraffin, stearyl alcohol, white wax, yellow wax, and high molecular weight PEGs) together to form a homogeneous solution. Fusion method is used when the base contains solids that have higher melting points (e.g., waxes, cetyl alcohol, or glyceryl monostearate). This process is employed only when the components are stable at fusion temperatures. Hydrophilic o/w emulsions (such as water-removable ointments and creams) are typically prepared by the fusion process. The hydrophobic components are melted together and added to the aqueous phase/water-soluble components containing an emulsifying agent with constant mixing until the mixture congeals.

Normally, drug substances are in fine powered forms before being dispersed in the vehicle. Levigation of powders into a small portion of base may be facilitated by the use of a melted base or a small quantity of compatible levigation aid, such as mineral oil or glycerin. Water-soluble salts of drugs are incorporated by dissolving them in a small volume of water and incorporating the aqueous solution into a compatible base.

23.3 CREAMS

Creams are semisolid dosage forms containing one or more drug substances dissolved or dispersed in a suitable o/w or w/o emulsion base. Creams are more fluid compared to other semisolid dosage forms, such as ointments and pastes. Creams have a whitish, creamy appearance, which is a result of scattering of light from their dispersed phases, such as oil globules. This distinguishes them from simple ointments, which are translucent.

Creams based on o/w emulsions are useful as water-washable bases, whereas w/o emulsions have emollient and cleansing action. As described earlier, an o/w cream with high water content is also known as a *vanishing cream*. Upon rubbing this cream on the skin, the external/continuous aqueous phase evaporates, leading to increased concentration of a water-soluble drug in the oily film that adheres to the skin. This increase in the concentration gradient of the drug across the stratum corneum promotes percutaneous absorption.

Creams based on w/o emulsions, such as *cold cream*, are useful as softening and cleansing agents. The name, cold cream, refers to the cooling sensation associated with the slow evaporation of the dispersed aqueous phase. A cold cream, typically, also contains scents and is used to remove makeup. Other common cold cream components include mineral oil, jojoba oil, lanolin, glycerin, alcohol, borax, and beeswax in addition to antimicrobial preservatives such as methylparaben and propylparaben.

The use of creams as drug delivery systems is associated with good patient acceptance. In addition to the general requirements for semisolid dosage forms, incorporation of drug in a cream requires that the drug should

- Be soluble in desired concentration.
- Have relatively wide therapeutic window since accurate dosing is difficult.
- Not crystallize upon evaporation of water.

23.4 GELS AND JELLIES

23.4.1 Gels

Gels are semisolid systems consisting of dispersions of small or large molecules in an aqueous liquid vehicle, which has been thickened with a gelling agent. Gels can be a single phase or a biphasic system.

- Single-phase gels use high molecular weight hydrophilic polymers as gelling agents. Examples of such polymers include carbomers (cross-linked acrylic acid polymers). These gels are considered to be one-phase systems because no definite boundaries exist between the dispersed macromolecules and the liquid.
- Biphasic gels could contain a gelatinous, cross-linked precipitate of one substance in the aqueous phase. For example, magma or milk of magnesia consists of a gelatinous precipitate of magnesium hydroxide.

Gelling agents in single-phase gels could be (a) synthetic macromolecules, for example, carbomer 934; (b) cellulose derivatives, such as carboxymethylcellulose; and (c) natural gums, for example, tragacanth. Carbomers are high molecular weight water-soluble polymers of acrylic acid cross-linked with allyl ethers of sucrose and/or pentaerythritol. Their viscosity depends on their polymeric composition. They are used as gelling agents at concentrations of 0.5%–2% w/w in water.

In addition to the gelling agent and water, gels may also contain a drug substance, cosolvents (such as alcohol and/or propylene glycol), antimicrobial preservatives (such as methylparaben and propylparaben, or chlorhexidine gluconate), and stabilizers (such as the chelating agent edetate disodium).

Gels can be classified based on their gelling agent as inorganic and organic. Inorganic gels use precipitates of inorganic salts, such as magnesium hydroxide, as gelling agents, whereas organic gels generally use a carbon-based hydrophilic polymer. Inorganic gels are generally two-phase systems, whereas organic gels are generally single-phase systems.

Based on the solvent phase of the gels, they may be classified as hydrogels or organogels. Hydrogels contain water as the main continuous phase solvent, whereas organogels may contain an organic liquid. Hydrogels contain significant amounts of water but remain as water insoluble.

The diffusion rate of a drug from a gel depends on the physical structure of the polymer network and its chemical nature. If the gel is highly hydrated, diffusion occurs through the pores. In gels of lower hydration, the drug dissolves in the polymer and is transported between the chains. Polymer cross-linking increases the hydrophobicity of a gel and reduces the diffusion rate of the drug.

Gels typically display non-Newtonian flow characteristics, that is, they show a nonlinear relationship between shear stress and strain rate, which can also be time dependent. Depending on their flow characteristics, gels may be shear thinning (pseudoplastic, i.e., viscosity decreases and flow increases on agitation), shear thickening (dilatant, i.e., viscosity increases and flow decreases on agitation), or thixotropic (e.g., requires decreasing stress to maintain a constant strain rate over time; or, in other words, viscosity decreases and flow increases over time under the same agitation rate). Inorganic gels consist of floccules of small particles, as found in aluminum hydroxide gel or bentonite magma. Such gels may be thixotropic, displaying higher viscosity and a semisolid state on standing and becoming low viscosity liquids on agitation.

23.4.2 Jellies

Jellies are semisolid gels of intertwining hydrophilic polymers that form a structurally coherent matrix and contain a high proportion of liquid, usually water, hydrogen bonded and associated with the hydrophilic polymer chains. Adding a thickening agent to an aqueous solution of a drug substance forms a jelly. The thickening agent could be natural gums, such as alginates, tragacanth, and pectin or synthetic derivatives of natural substances such as sodium carboxymethyl cellulose (CMC) and methyl cellulose (MC). The resultant product is usually a clear and uniform semisolid. Jellies, being aqueous, are prone to bacterial growth. Thus, antimicrobials are usually added as preservatives.

23.5 LOTIONS

A lotion is a low- to medium-viscosity medicated or nonmedicated topical preparation, intended for application to unbroken skin. Lotions are usually applied to external skin with bare hands, a clean cloth, cotton wool, or gauze. Solid particles incorporated in lotions should be in a finely divided state to avoid grittiness.

Most lotions are o/w emulsions, but w/o lotions are also formulated. The key components of a lotion are the aqueous and oily phases, an emulsifying agent to prevent separation of these two phases, and, if used, the drug substance or substances. A wide variety of other ingredients such as fragrances, glycerol, petroleum jelly, dyes, preservatives, and stabilizing agents are commonly added to lotions for improved organoleptic and preservation characteristics.

Lotions can be used for the topical delivery of medications such as antibiotics, antiseptics, antifungals, corticosteroids, antiacne agents, and soothing/protective agents (such as calamine). Aside from medical use and skin care, lotions are often used as accessories to aid massage, masturbation, or sex. Noncomedogenic lotions, products that do not block the natural pores of the skin, are recommended for use on pimples or acne-prone skin. These lotions are also termed as nonocclusive. Thus, they may reduce acne and/or reduce the incidence of pimples.

The same drug substance can be formulated into a lotion, cream, and ointment. Creams are the most convenient of the three but are inappropriate for application to regions of hairy skin such as the scalp; whereas a lotion is less viscous and may be readily applied to these areas. Many medicated shampoos are, in fact, lotions. Lotions also have an advantage that they may be spread thinly compared to a cream or ointment and may economically cover a larger area of skin.

23.6 PASTES

Pastes are semisolid dosage forms that contain a large proportion of solid component. They differ from ointments in their consistency, as they contain larger amounts of solids and consequently are thicker and stiffer. Pastes can be made either of fatty bases, such as petrolatum and hydrophilic petrolatum, or of aqueous gels, such as celluloses. Pastes may contain one or more drug substances intended for topical application.

Pastes are well adsorbed on the skin. Pastes can absorb watery solutions so that they can be used around oozing lesions. Pastes can be easily removed from skin and are water washable, which is an important consideration when they are applied on traumatized skin.

Pastes that contain hydrophobic components can be water impermeable and prevent dehydration. Examples of pastes include the commonly used toothpastes and zinc oxide paste. Toothpaste contains an abrasive solid for cleansing purposes and sometimes also includes a fluoride salt, such as sodium fluoride or stannous fluoride, as a medicament. Zinc oxide paste is typically composed of 25% w/w zinc oxide, 25% w/w starch, and 50% w/w white petrolatum.

Pastes can be formed from several bases, such as gelatin, starch, tragacanth, polyethylene glycol, pectin, or cellulose derivatives.

23.7 FOAMS

Stable foams are semisolid preparations that entrap air upon application to form a lightweight, flexible matrix with a large surface area of the liquid. Foams are sometimes used for topical application to areas that are otherwise difficult to reach, such as hairy scalp, or on sensitive skin, such as in acne.

For example, Luxiq® aerosol foam is a topical anti-inflammatory corticosteroid formulation that contains 0.12% w/w betamethasone valerate in a thermolabile hydroethanolic foam vehicle. This foam vehicle consists of ethanol (~60%), cetyl alcohol, stearic acid, polysorbate 80, potassium citrate, propylene glycol, purified water, and cetyl alcohol. It is pressurized with a hydrocarbon (propane/butane) propellant. The foam melts upon contact with warm skin and is intended for application to the scalp. Similarly, clobetasol 0.05% foam is an anti-inflammatory corticosteroid formulation intended for application to the scalp. Clindamycin phosphate foam 1% is a topical antibiotic preparation for use in acne.

Foams typically contain a hydrocarbon propellant in the packaging container to pressurize the drug solution. The drug is dissolved in a low boiling point vehicle, such as the one containing a high proportion of ethanol, which also has a surfactant and a base to dissolve the drug. The vehicle may also contain preservatives and buffering agents. Evaporation of ethanol upon aerosolization leads to expansion of liquid droplets and formation of foam by entrapment of air.

23.8 MANUFACTURING PROCESSES

23.8.1 Laboratory scale

Preparation of semisolid dosage forms on a laboratory or compounding pharmacy scale can be accomplished using one or more of the following techniques and principles:

- Geometric mixing using a spatula on a plate. This allows uniform incorporation of a small quantity of an ingredient into a large quantity of the other ingredient(s). Geometric mixing involves mixing a small quantity ingredient with the same volumetric or weight quantity of the larger quantity ingredient, followed by repeating this procedure with the small component mix until all the large quantity ingredient has been incorporated.
- Powder communition or particle size reduction by grinding in a pestle and mortar.
- Levigation by grinding the powder in a small quantity of suitable levigation aid in a pestle and mortar, followed by geometric mixing with the base using a spatula on a plate.
- Fusion by melting the components together on a water bath.
- Using pestle and mortar to prepare an emulsion concentrate using lower quantity of the external or continuous phase, followed by dilution of the emulsion concentrate to volume.

23.8.2 Industrial scale

Manufacture of semisolid dosage forms on a large scale presents challenges with respect to the inherent viscosity of the formulation, non-Newtonian flow characteristics, possibility of air entrapment, heat distribution within a vessel, variation in the volume of liquid components with changes in operating or ambient temperature, and the energy requirement for efficient mixing of viscous fluids.

On a pilot plant to a production scale, semisolid formulations are manufactured using one or more of the following equipment and techniques:

1. Electrically operated propeller mixer in a suitable mixing vessel.
2. Temperature control using jacketed mixing vessel, with the jacket having a supply of hot or cold water or steam. The mixing vessel also often has a mixer that sweeps close to the wall to prevent overheating and allow mixing of semisolid mass, which otherwise has low convective mixing rate.
3. Homogenization using a homogenizer mixer or a colloid mill.
4. Use of proportioning pump to allow simultaneous blending of phases.
5. Transfer of the semisolid material from one unit operation to another, or to the packaging line, in a container, gravity-facilitated, if feasible, or pumping through a tube.

The choice of technique depends on rheological properties of the formulation in addition to plant design and feasibility of equipment.

23.9 ANALYSIS OF SEMISOLID DOSAGE FORMS

The following quality attributes of semisolid dosage forms of drugs, such as ointments and creams, are evaluated:

1. Physical stability, in terms of nonseparation of emulsion phases, when applicable, and homogeneity of appearance/color.
2. Drug identity, purity, content, and uniformity of content. The content of drug per unit mass of the dosage form and impurities/related substances of the drug substance indicate its potency and purity.
3. Drug release rate using an *in vitro* test.
4. Viscosity of the formulation.
5. Minimum fill in the container and deliverable volume or doses.
6. Although these dosage forms are not required to be sterile, the microbial content of certain bacterial species, such as *Staphylococcus aureus* and *Pseudomonas aeruginosa*, is controlled.

REVIEW QUESTIONS

23.1 The following are semisolid topical preparations:
   A. Ointments
   B. Creams
   C. Lotions
   D. All of the above
23.2 The main difference between creams and ointments is
   A. Creams are thicker than ointments.
   B. Ointments are thicker than creams.
   C. Creams are emulsions, whereas ointments are suspensions.
   D. None of the above.
23.3 The presence of petrolatum-like bases renders them:
   A. Occlusive
   B. Greasy
   C. Water washable
   D. Occlusive and greasy
   E. All of the above
23.4 Select none, one, or more correct answers from the following for the subset of questions:
   A. Cold cream
   B. Vanishing cream
   C. Vaseline
   D. Calamine lotion
   E. Lanolin
   F. Hydrophilic ointment

G. Jelly

    i. Which of these are o/w emulsions?

    ii. Which of these are w/o emulsions?

    iii. Which of these are suspensions?

    iv. Which of these are solutions?

23.5 Select the one most appropriate answer from the following for the subset of questions:

A. An o/w emulsion

B. A w/o emulsion

C. A suspension

D. An aqueous solution

E. An oily solution

F. Mixture of PEG 400 and PEG 4,000

    i. Which of these will lead to a lasting cooling feeling upon application to skin?

    ii. Which of these would lead to a water-soluble drug deposition on the skin in a concentrated state?

    iii. Which of these is likely to be gritty?

    iv. Which of these is likely to be not water washable?

23.6 Which of the following ointment bases would be considered as the most suitable for the subset of application questions that follow?

A. A hydrocarbon/oleaginous base

B. An absorption base

C. An emulsion base

D. A water-soluble base

    i. Which base should be selected when water washability is the key requirement?

    ii. Which base should be selected for formulating a hydrophobic drug for transcutaneous absorption?

    iii. Which base is likely to be the most occlusive on the skin?

    iv. Which base is the most likely to cause skin dryness?

    v. Which base can be expanded with water as an external phase?

23.7 Which of the following are non-Newtonian flow types?

A. Pseudoplastic

B. Dilatant

C. Thixotropic

D. All of the above

FURTHER READING

Allen LV Jr. (2002) *The Art, Science, and Technology of Pharmaceutical Compounding*, 2nd ed., Washington, DC: American Pharmaceutical Association.

Allen LV, Popovich NG, and Ansel HC (2005) *Ansel's Pharmaceutical Dosage Forms and Drug Delivery Systems*, 8th ed., New York: Lippincott Williams & Wilkins.

Idson B and Zazarus J (1976) Semisolids. In Lachman L, Lieberman HA, and Kanig JL (Eds.) *The Theory and Practice of Industrial Pharmacy*, 2nd ed., Philadelphia, PA: Lea and Febiger.

Shah VP, Behl CR, Flynn GL, Higuchi WI, and Schaefer H (1992) Principles and criteria in the development and optimization of topical therapeutic products. *Pharm Res* 9: 1107–1112.

Singh SK, Nagpal K, and Saini S (2014) Semi-solid dosage forms, Chapter 11. In Dash AK, Singh S, and Tolman J (Eds.) *Pharmaceutics: Basic Principles and Application to Pharmacy Practice*. San Diego, CA: Elsevier.

Inserts, implants, and devices

24.1 INTRODUCTION

Inserts, implants, and devices represent pharmaceutical interventions in healthy and/or disease states that may be used to improve health and/or promote quality of life. Inserts, as the name implies, are drug delivery systems that are designed for insertion into one or the other body cavity, such as vagina, rectum, buccal cavity, or the cul-de-sac of the eye, in the patient. Suppositories are solid dosage forms that are used to administer drugs through the rectum or vagina. Implants, on the other hand, are designed for surgical placement inside the body, such as in the subcutaneous (SC) tissue, breast, penis, heart, bones, teeth, eye, or the ear.

Devices are recognized as relatively sophisticated drug delivery systems intended for a specific application, such as transdermal drug delivery, intra-uterine devices (IUDs), ventricular assist devices, and insulin pumps and pens. Transdermal patches are used for drug delivery across the skin. Aerosols and inhalation drug delivery devices are used for pulmonary drug delivery.

Inserts, implants, and devices may or may not be loaded with drug(s). Drug containing inserts, implants, and devices are used to deliver drugs for localized or systemic effects. Sometimes, the rate of drug release is controlled. In such cases, the drug may embed into biodegradable or nonbiodegradable materials forming a uniform matrix that allows slow release of the drug.

24.2 INSERTS

24.2.1 Ocular inserts

Drug administration to the eye commonly involves the use of eye drops, which can be formulated as a drug solution or suspension, or as semi-solid ointments. Tear turnover and drainage can quickly eliminate the administered drug, reducing the amount of drug absorbed into the eye. Less than 10% of a topically applied dose is usually absorbed into the eye. A part of the dose also passes into the nasal sinus and is absorbed through the highly vascular nasal mucosa into the bloodstream. This may result in unwanted systemic side effects. For example, topical administration of latanoprost (Xalatan®) eye drops, a prostaglandin PGF2α analogue used to treat glaucoma, can result in chest tightness in some patients. Similarly, the use of topical α-blockers, such as timolol, for glaucoma treatment can lead to systemic side effects, such as hypotension and bradycardia.

These safety concerns are addressed by the use of inserts that stay on the cornea for long duration of time. Inserts can be biodegradable or nonbiodegradable. Inserts can also be designed for immediate or controlled drug release. Drug containing inserts are placed on the cornea, sometimes hidden below the eyelid, by the patient. These inserts are designed to maintain drug concentration in the precorneal fluids at relatively steady levels over a prolonged period of time and allow drug diffusion across the cornea. Ocular inserts are less affected by nasolacrimal drainage and tear flow than conventional dosage forms. They can provide slow drug release and longer residence times in the conjunctival cul-de-sac. Ocular inserts (e.g., medicated contact lenses, collagen shields, and minidiscs) also reduce systemic absorption of topically applied drugs as a result of decreased drainage into the nasal cavity. In addition, contact lenses are becoming increasingly useful as potential drug delivery devices by presoaking them in drug solutions. The use of contact lenses can simultaneously correct vision and release drug.

The ophthalmic inserts can be insoluble or soluble. Insoluble inserts may or may not be erodible/biodegradable. Insoluble inserts are further classified as diffusional, osmotic, and contact lens. Biodegradable inserts consist of degradable polymers such as polyvinyl alcohol (PVA), hydroxypropylcellulose (HPC), polyvinylpyrrolidone (PVP), and hyaluronic acid.

Nonbiodegradable inserts are prepared from insoluble materials such as ethylene–vinyl acetate copolymers and styrene–isoprene–styrene block copolymers. Ocular inserts are exemplified by the following:

- Ocusert® consists of a drug reservoir (e.g., pilocarpine HCl in an alginate gel) sandwiched on both sides by a release-controlling membrane, which is made of ethylene–vinyl acetate copolymer. This system is encased in the periphery by a white ring, which allows positioning of the system in the eye (Figure 24.1). Ocusert provides slow release of pilocarpine HCl for the control of increased intraocular pressure in glaucoma.
- Lacrisert® is a soluble insert composed of HPC. It is useful in the treatment of dry eye syndrome. The device is placed in the lower fornix (below the lower eyelid), where it slowly dissolves over 6–8 h to stabilize and thicken the tear film.

24.2.2 Suppositories

A suppository is a solid dosage form designed for easy insertion into body orifices of rectum, vagina, or urethra. Once inserted, the suppository base melts, softens, or gets dissolved at body temperature, distributing its medication to the tissues of the region. Suppositories are used for local or systemic effects. Suppositories are also used to administer drugs to infants and small children, to severely debilitated patients, to geriatric patients who cannot take medications orally, and to those for whom both the oral and the parenteral routes may not be suitable. Vaginal or rectal suppositories are sometimes also termed as pharmaceutical pessaries (singular, pessary).

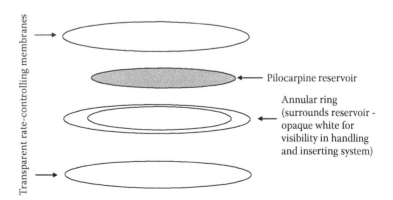

Figure 24.1 An illustration of design elements of an ocular insert device.

24.2.2.1 Types of suppositories

Based on their route of administration, suppositories can be rectal, vaginal, or urethral.

- *Rectal suppositories* are cylindrical or conical in shape. Suppositories containing a moisturizer or a vasoconstrictor are often used to relieve the pain, irritation, itching, and inflammation associated with hemorrhoids. Glycerin or bisacodyl suppositories are used as a laxative. They may also be used for systemic administration of drugs, such as opiate analgesics.

  Rectal suppositories are often intended for systemic drug action. Examples of such rectal suppositories include Thorazine® (chlorpromazine) and Phenergan® (promethazine HCl). The suppository dissolves at body temperature and gradually spreads over the lining of the lower bowel (rectum), from where it is absorbed into the bloodstream. The medicine is easily absorbed from the rectum, because there is a rich supply of blood vessels in this area. Addition of surfactants may increase the wetting and spreading of the molten mass, which tends to increase the extent of drug absorption. Surfactants, such as polyoxyethylene sorbitan monostearate and sodium lauryl sulfate, may also increase the permeability of the rectal mucosal membrane resulting in significant increase in drug absorption.
- *Vaginal suppositories* are available in ovoid, globular, or other shapes. They are employed as contraceptives, antiseptics in feminine hygiene, treatment of local vaginal infections (e.g., candidiasis), or for systemic delivery of hormones (e.g., progesterone), with high local concentration, especially in the uterus.
- *Urethral suppositories* are sometimes used for the treatment of severe erectile dysfunction. For example, alprostadil pellets that contain the vasodilator prostaglandin E1 is marketed under the trade name MUSE® (medicated urethral suppository for erection).

24.2.2.2 Suppository bases

Most suppositories consist of a drug substance dissolved or dispersed in a matrix, termed as a suppository base. The suppository base has a marked influence on the release of active constituents. Suppository bases can be either oleaginous or water-soluble bases.

- *Oleaginous bases* are exemplified by theobroma oil or cocoa butter and synthetic triglycerides, such as hydrogenated vegetable oils.
- Cocoa butter is a hard, amorphous solid at ambient temperature (15°C–25°C) and melts at 30°C–35°C into a bland, nonirritating oil.

This may necessitate refrigeration of suppositories in warm regions. Addition of certain drugs can change (lower) the melting point. Melting point may also be lowered if cocoa butter is heated above 35°C at which point it undergoes polymorphic transition into a lower melting metastable morph. These considerations limit the manufacturability with cocoa butter bases. Synthetic triglyceride bases, such as Fattibase®, Wecobee®, Suppocire®, Wtepsol®, Hydrokote®, or Dehydag®, do not exhibit polymorphism.

- *Water-soluble or water-miscible suppository bases* are exemplified by glycerinated gelatin and polyethylene glycols (PEGs). PEG suppository bases do not melt at body temperature but rather dissolve slowly in the body's fluids. Melting point of PEG is a function of its molecular weight. Higher the molecular weight, higher the melting point. Typically, a combination of lower and higher melting PEGs is used to make a suppository base.
- Factors affecting the bioavailability of suppository dosage forms include the retention time of the suppository in the cavity, the size and shape of the suppository, and its melting point. Drug release and the onset of drug action also depend on the liquefaction of the suppository base, dissolution of the drug in the local fluids, and drug diffusion across the mucosal layer.

24.2.2.3 Manufacturing process and formulation considerations

Drugs are usually dissolved or dispersed in a suitable suppository base. Other excipients that may be used include surfactants and preservatives. Hand rolling, compression molding, or fusion molding are the three processes commonly used to manufacture suppositories.

- *Hand rolling* is typically employed for cocoa butter-based suppositories. The base is triturated with the drug in a mortar. The mass is formed into a ball in the palm of the hands. The ball is rolled on a flat board or pill tile to form an elongated cylinder. The cylinder is cut into appropriate number of pieces, one end of each of which is rolled to produce a conical shape.
- *Compression molding* requires forcing a fixed quantity of suppository formulation into a special compression mold. The quantity of the formulation is calculated based on the prior determination of the volume of molds and the density of the formulation.
- *Fusion molding* involves melting the suppository base, followed by dissolving or dispersing the drug in the base, and pouring the molten mixture into a metallic suppository mold—where the mixture is allowed to congeal into shape.

Formulation considerations for suppository manufacturing include a careful consideration of density, because suppository molds are volume filled, whereas the formulation composition is weight based. The possible variation in drug loading that can result from the manufacturing process and potential variability in drug absorption due to loss with body fluids indicates that low therapeutic index medicaments may not be suitable for delivery via a suppository. Quality control of suppositories involves testing the melting range, liquefaction or softening time, physical integrity or breaking test, drug release rate testing, and stability determination for the physical (appearance and odor) and chemical (pH and drug degradation) attributes.

24.2.3 Vaginal rings

Vaginal rings, also known as V-rings or intravaginal rings, are *doughnut-shaped* polymeric drug delivery devices designed to provide controlled release of drugs to the vagina. They are manually placed in vagina and are held in place by the anatomy, usually close to the cervix.

- Nuvaring® is a contraceptive vaginal ring that contains etonorgestrel (progestogen) and ethinyl estradiol (estrogen). It is made using poly(ethylene-co-vinyl acetate) polymer and provides slow release of hormones over a period of 3 weeks.
- Estring® is a low-dose estradiol-releasing ring for treating vaginal atrophy.
- Femring® is a low-dose estradiol acetate-containing ring. It is used for vaginal atrophy and hot flashes. It can provide drug release over a period of 3 months.

24.3 IMPLANTS

An implant may be defined as a material that is securely placed (inserted or grafted) *into* the body. Most of the implants are surgically placed inside the body. A drug-containing implant is usually a sterile, solid dosage form prepared by compression or melting for drug delivery at a desired rate over a prolonged period of time.

24.3.1 Types of implants based on drug release mechanism

Drug-containing implants may be classified into following types:

1. Diffusion-controlled implants
2. Osmotic minipumps

These implants differ in the mechanism of control of drug release.

24.3.1.1 Diffusion-controlled implants

The rate of drug delivery from polymeric systems may be controlled by (a) drug diffusion and dissolution through an insoluble matrix and/or (b) the use of a rate-controlling membrane. Devices that use a rate-controlling membrane achieve controlled rate of drug delivery through diffusion across the membrane. These membrane systems contain a reservoir, which is in contact with the inner surface of the rate-controlling membrane. The reservoir contains the drug in a liquid, gel, colloid, semisolid, or solid matrix. For drug delivery systems that utilize diffusion and dissolution through a matrix for control of drug release rate, the matrix could be composed of hydrophilic or hydrophobic polymers, or a combination of the two to obtain optimum drug release. Depending on the nature of polymers used, the matrix implants could be biodegradable or nonbiodegradable. Drug release from biodegradable implants is a function of both the rate of drug diffusion and the rate of polymer degradation.

Kinetics of drug release from an implant is determined by its mechanism. Usually, a reservoir system gives a zero-order profile because the rate-controlling step is the process through which the drug must diffuse from a concentrated solution in the core. A matrix system usually provides a square root of time profile of drug release reflecting an erosion-limited drug absorption. If the rate of polymer degradation is slow compared to the rate of drug diffusion, drug release kinetics obtained with a biodegradable implant can become diffusion limited from a concentrated core and thus similar to nonbiodegradable implants.

Drug-containing implants are exemplified by the following:

- Zoladex® is an implant, which contains goserelin acetate dispersed in a matrix consisting of D,L-lactic and glycolic acid copolymer. Goserelin acetate is a potent synthetic decapeptide analogue of luteinizing hormone-releasing hormone (LHRH) and is a gonadotropin-releasing hormone (GnRH) agonist. Zoladex is implanted subcutaneously into the upper abdominal wall. It is used for palliative treatment of advanced carcinoma of the prostate, endometriosis, and advanced breast cancer.
- Vantas® implant contains histrelin, which is a synthetic analogue of GnRH agonist. It is a diffusion-controlled device that provides drug release for up to 12 months. It is used for treating prostate cancer by decreasing the production of certain hormones, which reduces testosterone levels.

24.3.1.2 Osmotic minipumps

In contrast to rate-controlling membranes (that use a porous membrane), osmotic minipumps use a membrane impermeable to the drug with well-defined openings for drug release. The opening may be a laser-drilled orifice on a tablet coating, for example. The core of these devices contains the drug alone or together with an osmotic agent, usually a salt. The membrane is permeable to solvent (water) but impermeable to solute (drug). Such a membrane is called semipermeable membrane. Penetration of water inside the device through the semipermeable membrane allows dissolution of salt (osmotic agent) and creation of high osmotic pressure inside the membrane (highly concentrated salt solution). This osmotic pressure facilitates the release of the drug through the orifice. When in contact with body fluids, the osmotic agent draws in water through the semipermeable membrane because of the osmotic pressure gradient and forms a saturated solution inside the device. The flow of saturated solution of the drug out of the device through the delivery orifice relieves the pressure inside. This process continues at a constant rate until the entire solid agent has been dissolved. The drug release rate is usually unaffected by the pH of the environment and essentially remains constant as long as the osmotic gradient remains constant. Thus, the kinetics of drug release is governed by the salt concentration and dosage form volume—which impact osmotic pressure, and the orifice diameter.

Polymers, such as cellulose acetate, ethylcellulose, polyurethane, polyvinyl chloride, and PVA, are used to prepare semipermeable membranes to regulate the osmotic permeation of water. A water-insoluble polymer impregnated with a small quantity of a water-soluble polymer allows the formation of micropores that allow solvent diffusion across the membrane, thus making a semipermeable membrane.

Oral osmotic pump is one of the commonly used devices. It is composed of a core tablet surrounded by a semipermeable coating. The coating membrane has a 0.3–4 mm diameter hole, which is produced by a laser beam, for drug exit. This system requires only osmotic pressure to be effective. The drug release rate is dependent on the surface area and nature of the membrane, and the diameter of the hole. When the dosage form comes in contact with water, water is imbibed, and the drug is released from the orifice at a controlled rate driven by the resultant osmotic pressure of the core.

Drug-containing implants are exemplified by the following:

- Alzet® miniosmotic pump (illustrated in Figure 24.2a) permits easy manipulation of drug release rate over a range of time periods (from 1 day to 6 weeks). These miniature infusion pumps are designed for continuous dosing of unrestrained laboratory animals.
- Osmotic minipump for human use is exemplified by Viadur®, which uses the DUROS® technology (illustrated in Figure 24.2b). Nondegradable, osmotically driven system is intended to enable delivery of small drugs, peptides, proteins, and DNA for systemic

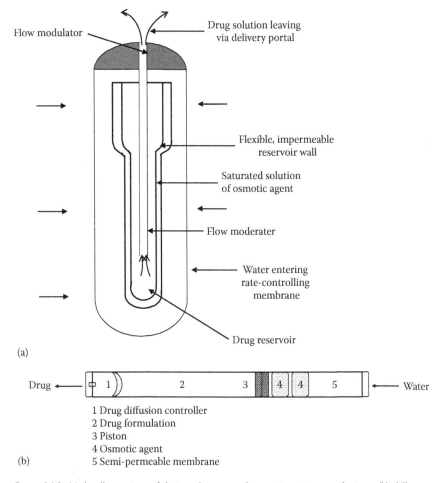

Figure 24.2 (a) An illustration of design elements of osmotic minipump devices. (b) different components of an osmotic pump.

or tissue-specific therapy. These implants are used for continuous therapy for up to 1 year. Viadur is a luprolide acetate-containing implant, once yearly, for the palliative treatment of advanced prostate cancer.

24.3.2 Types of implants based on clinical use

Implants may also be classified based on the organ in which the device is implanted. These implants can be drug-containing or nondrug-containing devices. Drugs may also be incorporated into or on the surface of devices used in routine clinical medicine, such as cardiac stents.

24.3.2.1 Cardiac implants

Cardiac implants are devices that are surgically placed in the heart for restoring and assisting regular heart function. For example, polymeric closure devices, such as Amplatzer® and CardioSEAL®, are used to close a hole or an opening between the right and the left side of the heart to correct birth defects located in the interatrial septum. The use of cardiac pacemakers and artificial heart valves is well known.

Drug loading into conventional cardiac implants can improve their clinical outcome. For example, *drug eluting stents* are used for the prevention of in-stent restenosis (fibrosis and thrombus-induced blockade of the stented artery). These stents contain drugs, such as sirolimus (Cypher®), paclitaxel (Taxus®), zotarolimus (Endeavour®), and everolimus (Xience V®). In addition, iontophoretic cardiac drug delivery system allows cardiac electrical pulse-induced drug release for the treatment of arrhythmias.

24.3.2.2 Dental implants

Dental implants, such as artificial tooth, fillings, and dentures, are fairly common in the practice of dentistry. Antibiotics and analgesic drugs are commonly used in medicated dental implants. Prophylactic antibiotic treatment is frequently practiced in dental implant placement surgery to minimize chances of infection at the implant site. Local release of the antibiotic from a polymeric matrix close to the implant has greater efficacy while minimizing systemic side effects.

Atridox® is a FDA-approved product designed for controlled delivery of the antibiotic doxycycline for the treatment of periodontal disease. When injected into the periodontal cavity, the formulation sets, forming a drug delivery depot that delivers the antibiotic to the cavity.

24.3.2.3 Urological and penal implants

Urological implants, such as urethral and ureteral stents and catheters, can be used for local drug therapy. Penile implants are surgically placed inside the penis for male infections and impotence.

Surface deposition of ionic and organic components (encrustation) can affect drug release from these devices. Encrustation is promoted by high urinary pH, which is common with urinary infection, and due to their prolonged contact with urine. Encrustation also leads to higher risk of infection. Glycosaminoglycans can act as crystal growth inhibitors. Therefore, surface coating of the glycosaminoglycan heparin on the stent has been proposed to minimize encrustation.

Infection of the implants is a relatively common problem that requires expensive and invasive replacement of the prosthesis. This problem can be overcome by the use of antibiotic-releasing implants. An antibiotic eluting implant, Inhibizone®, was introduced to minimize the risk of infection by providing a controlled release of antibiotics minocycline and rifampin in the microenvironment surrounding the implant.

24.3.2.4 Breast implants

Cosmetic breast enhancement implants are fairly common. Pain management with the implant involves the use of oral medication, including narcotic analgesics. Intraoperative administration of analgesics into the implant pocket facilitates early postoperative recovery and reduces the incidence of pain in patients undergoing surgery.

Capsular fibrosis is one of the most serious complications associated with silicone breast implants. Fibrosis is mediated by the transforming growth factor-β, which is inhibited by the drug halofuginone lactate. Surface modification of silicone breast implants with halofuginone lactate reduces the risk of fibrosis.

24.3.2.5 Ophthalmic implants

Local drug delivery by the use of ophthalmic implants provides higher local drug concentration and improves patient response compared to intravenous (IV) therapy. Vitrasert® is a ganciclovir intravitreal implant for the treatment of patients with AIDS-related cytomegalovirus (CMV) retinitis. Vitrasert contains ganciclovir embedded in a polymer matrix, which releases the drug over a period of 5–8 months.

Retisert® is a controlled-release intravitreal implant of the corticosteroid antiinflammatory agent fluocinolone acetonide used for the treatment of chronic noninfectious uveitis—a leading cause of blindness. This insert contains 0.59 mg drug, which is released over a period of about 30 months.

Intravitreal-controlled drug delivery can be achieved with the use of implantable devices. For example, I-vation® intravitreal implant is made of a nonferrous metal alloy, which is placed inside the vitreous humor of the eye using a 25 gauge needle. Drugs are delivered by coating onto its surface. The helical shape of this implant maximizes drug loading and release. The use of this device helps reduce frequent intraocular injections. Duration of drug release from this device could range from 6 months to 2 years.

24.3.2.6 Dermal or tissue implants

Drug implants in the SC region or within certain tissues are used for controlled/prolonged drug release for local or systemic action. These implants are exemplified by the following:

- SC contraceptive implants provide slow drug release over a prolonged period of time. Most of these implants contain a progestogen, such as levonorgestrel, etonorgestrel, nestorone, elcometrine, or nomegestrol acetate. The polymers used in these inserts are exemplified by ethylvin-ylacetate and polydimethyl/polymethyl-vinyl-siloxanes. These implants have a steroid load of 50–216 mg, are placed under the skin, and release the hormone at 30–100 µg/day over a period of 6 months to 7 years.

- Gliadel® wafer, which contains the antitumor agent carmustine in a biodegradable polyanhydride copolymer, is used for the treatment of malignant glioma (brain tumor) and recurrent glioblastoma multiforme by implantation in or close to the tumor site. Each wafer contains 7.7 mg carmustine and is 1 mm thick and 1.45 cm in diameter. It is used as an adjunct to surgery and radiation.
- Vantas® SC implant contains histrelin acetate and is indicated for palliative treatment of advanced prostate cancer by suppressing testosterone levels while requiring less frequent administration than other LHRH agonists. It releases 50 mg of drug over a period of 12 months.

24.4 DEVICES

Devices are specialized pharmaceutical drug delivery systems in which the desired drug delivery and targeting are achieved with the aid of the packaging container. Pulmonary delivery devices, transdermal devices, and IUDs exemplify them.

24.4.1 Inhaler devices for pulmonary drug delivery

Delivery devices play a major role in the efficiency of pulmonary drug delivery. Drug particles or solution are aerosolized and inhaled with the breath for delivery to the lung. An aerosol is a colloidal dispersion of a liquid or a solid in a gas. Aerosol device is a pressurized or breath-actuated dosage forms designed to deliver the containing solution or suspension of drug(s) to the lung by forming an aerosol at the time of administration. A pressurized aerosol device contains a liquid propellant in a pressurizable container, a valve that allows the pressurized product to be expelled from the container when the actuator is pressed, and a dip tube that conveys the formulation from the bottom of the container to the valve assembly. The propellant is a liquefied gas that expands readily upon release of pressure to provide the driving force for the delivery of the contents.

Formulation factors affecting pulmonary drug delivery include particle size and size distribution, shape, and density. Generally, particles in the size range of 1–5 µm are considered respirable (have significant lung deposition). Particle shape and density determine the proportion of inhaled particles that deposit in deep lung alveoli versus major airways. *Device factors* affecting pulmonary drug delivery include efficiency of spray, size and size uniformity of sprayed droplets, location of spray generation in the context of patient's anatomy, width of spray zone, and the speed of the aerosol. *User* or *patient factors* impacting drug delivery to the lung include coordination of inspiration time with device actuation and the strength, quantity, and consistency of air intake for breath-actuated devices.

Formulation considerations important for the development of aerosol dosage forms include uniformity of drug content, especially in the case of powders and suspension; particle size and size distribution, shape, and density (for powders and suspensions); flow through the nozzle, compatibility with the container components, emitted dose, and fine particle dose or fine particle fraction in the emitted dose.

The most commonly used devices for pulmonary drug delivery include nebulizers for liquid formulations, metered-dose inhalers (MDIs), and dry powder inhalers (DPIs). Figure 24.3 shows schematic of these inhalation devices. These devices vary as much in their sophistication as they do in their effectiveness. Each type of device has its own advantages, disadvantages, and limitations. The choice of device depends on the drug (such as solubility and stability in aqueous medium), the formulation (e.g., dry powder or aqueous solution), pathophysiology of the lungs (e.g., lung capacity), and the status of the patient (e.g., in patient or ambulatory self-use).

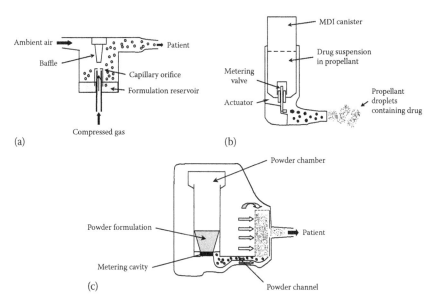

Figure 24.3 An illustration of design elements of inhalation devices: (a) nebulizer, (b) metered-dose inhaler, and (c) dry powder inhaler.

24.4.1.1 Nebulizers

Nebulizers convert aqueous solutions or micronized suspension of a drug into an aerosol for inhalation using compressed air (usually through a pump) and device design. The patient inhales normally while the aerosolized product is delivered to the patient through a mouthpiece adapter. Nebulizers require minimal patient coordination of breathing but are cumbersome, nonportable, and time consuming to use.

There are two main types of nebulizer:

- Air-jet (high velocity air stream-aided dispersion) nebulizers
- Ultrasonic (ultrasonic energy-aided dispersion) nebulizers

Air-jet nebulizers rely upon *compressed gas* to aerosolize a solution that is then available for inhalation by the patient. Ultrasonic nebulizers utilize ultrasonic vibrations for aerosolization. Both air-jet and ultrasonic nebulizers produce aerosol at a constant rate regardless of the respiration cycle. This leads to loss of approximately two thirds of the aerosol during the expiration and breath-holding phases.

Two improved nebulizers, the *breath-enhanced* nebulizers and *dosimetric* nebulizers, overcome this limitation. These inhalers direct the patient's inhaled air within the nebulizer to enhance aerosol volume during the inhalation phase and release aerosol exclusively during the inhalation phase, respectively.

All nebulizers require a solution or suspension-based formulation, which places stringent demands on the solubility and stability of the drug in aqueous media. For protein and peptide drugs, the stability of proteins and peptides upon shearing can pose additional limitation. Nebulization exerts high shear stress on these macromolecules, which can lead to their denaturation. This problem gets exacerbated because 99% of the droplets generated are recycled back into the reservoir to be nebulized during the next dosing cycle. Furthermore, the physical properties of drug solutions (e.g., ionic strength, viscosity, osmolarity, pH, and surface tension) may change with dosing and may affect the nebulization efficiency. The droplets produced by nebulizers are heterogeneous in size, which results in very poor drug delivery to the lower respiratory tract. They often require several minutes of use to administer the desired dose of medicine. Nebulizers are more effective for drug delivery to the major pathways of the respiratory track, such as the trachea, than deep lung alveoli. Thus, adrenergic agonists such as albuterol and steroids such as budesonide are commonly delivered by nebulizers. Recombinant human DNase, Dornase alfa (Pulmozyme® by Genentech in South San Francisco, CA) uses nebulizer for protein delivery to the respiratory tract. Pulmozyme reduces viscosity of the airway secretions by cleaving the extracellular fibrillar aggregates of DNA from autolyzing neutrophils in cystic fibrosis.

New nebulizer devices, such as the AERx (Aradigm, Hayward, CA) and Respimat (Boehringer, Germany), generate an aerosol mechanically and reduce the shear forces on the drug. In addition, *vibrating mesh* technologies such as AeroDose (Aerogen Inc., Mountain View, CA) have been used successfully to deliver proteins to the lungs.

24.4.1.2 Metered-dose inhalers

MDIs generate aerosol for inhalation by expelling a measured dose of pressurized liquid propellant containing drug via an orifice. They are portable, easy to use, and the most commonly used inhalation aerosol devices today. A typical MDI comprises of a canister, metering valve, actuator, spacer, and holding chamber. In addition, they may also have dose counters and content indicators. During MDI manufacturing, more aerosol formulation than claimed is commonly added, which is sufficient for additional 20–30 sprays. However, the last doses from the container are inconsistent and unpredictable. Therefore, the dose counter feature allows patients to track the number of actuations and avoids using the product beyond the recommended number of doses.

MDIs utilize propellants, such as chlorofluorocarbons (CFC) and hydrofluoroalkanes (HFAs), to emit the drug solution through a nozzle. A metering chamber within the valve measures individual doses volumetrically. High velocity of the generated aerosol spray causes substantial oropharyngeal deposition by impaction, which results in poor drug delivery to the lung. This can be avoided by adding a *spacer device*, which reduces aerosol velocity. The spacer also overcomes difficulties in the coordination of inhalation and actuation, especially for pediatric patients, resulting in improved dosing reproducibility.

MDI delivery efficiency depends on the patient's inspiratory flow rate, breathing pattern, and hand–mouth coordination. Increase in tidal volume (volume of air moved into or out of the lungs during normal breathing) and decrease in respiratory frequency increase peripheral drug deposition in the lung. Most patients need to be trained for proper use of the MDI.

24.4.1.3 Dry powder inhalers

DPIs are one of the most popular methods of protein delivery to the lungs. DPIs generate aerosols by drawing air through the loose dry powder of a drug formulation. These are usually capsule-based devices, wherein the dry powder formulation is filled and provided in a hard gelatin capsule. The device pierces the capsule and provides inspiratory air pathway that would fluidize the capsule and would enable release of its contents through the piercing. The drug particles form an aerosol in the inspired air upon breathing by the patient. DPIs are generally easier to use, compared to MDIs. However, DPIs require a rapid rate of inhalation to provide necessary energy for aerosolization, which may be difficult for pediatric or distressed patients, and in certain disease states such as asthma or chronic obstructive pulmonary disease (COPD). DPIs range from unit dose systems, employing only the patient's breath to generate the aerosol, to *multiple-dosing reservoir devices*, which actively impart energy to the powder bed to introduce aerosol particles into the patient's respiratory airflow. For stability reasons, *unit-dose* devices are preferred for protein delivery. Figure 24.3c shows the schematic design of a noncapsule-based DPI (Novolizer®).

Lung deposition of drug particles varies among different DPIs. DPIs are complex systems, and their performance depends on effective powder deagglomeration. Drugs in low doses are often combined with excipients that provide drug-binding sites on surface while serving as bulking agents. These help with uniformity of drug content and consistency of emitted dose. Carrier particles, such as lactose, are commonly added to avoid drug agglomeration due to cohesive forces among the micronized drug particles. When air is directed through the powder, turbulent airflow detaches small drug particles from the carrier particles. The smaller particle size drug migrates to the lung alveoli, whereas the larger particle size excipient deposits in the back of the throat and in the major respiratory airways. Thus, optimized performance of both the device and the formulation is critical to ensuring high and consistent lung deposition.

Most of the therapeutic dry powders for DPIs are currently made with particles of small aerodynamic particle diameter (e.g., 90% particles below 5 μm) and density of 1 ± 0.5 g/cm^3. Increased porosity of particles, such as when produced by spray drying, helps with deep lung penetration by improving the aerodynamic performance. Uniformity of particle size distribution, shape, and density are important for achieving efficient pulmonary delivery.

Drugs administered by inhalation are mostly intended to have a direct effect on the lungs. Inhaled drugs play a very prominent role in the treatment of asthma. This route has significant advantages over oral or parenteral administration, because lipid-soluble compounds are rapidly absorbed across the respiratory tract epithelium. Bronchodilators and corticosteroids are commonly used for treating asthma and COPD. Azmacort® (triamcinolone acetamide), Ventolin® HFA (albuterol sulfate), and Serevent® (salmeterol) are examples of commercially available aerosols for the treatment of asthma.

Proteins, oligonucleotides, and genes demonstrate poor oral bioavailability due to the harsh environment of the gastrointestinal tract and their hydrophilicity, large size, and rapid metabolism. In such cases, the pulmonary route enables higher rates of passage into systemic circulation than oral administration.

24.4.2 Intrauterine devices

IUDs, as the name suggests, are the devices that are placed in the uterus. These devices are mostly used for contraception by preventing the fertilization of the egg by the sperm, inhibiting tubular transport, and/or preventing the implantation of the blastocyst into the uterine endometrium. The hormone containing devices can be used for other hormonal effects such as in menorrhagia.

IUDs can be (a) inert, (b) copper based, or (c) hormone containing. Most IUDs are T shaped so that they are held in place in the uterus by the arms of the T shape. The copper surface of copper-based IUDs allows the release of copper in the uterine mucosal microenvironment, which aids contraception. A side effect of copper-based IUDs is increased uterine bleeding. The hormone-based IUDs mostly contain a progestogen. The use of these devices can provide much lower systemic and high local progestogen levels.

- Progestasert® device is designed for implantation into the uterine cavity, where it releases 65 µg progesterone per day to provide contraception for 1 year.
- Mirena® device, also known as the LNG-20 IUS (intrauterine system), contains levonorgestrel. It is designed to provide an initial drug release rate of 20 µg/day and is used to provide contraception for up to 5 years.

24.4.3 Subcutaneous devices

Parenteral drug administration is indicated for several drugs, especially the new biotechnology-based drug products such as monoclonal antibodies, and protein and peptide therapeutic agents. During the initial clinical development of these therapies, IV route of drug administration is adopted to allow dose flexibility and the ability to closely monitor and control drug exposure. However, the IV route of drug administration is not preferred for commercial use of these drugs by patients because IV drug administration requires the intervention of a health care professional and the use of a health care facility. In contrast, SC administration is often preferred because it can allow for patient self-administration at home and utilizes smaller size needle, which causes less pain at the injection site and is more patient friendly. The development of SC dosage forms of drugs has, therefore, gathered significant momentum in the recent years.

At the same time, SC delivery of several drugs is limited by the injection volume that can be administered in the SC space (usually about 1 mL, maximum 1.5 mL), the required dose of the drug, and the solubility of the drug in the injection vehicle (for a soluble drug product). Based on the dose and the solubility calculations, the required dosing volume of a drug can sometimes exceed the usual injectable volume in the SC space. In these cases, the use of a device for SC drug delivery can facilitate SC delivery of a drug that would otherwise not be possible.

The devices that can be used for SC delivery of a drug could be a syringe pump in an inpatient setting or a patch pump in an outpatient setting.

A syringe pump is a mechanical device that pumps the drug product through a syringe at a low enough rate to match the absorption of the fluid from the injected SC space. This allows higher volume of drug to be administered subcutaneously over a prolonged period of time. The use of a syringe pump, however, requires the patient site of drug administration to be immobilized for the duration of drug administration. Also, the operation and calibration of the syringe pump usually require the expertise of a health care professional. Thus, although syringe pumps can generally be used in an inpatient setting in a hospital and during clinical trials, their use is limited for outpatient clinical use.

SC devices for outpatient clinical use are generally smaller, battery-operated pumps that can be attached to the abdominal cavity and worn under the clothing. These allow patient self-administration and the patient can have mobility during drug administration. Such devices are exemplified by Roche's MyDose® device and West's SmartDose® electronic wearable injector. The improved quality of life and patient convenience associated with these devices provides better patient compliance and satisfaction with the therapy.

REVIEW QUESTIONS

24.1 Suppositories are solid dosage forms intended for insertion into body orifices and are used for
   A. Rectal and vaginal drug delivery
   B. Oral drug delivery
   C. Nasal drug delivery
   D. Skeletal drug delivery
   E. All of the above
24.2 The rate of drug release from an aerosol depends on
   A. The power of a compressed or liquefied gas to expel the container
   B. Particle size of the formulation
   C. The type of drug
   D. The type of container
   E. All of the above
24.3 Which of the following statements is not true about aerosols?
   A. Dry powders can be dispensed.
   B. Contamination is avoided.
   C. Emulsions cannot be dispensed.
   D. More patient compliance compared to injectables.
   E. None of the above
24.4 What are Ocuserts? Mention a marketed drug product using this dosage form.
24.5 Enlist factors that affect drug bioavailability from a suppository. What are the different kinds of suppository bases?

24.6 Identify clinical considerations important to the development of all implantable drug delivery systems. What are the differences in the principle of drug delivery between an osmotic minipump and a diffusion-controlled implant?

FURTHER READING

Akala EO (2004) Oral controlled release solid dosage forms. In Ghosh TK and Jasti BR (Eds.) *Theory and Practice of Contemporary Pharmaceutics*, Boca Raton, FL: CRC Press, pp. 333–366.

Bensinger R, Shin DH, Kass MA, Podos SM, and Becker B (1976) Pilocarpine ocular inserts. *Invest Ophthalmol* **15**: 1008–1010.

Cheng K and Mahato RI (2007) Biopharmaceutical challenges in pulmonary delivery of proteins and peptides. In Meibohm B (Ed.) *Pharmacokinetics and Pharmacodynamics of Biotech Drugs: Principles and Case Studies in Drug Development*, New York: John Wiley and Sons, pp. 209–242.

Niven R (1993) Delivery of biopharmaceutics by inhalation aerosols. *Pharm Technol* **17**: 72–81.

Owens DR, Grimley J, and Kirkpatrick P (2006) Inhaled human insulin. *Nat Rev Drug Discov* **5**: 371–372.

Patton J and Platz RM (1994) Pulmonary delivery of peptides and proteins for systemic action. *J Control Release* **28**: 79–85.

Scheindlin S (2004) Transdermal drug delivery: Past, present and future. *Mol Interv* **4**: 308–3122.

Chapter 25

Protein and peptide drug delivery

LEARNING OBJECTIVES

On completion of this chapter, the students should be able to

1. Describe the differences between the primary, secondary, and tertiary structure of proteins.
2. List types of physical instability of proteins and peptides.
3. Describe major pathways of protein degradation, their chemistry, and the corresponding stabilization strategies.
4. Identify key components of protein formulations.

25.1 INTRODUCTION

The use of therapeutic proteins to replace or supplement endogenous protein molecules has been a long established treatment for diseases such as diabetes, growth hormone deficiency, and hemophilia. The use of proteins and peptides as pharmaceutical products has increased significantly in recent years with the commercialization of monoclonal antibody (mAb)-based therapeutics such as immune-oncology agents Opdivo® and Yervoy®, and the antibody–drug conjugates (ADCs) such as Kadcyla® and Adcentris®.

Protein and peptide drugs are either natural in origin or synthetically produced using recombinant DNA technology or from transgenic animals. Recombinant DNA technology has allowed the large-scale production and biological characterization of several therapeutic proteins, including granulocyte macrophage colony-stimulating factor (GM-CSF), erythropoietin (EPO), interleukins, insulin-like growth factor-1 (IGF-1), human factors VIII and IX (involved in blood coagulation and useful for hemophilia), mAbs, and tissue plasminogen activator (t-PA). Table 25.1 lists some of the FDA-approved marketed products of therapeutic proteins.

Table 25.1 List of some commercial products of therapeutic proteins

Protein type	Protein names	Description	Indication
Polyclonal antibodies (lyophilized)	Sandoglobulin	Human immune globulin for intravenous administration. It is a polyvalent antibody product that contains all IgG antibodies, which regularly occur in the donor population in a concentrated form. It is prepared by fractionation of the plasma of volunteer donors. It is a lyophilized preparation.	Primary immune deficiencies such as severe combined immunodeficiency (SCID), common variable immunodeficiency, X-linked agammaglobulinemia, and immune thrombocytopenic purpura (ITP)
Polyclonal antibodies (solution)	Gammagard	Concentrated human IgG antibodies similar to that of normal plasma. It is manufactured from pooled human plasma from donors. It is available as a 10% ready-to-use sterile liquid formulation.	Primary immunodeficiencies
Monoclonal antibodies	Rituximab	Monoclonal antibody that recognizes specific proteins on the surface of some lymphoma cells and triggers body's immune system.	Combination therapy for tumors such as non-Hodgkin's lymphoma (NHL) and chronic lymphocytic leukemia (CLL), and autoimmune diseases such as rheumatoid arthritis
Radioactively tagged antibodies	Ibritumomab tiuxetan	Monoclonal antibody radioimmunotherapy. It is prepared from monoclonal mouse IgG1 antibody ibritumomab and uses the chelator tiuxetan, which has a radioactive isotope (yttrium-90 or indium-111).	B-cell non-Hodgkin's lymphoma

(Continued)

Table 25.1 (Continued) List of some commercial products of therapeutic proteins

Protein type	Protein names	Description	Indication
Mouse antibodies	Tositumomab	IgG2 anti-CD20 monoclonal antibody of murine origin. Also available as radioactively labeled ^{131}I-tositumomab, which has covalently bound iodine-131.	Follicular lymphoma
Chimeric antibodies	Infliximab	Monoclonal antibody against TNFα.	Psoriasis, Crohn's disease, ankylosing spondylitis, psoriatic arthritis, rheumatoid arthritis, and ulcerative colitis
Humanized antibodies	Daclizumab	Monoclonal antibody against the α subunit of IL-2 receptor on T-cells.	To prevent the rejection in organ transplantation, especially in kidney transplantation
Fusion proteins	Abatacept	Fusion protein that is composed of human Ig fused to the extracellular domain of cytotoxic T-lymphocyte-associated protein 4 (CTLA-4), a molecule involved in T-cell stimulation.	Rheumatoid arthritis
Physiological proteins	Erythropoietin	Glycoprotein hormone that controls erythropoiesis (red blood cell production). It is available as a lyophilized preparation.	Kidney diseases, anemia, and cancer
Antibody–drug conjugates	Kadcyla®	Human immunoglobulin conjugated to a small molecule antimicrotubule drug.	Breast cancer

The physical and chemical instabilities of proteins and peptides, arising from their large molecular weight and complex structure, pose many challenges for pharmaceutical formulation development. Clinical applications of protein drugs are limited by their inadequate concentration in blood, poor oral bioavailability, high manufacturing cost, chemical or biological instability, and/or rapid hepatic metabolism. In addition, most protein drugs do not efficiently pass through biological membranes and enter their target cells. These limitations lead to their high dose and/or need for frequent administration, which can cause undesirable side effects. Also, proteins can elicit host immune response following repeated use due to the development of neutralizing antibodies or hypersensitivity reactions.

Proteins and peptides are rapidly degraded in the gastrointestinal tract due to the harsh pH and enzymatic environment, resulting in poor oral bioavailability. Therefore, proteins are primarily administered parenterally by intravenous (IV), subcutaneous (SC), and/or intramuscular (IM) injection. Thus, the development of a protein formulation is primarily focused on a sterile solution or a sterile, lyophilized powder for reconstitution prior to administration.

25.2 STRUCTURE

Proteins and peptides consist of simple building blocks called amino acids, which are linked together by peptide bonds. A peptide bond is formed by the nucleophilic addition of the primary amine of one amino acid to the electropositive carboxylate carbon of the other amino acid (Figure 25.1). Two amino acids linked together by a peptide bond form a dipeptide; three amino acids linked together by a peptide bond form a tripeptide; and so on. Polypeptides consist of a linear chain of several amino acids. Long chains of amino acids tend to self-associate and fold into three-dimensional

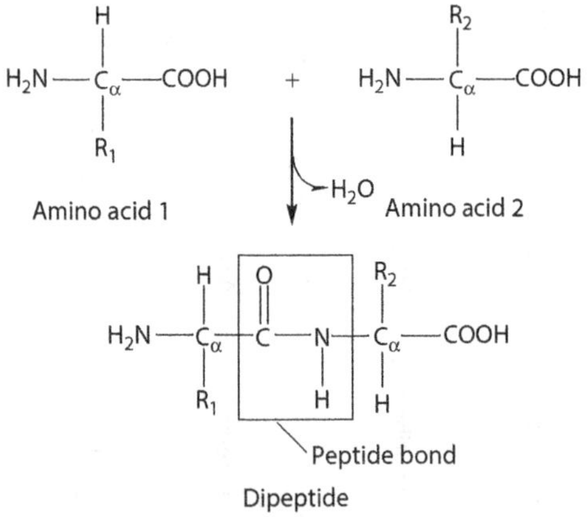

Figure 25.1 Chemical structure of a typical peptide bond. Polypeptides consist of a linear chain of amino acids successively linked via peptide bonds.

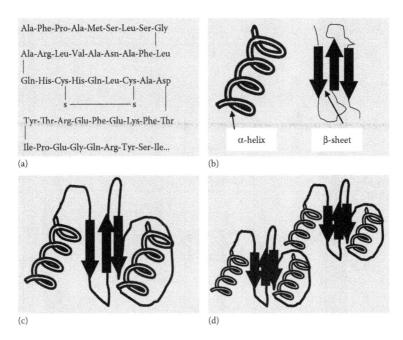

Ala-Phe-Pro-Ala-Met-Ser-Leu-Ser-Gly
|
Ala-Arg-Leu-Val-Ala-Asn-Ala-Phe-Leu
|
Gln-His-Cys-His-Gln-Leu-Cys-Ala-Asp
        | |
        s —————————— s |
Tyr-Thr-Arg-Glu-Phe-Glu-Lys-Phe-Thr
|
Ile-Pro-Glu-Gly-Gln-Arg-Tyr-Ser-Ile...

(a)

α-helix   β-sheet

(b)

(c)   (d)

Figure 25.2 Illustration of protein structures: (a) primary structure (amino acid sequence), (b) secondary structure (α-helix and β-sheets), (c) tertiary structure (further folding of the secondary structurally folded protein), and (d) quaternary structure (combination of polypeptides).

conformations depending on their unique amino acid sequence. Specific functions of proteins are often a function of their unique amino acid sequence and the resulting conformation that makes up the protein.

As shown in Figure 25.2, a chain of amino acids forming a polypeptide through covalent linkages constitutes a protein or peptide's *primary structure*. Spatial folding of a polypeptide chain through noncovalent interactions of *neighboring* amino acids results in the *secondary structure*, which consists of patterns of structural domains such as α-helices and β-sheets. The surfaces of polypeptide chains, organized into these domains, can further bond with each other through noncovalent interactions of *distant* amino acids, which give the overall structure to one polypeptide chain, called the *tertiary structure*. Spatial interaction of more than one polypeptide chain to form protein is termed the *quaternary structure*.

25.2.1 Amino acids

There are 20 naturally occurring amino acids that form the structural basis of all the proteins and peptides. The chemical structures of these amino acids, along with their abbreviated and one-letter designations, are presented in Figure 25.3. Each amino acid possesses unique physicochemical properties governed by their chemical structure.

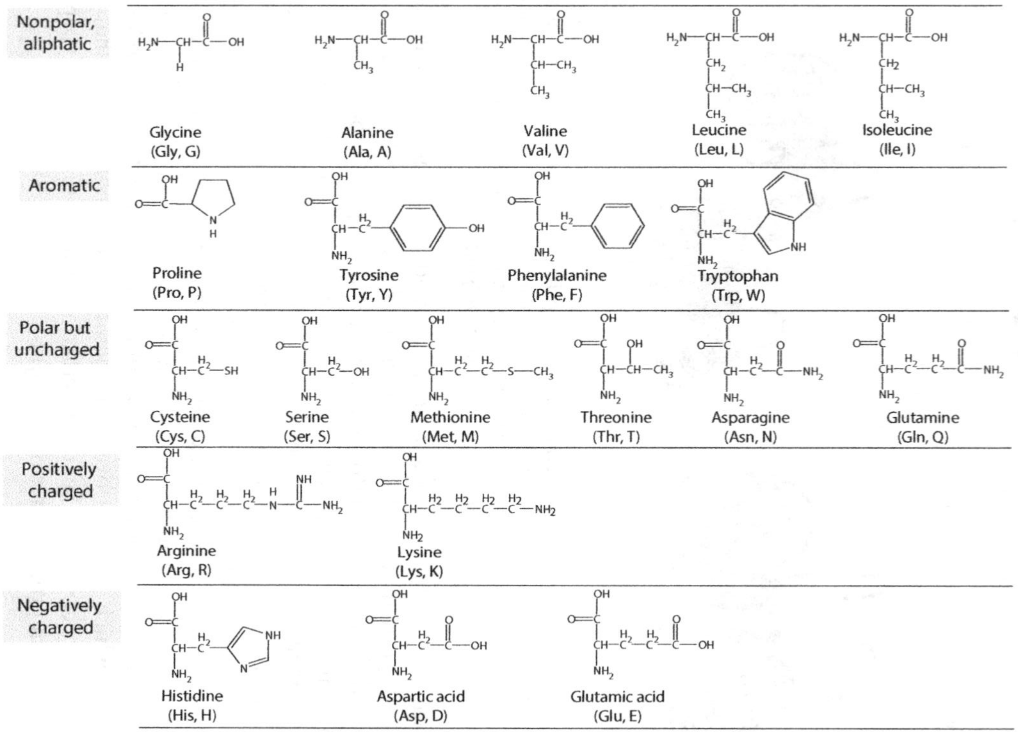

Figure 25.3 Chemical structure of the 20 amino acids commonly found in proteins. The amino acids may be subdivided into five groups on the basis of side-chain structure. Their three- and one-letter abbreviations are also listed.

Table 25.2 Hydrophobicity and acidity of amino acids

Amino acid	Three letter abbreviation	One letter designation	Log P value	pK_a value of carboxylate group	pK_a value of amino group	pK_a value of side chain
Alanine	Ala	A	−2.83	2.35	9.87	—
Arginine	Arg	R	−1.43	2.01	9.04	12.48
Asparagine	Asn	N	−2.33	2.02	8.80	—
Aspartic acid	Asp	D	−1.67	2.10	9.82	3.86
Cysteine	Cys	C	−0.92	2.05	10.25	8.00
Glutamic acid	Glu	E	−1.39	2.10	9.47	4.07
Glutamine	Gln	Q	−2.05	2.17	9.13	—
Glycine	Gly	G	−1.39	2.35	9.78	—
Histidine	His	H	−1.67	1.77	9.18	6.10
Isoleucine	Ile	I	0.41	2.32	9.76	—
Leucine	Leu	L	−1.62	2.33	9.74	—
Lysine	Lys	K	−1.15	2.18	8.95	10.53
Methionine	Met	M	−0.56	2.28	9.21	—
Phenylalanine	Phe	F	−1.49	2.58	9.24	—
Proline	Pro	P	−0.40	2.20	10.60	—
Serine	Ser	S	−1.75	2.21	9.15	—
Threonine	Thr	T	−1.43	2.09	9.10	—
Tryptophan	Trp	W	−1.07	2.38	9.39	—
Tyrosine	Tyr	Y	−2.15	2.20	9.11	10.07
Valine	Val	V	−0.01	2.29	9.72	—

- Nineteen amino acids contain an amino ($-NH_2$) and carboxyl ($-COOH$) group attached to a carbon atom to which various side chains (–R) are connected. This carbon atom is termed α-carbon because it is next to the carboxylate group in the structure. The amino acid *proline* is unusual; in that, its side chain forms a direct covalent bond with the nitrogen atom of amino group. This is indicated in the higher hydrophobic character of proline (higher log P, Table 25.2) compared to most other amino acids.
- The α-carbon has four different groups attached to it and is chiral, except in the case of glycine. This chirality can lead to two optical isomers, L- and D-amino acids, which would be mirror images of each other. Natural amino acids are exclusively L-amino acids.

Amino acids are classified by the acidity (or basicity) and polarity (or hydrophilic/hydrophobic nature). The acidity/basicity is indicated by their ionization constant, the pK_a, whereas the polarity is indicated by log P.

These constants are defined by the following equations and are listed for each amino acid in Table 25.2.

An amino acid backbone can involve the ionization of the acid and/or base:

$$R\text{-COOH} \rightleftharpoons R\text{-COO}^- + H^+$$

$$R\text{-NH}_2 + H^+ \rightleftharpoons R\text{-NH}_3^+$$

$$k_a = \frac{[R\text{-COO}^-][H^+]}{[R\text{-COOH}]} \quad \left(\begin{array}{l}\text{ionization constant for the acid}\\ \text{functional group}\end{array}\right)$$

$$k_b = \frac{[R\text{-NH}_3^+]}{[R\text{-NH}_2][H^+]} \quad \left(\begin{array}{l}\text{ionization constant for the base}\\ \text{functional group}\end{array}\right)$$

$$pK_a = -\log K_a$$

$$pK_b = -\log K_b$$

$$pK_a + pK_b = pK_w = 14, \quad \text{where } pK_w \text{ is the ionization constant of water}$$

and

$$\log P = \log\left(\frac{[\text{solute}]_{\text{Octanol}}}{[\text{solute}]_{\text{Water}}}\right)$$

Amino acids with low pK_a values are acidic, whereas amino acids with high (>7) pK_a values (which would correspond to low pK_b values) are basic. Amino acids typically have an acidic carboxylate group and a basic amino group, which contribute to its acidity or basicity. In addition, the side chain may also ionize. Thus, there are multiple pK_a values associated with an amino acid. However, in a polypeptide chain, the carboxylate and amino groups are covalently bonded to neighboring amino acids (except for the terminal amino acids). In addition, the electron density on the side chain is influenced by the side chains of other spatially close amino acids. Thus, ionization constants of amino acids in a protein are different than those of pure amino acids. Aspartic and glutamic amino acids are considered acidic because of the presence of ionizable carboxylic acid functional groups. Arginine, histidine, and lysine contain basic ionizable side chains and are referred to as basic amino acids.

The hydrophobic character of amino acids as individual molecules is indicated by their $\log P$ value (Table 25.2). These values are predominantly influenced by the ionizable carboxylate and amino functional groups. In a protein structure, these functional groups are covalently bonded. Hydrophobicity, in the context of protein surface, is primarily influenced

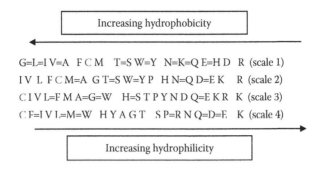

Figure 25.4 Relative hydrophobicity of different amino acids estimated based on either their side-chain sequence (scales 1 and 2) or their typical location in a globular protein structure (scales 3 and 4).

by the protein structure and the interactions of side chains of amino acids with water (Figure 25.4).

Thus, the hydrophobic character of amino acids depends on their microenvironment in the specific protein. In general, the hydrophobic character of an amino acid has been defined by either (a) physicochemical properties of amino acid side chains while ignoring the effects of the carboxylate and the amino groups (scales 1 and 2 in Figure 25.4), or (b) scaling the probability for an amino acid to be found inside or outside a protein structure by examining three-dimensional structures of known proteins (scales 3 and 4 in Figure 25.4). The scaling criteria inherently result in different predictions. For example, cysteine typically forms disulfide bonds in proteins, and stable disulfide bonds are generally present on the hydrophobic interior of a globular protein. Thus, cysteine is relatively more hydrophobic by the scaling criterion of its location in a protein.

25.2.2 Primary structure

The primary structure of a protein refers to the sequence of amino acids and the location of disulfide bonds in the constituent polypeptide chain(s) (Figure 25.2). Primary structure determines a protein's folding and higher levels of structural organization. However, the primary structure generally cannot predict the three-dimensional structure and shape of the proteins in solution.

25.2.3 Secondary structure

Secondary structure can be described as the local spatial conformation of a polypeptide's backbone, excluding the constituent amino acid's side chains. Common secondary structural forms are the α-helix and β-sheets (Figure 25.2). The α-helix results from the helical coiling of a stretch of hydrophobic amino acids with the hydrophobic groups facing inside and the hydrophilic groups facing outside the helix. β-sheets, on the other hand, are characterized by side-by-side hydrogen bonding either within the same chain or between two different chains, thus exposing the amino acid functional groups to the solvent medium. The chain folding of the secondary structures often arises from cross-linking through hydrogen bonding or disulfide bridges. Generally, α-helices are present in membrane proteins, whereas secreted proteins mostly have β-sheet or irregular structure.

25.2.4 Tertiary structure

Tertiary structure of a protein refers to the exact three-dimensional structure of its constituent polypeptide chain(s) (Figure 25.2). The spatial proximity of secondary structural elements determines the tertiary structure of a polypeptide. Spatially close amino acids on the folded (secondary structure) polypeptide chains can form attractive hydrogen bond, ionic, or hydrophobic interactions, resulting in stabilization of the tertiary structure. Proteins under physiological conditions assume their distinctive tertiary structure of minimum free energy, which is a prerequisite for their biological function.

25.2.5 Quaternary structure

Quaternary structures are the highest level of protein organization that can be achieved by proteins that have more than one noncovalently linked constituent polypeptide chain (Figure 25.2). These polypeptide chains can associate to form dimers, trimers, and oligomers, which constitute the quaternary structure of a protein. Almost all proteins that are greater than 100 kDa have a quaternary structure. For example, hemoglobin consists of nonidentical subunits that associate to form a dimer (heterodimer) or a tetramer (heterotetramer); glutathione-S-transferase consists of homotetramer (all subunits identical); collagen is a homotrimeric protein; and the enzyme reverse transcriptase is a heterodimer.

The stabilization of higher orders of protein structure by multiple weak bonds is responsible for the flexibility of structure, which is often required for its functionality. For example, enzymes change conformation upon binding of an agonist, and membrane ion channels change conformation to facilitate transport upon ion binding on their surface.

25.3 TYPES OF PROTEIN AND PEPTIDE THERAPEUTICS

25.3.1 Antibodies

Antibody is a protein produced by β-lymphocytes in response to substances recognized as foreign (*antigens*). Antibodies recognize and bind to antigens, resulting in their inactivation or opsonization (binding of antibody to the membrane surface of invading pathogen, thus marking it for phagocytosis) or complement-mediated destruction. Antibodies are also known as immunoglobulins (abbreviated Ig) because they are immune-response proteins that are *globular* proteins (compact with higher orders of structure and hydrophilic surface making them soluble; as against *fibrous* proteins, which have predominantly secondary structure and are insoluble). Of the five major types of antibodies (Table 25.3), IgG is preferred for therapeutic application due to its wide distribution and function. Structurally, Ig is commonly represented in a typical Y-arm structure (Figure 25.5) consisting of two large/heavy and two small/light polypeptide chains joined by disulfide bridges. Antibody fragments consist of a constant region (designated, Fc) and a variable, antigen-binding region (designated, Fab). Antibodies that recognize multiple sites of an antigen are termed *polyclonal*, whereas antibodies that target only a specific site are *monoclonal*. Identical immune cells make monoclonal antibodies, whereas polyclonal antibodies are produced by a mass of immune cells that may produce antibodies against different regions of the antigen. In industrial application, monoclonal antibodies are prepared by recombinant DNA technology in cell cultures. For human clinical applications, generally monoclonal antibodies are preferred. Polyclonal antibodies are utilized for diagnostic and lab use such as immunohistochemistry.

Table 25.3 Types of antibodies

Antibody	Proportion of total antibodies	Where found in the body	Function	Size
IgA	10%–15%	Nose, breathing passages, digestive tract, ears, eyes, saliva, vagina, tears, and blood	Protection on the mucosal surfaces of the body exposed to the outside environment	
IgG	75%–80%	All body fluids. Smallest and the most common	Fighting bacterial and viral infections. Only type of antibody that can cross placenta	Smallest
IgM	5%–10%	Blood and lymph	First type of antibody made in response to infection. Stimulate other immune cells	Largest
IgE	Small amounts	Lungs, skin, mucous membranes	React to pollen, fungal spores, and animal dander. May be involved in allergic reactions	
IgD	Small amounts	Tissue that lines belly or chest	Not clear	

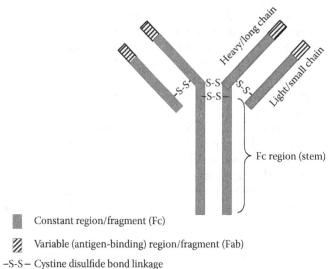

Constant region/fragment (Fc)

Variable (antigen-binding) region/fragment (Fab)

−S-S− Cystine disulfide bond linkage

Figure 25.5 Typical structure of an antibody.

A number of immunoglobulin (Ig) G products have been developed for therapeutic use in various immune disorders (Table 25.1). Due to their specificity, there is a growing interest in the use of monoclonal antibodies and their modifications as therapeutics. For example, antibodies whose Fab fragment segments have been reduced in size to the smallest known antigen-binding fragments are known as *domain antibodies*. Also, antibodies that can bind two different antigens are called *bispecific antibodies*.

The usefulness of antibodies was limited by the immune response generated by the host to the administered antibodies, especially when the antibodies were generated by antigen injection in foreign animal species, such as mouse. *The antibodies generated in mouse were named with the suffix ~momab*. The use of humanized/human monoclonal antibodies with the use of recombinant DNA technology has helped to overcome these limitations.

- *Chimeric and humanized antibodies* are the antibodies produced from nonhuman species whose protein sequences have been modified to increase their similarity to the antibody variants that are naturally found in humans.
  - *Chimeric antibodies* consist of murine variable regions fused with human constant regions, resulting in ~65% human amino acid sequence. This reduces immunogenicity and increases plasma half-life. *These antibodies are named with the suffix ~ximab*. For example, rituximab is a chimeric antibody.

- *Humanized antibodies* are made by grafting the murine variable amino acid domains (which determine antigen specificity) onto human antibodies, resulting in ~95% human amino acid sequence. These, however, have lower antigen-binding affinity than murine antibodies. *These antibodies are named with the suffix ~zumab.* For example, bevacizumab (Avastin®) is a humanized antibody that targets vascular endothelial growth factor (VEGF) and is recommended as first-line therapy in advanced colorectal cancer in combination with other drugs.
- *Human monoclonal antibodies* can be produced using phage display or transgenic mice. Transferring the human Ig genes into the mouse genome can produce these antibodies. *These antibodies are named with the suffix ~mumab.* For example, ipilimumab is a human mAb that inhibits the checkpoint receptor cytotoxic T lymphocyte-associated antigen 4 (CTLA4) and is recommended for advanced-stage melanoma.

Most therapeutic antibodies exert their therapeutic effects by binding to selected cellular targets, which are then destroyed by physiological mechanisms activated by the effector functions of the antibody. In addition, antibodies can also be used as drug delivery and targeting vehicles. Active research and development is being pursued on customized antibodies conjugated to toxins, radioisotopes, small drugs, enzymes, and genes for selectively destroying harmful cells in the body. For example, several ADCs have been developed for the treatment of cancer that utilizes a toxin, which is a small molecule attached to an antibody. For example, Adcetris® and Kadcyla® are ADCs for tumor treatment.

25.3.2 Hormones and physiological proteins

Protein therapeutics to replace or supplement endogenous protein molecules are used for several diseases such as diabetes (insulin), growth hormone deficiency (growth hormone), and hemophilia (factors VIII and IX). Table 25.1 lists some protein therapeutics and their clinical applications.

25.3.3 Chemically modified proteins and peptides

Chemical modifications of proteins are carried out to either

- Increase target specificity, for example, abatacept (Table 25.1) and conjugation to sugars.
- Increase therapeutic ability, for example, radiolabeled antibodies and ADCs (Table 25.1).
- Increase plasma half-life, for example, by PEGylation of antibodies.

25.3.3.1 Conjugation with sugars

Conjugation of sugars, such as sucrose, mannose (mannosylation), or lactose (lactosylation), to proteins can be used to provide targeted delivery of proteins. For example, receptors for carbohydrates, such as the asialoglycoprotein receptor on hepatocytes, and the mannose receptor on macrophages, such as Kupffer cells, recognize corresponding sugars. Mannosylated bovine serum albumin (Man-BSA) and galactosylated BSA (Gal-BSA) preferentially bind to alveolar macrophages and hepatocytes, respectively. Galactosylated and mannosylated recombinant human superoxide dismutase (Gal-SOD, Man-SOD) exhibited inhibitory effects superior to native SOD against hepatic ischemia-perfusion injury.

25.3.3.2 PEGylation

Proteins may be conjugated to Polyethylene glycol (PEG), a nonimmunogenic, nontoxic, and FDA-approved polymer, to increase their plasma half-life. The process of conjugation with PEG is called *PEGylation*, and the protein after the conjugation is called the *PEGylated* protein. PEG consists of a flexible polyether chain that provides a hydrophilic surface, thus shielding hydrophobic groups and minimizing nonspecific interactions. Attachment of PEG on protein surface also increases the hydrodynamic diameter of proteins. Either straight chain or branched PEG can be used for PEGylation. The flexibility of the side chain allows the PEGylated protein to interact with the target.

PEGylation can increase biocompatibility, reduce immune response, increase *in vivo* stability, delay clearance by the reticuloendothelial system, and prevent protein adsorption on the surface of the delivery device, such as syringe.

25.3.3.2.1 Applications

Interferon (IFN)-2α has a low plasma half-life and needs daily injections. However, IFN-2α conjugated to *branched PEG* 40 (i.e., PEG of 40 kDa average molecular weight) provides sustained plasma concentrations upon once a week injection. Other examples of PEG-modification to modulate clearance rate of proteins include PEG-adenosine deaminase (PEG-ADA), PEG-asparaginase, PEG-rIL2, and PEG-interferon. Native ADA is not effective due to its short half-life (<30 min) and is immunogenic due to bovine source, whereas PEGylated ADA (Adagen®) is quite effective, has long half-life, and is nonimmunogenic.

25.3.3.2.2 Chemistry

PEG has two hydroxyl groups at each end of the linear chain. PEGylation is often done by creating a reactive electrophilic intermediate with succinimide (thus producing N-hydroxysuccinimide, NHS), which undergoes electrophilic

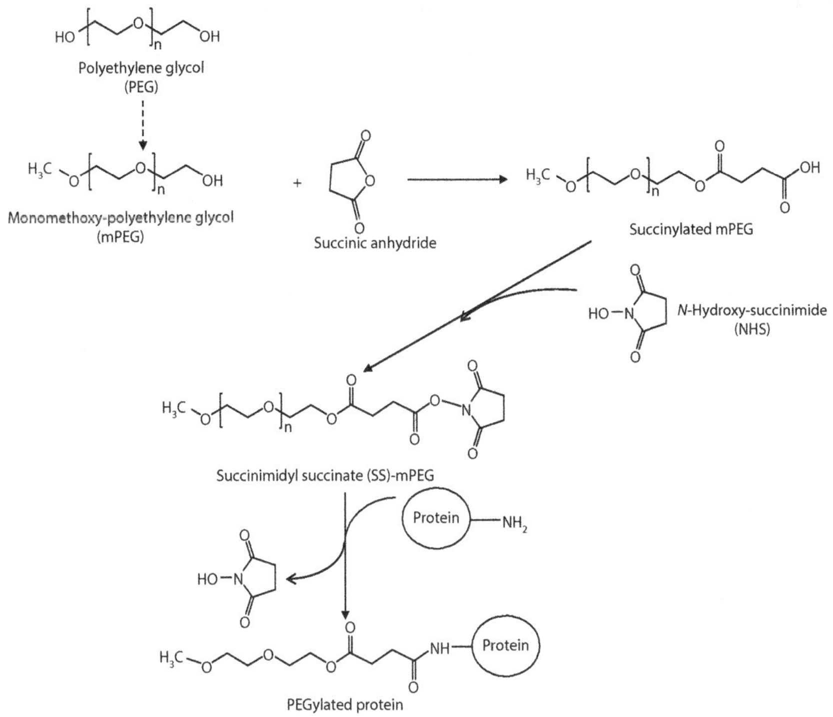

Figure 25.6 PEGylation of proteins using *N*-hydroxysuccinimide (NHS) derivative of methoxy PEG.

substitution by an amine group of the protein (Figure 25.6). The NHS ester groups primarily react with the α-amines at the N-terminals and the ε–amines of lysine side chains. Two hydroxyl groups—one at either end—make the natural PEG bifunctional. To prevent the potential for cross-linking and polymerization with the natural bifunctional polymer, monofunctional PEG polymer can be used. To make PEG monofunctional, one end of the chain is blocked with a methyl ether (methoxy) group. Such a monofunctional PEG is termed monomethoxyPEG (mPEG). Thus, mPEG contains only one hydroxyl group per chain, thus limiting activation and coupling to one site.

25.3.3.2.3 Limitations

PEGylation usually reduces binding affinity of the protein to its target. PEGylation also increases the viscosity of protein formulations, which may limit the development of concentrated solutions for injection. Protein reaction with PEG generally has low efficiency and is difficult to optimize. In addition, PEG often contains peroxide impurities, which can lead to oxidative protein degradation during shelf life storage.

25.3.3.3 Other protein conjugation approaches

Proteins can also be conjugated to hydroxyethyl starch (HESylation) or to polysialic acid (PSAylation) using similar chemistry to increase their plasma half-life. PEGylation remains the most common protein modification.

25.3.3.4 Antibody drug conjugates

In recent years, several mAb-based therapeutics that have a small molecule conjugated to the antibody—the ADCs—have been commercialized, such as Kadcyla® and Adcentris®. Most current ADCs are developed for oncology indications and utilize a high potency cytotoxic drug called payload attached through a covalent linker to a monoclonal antibody that serves as a targeting moiety. The discovery and development of ADCs follow unique paradigms that overlap both small and large molecule drug discovery and development but have unique distinctions. For example, the attachment of hydrophobic drug on the mAb changes mAb surface properties and conformational stability. It can increase protein aggregation and surface hydrophobicity.

Chemistry of conjugation of small molecule drug to the antibody is constantly evolving. In general, the conjugation can be random (through, e.g., lysine or cysteine residues) or site specific (through, e.g., engineered antibodies that have specific amino acid residues). One needs to pay attention to the selection of mAb, payload, and linker for an effective ADC. Currently, several ADCs are in clinical trials as monotherapies or in combination with other anticancer drugs.

25.4 PROTEIN CHARACTERIZATION

25.4.1 Biophysical characterization

Therapeutic applications of proteins require an understanding of fully elucidated structure, pharmacology, and mechanism of action. In addition, protein behavior in solution and the impact of chemical properties and components of solutions on the physical properties of solutions (termed biophysical characterization) need to be well defined. Biophysical characterization of proteins includes the determination of *size, shape,* and *solution properties* of proteins through direct and indirect techniques that include the following:

- Hydrodynamic protein size measurement by analytical ultracentrifugation, gel filtration chromatography, gel electrophoresis, and/or viscometry.
- Thermodynamic methods such as microcalorimetry and surface plasma resonance can help delineate the state of protein association and interactions with other molecules in solution.

- Particulate formation by protein self-association or interaction with other components in solution by dynamic light scattering (DLS).
- Spectroscopic methods such as circular dichroism (CD) and thermal melt fluorescence spectroscopy can help determine the stability of protein *conformation* in solution.

25.4.2 Physicochemical characterization

25.4.2.1 Solubility

Under physiological conditions, solubility of proteins can vary from the very soluble to the virtually insoluble. Water solubility of a protein requires interactions, such as hydrogen bonding and electrostatic interactions, of protein surface with the aqueous medium. The hydrophilic interactions, which are stronger and predominant in aqueous conditions, are enhanced by the ionization of functional groups on proteins such as amines and carboxylates. Ionization of these functional groups is pH dependent. Thus, the solubility of proteins and peptides is dependent on the pH of the solution.

The overall charge on a protein can be either positive or negative, depending on the ionization status of all of its functional groups. A protein is usually positively charged at a low pH and negatively charged at a high pH. Protein solubility increases as the pH of the solution moves away from the *isoelectric point* (IEP) (Figure 25.7), which is the pH at which the molecule is ionized but has a net zero charge and does not migrate in an electric field (determined by gel electrophoresis). The presence of both positive and negative charges on the protein at its IEP leads to a *greater tendency for self-association*. As the net charge on the protein changes in any one direction (positive or negative) with a change in solution pH, the affinity of the protein for the aqueous environment increases and the protein molecules also exert greater electrostatic repulsion among each other, thus

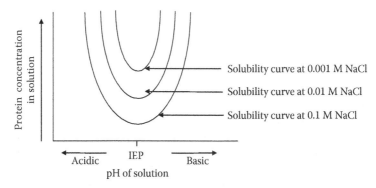

Figure 25.7 A typical profile of protein solubility in solution as a function of solution pH and salt concentration. IEP, isoelectric point.

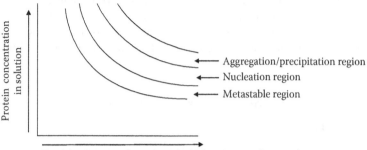

Precipitating agent (concentration)/phenomenon (intensity)
for example nonsolvent concentration and temperature

Figure 25.8 Phase behavior of proteins in solution formulation. Typical phases of physical instability of protein in solution with the addition of a precipitating agent (such as salt) or change of a precipitation inducing phenomenon (such as temperature).

preventing them from self-associating. This increases their aqueous solubility. However, extremes of pH can cause protein unfolding with the exposure of hydrophobic groups and protein self-association at their exposed hydrophobic regions leading to precipitation.

The *phase behavior of protein solutions*, that is, whether protein solution is a single-phase solution or has protein separation (two phases—solid and liquid), is affected by pH, ionic strength, and temperature (Figure 25.8). Generally, protein solubility decreases with increasing ionic strength of salts, such as NaCl and KCl (Figure 25.7). This phenomenon is called the *salting out effect*. This phenomenon is used to concentrate dilute solutions of proteins and to separate a mixture of proteins (if one of the proteins salts out at a lower salt concentration than the other). The added salt can then be removed by dialysis.

Organic solvents tend to decrease the solubility of proteins by lowering the dielectric constant of the solution (Figure 25.8). The presence of other highly water-soluble polymers in the solution (cosolutes) also tends to reduce protein solubility by their interactions with solvent molecules, thus tying up the solvent and reducing protein–solvent interactions. This phenomenon is known as the *volume exclusion effect*.

25.4.2.2 Hydrophobicity

Different amino acids have different degrees of hydrophobicity (Figure 25.4). Overall hydrophobicity or hydrophilicity of a protein is determined by the nature of functional groups exposed on the surface of the protein. These are the groups that contribute to protein–solvent and protein–protein interactions.

In an aqueous solution, hydrophobic regions of a polypeptide tend to point away from the hydrophilic aqueous environment to achieve the thermodynamically least energy state of greatest stability. In doing so, the hydrophobic surfaces of a protein tend to cluster together on the inside of the protein and form multiple weak van der Waals interactions. These multiple simultaneous weak hydrophobic interactions are the single most important stabilizing influence of protein native structure, which also provide flexibility of protein conformation depending on its solution environment. Thus, in addition to the stabilizing interactions with the solvent on the surface, including electrostatic, van der Waals, hydrogen bonds, and ionic interactions, hydrophobic interactions within and among a protein's polypeptide chains stabilize native protein native structure. For example, if alternating hydrophilic and hydrophobic amino acid sequences in synthetic peptides are at the optimum distances in space, the molecules coil with the hydrophobic amino acids on the inside of each coil and the hydrophilic ones to the outside. Thus, secondary, tertiary, and quaternary structures of polypeptide chains are important in determining the net hydrophobic or hydrophilic nature of the protein.

25.5 INSTABILITY

Protein pharmaceuticals commonly exhibit both physical and chemical instability. Physical instability refers to changes in the higher order structure that does not include covalent bond cleavage or formation, whereas chemical instability refers to modification of proteins via bond formation (e.g., oxidation) or bond cleavage (e.g., deamidation), yielding a new chemical entity. Physical instability often results in protein denaturation (loss of natural conformation), which can lead to adsorption to surfaces, aggregation, and precipitation.

25.5.1 Physical instability

Protein denaturation is a result of change in higher order folding or conformation that commonly exhibits as a change in the surface exposure of functional groups. Increase in surface hydrophobicity due to protein denaturation can lead to aggregation, precipitation, and/or adsorption to the surface of the container or closure.

25.5.1.1 Denaturation

Protein native structure represents the least overall thermodynamic free energy of interaction of different residues of the polypeptide(s) with the solvent (water) and with themselves. This determines the *native state* of protein structure. The three-dimensional structure of a protein is held together by weak noncovalent interactions, is flexible, and relatively unstable. It can be modified by environmental factors, such as solution composition and temperature. For example, a change in the solvent medium can result in a different structure being the lower, thermodynamically least free energy state of protein conformation. For example, addition of salt or organic solvent would reduce the propensity for hydrophilic interactions on the protein surface.

If the enthalpy barrier from the native state to the altered lower thermodynamic free energy state can be met (e.g., by heating the protein solution), the protein conformation might change to the new form of thermodynamically least free energy. This loss of natural, or native, state of a protein is termed *denaturation*. Protein denaturation refers to disruption of the tertiary and secondary structure of a protein or peptide. It can be caused by heating, cooling, freezing, extremes of pH, and contact with organic chemicals. Protein denaturation is often associated with increased hydrophobic surface of a protein. In such cases, several protein molecules in solution might self-associate and exclude the solvent. This phenomenon is termed *aggregation*. If the aggregates separate from the solution and become visible, the phenomenon is called protein *precipitation*.

Protein denaturation can also lead to protein unfolding. It can be reversible or irreversible. Reversible denaturation can be caused by temperature or exposure to *chaotropic agents*, such as urea and guanidine hydrochloride. The chaotropic agents interfere with stabilizing intramolecular noncovalent interactions in proteins, including hydrogen bonding, van der Waals forces, and hydrophobic effects. In the case of *reversible denaturation*, if the denaturing condition is removed, the protein will regain its native state and maintain its activity. *Irreversible denaturation* implies that the unfolding process disrupted the native protein structure to the extent that the native structure cannot be regained simply by changing the denaturing condition (such as temperature). The ease of protein denaturation depends on the strength and number of intermolecular interactions that keep the protein in its native conformation.

25.5.1.2 Aggregation and precipitation

Aggregation of proteins refers to nonreversible interaction and clustering of two or more protein molecules. Protein aggregates *may be soluble or insoluble*. Protein aggregation is driven by the unfolding process, which exposes the interior hydrophobic region to the solvents, usually water, leading to thermodynamically unfavorable surroundings of the hydrophobic protein. This drives intermolecular interactions between exposed hydrophobic regions of different protein molecules, leading to association and, thus, aggregation.

Several factors may lead to protein aggregation. For example:

- *Shear forces*: Shearing and shaking of protein solutions during formulation and shipment may lead to aggregation.
- *Temperature*: An increase in temperature results in greater flexibility of proteins and an increased tendency to form aggregates.

- *Ionic strength*: An increase in the ionic strength may lead to neutralization of the surface charge of the protein molecules, which may lead to aggregation.
- *pH*: Charge neutralization and subsequent aggregation can also occur when the pH of the solution approaches the IEP of the protein.
- *Moisture*: An optimal residual moisture level is required to maintain the stability of lyophilized protein formulations, the absence of which may lead to protein aggregation. Thus, hydration in formulated proteins must be ensured by either increasing residual moisture content or by adding water-substituting excipients.

When *insoluble protein aggregates* are visually evident, the protein is said to have precipitated. Protein precipitation is a macroscopic process producing a visible change of the protein solution, such as turbidity/clouding of the solution or formation of visible particulates. Accumulation of *soluble protein aggregates*, on the other hand, is evident by the changes in solution properties of proteins, such as viscosity.

Native, folded proteins may precipitate under certain conditions, most notably salting out and isoelectric precipitation. Protein precipitation can be a result of both covalent and noncovalent aggregation pathways.

25.5.1.3 Surface adsorption

The adsorption of proteins and peptides to the surfaces of the container, closure, or filter results from protein surface interaction with nonpolar surfaces. This can cause proteins to expose their hydrophobic interior, leading to adherence or adsorption to the surfaces of the containers. Alterations in the pH and ionic strength of the media can significantly enhance or reduce the protein's tendency to adsorb. Protein adsorption to neutral or slightly charged surface is greatest at its IEP.

The effect of surface adsorption on the amount of administered drug can be substantial when the initial concentration of the protein in solution is low, leading to a high proportion of drug loss due to adsorption. The extent and reversibility of protein adsorption are dependent on the conformational state of the protein, the pH and ionic strength of the solution, the nature of the exposed surface, surface area, and time of exposure. Poly(oxyethylene oxide) (Teflon)-coated surfaces and siliconized rubber stoppers for vials can minimize the likelihood of protein adsorption at the surface. Certain formulation strategies, such as increase in the concentration of surfactant, and prerinse of the IV administration tube-set and filter with the diluent can also minimize protein adsorption to surfaces.

25.5.2 Chemical instability

Chemical instability of proteins and peptides generally involves one or more of the following chemical reactions.

25.5.2.1 Hydrolysis

Proteolysis is the hydrolysis of the peptide bond between amino acids in a peptide or protein. At an extreme pH and temperature, the peptide bond can undergo rapid proteolysis resulting in protein degradation and/or fragmentation. The most commonly observed proteolytic reactions in proteins and peptides involve the side-chain amide groups of asparagine (Asn) and glutamine (Gln), and the peptide bond on the C-terminal side of an aspartic acid (Asp) or a proline (Pro) residue. Several therapeutic proteins are known to degrade through hydrolysis. These include luteinizing hormone-releasing hormone (LHRH), macrophage colony-stimulating factor (M-CSF), human growth hormone, and vasoactive intestinal peptide (VIP).

Hydrolysis leading to protein fragmentation generally compromises protein efficacy and may produce toxicities or immunogenicity as well.

Protein degradation by hydrolysis can be observed during stability testing by the formation of charge variants (by isoelectric focusing) or size variants (by size exclusion chromatography [SEC]). Isolation of these new peaks followed by their size determination by mass spectroscopy (MS) and/or composition determination by tryptic peptide mapping (TPM) helps identify the exact size and sequence of degradants, and the location of hydrolysis.

25.5.2.2 Deamidation

Deamidation is one of the main chemical degradation pathways of proteins in which the side-chain linkage in a glutamine (Gln) or asparagine (Asn) residues is hydrolyzed to form a carboxylic acid. The hydrolysis changes the asparaginyl residue into an aspartyl or isoaspartyl residue. The deamidation of Asn and Gln residues of proteins is an acid and base-catalyzed hydrolysis reaction, which can occur rapidly under physiological conditions.

Deamidation may or may not impact protein efficacy, safety, and immunogenicity. Thorough characterization and understanding of the sites and extent of deamidation, nonetheless, are critical to clinical comparability of the dosed drug substance.

Deamidation is generally detected by the change in the size and charge variants, and the location of deamidation is confirmed by TPM.

Solution pH optimization and lyophilization are frequently used to minimize deamidation in proteins. However, residual moisture present in the lyophilized formulation can still allow deamidation to take place. In some cases, protein engineering to replace Asn residue with Ser can be used if it does not affect protein conformation and biological activity.

25.5.2.3 Oxidation

Oxidation is one of the major causes of chemical degradation in proteins and peptides. The *functional groups* in proteins that can undergo oxidation include the following (Figure 25.9):

- Sulfhydryl in cysteine (Cys)
- Imidazole in histidine (His)
- Thiolether in methionine (Met)
- Phenol in tyrosine (Tyr)
- Indole in tryptophan (Trp)

Factors that increase oxidative degradation in proteins include the following:

- Atmospheric oxygen, which alone can lead to oxidation of Met residues, producing the corresponding sulfoxide.
- Peroxides, such as hydrogen peroxide, can modify indole, sulfhydryl, disulfide, imidazole, phenol, and thioether groups of proteins at neutral or slightly alkaline pH. The source of peroxides in formulation is often the hydrophilic polymeric excipients used.
- Oxidation can be catalyzed by metal contaminants (e.g., Fe^{2+}/Fe^{3+} and Cu^+/Cu^{2+}), light, acid/base, and free radicals.
- Solution pH, nature of buffers, presence of metal ions and metal chelators, and neighboring amino acid residues of susceptible amino acids influence oxidation in solution.
- Light, which may photoactivate triplet ground state oxygen to the excited, more reactive singlet state.

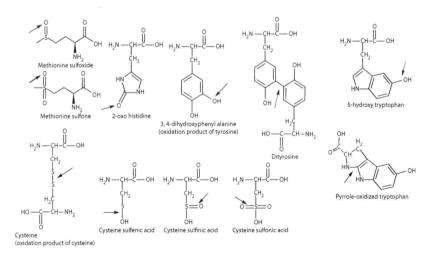

Figure 25.9 Side-chain oxidation products of oxidizable amino acid residues in a protein.

Stabilization strategies to prevent or minimize oxidative degradation of proteins include the following:

- Low temperature storage or refrigeration to reduce reaction rates.
- Nitrogen overlay in packaging to minimize the impact of headspace air/oxygen exposure.
- Protection from light by the use of amber glass containers for storage.
- pH optimization.
- Use of antioxidants and chelating agents. Antioxidants terminate free-radical reactions. Chelating agents sequester free metals, such as iron and copper from the formulations.
- Lyophilization.
- Certain sugars might prevent or minimize protein oxidation by complexation with metal ions or hydrogen bonding on the protein surface to preserve its native conformation.

25.5.2.4 Racemization

Racemization can affect protein conformation. All amino acid residues except glycine (Gly) are chiral at the carbon atom bearing the side chain and are subject to base-catalyzed racemization. The rate of racemization depends on the particular amino acids and is influenced by temperature, pH, ionic strength, and metal ion chelation. Aspartic acid and serine residues are most prone to racemization.

25.5.2.5 Disulfide exchange

Disulfide bonds provide covalent structural stabilization in proteins. Cleavage and subsequent rearrangement of these bonds can alter the tertiary structure, thereby affecting protein conformation, stability, and biological activity. Disulfide exchange is catalyzed by thiols, which can arise by initial reduction of disulfide bond, or β-elimination in neutral or alkaline media. Disulfide thiol exchange reactions can be inhibited by the addition of efficient thiol scavengers, such as *p*-mercuribenzoate and *N*-ethylmaleimide. Figure 25.10 illustrates a cysteine–disulfide exchange reaction.

25.5.2.6 Maillard reaction

The use, or presence as impurities, of *reducing sugars* (e.g., glucose, lactose, fructose, maltose, xylose) in a protein formulation can result in the Maillard *browning reaction*, which involves nonenzymatic glycation of the protein at the basic protein residues such as lysine, arginine, asparagine, and glutamine. Reducing sugars have an open chain (with an aldehyde or ketone group) and a closed chain (cyclic oxygen) form coexisting in solution in equilibrium. The presence of the aldehyde or the ketone group in the

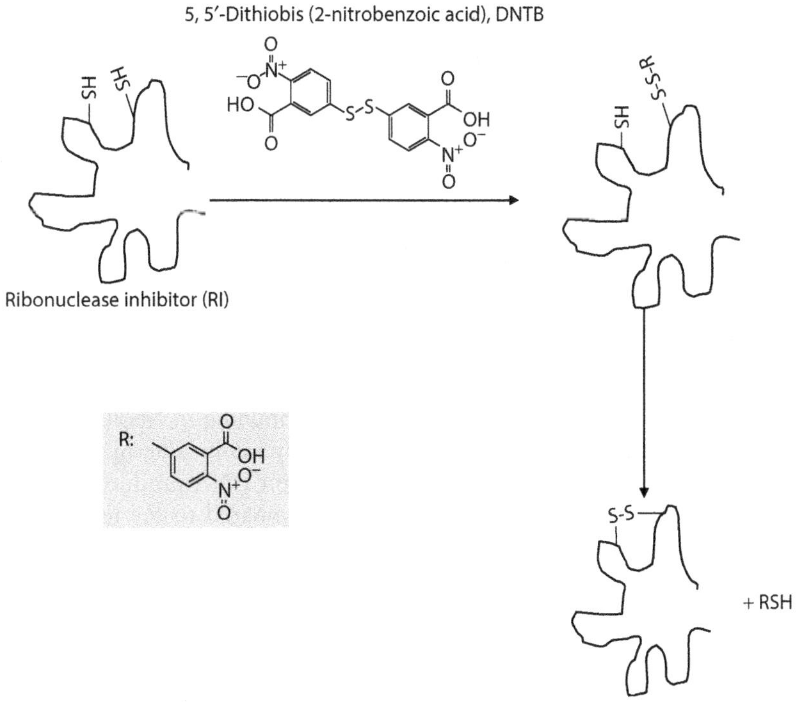

Figure 25.10 An illustration of the effect of cysteine disulfide exchange on protein conformation.

open chain allows nucleophilic attack of the amine group of a basic amino acid side chain on the carboxyl carbon. Sucrose, even though a nonreducing sugar, is a disaccharide and can get hydrolyzed at acidic pH into reducing sugars.

Maillard reaction results in the formation of a Schiff base ($R_1R_2C=N-R_3$), which can further rearrange to form products with π-electron cloud conjugation, which are colored products—hence the name *browning reaction.* Maillard reaction could be minimized or prevented by removing reactive substrate (reducing sugars), pH adjustment, chelation of trace metals, use of antioxidant, reducing water content (thus minimizing the plasticity and solute reactivity in the lyophilized solid matrix), and storage at low temperatures.

25.6 ANTIGENICITY AND IMMUNOGENICITY

The ability of a protein to generate an immune response, triggering the production of antibodies, is referred to as *immunogenicity.* Sometimes, the first administration of a protein does not elicit an immune response due to low concentration or longer time required for the humoral and cellular immune processes. Repeated protein administration may often lead to the formation of antibodies, causing an immune reaction.

Antigenicity, on the other hand, refers to the ability of specific sites (*epitopes*) on the protein to recognize antibodies in the host immune system. Thus, the first administration of an antigenic protein would lead to an immune reaction if the host immune system has antibodies against the foreign protein epitope. When antibodies are developed upon repeated protein administration (immunogenicity), the protein may not be antigenic when administered first but becomes antigenic upon subsequent administration when the host has formed mature antibodies against the protein.

Although proteins made in a particular organism are recognized by the immune system as *self*-protein and normally do not elicit an immune response, misfolded or denatured forms of *self*-proteins may be immunogenic. Thus, immunogenicity may be prevented by maintaining the molecule in the properly folded native conformation as well as by minimizing or preventing protein self-association. In general, the recombinant DNA-produced proteins are likely to be relatively more immunogenic compared to the natural proteins. Approaches toward humanizing antibodies or adding specific human sequences to murine antibodies to make chimeras have greatly improved their therapeutic potential by reducing or eliminating their immune response.

25.7 DRUG PRODUCT FORMULATION AND PROCESS

Most proteins and peptides are not absorbed to any significant extent by the oral route. Therefore, most commercially available protein pharmaceuticals are administered by parenteral routes. Parenteral protein formulations are typically administered by IV, IM, or SC injection. In addition, some protein drugs, for example, insulin, can also be delivered by inhalation for absorption through the alveolar mucosal membrane.

Parenterally administered proteins are rapidly cleared from circulation by the reticuloendothelial systems (RES). Proteins are metabolized by peptidases, leading to rapid loss of their biological activity. Pharmacokinetics of proteins after parenteral administration can be improved by covalent conjugation with a hydrophilic polymer such as PEG (PEGylation), as discussed earlier. Protein bioavailability from the SC route is generally low (~30%–70%). In addition, immunogenicity potential of proteins is higher when administered by the SC route. At the same time, SC administration is preferred because it allows for patient self-administration—as compared to IV administration, which must be carried out by a health care provider.

25.7.1 Route of administration

Selection of the appropriate route of administration for a protein drug depends on several factors, including the disease state, the desired onset and duration of drug absorption/action, drug dose, frequency of administration, patient compliance, and the physicochemical properties of the drug. For example:

- IV route is preferred for rapid onset of administration, whereas the SC route can be used for sustained drug delivery devices. Thus, sustained drug delivery devices such as poly(lactide-co-glycolide) (PLGA)-entrapped drugs are often designed for SC administration.
- Compared to the SC route, IM injection is exposed to much greater blood supply and, thus, faster absorption.
- Higher injection volumes may be administered by the IM (2–5 mL) than the SC (up to 1 mL) route.
- In cases where patient self-administration of a drug is required, IM or SC injections are needed over IV.

In terms of formulation requirements, the needed volume of injection is determined by the drug dose and solubility. If solubility is inadequate, solubilization approaches may be needed. Preparation of concentrated protein solutions can, however, lead to high viscosity—which could make deaeration upon agitation and injectability through a syringe difficult. For example, SC injections typically use lower diameter (25–30G) needles compared to IM injections (20–22G).

25.7.2 Type of formulation

Selection of the type of protein formulation depends on several factors, such as follows:

- *Disease condition*: For example, requirement of patient self-administration (SC route preferred) versus administration by a health care professional in a hospital setting (IV route preferred) might depend on whether typically the patients are hospitalized or outpatients. The SC formulation typically has limitations on the number of injections per dose and per day as well as on the volume per injection. IV injection or infusion in a hospital setting generally does not have such a limitation.
- *Drug half-life*: Rapidly cleared drugs must be administered as an IV infusion to obtain sustained plasma concentrations.
- *Patient population*: Age of the patient may determine the kind of delivery devices that may be the most suitable. For example, slow infusion pump or autoinjector may be preferred for a geriatric population for drug self-administration over vial of lyophilized drug and syringe due to the dexterity required to reconstitute the lyophilized powder and fill the syringe from a vial before injection.
- Route of delivery, such as IM, IV, SC, intraperitoneal, topical, inhalation, or nasal. IV formulations can further be IV bolus or IV infusion. Inhalation formulations can be dry powder based or solution based.
- Drug dose, solubility, stability, and other physicochemical properties.

Proteins and peptides for parenteral administration are typically formulated as ready-to-use aqueous solutions or as lyophilized solid mass that is reconstituted into a protein solution by dilution with water, isotonic dextrose solution, or isotonic sodium chloride solution immediately before administration. Proteins and peptides for inhalation and nasal routes of administration are typically formulated as dry powders. The details of dry powder formulations will not be discussed in this chapter.

25.7.3 Formulation components

The development of a suitable pharmaceutical formulation of a protein usually involves the screening of a number of physiologically acceptable buffers, salts, chelators, antioxidants, surfactants, cosolvents, and preservatives (Table 25.4). Formulation components are selected to address one or more requirements for protein formulations, such as follows:

Table 25.4 Typical excipients in protein formulations

Category	Type	Functionality	Examples
Buffering agents	Nonamino acid buffers	Ensure optimal pH control	Acetate, citrate, carbonate, HEPES, maleate, phosphate, succinate, tartrate, TRIS
	Amino acids		glycine, histidine
Tonic agents	Salts	Stabilization from aggregation, isotonicity	Sodium chloride, potassium chloride, calcium chloride, magnesium chloride, sodium gluconate, sodium sulfate, ammonium sulfate, magnesium sulfate, zinc chloride
Hydrophilic additives	Sugars	Conformation stabilizing agents, especially in lyophilized formulations, and isotonicity	Glucose, fructose, lactose, maltose, mannitol, sorbitol, sucrose, trehalose, inositol
	Other polyols		Glycerol, cyclodextrins
	Amino acids	Buffering action and nonspecific interactions	Alanine, arginine, aspartic acid, lysine, proline
	Hydrophilic polymers	Polymer matrix in solution	Dextran, PEG
Solubilizers	Surfactants	Reduce surface tension, solubilization	Polysorbate, poloxamer, sodium lauryl sulfate
	Cosolvents	Increase protein solubility	Ethanol
Preservatives	Antioxidants	Preferentially oxidized over the protein substrate	Ascorbic acid, citric acid, glutathione, methionine, sodium sulfite
		Heavy metal binding	EDTA, DTPA, EGTA
	Antimicrobial preservation	Antimicrobial agents	Benzyl alcohol, benzoic acid, chlorobutanol, m-cresol, methylparaben, propylparaben

Abbreviations: EDTA, ethylenediamine tetraacetic acid; DTPA, diethylene triamine pentaacetic acid; EGTA, ethylene glycol tetraacetic acid.

- Increasing protein solubility by the use of surfactants and/or cosolvents and pH adjustment.
- Using pH of optimum stability by the use of buffering agents. Selection of an appropriate buffer type and strength is carried out to minimize specific/general acid/base degradation of the protein.
- Physical stability improvement by the addition of polyhydric alcohols, carbohydrates, and amino acids. Addition of these components to aqueous solutions of proteins leads to their hydrogen bonding on the protein surface, thus stabilizing the native protein conformation.
- Stabilization of protein conformation by the addition of cosolvents such as glycerol or PEG, which may decrease the protein surface area in contact with the solvent.
- Electrostatic interactions in proteins may be modulated by the alteration of the solvent polarity and dielectric constant to change protein electrostatic interactions in solution, which may reduce the association tendency of a protein.
- Antimicrobial agents may be added to large volume parenteral (LVP) solutions or multidose vials where repeated puncturing for dose withdrawal is expected to preserve aqueous solutions of proteins against bacterial and fungal growth.
- Chelating agents and antioxidants may be added to prevent metal and/or oxidation-induced chemical instability.
- Osmolarity control is required for parenteral formulations. This is often achieved by the use of salts, buffers, and sugars.

25.7.4 Manufacturing processes

25.7.4.1 Protein solution

A typical manufacturing process of protein solution involves

1. Freeze ad thaw of the bulk drug substance (therapeutic protein).
2. Formulation (dilution and addition of excipients).
3. Filtration for removing any particulate matter and/or sterilization.
4. Filling of drug product in vials or syringes.
5. Inspection of filled vials or syringes for the presence of any particulate matter.
6. Labeling and packaging.
7. Storage and shipment of drug product.
8. Use of a delivery device for drug administration to the patient.

Many of these processes may affect formulation stability. For example:

- Exposure to light and shear during inspection and transportation can lead to the formation of microbubbles in the formulation, which can increase the propensity for aggregation and oxidation.

- Protein may interact with the silicone oil typically used in syringes for smooth barrel movement, leading to instability of syringe-filled protein formulations.
- Protein loss may occur due to adsorption to manufacturing equipment and filter membranes. In addition, leaching of metal ions from manufacturing vessels into the protein formulation can lead to protein instability. Modern day manufacturing practices utilize single-use plastic liners in process tanks to avoid the risk of metal leaching and also to eliminate cleaning verification needs and cross-contamination risk.

25.7.4.2 Lyophilization

Many proteins are very unstable in solution and may not yield acceptable shelf life, even under refrigerated (2°C–8°C) storage conditions. In such cases, freeze-drying or lyophilization is often employed to minimize the kinetics of degradation processes that occur in solutions. Many variables impact the stability of lyophilized drug product. For example, high concentration of reacting species in the protein microenvironment can be detrimental. Further, careful optimization of residual water and protein-binding sugar concentration is required to ensure cake integrity and rapid reconstitution.

The role of *residual moisture* in the lyophilized formulations on proteins stability can be complex. The amount of moisture adsorbed on each protein as a monolayer can be determined by the Brunauer–Emmett–Teller (BET) method. Lyophilized protein product needs closely bound water layer on the protein to shield its highly polar groups, which would otherwise be exposed leading to aggregation and cause opalescence upon reconstitution. High moisture content, on the other hand, could increase plasticity in the system leading to high reactivity and compromise the physicochemical stability. For example, insulin, tetanus toxoid, somatotrophin, and human albumin aggregate in the presence of moisture, which can lead to reduced activity, stability, and diffusion.

Formulation components that stabilize the protein during and after lyophilization depend on the particular protein. For example, polysorbate 80, hydroxypropyl β-cyclodextrin, and human serum albumin-stabilized human IL-2. Mannitol in combination with dextran, sucrose, and trehalose reduced aggregation in lyophilized TNF-α. Sugars stabilize most proteins during lyophilization by protecting against dehydration. Polyvinylpyrrolidone (PVP) and BSA protect some tetrameric enzymes, such as asparaginase, lactate dehydrogenase, and phosphofructokinase, during lyophilization and rehydration by preventing protein unfolding.

Lyophilization is a high cost, long (several days), batch unit operavtion that requires careful formulation and process cycle development.

Lyophilization process involves freezing of a protein solution to a very low temperature (such as $-40°C$), followed by primary drying (removal of water from the frozen state) under vacuum at a higher temperature (such as $-30°C$) and then secondary drying at an even higher temperature (such as $-20°C$). Sometimes, an annealing step is inserted before primary drying and after initial freezing of solution. Annealing involves short-term increase of product temperature to allow reorientation of polymeric proteins and other components with excipients and provides better cake performance. Completion of each drying stage of lyophilization is ascertained through changes in the humidity in the lyophilization chamber (which indicates the rate of evaporation of water), change in the condensate weight or volume at the vacuum pump (which indicates amount of water removed), and/or change in product temperature through temperature probes inserted in vials (which indicates changes in the heat of sublimation).

REVIEW QUESTIONS

25.1 Most protein drugs have poor oral absorption because of their
   A. Large molecular size and high hydrophilicity
   B. Poor transport via the paracellular route
   C. Negligible passive diffusion
   D. Degradation in the GI tract
   E. All of the above
25.2 Protein aggregation in an oral solution formulation can be minimized by
   A. Conjugation to PEG
   B. pH optimization
   C. Addition of certain polymers
   D. Addition of sugars to the formulation
   E. All of the above
25.3 Which of the following enzymes is most responsible for protein metabolism and degradation?
   A. Proteases
   B. Kinases
   C. Oxidases
   D. Phosphorylases
25.4 Which of the following antioxidant is not a metal chelator?
   A. EDTA
   B. EGTA
   C. DTPA
   D. Ascorbic acid
   E. Citric acid

25.5 Which of the following level of protein structure is only possible for proteins that have more than one polypeptide chain?
A. Primary structure
B. Secondary structure
C. Tertiary structure
D. Quaternary structure
E. All of the above

25.6 Which of the following antibody is expected to be least antigenic and immunogenic?
A. Antihuman CD31 mouse monoclonal antibody
B. Antihuman CD31 humanized mouse monoclonal antibody
C. Antihuman CD31 human monoclonal antibody
D. Antihuman CD31 mouse domain antibody
E. Antihuman CD31 mouse-human chimeric monoclonal antibody

25.7 Aqueous protein solubility is least likely to depend on
A. pH
B. Salt concentration
C. Isoelectric point
D. Cosolvent content
E. Preservative content

25.8 Which of the following does not represent an advantage of lyophilization?
A. Improving chemical stability of the protein
B. Ease of handling and transportation
C. Economically cheaper option of protein formulation
D. Reducing the kinetics of degradation reactions

25.9 Which of the following amino acid residues are sensitive to deamidation? Select all that apply.
A. Asparagine
B. Cysteine
C. Glutamine
D. Histidine
E. Proline

25.10 Which of the following amino acid residues are sensitive to disulfide exchange? Select all that apply.
A. Asparagine
B. Cysteine
C. Glutamine
D. Histidine
E. Proline

25.11 Which of the following amino acid residues are sensitive to oxidation? Select all that apply.
A. Asparagine
B. Cysteine

    C. Glutamine

    D. Histidine

    E. Proline

25.12 Which statements are TRUE and which ones are FALSE?

    A. The secondary structure of proteins refers to the conformation of the polypeptide backbone.

    B. Oxidation of methionine to methionine sulfoxide can be reversed with a suitable reducing agent.

    C. The peptide bond between aspartic acid and proline are susceptible to hydrolysis at acidic pH.

    D. The amide groups of asparaginyl and glutaminyl residues are labile at acidic pH.

    E. High residual moisture content may induce aggregation of the lyophilized proteins.

    F. Exposure to hydrophobic surfaces may promote protein aggregation.

25.13 Protein denaturation

    A. Can be either reversible or irreversible

    B. Can be caused by exposure to hydrophobic surfaces

    C. Can be induced extreme pH

    D. All of the above

25.14 Name three functional groups in proteins, which can be used for conjugation.

25.15 Why to PEGylate protein drugs? How will you PEGylate insulin?

FURTHER READING

Berange J (2000) Physical stability of proteins. In Frokjaer S and Hovgaard L (Eds.) *Pharmaceutical Formulation Development of Peptides and Proteins*, London: Taylor & Francis Group, pp. 89–112.

Florence AT and Attwood D (2006) *Physicochemical Principles of Pharmacy*, 4th ed., London: Pharmaceutical Press.

Goolcharan C, Khossravi M, and Borchardt RT (2000) Chemical pathways of peptide and protein degradation. In Frokjaer S and Hovgaard L (Eds.) *Pharmaceutical Formulation Development of Peptides and Proteins*, London: Taylor & Francis Group, pp. 70–88.

Ha E, Ganguly M, Li Xm Jasti BR, and Kompella UB (2004) Delivery of peptide and protein drugs. In Ghosh TK and Jasti BR (Eds.) *Theory and Practice of Contemporary Pharmaceutics*, Boca Raton, FL: CRC Press, pp. 525–547.

Lee VHL (Ed.) (1991) *Peptide and Protein Drug Delivery*, New York: Marcel Dekker.

Walsh G (2002) *Proteins: Biochemistry and Biotechnology*, New York: John Wiley & Sons.

Woodbury CP Jr. (2006) Proteins. In Groves MJ (Ed.) *Pharmaceutical Biotechnology*, 2nd ed., London: Taylor & Francis Group, pp. 5–29.

Chapter 26

Biotechnology-based drugs

LEARNING OBJECTIVES

On the completion of this chapter, the students should be able to

1. Define transcription and translation.
2. Define gene, gene expression, antisense oligonucleotides, and gene therapy.
3. Describe the three basic components of gene medicines.
4. Discuss the characteristics of viral and nonviral gene therapy.

26.1 INTRODUCTION

Almost all human diseases are the result of inappropriate protein production or due to some structural disorder that impacts protein performance. *Traditional small molecule drugs* are designed to interact with protein molecules that support or cause diseases. *Protein drugs* seek to replace the defective protein in cases where missing or defective protein is the cause of the disease. *Enzyme replacement therapy* is a common example of protein drugs that seek to replace inherently defective enzymes. Many severe and debilitating diseases (e.g., diabetes, hemophilia, and cystic fibrosis) and several chronic diseases (i.e., hypertension, ischemic heart disease, asthma, Parkinson's disease, motor neuron disease, and multiple sclerosis) remain inadequately treated by conventional small molecular weight and protein drugs.

Compared to conventional small molecular weight and protein drugs, *nucleic acid medicines* are designed to suppress or generate endogenous proteins by acting on or with the transcription and translation mechanisms of formation of proteins from the genetic code. These medicines can be administered to patients by conventional routes, such as direct injection, inhalation, or intravenous injection. Several different approaches are used for turning nucleic acids into therapeutics. Among them, antisense oligonucleotides (ODNs), RNA interference (RNAi) technologies, plasmid DNA and virus-based *gene therapy* approaches are most widely studied. Antisense ODNs and small interfering RNAs (siRNAs) aim at inhibiting aberrant protein production, whereas gene therapy aims at using the patient's somatic cells to produce therapeutic proteins needed for treating genetic or acquired diseases. The promise of these nucleic acid drugs is to allow either the production of therapeutic proteins that may be difficult to administer exogenously or the inhibition of abnormal protein production.

26.2 GENE AND GENE EXPRESSION

The information necessary to produce proteins in cells is encoded in the genetic material—the chromosomal strands, which are made of deoxyribonucleic acid (DNA). A *gene* is a sequence of the chromosome that codes for a specific protein. Thus, genes are made of DNA and contain information to produce specific proteins.

Transcription is a nuclear process whereby information from DNA is transferred to messenger ribonucleic acid (mRNA). In this process, the two complementary strands of the DNA partly uncoil. The *sense* strand separates from the *antisense* strand. The *antisense* strand of DNA is used as a template by the transcribing enzymes to assemble complementarity-defined mRNA. Then, mRNA migrates into the cytoplasm, where ribosomes transfer the encoded information in mRNA's base sequence in a complementarity-defined manner to assemble proteins. This process is called *translation*. The long strings of amino acids (called polypeptide chains) thus generated can fold by themselves or assemble with other polypeptide chains to form a specific protein. The covalently linked amino acids constitute the primary structure of proteins. These folding and assembly processes happen through multiple noncovalent (such as hydrophobic, hydrogen bond, and ionic) interactions, leading to the secondary, tertiary, and quaternary structure of proteins. Some proteins additionally undergo posttranslational modifications, such as the attachment of a glycan (carbohydrate) moiety to the protein by a cytoplasmic enzyme. The assembled, functional proteins then migrate to their site of action—which can be membrane, cytoplasmic, intraorgnelle (such as intranuclear or intramitochondrial), or extracellular (secreted proteins, such as hormones).

26.3 GENE SILENCING

Antisense drugs inhibit the existing but abnormally expressed genes by blocking the transcription of DNA or the translation of mRNA.

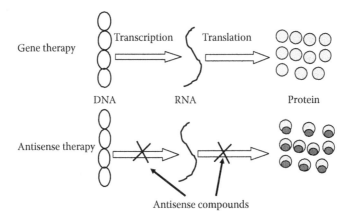

Figure 26.1 Mode of action of nucleic acids. Gene therapy aims at producing therapeutic proteins, whereas antisense therapy aims at blocking the production of aberrant proteins.

Figure 26.1 illustrates the different modes of action of antisense compounds. Overexpression of a particular protein can lead or contribute to a disease state, such as fibrosis and cancer. Antisense drugs are used to stop the production of these aberrant proteins. Antisense drugs work at the genetic level to interrupt the process by which disease-causing proteins are produced. This is true of both host diseases (such as cancer) and infectious diseases (such as acquired immune deficiency syndrome [AIDS]).

26.4 GENE SILENCING TECHNOLOGIES

Gene silencing technologies can be exemplified by antisense ODNs, peptide nucleic acids (PNAs), antisense RNA, aptamers, ribozymes, and double-stranded siRNA.

26.4.1 Antisense oligonucleotides

To create antisense drugs, nucleotides are linked together in short chains, called *ODNs*. When deoxyribonucleotides are linked in small chains, these are called *oligodeoxyribonucleotides*. The sequence of nucleotides in the antisense drugs is complementary to small segments of mRNA. Each antisense drug is designed to bind to a specific sequence of nucleotides in its mRNA target to inhibit production of the protein encoded by the target mRNA. By acting at this earlier stage in the disease-causing process to prevent the production of a disease-causing protein, antisense drugs have the potential to provide greater therapeutic benefit than traditional drugs—which do not act until the disease-causing protein has already been produced. Antisense drugs also have the potential to be much more selective or specific than traditional drugs, and therefore more effective, because they bind to specific mRNA targets through multiple points of interaction at a single binding site. An oligomer of about 15–20 nucleotides in length is considered to be the best because this corresponds to both the appropriate length of a single unique target site in mRNA and the length required for effective hybridization. Cellular uptake of ODNs occurs by means of fluid-phase pinocytosis and/or receptor-mediated endocytosis.

Figure 26.2 shows the structures of antisense compounds. Unmodified ODNs are polyanions with a phosphodiester backbone. They are rapidly degraded under physiological conditions by enzymes called nucleases, primarily 3'-exonucleases. Because of this, ODN modifications have been designed to prevent or reduce the rate of degradation. The phosphorothioate modification of the ODN backbone, in which a sulfur atom replaces one of the nonbridging oxygen atoms in the phosphate group, produces ODNs that are relatively resistant to cellular and serum nucleases. Methylphosphonate ODNs have no net charge, which prevents nuclease digestion, but also decreases water solubility (Figure 26.2a).

26.4.2 Triplex-forming oligonucleotides

In contrast to antisense ODNs, triplex-forming oligonucleotides (TFOs) inhibit gene transcription by forming DNA triple helices in a sequence-specific manner on polypurine–polypyrimidine tracts. Targeting TFOs to

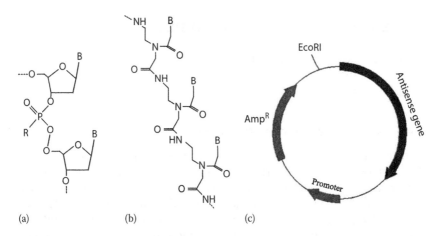

(a)   (b)   (c)

Figure 26.2 Structures of antisense compounds. (a) antisense oligonucleotide (ODN), (b) peptide nucleic acid (PNA), and (c) antisense RNA. R: −O, phosphodiester oligonucleotide; −S, phosphorothioate oligonucleotide; −CH$_3$, methylphosphonate oligonucleotide.

the gene itself presents several advantages compared to antisense ODNs, which are directed to mRNA. There are only two copies of targeted gene, whereas there are thousands of copies of mRNA. Blocking mRNA translation does not prevent the corresponding gene from being transcribed, which continuously repopulates the RNA pool. In contrast, prevention of gene transcription can bring down the mRNA concentration in a more efficient and long-lasting way.

DNA normally exists in a duplex form (two strands coiled around each other). However, under some circumstances, DNA can assume triple helix structures. Triplex helix formation may then prevent the interaction of various transcription factors, or it may physically block the initiation or elongation of the transcription complex. This process is used in the application of TFOs. The TFOs are specific DNA duplex binding ODNs that can bind to DNA duplex, leading to triple helix formation and prevention of the transcription process.

26.4.3 Peptide nucleic acids

PNA has a chemical structure similar to DNA and RNA but differs in the composition of its backbone. DNA and RNA have a deoxyribose and ribose sugar backbone, respectively, whereas PNA's backbone is composed of repeating N-(2-aminoethyl)-glycine units linked by peptide bonds. The various purine and pyrimidine bases are linked to the backbone by methylene carbonyl bonds. PNAs are depicted similar as peptides, with the N-terminus at the first (left) position and the C-terminus at the right (Figure 26.2b). As the backbone of PNA contains no charged phosphate groups, the binding between PNA/DNA strands is stronger than between DNA/DNA strands due to the lack of electrostatic repulsion. PNAs are resistant to both nucleases and proteases (enzymes that digest proteins). PNAs can bind to DNA and RNA targets in a sequence-specific manner to form PNA/DNA and PNA/RNA Watson–Crick double helical structures. Therefore, PNAs can be used as antisense medicines similar to ODNs.

26.4.4 Antisense RNA

The antisense mRNA strategy relies on the transfection and subsequent expression of a plasmid vector whose gene expression cassette carries the cDNA of the gene of interest subcloned into the vector in an antisense orientation (Figure 26.2c). After *transfection* (the process of introducing foreign genetic material into cells) into the cells, the plasmid expresses the antisense mRNA within the cell cytoplasm. This antisense mRNA can hybridize exclusively with the mRNA of the gene of interest and can block protein synthesis (translation). Hence, antisense mRNA gene medicines require expression vectors and delivery systems similar to gene therapy medicines (discussed later in this chapter).

26.4.5 MicroRNA

MicroRNAs (miRNAs) are endogenous, noncoding, small (~22 nucleotides) double-stranded RNAs that participate in gene silencing and posttranscriptional regulation of gene expression. The miRNAs regulate several cellular processes to impact overall outcomes in areas such as cell survival and fat metabolism. Each miRNA has multiple targets and makes global changes in the cellular systems. Changes in the expression level of miRNAs are associated with phenotypic or performance differentiation among different cells. For example, high and low titer producing chinese hamster ovary (CHO) cell lines have shown differences in the concentrations of specific miRNAs. Manipulating the levels of miRNAs in CHO cells leads to increased recombinant protein production by influencing cell proliferation, increasing cellular resistance to apoptosis, and increasing specific productivity.

Gene medicines that utilize the miRNA pathway might either be exogenously administered miRNAs or gene silencing therapies that block the production of endogenous miRNAs.

26.4.6 Aptamers

Aptamers are single-stranded or double-stranded nucleic acids that can bind proteins involved in the regulation and expression of genes (i.e., transcription factors). In addition, they also bind to proteins that perform other regulatory functions. For example, a 15-mer (i.e., 15 nucleotide long) DNA aptamer binds to human thrombin and prevents thrombin-catalyzed blood coagulation. In this approach, the target site is extracellular, and hence the aptamer nucleic acid does not have to cross the cell membrane to be effective, after parenteral administration.

26.4.7 Ribozymes

Ribozyme, also known as RNA enzyme or catalytic RNA, is an RNA molecule having catalytic enzyme activity that uses either transesterification or a hydrolysis mechanism to cleave a unique phosphodiester bond in a single-stranded RNA molecule in a sequence-dependent manner. This process, therefore, leads to interference with the translation process.

26.4.8 RNA interference

RNAi is a phenomenon in which double-stranded RNA (dsRNA) molecules efficiently and specifically inhibit gene expression at a posttranscriptional level. Endogenous mRNA exists as a single strand. Mammalian cells have a specific enzyme, called Dicer, which recognizes the double-stranded RNA and chops it up into small fragments of between 21–25 base pairs in length. Such a short double-stranded RNA fragment is called siRNA. The siRNA can bind certain cellular proteins to form the RNA-induced silencing complex (RISC). The RISC gets activated when the siRNA unwinds. The activated complex binds to the mRNA corresponding to the antisense RNA. Thus, siRNA silences a target gene by binding to its complementary mRNA and by triggering its degradation. The mechanisms of RNAi are illustrated in Figure 26.3.

Potent knockdown of the target gene with high sequence specificity makes siRNA a promising therapeutic strategy. Three different ways are commonly used for producing siRNA: chemical synthesis, administration of plasmid DNA, and viral vectors encoding small hairpin RNA (shRNA) expression cassette. The transcription of genetic sequence in the plasmid DNA or viral vector leads to the production of an mRNA that has internal self-complementarity, which leads to the formation of double strand with a closed hairpin-like loop at one end (shRNA). The shRNA becomes a substrate for the Dicer, leading to endogenous formation of siRNA.

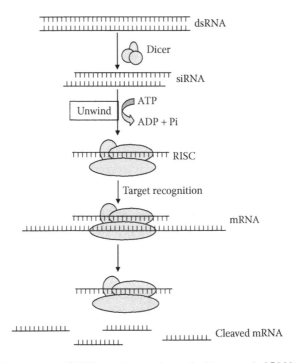

Figure 26.3 Mechanisms of RNA interference. Long double-stranded RNA is cleaved by Dicer into fragments of 21–23 nucleotide siRNAs. Following unwinding, the antisense strand of duplex siRNA is incorporated into RNA-induced silencing complex (RISC) protein. Subsequently, the incorporated siRNA stand guides RISC to its homologous target mRNA for endonucleolytic cleavage.

26.5 GENE THERAPY

Gene therapy is a method for the treatment or prevention of disease that uses genes to provide the patient's somatic cells with the genetic information necessary to produce specific therapeutic proteins needed to correct or to modulate a disease. The promise of gene therapy is to overcome limitations associated with the administration of therapeutic proteins, including low bioavailability, inadequate pharmacokinetic profiles, and high manufacturing cost. Gene therapy approaches are utilized for treating genetic and acquired diseases.

Two approaches are currently used for gene transfer: viral and nonviral. Viral gene transfer utilizes a genetically modified natural virus with a part of the viral genome replaced by a therapeutic gene (called *transgene*) and making the virus replication deficient. Viruses have evolved to efficiently penetrate cells and transfer their genetic material into host cells (a process called *transduction*). Thus, a viral vector efficiently transfers the desired genetic material into target cells, leading to transgene expression. Several different viral vectors have been developed for gene therapy, including retrovirus, adenovirus, adeno-associated virus (AAV), and herpes simplex virus (HSV). The advantages and disadvantages of different viral vectors are listed in Table 26.1.

AAVs are the most common viral vectors used in the clinic. This is mainly due to their low safety risk compared to other viral vectors. Adenoviruses and AAVs do not integrate their genes into the host cell genome. Retroviruses, on the other hand, integrate their genetic material with the host cell genome. Thus, the duration of gene expression is much longer with retroviruses (weeks to months) as compared to adenoviruses and AAVs (days to weeks).

26.5.1 Retroviral vector

Retroviral vectors are RNA viruses (i.e., their genome is RNA) possessing the main feature of reverse transcribing their viral RNA genome into a double-stranded viral DNA. Retroviral vectors can stably insert into the host DNA. The retroviral genome consists of three encoding regions (portions of DNA strand that code for specific functional proteins) responsible for viral replication: (1) *gag* region, encoding group-specific antigens and proteins; (2) *pol* region, encoding reverse transcriptase; and (3) *env* region, encoding viral envelope protein. These regions are flanked on either side by a long terminal repeat (LTR) region. LTRs are responsible for regulation and expression of the viral genome.

These vectors can carry foreign genes of <8 kb (kilo base pair length). They carry an inherent risk of mutagenesis by inserting their genome (called insertion mutagenesis) within a functional gene, which can compromise the functionality of a critical normal human protein.

Table 26.1 Characteristics of viral vectors

Viral vector classification	DNA/ RNA	Insertion size (kb)	Expression	Advantages	Disadvantages
Retrovirus (MMLV)	RNA	9.0	Stable	Integrates, no immune response	Low viral titer, transduces only dividing cells, insertional mutations
Retrovirus (lentivirus)	RNA	9.0	Stable	Transduces nondividing cells	Low viral titer, transduces only dividing cells, insertional mutations
Adenovirus	dsDNA	7.5	Transient	High viral titer, transfects both dividing and nondividing cells	Immunogenic, transient expression
Adeno-associated virus (AAV)	ssDNA	4.5	Stable	Little immunogenicity, integrates	Low transfection efficiency
Herpes simplex virus (HSV)	dsDNA	30	stable	Can target neuronal tissues	Immunogenic (?)

Note: MMLV, Moloney murine leukemia virus.

Defective retroviral vectors are devoid of the genes encoding viral proteins but retain the ability to infect cells and insert their genes into the chromosomes of the target cells. Members of this class include the Moloney murine leukemia viruses (MuLVs) and the lentiviruses.

26.5.1.1 MuLV

MuLV consists of three functional genes: *gag, pol,* and *env,* flanked by the viral LTRs. Removing these structural genes and inserting therapeutic genes in their place make muLV-based vectors. MuLV-derived vectors integrate exclusively in dividing cells.

26.5.1.2 Lentiviruses

The human immunodeficiency virus (HIV) is a lentivirus and is known to cause AIDS. Their special ability to infect and integrate into nondividing cells has application for the construction of lentiviral vectors for gene delivery into nondividing, terminally differentiated cells such as neuronal tissue, hematopoietic cells, and myofibers.

26.5.2 Adenoviral vectors

Adenoviruses are nonenveloped DNA viruses carrying linear double-stranded DNA of about 35 kb length. The base pair length genome carrying capacity of a virus is limited by the size of genome that can be accommodated within the capsid. Adenoviral vectors infect both dividing and nondividing cells. Adenoviral vectors do not integrate into the host cell chromosomes. Genes introduced into cells using adenoviral vectors are maintained extrachromosomally in the nucleus and provides transient transgene expression.

For producing transgene-containing therapeutic adenoviruses, their genome is modified by deletion of the viral replication specific gene known as early gene 1 (*E1A*). This also creates space for the insertion of the desired gene. Adenoviral vectors are based on natural adenoviruses of serotypes 2 and 5. In these first generation adenoviral-vectors, additional partial deletions of *E1B* and *E3* genes can be made to create more space for transgene insertion.

An advantage of adenoviruses over retroviral vectors is achievement of very high viral titers. This suggests efficient gene transfer. Their key disadvantages are their episomal (extrachromosomal) status in the host cell that permits only transient expression of the therapeutic gene. Furthermore, expression of the E2 viral protein provokes inflammatory reactions and toxicities that limit repeated application of adenoviral vector for therapeutic benefit.

26.5.3 Adeno-associated virus vectors

AAV is a single-stranded DNA virus and belongs to the family of parvoviruses. For example, AAV-2 is a nonpathogenic human virus, and the wild type AAV-2 genome establishes a latent infection in human cells, where the viral genome integrates into the chromosomal DNA in a site-specific manner. AAV requires an adenovirus or a herpes virus for viral replication. Compared to adenoviruses, AAV has low immunogenicity. It has a limited capacity for insertion of foreign genes ranging only from 4.1 to 4.9 kb. For construction of rAAV-based vectors, the *rep* and *cap* genes (which are responsible for the production of proteins that would *rep*licate the virus or produce structural proteins for the *cap*sid) are replaced by the therapeutic genes.

26.5.4 Herpes simplex virus vectors

Herpes simplex virus 1 (HSV-1) is a DNA virus possessing a double-stranded linear genome of 150 kb. The HSV affords large packaging capacity for insertion of foreign genes. HSV-1 can infect both dividing and nondividing cells. HSV has natural tropism toward neuronal cells, and this property can be exploited for gene therapies for neuronal tumors. HSV-1 particles are relatively stable and can be concentrated to high virus titers, which are valuable for low-volume administration of a large number of viral particles. The virus does not integrate into the host genome, and, therefore, exhibits transient gene expression in infected cells.

26.5.5 Nonviral gene expression system: Plasmid vectors

Unlike viral vectors, which have many inherent risks, such as inflammation and the potential to generate host immune response (both cellular and humoral), plasmid-based nonviral vectors are fairly safe. As illustrated in Figure 26.4, *three essential components of gene medicines* are a therapeutic gene that encodes a specific therapeutic protein; a gene expression system that recruits host cell machinery to allow the transcription of the encoding gene within a target cell; and a gene delivery system that translocate the expression system to specific location within the body and across the cell membrane barrier.

The gene and the gene expression system are the components of plasmid DNA, which is a circular double-stranded DNA molecule. Basic components of a gene expression plasmid are illustrated in Figure 26.5. Plasmid-based gene expression systems contain a cDNA sequence coding for a therapeutic gene and several other genetic elements, including introns, polyadenylation sequences, and transcript stabilizers to control transcription, translation, and protein stability. Optional components can be added

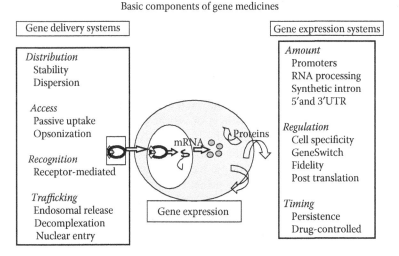

Figure 26.4 Basic components of a nonviral gene medicine. Therapeutic gene, gene delivery system, and gene expression plasmid are the three basic components of a nonviral gene medicine.

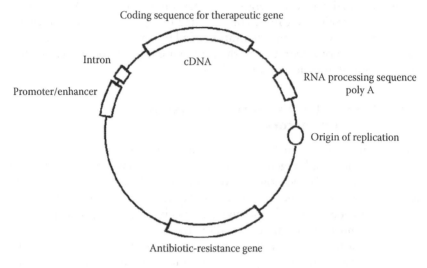

Figure 26.5 Basic components of a gene expression plasmid.

to an expression plasmid, such as a gene switch, which enables expression of the therapeutic protein to be turned on or off after oral administration of a specific low molecular weight drug. The gene delivery system distributes the plasmid to the desired target cells and promotes its internalization into the cells. Once inside the cytoplasm, the plasmid can then translocate to the nucleus, where gene expression begins through the natural cellular processes of transcription and translation.

26.5.6 Gene delivery systems

Plasmid DNA is a long polyanionic polymer. Depending on the number of base pairs, its hydrodynamic size can range from 100 to 200 nm. As the cell membrane is negatively charged, electrostatic repulsion is a barrier to the cell membrane translocation of the plasmid DNA. This barrier is commonly overcome through the use of polycationic lipids, polymers, or lipopolymers utilized as gene delivery systems. Most commonly used synthetic gene carriers are cationic polymers and lipids, which condense plasmids into small particles and protect them from degradation by nucleases. These positively charged lipids, polymers, or lipopolymers interact with the negatively charged plasmid DNA in aqueous solution to form condensed colloidal particles with low hydrodynamic diameter and an overall positive charge, which have higher cellular uptake. The positively charged gene delivery systems can have interactions with other physiological proteins that are negatively charged, leading to toxicities. The apparent potency of a plasmid is further reduced by its chemical, enzymatic, and colloidal instability, sequestration by cells of the immune system, uptake and adsorption by nontarget cells and structures, access to target tissues, cellular uptake, and trafficking to the nucleus of the cells. Therefore, there is a growing need for novel delivery systems, which should be safe for repeated administration.

26.5.6.1 Lipid-based gene delivery

Plasmids may be incorporated into cationic or neutral liposomes. With the right selection of lipids for making liposomes, the liposomes can be made pH sensitive so that they are fusogenic (i.e., fuse with the cell membrane) at acidic pH. This feature has been used to facilitate the endosomal disruption and subsequent release of plasmids in the cytoplasm. The cellular uptake process involves incorporation of a plasmid in an intracellular organelle called endosome. The endosome slowly acidifies its contents in an attempt to degrade its contents. The acidified endosome is called a lysosome. The fusogenic lipids fuse with the lysosomal cell membrane as the pH becomes acidic, thus disrupting the lysosome and releasing its cargo. Thus, the plasmid DNA escapes into the cytoplasm without getting degraded within the lysosome.

Transfection reagent Lipofectin™, for example, is a cationic liposome composed of 1:1 w/w mixture of the cationic lipid N[1-(2,3-dioleyloxy) propyl]-N,N,N-trimethylammonium chloride (DOTMA) and the colipid dioleoyl phosphatidylethanolamine (DOPE). Cationic lipids interact electrostatically with the negatively charged phosphate backbone of DNA, neutralizing the charges and promoting the condensation of DNA into a more compact structure. Usually, cationic lipids are mixed with a zwitterionic or neutral colipid such as DOPE or cholesterol, respectively to form liposomes or micelles. The cationic lipid and colipid are mixed together in chloroform, which is then evaporated to dryness. Water is added to the dried lipid film, and the hydrated film is then either extruded or sonicated to form cationic liposomes. Cationic liposomes have also been prepared by an ethanol injection technique, whereby lipids dissolved in ethanol at a high concentration are injected into an aqueous solution to form liposomes.

As shown in Figure 26.6, the general structure of a *cationic lipid* has three parts: (1) a *hydrophobic lipid anchor group*, which helps in forming liposomes (or micellar structures) and can interact with cell membranes; (2) a positively charged *headgroup*, which interacts with plasmid, leading to its condensation; and (3) a *linker group* that connects the lipid anchor with the charged headgroup. The net charge of the complex has a significant effect on transection efficiency (i.e., efficiency of cellular transfer of plasmid DNA) and DNA stability. Usually, positively charged complexes show high transfection efficiency *in vitro*. The relative proportions of each component and the structure of the head group influence the physicochemical properties of liposome/plasmid complexes.

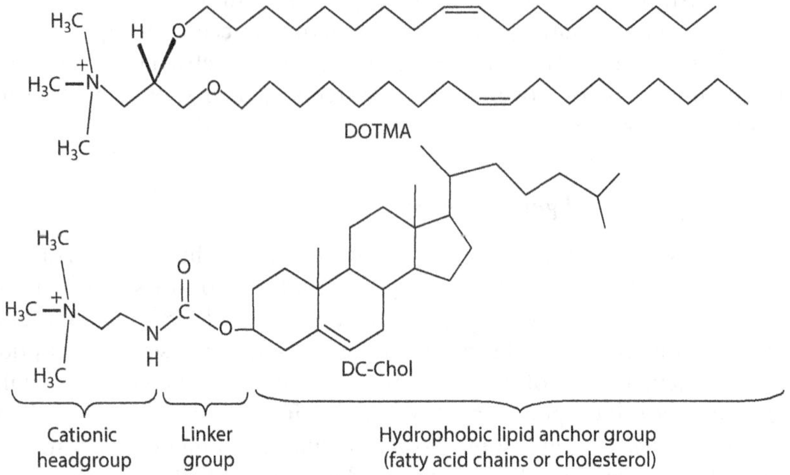

Figure 26.6 Basic components of a cationic lipid. (a) hydrophobic lipid group, (b) linker group, and (c) cationic headgroup.

26.5.6.2 Peptide-based gene delivery

For site-specific delivery of plasmids, positively charged macromolecules, such as poly(L-lysine) (PLL), histones, protamine, or poly(L-ornithine) may be linked together to a cell-specific ligand and then complexed to plasmids via electrostatic interaction. The resulting complexes retain their ability to interact specifically with target cell receptors, leading to receptor-mediated internalization of the complex into the cells. Receptor ligands currently being investigated include glycoproteins, transferrin, polymeric immunoglobulin, insulin, epidermal growth factor (EGF), lectins, folate, malaria circumsporozoite protein, α2-macroglobulin, sugars, integrins (asp-gly-asp [RGD] peptides), thrombomodulin, surfactant protein A and B, mucin, and the c-kit receptor.

Site-specific gene delivery and expression are influenced by the extent of DNA condensation, the method of complexation, the molecular weights of both polycations and plasmids, and the number of ligand residues bound per polycation molecule. To avoid high cytotoxicity, molecular heterogeneity, and possible immunogenicity of PLL and polyethylenimine (PEI), molecularly homogenous lysine and arginine-rich peptide-based gene delivery systems are being investigated. Peptides with moieties that provide cooperative hydrophobic behavior of the alkyl chains of cationic lipids would improve the stability of the peptide-based DNA delivery systems. Short synthetic peptides containing the first 23 amino acids of the HA2 subunit of influenza hemagglutinin protein (HA) are attractive because of their pH-dependent lytic properties, with little activity at pH 7 but greater than or equal to a 100-fold increase in transfection efficiency at pH 5.

26.5.6.3 Polymer-based gene delivery

Polymeric biomaterials can be tailored to interact more on cellular and protein levels to achieve high degrees of specificity, activity, and functionality. These polymeric materials include (1) polymers that can be covalently attached to proteins and antibodies to form drug conjugates, (2) stimuli sensitive polymers, (3) polymer/cell matrix, (4) functional biodegradable polymers, and (5) polymeric gene carriers. These polymers are being utilized for the delivery of proteins, ODNs, and genes. Noncondensing polymers, such as polyvinylpyrrolidone (PVP) and pluronics, can also be used for delivery of nucleic acids to muscles and tumors. These polymer-based DNA formulations are hyperosmotic and result in an improved dispersion of plasmids through the extracellular matrix of solid tissues, such as muscles or solid tumors, possibly by protecting plasmids from nuclease degradation, dispersing plasmids in the muscle, and facilitating their uptake by muscle cells. Synthetic polymers offer a wide array of choices in influencing different aspects of DNA condensation, targeting, cellular uptake, intracellular release, and bioactivity.

REVIEW QUESTIONS

26.1 The term gene therapy refers to a method
  A. For the treatment or prevention of disease by allowing the patient's cells to produce specific therapeutic proteins
  B. For the treatment of genetic as well as acquired or chronic diseases
  C. Which allows production of therapeutic protein or inhibition of abnormal protein production
  D. Which allows somatic or germ-line cells to produce therapeutic/reporter proteins
  E. All of the above

26.2 Gene therapy has a great potential because
  A. It can control the intracellular production of a gene product in response to a disease
  B. It can restrict the availability of any gene product to specific sites in the body
  C. It can deliver sustained therapeutic protein levels over a prolonged period
  D. Gene expression can be turned on or off in response to the benefits of treatment
  E. All of the above

26.3 Define gene, gene expression, transcription and translation. What are the three basic components of a gene medicine?

26.4 What is antisense therapy? What are the types of antisense compounds? What are the types of antisense oligonucleotides?

26.5 What is RNA interference? Define siRNA, shRNA, and miRNA.

26.6 Describe the essential feature of a gene expression system.

26.7 Describe the influential factors for the development of nonviral gene therapy products.

FURTHER READING

Cheng K and Mahato RI (Eds.) *Advanced siRNA Delivery* (2013) Chichester, UK: John Wiley & Sons.

Cheng K and Mahato RI (Eds.) (2011) Biological and therapeutic applications of small RNAs. *Pharm Res* **28**: 2961–2965.

Crooke ST (2004) Progress in antisense technology. *Annu Rev Med.* **55**: 61–95.

Khalia IA, Kagure K, Akita H, and Harashima H (2006) Uptake pathways and subsequent intracellular trafficking in nonviral gene delivery. *Pharmacol Rev* **58**: 32–45.

Li F and Mahato RI (Eds.) (2015) miRNAs as targets for cancer treatment: Therapeutics design and delivery. *Adv Drug Del Rev* **8**: 1–198.

Mahato RI (Ed.) *Biomaterials for Delivery and Targeting of Proteins and Nucleic Acids*, Boca Raton, FL: Taylor & Francis Group, 2005.

Mahato RI and Kim SW (Eds.) (2002) *Pharmaceutical Perspectives of Nucleic Acid-based Therapeutics*, London: Taylor & Francis Group.

Mahato RI, Cheng K, and Guntaka RV (2005) Modulation of gene expression by antisense and antigene oligodeoxynucleotides and small interfering RNA. *Expert Opin Drug Deliv* **2**: 3–28.

Mahato RI, Smith LC, and Rolland A (1999) Pharmaceutical perspectives of non-viral gene therapy. *Adv Genet* **41**: 95–156.

Smith AE (1995) Viral vectors in gene therapy. *Annu Rev Microbiol* **49**: 807–838.

Answers to review questions

CHAPTER 1

1.1 D.
1.2 B.
1.3 A, C.
1.4 E.
1.5 B.
1.6 C.

CHAPTER 2

2.1 D.
2.2 A. False
    B. False
    C. True
    D. False
    E. False
    F. False
    G. True
2.3 A. FDA: Food and Drug Administration; IND: investigational new drug application; NDA: new drug application; CDER: Center for Drug Evaluation and Research; Biologics: viruses, therapeutic serum, toxin, antitoxin, vaccines, blood, blood components or derivatives, allergic products, or analogous products, applicable to the prevention, treatment, or cure of a disease or condition of a human being.
    B. Refer to the chapter.
    C. Healthy subjects are evaluated in phase I clinical trials of drug product development.
    D. A lead compound is the one that shows high bioactivity and low toxicity.

2.4 A. The CDER evaluates prescription, generic, and OTC drug products for safety and efficacy before they can be marketed. It also monitors all human drugs and biopharmaceuticals once they are in the market, and removes products from the market that may not be manufactured properly or may cause harm to patients. The CBER regulates biologics not reviewed by the CDER, such as vaccines, blood and blood products, gene therapy products, and cellular and tissue transplants.

B. Refer to the chapter.

C. Refer to the chapter.

D. Postmarketing surveillance is necessary as it may contribute to the understanding of the drug's mechanism or scope of action, indicate possible new therapeutic uses, and/or demonstrate the need for additional dosage strengths, dosage forms, or routes of administration. Post marketing surveillance studies may also reveal additional side effects, and rare, serious and unexpected adverse effects.

CHAPTER 3

3.1 B. According to the pH-partition theory, absorption of a weak electrolyte drug depends on the extent to which the drug exists in its unionized form at the absorption site. However, the pH-partition theory often does not hold true, as most weakly acidic drugs are well absorbed from the small intestine because of the large epithelial surface areas of the organ.

3.2 A. The Henderson–Hasselbalch equation for a weak acid and its salt is represented as pH = pKa + log [salt]/[acid], where pKa is the negative log of the dissociation constant of a weak acid and [salt]/[acid] is the ratio of the molar concentration of salt and acid used to prepare a buffer.

3.3 A. False

B. True

C. False

D. True

3.4 A. *Adsorption* is different from *absorption*, which implies penetration through organs and tissues. The degree of adsorption depends on the chemical nature of the adsorbent and the adsorbate, surface area of the adsorbent, temperature, and partial pressure of the adsorbed gas. Adsorption can be physical or chemical in nature.

B. The *pH-partition theory* states that drugs are absorbed from the biological membranes by passive diffusion, depending on the fraction of unionized form of the drug at the pH of that biological membrane. Their degree of ionization depends on both their pK_a and the solution pH. The GI tract acts as a lipophilic barrier and thus ionized drugs will have minimal membrane permeability compared to unionized form of the drug. The solution pH will affect the overall partition coefficient of an ionizable substance. The pK_a of the molecule is the pH at which there is a 50:50 mixture of conjugate acid–base forms. The conjugate acid form will predominate at a pH lower than the pK_a, and the conjugate base form will be present at a pH higher than the pK_a. The following *Henderson–Hasselbalch equations* describe the relationship between ionized and nonionized species of a weak electrolyte:

Weakly Acidic Drugs   Weakly Basic Drugs

$$pH = pK_a + \log^{[A^-]/[HA]} \qquad pH = pK_a + \log^{[B^-]/[BH^+]}$$

The *pH-partition theory often does not hold true.* For example, most weak acids are well absorbed from the small intestine, which is contrary to the prediction of the pH-partition hypothesis. Similarly, quaternary ammonium compounds are ionized at all pHs but are readily absorbed from the GI tract. These discrepancies arise because the *pH-partition theory does not take into account the following:*

- Large epithelial surface areas of the small intestine compensate for ionization effects.
- Long residence time in the small intestine also compensates for the ionization effects.
- Charged drugs, such as quaternary ammonium compounds, may interact with oppositely charged organic ions, resulting in a neutral species, which is absorbable.
- Some drugs are absorbed via active pathways.

   C. i. A highly water-soluble compound will be absorbed poorly as compared to lipophilic compounds.

     ii. A low-molecular weight compound will be absorbed better than a high-molecular weight compound.

3.5 A. Besides providing the mechanism for the safe and convenient delivery of accurate dosage, we need to formulate a drug into pharmaceutical dosage forms for the following additional reasons: (i) to protect the drug substance from the destructive influence of atmospheric oxygen or humidity;(ii) to protect the drug substance from the destructive influence of gastric acid after oral administration;

(iii) to conceal the bitter, salty, or offensive taste or odor of a drug substance; (iv) to provide liquid preparations of substances that are either insoluble or unstable in the desired vehicle; (v) to provide rate-controlled drug action; and (vi) to provide site-specific drug delivery.

B. A drug's partition coefficient is a measure of its distribution in a lipophilic–hydrophilic phase system and indicates its ability to penetrate biological membranes. The octanol–water partition coefficient is used in the formulation development and is defined as P = (concentration of drug in octanol or nonpolar solvent)/(concentration of drug in water polar solvent). The logarithm of partition coefficient (P) is known as log P. The value of log P is a measure of lipophilicity and is used widely because many pharmaceutical and biological events depend on lipophilic characteristics.

C. Nonelectrolytes are substances that do not form ions when dissolved in water. Their aqueous solutions do not conduct an electric current. Electrolytes are substances that form ions in solution. As a result, their aqueous solutions conduct electric current. Electrolytes are characterized as strong or weak. Strong electrolytes (e.g., sodium chloride and hydrochloric acid) are completely ionized in water at all concentrations. Weak electrolytes (e.g., aspirin and atropine) are partially ionized in water.

3.6　A. Eight intrinsic characteristics of a drug substance that must be considered before the development of its pharmaceutical formulation are the following:

- *Drug solubility and pH*: A drug substance must possess some aqueous solubility for systemic absorption and therapeutic response. Enhanced aqueous solubility may be achieved by forming salts or esters, by chemical complexation, or by reducing the drug's particle size. The pH affects solubility and stability. Cosolvents, complexation, micronization, and solid dispersion are used to improve aqueous solubility.
- *Partition coefficient*: The partition coefficient of a drug is a measure of its distribution in a lipophilic–hydrophilic phase system and indicates its ability to penetrate biological membranes.
- *Dissolution rate*: The speed at which a drug substance dissolves in a medium is called its dissolution rate.
- *Polymorphism*: Polymorphic forms exhibit different physicochemical properties, including melting point and solubility, which can affect the dissolution rate and thus the extent of its absorption.

- *Stability*: The chemical and physical stability of a drug substance alone, and when combined with formulation components, is critical to prepare a successful pharmaceutical product. For drugs susceptible to oxidative decomposition, the addition of antioxidant stabilizing agents to the formulation may be required to protect potency. For drugs destroyed by hydrolysis, protection against moisture in formulation, processing, and packaging may be required to prevent decomposition.
- *Membrane permeability*: To produce a biological response, the drug molecule must first cross a biological membrane. The biological membrane acts as a lipid barrier to most drugs and permits the absorption of lipid-soluble substances by passive diffusion, whereas lipid-insoluble drugs can diffuse across the barrier only with considerable difficulty.
- *Partition coefficient*: The octanol–water partition coefficient is used in formulation development. P = (concentration of drug in octanol)/(concentration of the drug in water).
- pK_a/*dissociation constants*: The extent of ionization or dissociation is dependent on the pH or the medium containing the drug.

B. The pH-partition theory often does not hold true, as most weakly acidic drugs are well absorbed from the small intestine, possibly because of the large epithelial surface areas of the organ. Drugs have a relatively long residence time in the small intestine, which also compensate for ionization effects.

3.7 According to the pH-partition theory, absorption of a weak electrolyte drug depends on the extent to which the drug exists in its unionized form at the absorption site. According to the Henderson–Hasselbalch equation

$$pK_a = pH + \log \frac{[HA]}{[A^-]}$$

$$\log \frac{C_u}{C_i} = pK_a - pH$$

$$\log \frac{C_u}{C_i} = 3.5 - 2 = 1.5$$

where:
C_u is the concentration of unionized drug
C_i is the concentration of ionized drug

$$\frac{C_u}{C_i} = \text{antilog } 1.5 = 31.62:1$$

In the plasma

$$pK_a = pH + \log\frac{[HA]}{[A^-]}$$

$$\log\frac{C_u}{C_i} = pK_a - pH$$

$$\log\frac{C_u}{C_i} = 3.5 - 7.4 = -3.9$$

$$\frac{C_u}{C_i} = \text{antilog}(-3.9) = 0.00125$$

Therefore, most of the administered aspirin remains unionized in the stomach and thus it is rapidly taken up by the stomach, leading to gastric bleeding.

3.8 Six physicochemical properties of a drug that influence drug absorption are (1) molecular weight, (2) drug solubility, (3) pK_a, (4) log P, (5) polymorphism, and (6) stability. Physicochemical properties of a drug can be improved by salt formation, bioconjugation, the use of cosolvents, and its use as a prodrug.

3.9　$pK_a = pH + \log\dfrac{[BH^+]}{[B]}, \rightarrow 7.15 = 7.4 + \log\dfrac{[BH^+]}{[B]}$

$$\frac{[B]}{[BH^+]} = \text{antilog}\,0.25 = 1.78$$

$$[B]\% = \frac{[B]}{([B]+[BH^+])} * 100\% = \frac{1}{(1+1.78)} * 100\% = 36.0\%$$

3.10　$pK_b + pK_a = pK_w, \rightarrow pK_a = pK_w - pK_b, pK_a = 14.0 - 5.6 = 8.4$

At pH 4.5,

$$pK_a = pH + \log\frac{[BH^+]}{[B]}, \rightarrow 8.4 = 4.5 + \log\frac{[BH^+]}{[B]}$$

$$\frac{[BH^+]}{[B]} = \text{antilog}(-3.9) = 0.000126$$

$$[B]\% = \frac{[B]}{([B]+[BH^+])} * 100\% = \frac{1}{(1+0.000126)} * 100\% = 99.99\%$$

At pH 8.0,

$$pK_a = pH + \log\frac{[BH^+]}{[B]}, \rightarrow 8.4 = 8.0 + \log\frac{[BH^+]}{[B]}$$

$$\frac{[B]}{[BH^+]} = \text{antilog}(-4.0) = 0.398$$

$$[B]\% = \frac{[B]}{([B]+[BH^+])} * 100\% = \frac{1}{(1+0.0398)} * 100\% = 71.53\%$$

3.11 $$pH = pK_a + \log\frac{[salt]}{[acid]}, \rightarrow 5.0 = 6.0 + \log\frac{[salt]}{[acid]}$$

$$\frac{[acid]}{[salt]} = \text{antilog}(1.0) = 10{:}1$$

CHAPTER 4

4.1 B.

4.2 B.   In passive transport, a drug travels from high concentration to a low concentration, whereas active transport moves drug molecules against a concentration gradient and requires energy.

4.3 C.   Fick's first law of diffusion states that the amount of material flow through a unit cross section of a barrier in unit time, which is known as the flux, is proportional to the concentration gradient. Fick's first law of diffusion describes the diffusion process under steady-state conditions when the concentration gradient does not change with time.

4.4 D.   The Noyes–Whitney equation describes the rate of drug dissolution from a tablet. Fick's first law of diffusion is similar to the Noyes–Whitney equation in that both equations describe drug movement due to a concentration gradient. The Michaelis–Menten equation involves enzyme kinetics, whereas Henderson–Hasselbalch equations are used for determination of pH of the buffer and the extent of ionization of a drug molecule.

4.5 D.   Diffusion coefficient is not a constant. It is affected by changes in the concentration, temperature, pressure, solvent properties, and chemical nature of the diffusant.

4.6 B.   According to the Noyes–Whitney equation, the rate of drug dissolution from a solid dosage form will increase with increase in surface area, which will increase with decrease in particle size or molecular weight of a drug.

4.7 C.   The permeability of a weak electrolyte through a biological membrane depends on the degree of its ionization; the more lipophilic drug will permeate more, which is possible when its partition coefficient increases.

4.8 A. True

B. True

C. True

4.9 Fick's first law of diffusion states that the amount of material (M) flowing through a unit cross section (S) of a barrier in unit time (t) is proportional to the concentration gradient (dC/dx).

$$J = \frac{1}{S}\frac{dM}{dt} = -D\frac{dC}{dx} = \frac{D(C_1 - C_2)}{h}$$

Because $K = C_1/C_d = C_2/C_r$, we can rewrite this equation as

$$\frac{dM}{dt} = \frac{DSK(C_d - C_r)}{h} = \frac{DSKC_d}{h}$$

The rate at which a solid dissolves in a solvent can be determined using the
Noyes–Whitney equation:

$$\frac{dM}{dt} = kS(C_s - C)$$

Under *sink conditions*, when the drug concentration (C) is much less than the solubility of the drug (Cs), we can ignore C ($C \to 0$). A simplified *Noyes– Whitney equation* can be used to measure dissolution rates:

$$\frac{dM}{dt} = KSC_s = \frac{DSC_s}{h}$$

or

$$\frac{dC}{dt} = \frac{kSC_s}{V} = \frac{DSC_s}{Vh}$$

$$\frac{dM}{dt} = \frac{DS(C_d - C_r)}{h}$$

$$D = 5 \times \frac{10^{-4}}{175} \times 1$$

$$D = 2.86 \times 10^{-6}\,cm^2/s$$

4.10 $dM/dt = S \times D \times (C_1 - C_2)/h$, $dM/dt = (2.5 \times 10^3) \times (1.75 \times 10^{-7})$

$$\times (0.35 - 2.1 \times 10^{-4})/(1.25 \times 10^{-4}) = 1.225\,mg/s$$

4.11 $Q/t = kDC_0/h = (6.8 \times 10^{-3})(8 \times 10^{-9}\,cm^2/s)$

$$(0.02\,g/cm^3)/(1.40 \times 10^{-2}\,cm) = 7.77 \times 10^{-11}\,gcm^{-2}s^{-1}$$

To obtain the results in micrograms per day, one must multiply the result by 10^6 µg/g and 86,400 s/24 h day.

$$\frac{Q}{t} = (7.77 \times 10^{-11})\,gcm^{-2}\,s^{-1}(10^6\,\text{µg/g})(86,400\,s/\text{day})$$

$$= 6.71\,\text{µg/cm}^2\text{day}$$

4.12 $C_s \gg C, \dfrac{dM}{dt} = kS(C_s - C) = kSC_s$

$$k = \frac{(dM/dt)}{(SC_s)} = \frac{(\Delta M/\Delta t)}{(SC_s)} = \frac{(0.15/2)}{(0.3 \times 10^4 \times 1.2 \times 10^{-3})} = 0.021\,cm/min$$

CHAPTER 5

5.1 A. 12.5 g × 1,000 mg/1 g = 12,500 mg × mL/50 mg = 250 mL
  B. 400 mg × 5 mL/250 mg = 8 mL
  C. 0.15 × 23 = 3.45 mg
  D. 250/0.15 = 1200/x, x = 4.8 mEq of sodium = 4.8 × 23 = 110.4 mg

5.2 A. 5 mg/kg× 175 lb × 1 kg/2.20 lb = 398 mg daily dose
  B. Number of tablets per day = 398/50 = 8 tablets, Tablets for 10 days = 8 × 10 = 80 tablets

5.3 Child's dose = Adult dose × $\dfrac{\text{Child's BSA in m}^2}{1.73\text{m}^2}$

$$= 8181.81 \times \frac{1.15}{1.73} = 5438\,mg$$

Child's dose = Adult dose × $\dfrac{\text{Child's age in months}}{150\text{ months}}$

$$= 8181.81 \times \frac{96}{150} = 5236.35\,mg$$

$$\text{Child's dose} = \text{Adult dose} \times \frac{\text{Child's age in years}}{\text{Child s age in years} + 12 \text{ years}}$$

$$= 8181.81 \times \frac{8}{8} + 12 = 3272.24 \text{ mg}$$

$$\text{Child's dose} = \text{Adult dose} \times \frac{\text{Child's weight in pounds}}{150 \text{ lb}}$$

$$= 8181.81 \times \frac{80}{150} = 4363.63 \text{ mg}$$

5.4   Sodium chloride equivalent = Molecular weight of sodium chloride × i factor of drug/i factor of sodium chloride × molecular weight of drug

A. E value = 58.5 × 1.8/220 × 2.6 = 0.18
   E value = 58.5 × 1.8/180 × 1 = 0.58
   E value = 58.5 × 1.8/140 × 1.9 = 0.40

B.   Sodium chloride equivalent = (40 × 0.18 + 25 × 0.58 + 100 × 0.40) × 10 − 3 = 0.0617 sodium chloride equivalent

C.   Amount of NaCl equivalent = (0.09 − 0.0617) = 0.0283 g

D. E value of dextrose = 58.5 × 1.8/180 × 1.9 = 0.31

$$\frac{1}{0.31} = \frac{x}{0.0283} x = 0.091 \text{g of dextrose}$$

5.5   A.   4.6 mL × 1.26 g/cm^3 = 5.796 g

B.   25 mL × 5 g/100 mL = 1.25 g
   1.25 g/0.78 = 1.60 mL

C.   25 mL × 5/100 = 1.25 g
   1.25 g/0.78 = 1.60 mL

5.6   A.   111/2 = 55.5 g

B.   2.775 g

C.   1.3875 g

D.   Total dose = 5 mg/kg/day × 150 lb × 0.455 kg/lb = 340 mg/day
   Impurity = 340 mg/day × 40 mg/kg = 0.0136 mg/day

E.   Mole fraction = Moles of NaCl/(Moles of NaCl + Moles of water) = 0.015/(0.015 + 5.50) = 0.0027

F. Concentration given   Proportional parts required

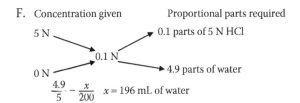

$\frac{4.9}{5} = \frac{x}{200}$ $x = 196$ mL of water

G. Concentration given   Proportional parts required

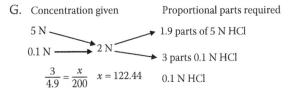

$\frac{3}{4.9} = \frac{x}{200}$ $x = 122.44$ 0.1 N HCl

5.7 A. i. Mean = 3.08
         Median = 3.2
         Variance = 0.95
         Standard deviation = 0.97
      ii. Mean = 13.22
         Median = 13.2
         Variance = 5.48
         Standard deviation = 2.34
      iii. Mean = 9.32
         Median = 9.3
         Variance = 30.60
         Standard deviation = 5.53
     B. Dataset c has the highest spread around the central tendency.
     C. Dataset a has the least spread around the central tendency.
     D. Differences between the means of datasets a and b are most likely
         to be statistically significant.
     E. Differences between the means of datasets b and c are least likely
         to be statistically significant.

CHAPTER 6

6.1 A. Pentaamminebromocobalt(III) sulfate
     B. Hexaammineiron(III) hexacyanochromate (III)
     C. Pentaamine cholorocobalt (III) sulfate

    D. pentaaquahydroxoiron (III) ion

    E. Cyclopentadienyliron dicarbonyl dimmer

6.2 A. $[Fe(NH_3)_6](NO_3)_3$

     B. $(NH_4)_2[CuCl_4]$

     C. $Na_3[FeCl(CN)_5]$

     D. $K_3[CoF_6]$

6.3 B.

6.4 A, B, C.

6.5 E.

6.6 A, B, C.

6.7 The factors affecting plasma protein binding of drugs are as follows:

- The extent of protein binding of many drugs is a linear function of partition coefficient P.
- Plasma protein binding may determine the characteristics of drug action or transport.
- Protein binding changes with drug concentration and protein concentration.
- On increasing the drug/protein ratios, saturation of some sites can occur and there may be a decrease in binding.

6.8 Only a free drug is able to cross the capillary endothelium. When protein binding occurs with high affinity and the total amount of drug in the body is low, the drug will be present almost exclusively in the plasma. As plasma proteins are large molecules, drugs that are bound to proteins cannot pass out of vascular space. Thus, plasma protein binding will control the distribution of drugs. As plasma protein binding increases, the extent of distribution decreases. However, some drugs may exhibit both a high degree of plasma protein binding and a large volume of distribution. Binding of drugs to plasma proteins is a dynamic equilibrium. If the unbound (or free) drug is able to cross biological membranes, the drug may exhibit an extensive volume of distribution, despite a high degree of protein binding. As a free drug moves across the membranes and out of vascular space, the equilibrium will shift, in essence drawing the drug off the plasma protein to *replenish* the free drug lost from vascular space. This free drug is now also able to traverse membranes and leave vascular space. In this way, a drug with a very low free fraction (i.e., a high degree of plasma protein binding) can exhibit a large volume of distribution. Disease states that alter plasma protein concentration may alter the protein binding of drugs. If the concentration of protein in plasma is reduced, there may be an increase in the free fraction of drugs bound to that protein. Similarly, if pathological changes in binding proteins reduce the affinity of the drug for the protein, there will be an increase in the free fraction of the drug.

CHAPTER 7

7.1 A. Second order
    B. First order
    C. Zero order
    D. First order
    E. Zero order
    F. Second order
    G. First order
    H. Zero and first order
    I. Second order

7.2 B. Stability at room temperature can be predicted from accelerated testing data by the Arrhenius equation: $\log (k_2/k_1) = E_a (T_2 - T_1)/(2.303 \, RT_2T_1)$, where k_2 and k_1 are the rate constants at the absolute temperatures T_2 and T_1, respectively; R is the gas constant; and E_a is the energy of activation. Stokes' equation is used to determine the sedimentation rate of a suspension, whereas the Noyes–Whitney equation is used to determine the dissolution rate.

7.3 I.

7.4 C.

7.5 A. $2H_2O_2 \rightarrow 2H_2O + O_2$

For a first-order reaction,

$$k = \frac{2.303}{t} \log \frac{C_0}{C}, \; k = \frac{2.303}{50 \, \text{min}} * \log \left(\frac{72.6 \text{mL}}{10.6 \text{mL}} \right) = 0.0385 \, \text{min}^{-1}$$

    B. $C = C_0 10^{-kt/2.303} = 72.6 * 10^{-0.0385*30/2.303} = 22.885 \, \text{mL}$

7.6 To calculate k, therefore diethyl acetate and potassium hydroxide have same initial concentration, $x/a(a - x) = kt$, $x =$ amount of reacted potassium hydroxide, $a =$ the initial amount, $\therefore k = x/(t * a * (a - x)) = 0.0088 \, \text{mol/L}/(35 \, \text{min} * 0.05 \, \text{mol/L} * (0.05 \, \text{mol/L} - 0.0088 \, \text{mol/L}))$,

$$k = 0.122 \, \text{L, mol}^{-1} \text{min}^{-1}$$

To calculate $t_{1/2}$, i.e., $x = 1/2a$,
$\therefore t = t_{1/2} = 1/2a/(ak(a. - 1/2a)) = 1/(ak) = 1/(0.05 \, \text{mol/ L} * 0.122$
$\text{L mol}^{-1} \text{min}^{-1}) = 163.86 \, \text{min}$

7.7 $\because \log \dfrac{k_2}{k_1} = \dfrac{E_a}{2.303R}\left(\dfrac{(T_2 - T_1)}{T_2 T_1}\right)$

$\log\left(\dfrac{3.8}{2.0}\right) = \dfrac{E_a}{2.303} * (423 - 383)/(423 * 383)/1.987$

$\therefore E_a = 5166.4\,\text{cal/mol};$

for the frequency factor A, $\log k = \log A - (E_a/2.303RT)$,

$\therefore \log\left(\dfrac{3.8}{3600\ \text{s}^{-1}}\right) = \dfrac{\log A - 5166.4}{(2.303 * 1.987 * 423)}$

$A = 0.493\ \text{s}^{-1}$

7.8 According to $t_{90(T_2)} = t_{90(T_1)}/Q_{10}^{\Delta T/10}$, using Q_{10} of 5,

life at 37°C = $21\,\text{days}/5^{[(37-5)/10]} = 0.12$ day or 2.92 h

CHAPTER 8

8.1 A. Chemical absorption is an irreversible specific process and may require activation energy, whereas physical adsorption is reversible and associated with van der Waals forces.

8.2 Adsorption is different from absorption, which implies penetration through organs and tissues. Physical adsorption is associated with van der Waals forces and is reversible. Removal of the adsorbate from the adsorbent is known as desorption. Physical adsorption is rapid, relatively weak, and nonspecific. Chemical adsorption (also known as *chemisorption*) is irreversible and in this the adsorbate is attached to the adsorbent by chemical bonds. Chemical adsorption is specific. It may require activation energy and therefore be slow, and only a monomolecular chemisorbed layer is possible.

8.3 The relationship between the amount of gas physically adsorbed on a solid and the equilibrium pressure or concentration is known as the *adsorption isotherm*. The isotherms are classified into five types. Both Freundlich and Langmuir isotherms are of type I, whereas BET is a type II isotherm. Type I isotherms show a fairly rapid rise in the amount of adsorption with increasing pressure, and adsorption is restricted to a monolayer. Type II isotherms are frequently encountered, and represent multilayer physical adsorption on nonporous solids. They are often referred to as *sigmoid isotherms*. Isotherm IV is typical of adsorption onto porous solids. Types III and V isotherms are produced in a relatively few instances in which the heat of adsorption of the gas in the first layer is less than the latent heat of condensation of successive layers.

8.4 Because gases have negligible intermolecular attractions, whereas liquids have significant attractive forces between the liquid molecules.

8.5 A wetting agent lowers the contact angle and aids in displacing an air phase at the surface and replacing it with a liquid phase. There are three types of wetting agents used in suspension formulations: (1) surfactants, (2) hydrophilic colloids, and (3) solvents.

8.6 According to $= 1/2\,rh\rho g$, $\gamma = 1/2 * 0.02 * 6.60 * 1.008 * 981$

$$= 65.3 \text{ dyn/cm}$$

8.7 $W_c = 2\gamma_L = 2 * 25 = 50 \text{ erg/cm}^2$

$$W_a = \gamma_L + \gamma_S - \gamma_{LS} = 25 + 72.8 - 30 = 67.8 \text{ erg/cm}^2$$

CHAPTER 9

9.1 B. Most lyophilic colloids are polymeric molecules including gelatin and acacia; they spontaneously form colloidal solution, and tend to be viscous. Dispersion of lyophilic colloids is stable in the presence of electrolytes.

9.2 D. Surfactants accumulate at the interface and lower the interfacial tension between oil and water phases.

9.3 A. False
 B. True
 C. False
 D. False
 E. False

9.4 Based on their particle size, colloidal systems are classified into molecular dispersions, colloidal dispersions, and coarse dispersions. Only coarse dispersions are visible to the naked eye.

9.5 A. Most substances acquire a surface electric charge when brought into contact with a polar medium, the possible charging mechanisms being ionization, ion adsorption, and ion dissolution.
 • *Ionization*: If the charge arises from ionization, the charge on the particles will be the function of pH and pK_a. Proteins acquire their charge mainly through the ionization of carboxyl and amino groups to give COO^- and NH_3^+ ions. Ionization depends strongly on pH of the medium. At low pH, a protein molecule will be positively charged, $-NH_2 \rightarrow NH_3^+$, and at high pH it will be negatively charged, $-COOH \rightarrow COO^-$. At a certain pH, specific for each individual protein, the total number of positive charges will be equal to the total number of negative charges, and the net charge will be zero. This pH is termed the *isoelectric point* of the protein.

- *Ion adsorption*: A net surface charge can be acquired by the unequal adsorption of oppositely charged ions. Surfaces that are already charged have a tendency to adsorb counterions, which may reverse the surface charge.
- *Ion dissolution*: Ionic substances can acquire a surface charge by virtue of unequal dissolution of the oppositely charged ions of which they are composed. For example, the particle of silver iodide in a solution with excess [I⁻] will carry a negative charge, but the charge will be positive if excess [Ag⁺] is present.

B. i. –COOH and NH$_2$

ii. –COOH has 2.35 and NH$_2$ has 9.69

iii. Low pH, NH$_3^+$; median pH 7, both groups ionized; basic pH, COO⁻

9.6 A. Zeta potential is defined as the difference in potential between the surface of the tightly bound layer of solvent/shear plane and the electroneutral region of the solution.

B. Electrophoretic properties are affected by the net charge on a particle, which includes that of an immobile solvent layer.

9.7 A. When the particles adhere by stronger forces, the phenomenon is called *aggregation*. Because of the large surface free energy of the dispersed-phase particles in emulsions, they tend to associate together by weak van der Waals forces forming light, fluffy conglomerates. This phenomenon is called *flocculation*. *Coagulation* is the condition when the dispersed-phase particles merge with each other to form a single phase.

B. Coagulation is an irreversible process and leads to caking, whereas flocculation is the process of forming light fluffy conglomerates, which are reversible on shaking.

9.8 A. Stokes' law defines the velocity of sedimentation as a function of the viscosity of the medium and the radius and the density of particles as per the following equation:

$$V = \frac{D^2(\rho - \rho_0)g}{18\eta_0'}$$

B. Creaming is the upward movement of dispersed droplets relative to the continuous phase, whereas sedimentation, the reverse process, is the downward movement of particles. These processes take place because of the density differences in the two phases and can be reversed by shaking. However, creaming is undesirable because it provides the possibility of inaccurate dosing and increases the likelihood of some coalescence, which may take place owing to the close proximity of the globules in the cream.

Factors that influence the rate of creaming are similar to those involved in the sedimentation rate as indicated by Stokes' law:

$$v = \frac{d^2(\rho_s - \rho_0)g}{18\eta_0}$$

where:
  v is the velocity of creaming
  d is the globule diameter
  ρ_s and ρ_0 are the densities of disperse phase and dispersion medium
  $\dot{\eta}_0$ is the viscosity of the dispersion medium (poise)
  g is the acceleration of gravity (981 cm/s^2)

C. According to Stokes' equation, we can minimize the rate of creaming and sedimentation by (i) reducing the globule size, (ii) decreasing the density difference between the two phases, and (iii) increasing the viscosity of the continuous phase. This may be achieved by homogenizing the emulsion to reduce the globule size and increasing the viscosity of the continuous phase by the use of thickening agents such as tragacanth or methylcellulose for o/w emulsions and soft paraffin for w/o emulsions.

D. The rate of sedimentation is directly proportional to the diameter of particles if density/shape is the same.

E. Water-soluble compounds will dissolve while being processed, causing increase in viscosity of the medium and reduction in diameter. According to Stokes' law, viscosity increase will affect the results.

9.9 D.

CHAPTER 10

10.1 E. Surface-active agents facilitate emulsion formation by lowering the interfacial tension between the oil and the water phases. Adsorption of surfactants on insoluble particles enables these particles to be dispersed in the form of a suspension.

10.2 C. Increasing the surfactant concentration above the critical micellar concentration will result in no change in surface tension.

10.3 D. Benzalkonium chloride is a cationic surfactant and can interact with bile salts.

10.4 D. Most substances acquire a surface charge by ionization, ion adsorption, and ion dissolution. At the isoelectric point, the total number of positive charges is equal to the total number of negative charges.

10.5   A. Surfactants are used as emulsifying agents, solubilizing agents, detergents, and wetting agents.

B. Anionic, cationic, and nonionic surfactants.

C. Because a surfactant having a HLB value of 18 is highly hydrophilic and does not deposit on the interface.

10.6   *Micelles* are the aggregates of surface-active agents in solution, which may contain 50 or more monomers. Micelles are small spherical structures composed of both hydrophilic and hydrophobic groups. The concentration of monomer at which micelles are formed is called the *critical micellization concentration* (CMC). The number of monomers that aggregate to form a micelle is known as the *aggregation number* of the micelle.

Micelles can be of three types: (1) *Normal micelles* have the lipophilic parts of the surfactant toward the core and hydrophilic parts toward the periphery, or solvent. These micelles are formed in water. (2) *Reverse micelles* have the hydrophobic groups toward the outside and the hydrophilic parts toward the core. These form in nonaqueous solvents. (3) *Lamellar micelles* form at concentrations much higher than the CMC. These are present in the physiological membranes (refer to Figure 9.3).

10.7   Sodium lauryl sulfate is a surfactant, glucose is not. Glucose will show very little change in surface tension, whereas sodium lauryl sulfate will display a sharp fall at CMC but an increase in conductivity (refer to Figure 9.4).

10.8   Cloud point is the temperature above which some surfactants begin to precipitate, whereas Krafft point is the temperature above which the solubility of a surfactant becomes equal to the CMC. Three factors affecting the cloud point are organic solubilisates, aliphatic hydrocarbons, and aromatic hydrocarbons. Organic solubilisates generally decrease the cloud point of nonionic surfactants. Aliphatic hydrocarbons tend to raise the cloud point. Aromatic hydrocarbons or alkanols may raise or lower the cloud point depending on the concentration.

Below the Krafft point, it is possible that even at the maximum solubility of the surfactant, the interface may not be saturated and, therefore, there is no reason for micelles to form. The surfactant has a limited solubility, and below the Krafft point, the solubility is insufficient for micellization. As the temperature increases, solubility slowly increases. At the Krafft point, surfactant crystals melt and are incorporated into the micelles. Above the Krafft point, micelles will form and, owing to their high solubility, there will be a dramatic increase in surfactant solubility.

10.9   Micelles can be used to increase the solubility of materials that are insoluble or poorly soluble in the dispersion medium used. This phenomenon is known as *solubilization* and the incorporated substance is referred to the solubilisate. The three factors affecting micellar solubilization are the nature of surfactants, the nature of solubilisates, and the temperature. For a hydrophobic drug solubilized in a micelle core, an increase in the lipophilic alkyl chain length of the surfactant enhances solubility, whereas an increase in the alkyl chain length results in an increase in the micellar radius, reducing pressure, resulting in an increase in the entry of the drug into the micelle. For ionic surfactant micelles, increase in the radius of the hydrocarbon core is the main way to enhance solubilization.

10.10 A. Cationic
    B. Cationic
    C. Anionic
    D. Ampholytic
    E. Nonionic
    F. Nonionic

CHAPTER 11

11.1 C.

11.2 E.

11.3 A. *Biomaterials and biocompatibility*: A biomaterial is a natural or synthetic polymer used as a device or carrier, intended to interact with biological systems. Biocompatibility is the ability of a material to perform with an appropriate host response in a specific application.

    B. *Block and graft copolymers*: Two or more monomers are employed for synthesizing copolymers. In copolymers, the monomeric units may be distributed randomly (random copolymer), in an alternating fashion (alternating copolymer), or in blocks (block copolymer). A graft copolymer consists of one polymer branching from the backbone of the other.

    C. *Repeating unit and end group*: The structural unit enclosed by brackets or parentheses is referred to as the *repeating unit* (or *monomeric unit*). To accent the repetition, a subscript n is frequently placed after the closing bracket, for example, $-[-CH2CH2-]n-$. *End groups* are the structural units that terminate polymer chains.

    D. *Monomer and oligomer*: Polymers are synthesized from simple molecules called *monomers* by a process called *polymerization*. If only a few monomer units are joined together, the resulting low-molecular weight polymer is called an *oligomer*.

11.4 A. Molecular weight methylmethacrylate $= 100$, DP$= 50,000/100 = 500$; molecular weight tetramethylene-*m*-benzenesulfonamide $= 211$, DP $= 26,000/211 = 123$

$$Mn = (15,000 * 9 + 25,000 * 5)/(9 + 5) = 18571.4$$

CHAPTER 12

12.1 A. True

B. True

12.2 According to Newton's law of flow, the rate of flow (D) is directly proportional to the applied stress (τ). That is, $\tau = \acute{\eta} \cdot D$, where $\acute{\eta}$ is the viscosity. Simple fluids, which obey this relationship, are referred to as *Newtonian fluids* and the fluids which deviate are known as *non-Newtonian fluids* (refer to Figure 11.1).

12.3 Thixotropy is the property of non-Newtonian pseudoplastic fluids to show a time-dependent change in viscosity. Many gels and colloids are thixotropic materials, exhibiting a stable form at rest but becoming fluid when agitated. Thixotropic flow is a reversible gel–sol gel transformation. On setting, a network gel forms and provides a rigid matrix that will stabilize suspensions and gels. When sheared by simple shaking, the matrix relaxes and forms a solution with the characteristics of a liquid dosage form for ease of use. On standing, the particles collide, flocculation occurs, and the gel is reformed. The shearing force on the injection as it is pushed through the needle ensures that it is fluid when injected; however, the rapid resumption of the gel structure prevents excessive spreading in the tissues, and consequently, a more compact depot is produced than with the non-thixotropic suspensions (refer to Figure 11.2).

CHAPTER 13

13.1 C.

13.2 C.

13.3 A.

13.4 C.

13.5 A.

13.6 D.

13.7 C.

13.8 The rate equation of decay of a radioactive compound, which follows a nonlinear first-order exponential rate kinetics given by

$$N(t) = N(0)e^{-\lambda t}$$

Here, radioactivity, typically represented by curie, C_i, can also be expressed as disintegrations per minute (dpm), since one $C_i = 3.7 \times 10^{10}$ Bq or dps $= 2.22 \times 10^{12}$ dpm.

$N(0) = 15$ dpm

$t = ?$ years

$t_{1/2} = 5,700$ years

$N(t) = 3$ dpm

Where, λ, the rate of decay can be calculated as

$$\lambda = \frac{\ln 2}{t_{1/2}} = \frac{0.693}{t_{1/2}} = \frac{0.693}{5700} = 1.216 \times 10^{-5} \text{ year}^{-1}$$

Thus,

$$3 = 15e^{-1.216 \times 10^{-5} t}$$

or

$$\ln 3 = \ln 15 - 1.216 \times 10^{-5} \times t$$

or

$$t = \frac{\ln 15 - \ln 3}{1.216 \times 10^{-5}} = 132{,}378 \text{ years}$$

13.9   The rate equation of decay of a radioactive compound, which follows a nonlinear first-order exponential rate kinetics given by

$$N(t) = N(0)e^{-\lambda t}$$

$N(0) = 600$ mCi/g
$t = 9$ days
$t_{1/2} = 6$ days
$N(t) = ?$ mCi/g

Where, λ, the rate of decay can be calculated as

$$\lambda = \frac{\ln 2}{t_{1/2}} = \frac{0.693}{t_{1/2}} = \frac{0.693}{6} = 0.1155 \text{ day}^{-1}$$

$$\ln N(t) = \ln N(0) - \lambda t$$

$$\ln N(t) = \ln 600 - 0.1155 \times 9 = 5.3574$$

$$N(t) = e^{5.3574} = 212.18.$$

Thus, at the end of 9 days, the radioactivity left in the compound is ~212 mCi/g.

Therefore, to dispense 30 mCi of radioactivity to the patient, the pharmacist would need to dispense, 30/212 = 0.14 g of compound.

13.10 B.

13.11 C.

13.12 C.

CHAPTER 14

14.1 E.

14.2 C.

14.3 A.

14.4 D.

14.5 PEGylation or reduction in size increases the retention time of liposomes in the bloodstream. Inclusion of PEG-lipid conjugates, such as polyethylene glycol–phosphatidylethanolamine (PEG–PE) reduces the uptake of liposomes by cells of the reticuloendothelial systems (RES), leading to their prolonged circulation half-life.

14.6 Polymeric *micelles* are small spherical structures composed of both hydrophilic and hydrophobic groups. The micelles are in dynamic equilibrium with free molecules (monomers) in solution; that is, the micelles are continuously breaking down and reforming. This fact distinguishes micellar solutions from liposomes, which are microscopic phospholipid vesicles composed of uni- or multilamellar lipid bilayers surrounding aqueous compartments.

14.7 The low oral bioavailability of peptide and protein drugs is primarily due to their large molecular size and vulnerability to proteolytic degradation in the GI tract. Most protein and peptide drugs are susceptible to rapid degradation by digestive enzymes. Furthermore, most peptide and protein drugs are rather hydrophilic and thus are poorly partitioned into epithelial cell membranes, leading to their absorption across the GI tract through passive diffusion.

14.8 A *microcapsule* is a reservoir-type system in which the drug is located centrally within the particle, whereas a *microsphere* is a matrix-type system in which the drug is dispersed throughout the particle. Microcapsules usually release their drug at a constant rate (zero-order release), whereas microspheres typically give a first-order release of drugs.

14.9 *Peyer's patches* belong to gut-associated lymphoid tissues (GALT) of the small intestine. Peyer's patches are capable of internalizing particulate matter, bacteria, and marker proteins. Localization of mucoadhesive polymeric delivery systems at or around Peyer's patches has the potential of favoring the absorption of peptides and proteins.

CHAPTER 15

15.1 A.

15.2 D.

15.3 C.

15.4 D.

15.5 A.

15.6 B.
15.7 D.
15.8 D.
15.9 A.
15.10 C.

CHAPTER 16

16.1 A. True
  B. False
  C. True
  D. True
  E. True
16.2 B. In flocculated systems, forces of attraction are predominant over repulsive forces.
16.3 B. According to Stokes' law, sedimentation rate will increase with an increase in the particle density.
16.4 Flocculated suspension has dispersed phase as loose, light, fluffy flocs (associations of particles) held together by weak van der Waals forces. Deflocculated suspension has dispersed phase in the form of aggregates, which are formed by growth and fusion of crystals in the precipitates to form a solid cake.

Flocculation is an acceptable characteristic for pharmaceutical suspension dosage forms as flocculated suspension form loose flocs, which are easy to redisperse at the time of dose administration as compared to redispersion of hard cake in a deflocculated suspension.
16.5 Because

$$V = \frac{2gr^2(\rho_1 - \rho_2)}{9\eta}, V = 2*981*(100/1000/2)2*\frac{(2.44 - 1.010)}{(9*27)}$$

$$= 0.0288 \text{ cm/s}$$

16.6 $V = \frac{2gr^2(\rho_1 - \rho_2)}{9\eta}, V = 2*981*(100/1000/2)2*\frac{(2.5 - 1.1)}{(9*5)}$

$$= 1.53 \times 10^5 \text{ cm/s}$$

16.7 A, B, C, and D.
16.8 C.
16.9 A, B, and C.
16.10 D.

CHAPTER 17

17.1 A.

17.2 A.

17.3 A. *Creaming and breaking*: *Creaming* is the upward movement of dispersed droplets relative to the continuous phase and it is a reversible process. In contrast, breaking is irreversible. When breaking occurs, simple mixing fails to resuspend the globules in a stable emulsified form, since the film surrounding the particles has been destroyed and the oil tends to coalesce.

B. *Creaming and sedimentation*: *Creaming* is the upward movement of dispersed droplets relative to the continuous phase, whereas *sedimentation* is the downward movement of particles.

C. *Coalescence and aggregation*: *Coalescence* is the process by which the emulsified particles merge with each other to form large particles. Coalescence is an irreversible process because the film that surrounds the individual globules is destroyed. In aggregation, dispersed droplets come together but do not fuse. Aggregation is to some extent reversible.

D. *Phase inversion*: An emulsion is said to invert when it changes from an o/w to a w/o emulsion or vice versa. Phase inversion can occur by the addition of an electrolyte or by changing the phase:volume ratio. Monovalent cations tend to form o/w emulsions, whereas divalent cations tend to form w/o emulsions. An o/w emulsion stabilized with sodium stearate can be inverted to a w/o emulsion by adding calcium chloride to form calcium stearate.

17.4 *Creaming* is the upward movement of dispersed droplets relative to the continuous phase, whereas sedimentation is the downward movement of particles. Factors that influence the rate of creaming are similar to those involved in the rate of sedimentation. According to Stokes' law,

$$v = \frac{d^2(\rho - \rho_0)g}{18\eta_0}$$

where:

v is the velocity of creaming

d is the globule diameter

ρ and ρ_0 are the densities of dispersed phase and dispersion medium, respectively

η is the viscosity of the dispersion medium (poise)

g is the acceleration of gravity ($981\ cm/s^2$)

According to this equation, we can minimize sedimentation and creaming phenomena by
- A reduction in the globule size
- A decrease in the density difference between the two phases
- An increase in the viscosity of the continuous phase

17.5 A. Increased free energy at the interface occurs because the increase in surface area of dispersed phase is responsible for the instability of the emulsion. The surfactants deposit on the interface between the two liquid phases and reduce the interfacial tension and free energy at the interface.

B. HLB value of surfactant and relative concentration of the two phases.

17.6 Emulsifying agents form a film around the dispersed globules to prevent coalescence and thus avoid the separation of two immiscible liquids used for emulsion formation. Emulsifying agents aid in forming emulsions through three different approaches: (1) reduction of interfacial tension, (2) formation of a rigid interfacial film, and (3) formation of an electrical double layer. The film can act as a mechanical barrier to the coalescence of the globules. An electrical double layer minimizes coalescence by producing electrical forces that repulse approaching droplets. Emulsifying agents can be divided into three groups: (1) surfactants, (2) hydrophilic colloids, and (3) finely divided solid particles.

- *Surfactants* are adsorbed at oil–water interfaces to form *monomolecular films* and reduce interfacial tensions.
- *Hydrophilic colloids* are used as emulsifying agents. These include proteins (gelatin and casein) and polysaccharides (acacia, cellulose derivatives, and alginates). These materials adsorb at the oil–water interface and form *multilayer films* around the dispersed droplets of oil in an o/w emulsion. Hydrated lyophilic colloids differ from surfactants as they do not cause an appreciable lowering in interfacial tension.
- *Finely divided solid particles* are adsorbed at the interface between two immiscible liquid phases and form a *film of particles* around the dispersed globules. Finely divided solid particles are concentrated at the interface, where they produce a particulate film around the dispersed droplets so as to prevent coalescence.

17.7 In general, o/w emulsions are formed when the HLB of the surfactants is within the range of about 9–12, and w/o emulsions are formed when the range is about 3–6.

17.8 A surfactant with a high HLB value (~9–12) is used as an emulsifier to form o/w emulsions; and a surfactant of low HLB value (~3–6) to form w/o emulsions.

17.9 A. SEDDS
     B. SEDDS
     C. SMEDDS
     D. SMEDDS
     E. SMEDDS
17.10 D.
17.11 B.

CHAPTER 18

18.1 A. True
     B. True
     C. False
     D. True
     E. False
18.2 A and C.
18.3 *Pharmaceutical solutions* are homogeneous mixtures of one or more solutes dispersed in a suitable solvent or a mixture of mutually miscible solvents:
- *Syrups* are aqueous solutions containing a sugar or sugar substitute with or without added flavoring agents and drugs.
- *Elixirs* are sweetened hydroalcoholic (combinations of water and ethanol) solutions.
- *Spirits* are hydroalcoholic solutions of aromatic materials.
- *Tinctures* are alcoholic or hydroalcoholic solutions of chemical or soluble constituents of vegetable drugs.

18.4 D.
18.5 A, B, and D.

CHAPTER 19

19.1 Size and shape
19.2 Mixing and granulation
19.3 A. Bulk < Tapped < True
     B. Bulk and true density
     C. Bulk density
     D. Bulk density
     E. Bulk and true density
19.4 Sticking to the tablet tooling, electrostatic charge, and flow.

19.5 A. $d[2,0]$, $d[3,2]$
     B. $d[3,0]$, $d[4,3]$
     C. $d(90)$, $d(50)$, $d(10)$
     D. Laser diffraction and focused beam reflectance measurement
     E. Microscopy

19.6 A.

19.7 C.

19.8 A.

19.9 A.

19.10 E.

CHAPTER 20

20.1 C. The ionized, or salt, form of a drug is generally more water soluble and therefore dissolves more rapidly than the nonionized (free acid or free base) form of the drug. According to the Noyes–Whitney equation, the dissolution rate is directly proportional to the surface area and inversely proportional to the particle size. Therefore, an increase in the particle size or a decrease in the surface area slows the dissolution rate. Use of sugarcoating around the tablet will also decrease the dissolution rate.

20.2 A. Disintegrating agents are added to the tablets to promote breakup of the tablets when placed in the aqueous environment. Lubricants are required to prevent adherence of the granules to the punch faces and dies. Binding agents are added to bind powders together in the granulation process. Glidants are added to tablet formulations to improve the flow properties of the granulations.

20.3 C. Enteric-coating materials include cellulose acetate trimellitate, poly(vinyl acetate)phthalate, hydroxypropyl methylcellulose phthalate, and cellulose acetate phthalate.

20.4 D. An enteric-coated tablet has a coating that remains intact in the stomach but dissolves in the intestine when the pH exceeds 6.

20.5 A.

20.6 B.

20.7 A.

20.8 B.

20.9 A. Increase in lubricant concentration.
     B. Decrease in dissolution rate. The diffusion coefficient across the membrane increases because lubricants are hydrophobic.

20.10 Glidant is used for improving the flow properties of the solids/granules, whereas lubricant serves to prevent adhesion of the tablet to dies and punches.

20.11 A. Disintegration is the process of breaking up of a tablet/capsule dosage form into the constituent granules. Dissolution is the process whereby the solid drug in a dosage form turns into a solution in the surrounding liquid media. Absorption is the process of the dissolved drug crossing the cellular membrane barrier to enter the systemic circulation.

B. (i) Increase the volume of the media, (ii) add a surfactant to the media, (iii) use a cosolvent such as alcohol, (iv) use biphasic solution such as water and chloroform, (v) increase paddle rpm, and (vi) change from basket to paddle apparatus.

CHAPTER 21

21.1 C.
21.2 A.
21.3 A.
21.4 C.
21.5 B.

CHAPTER 22

22.1 D.
22.2 D. The bacterial endotoxin test determines the level of bacterial endotoxin only from gram-positive bacteria. This test cannot determine the fever-producing potential of endotoxins. Gram-negative bacteria do not have to be alive for the endotoxin to produce an effect.
22.3 A. True
     B. True
     C. True
     D. True
     E. True
     F. False
     G. True
     H. True

CHAPTER 23

23.1 D.
23.2 B.

23.3 D.
23.4 i. Vanishing cream and hydrophilic ointment
ii. Cold cream and lanolin
iii. Alamine lotion
iv. Jelly
23.5 i. D.
ii. F.
iii. C.
iv. A and E.
23.6 i. D.
ii. B.
iii. A.
iv. C.

CHAPTER 24

24.1 A.
24.2 E.
24.3 C.
24.4 Occusert™ is a device consisting of a drug reservoir (e.g., pilocarpine HCl in an alginate gel) enclosed by two release-controlling membranes made of ethylene–vinyl acetate copolymer and enclosed by a white ring, allowing positioning of the system in the eye.
21.5 Factors affecting the bioavailability of suppository dosage forms include the retention time of the suppository in the cavity, the size, and the shape of the suppository, and its melting point.
Types of suppository base: Oleaginous bases
Water-soluble or water-miscible suppository bases
24.6 Diffusion-controlled implants and osmotic minipumps differ in mechanism of drug release. The rate of drug delivery from diffusion-controlled implant is controlled by drug diffusion or dissolution through an insoluble matrix and/or the use of a rate-controlling membrane. Minipumps, on the other hand, are osmotically-controlled devices consisting of an impermeable membrane with well-defined openings for drug release, encasing a drug-containing core.

CHAPTER 25

25.1 E.
25.2 E.

25.3 A.

25.4 D.

25.5 D.

25.6 C.

25.7 E.

25.8 C.

25.9 A and C.

25.10 B.

25.11 B and D.

25.12 A. True
       B. True
       C. True
       D. True
       E. True
       F. True

25.13 D.

25.14 Amino groups of N-terminal amino acid or ε-amino groups of lysine (–NH2), carboxyl group of aspartic and glutamic acids (–COOH), and sulfhydryl group of cysteine (–SH).

25.15 Proteins are often conjugated to polyethylene glycol (PEG), which is nonimmunogenic and nontoxic. In protein molecules, N-hydroxysuccinimide (NHS) ester groups primarily react with the α-amines at the N-terminals and the ε-amines of lysine side chains. PEGylation can provide increased biocompatibility, reduce immune response, increase *in vivo* stability, delayed clearance by the reticuloendothelial system, and prevent protein adsorption on the surface.

CHAPTER 26

26.1 E.

26.2 E.

26.3 Genes are made of DNA, which contains information about when and how much of which protein to produce, depending on the functions to be performed. Gene expression is the process of transcription of DNA into RNA and translation of mRNA into proteins. The *antisense* strand of DNA is used as a template for transcribing enzymes that assemble mRNA, a process called *transcription*. Then, mRNA migrates into the cytoplasm, where ribosomes read the encoded information, its mRNA's base sequence and, so doing, string together amino acids to form a specific protein. This process is called *translation*.

26.4 *Antisense therapy* aims at inhibiting the existing but abnormally expressed genes by blocking the transcription of DNA or translation of mRNA into harmful proteins. The types of antisense compounds include antisense oligonucleotides (ODNs), peptide nucleic acids (PNAs), antisense RNA, aptamers, ribozymes, and siRNA. The types of antisense oligonucleotides include phosphodiester ODNs, phosphorothioate ODNs, and methylphosphonate ODNs.

26.5 A gene expression plasmid is formed of circular double-stranded DNA molecules and contains a cDNA sequence coding for a therapeutic gene and several other genetic elements, including introns, polyadenylation sequences, and transcript stabilizers, to control transcription, translation, and protein stability.

26.6 The three important factors in the development of nonviral gene therapy products include the therapeutic gene, the gene expression system, and the gene delivery system. A *therapeutic gene* encodes a specific therapeutic protein, a *gene expression system* controls the functioning of a gene within a target cell, and a *gene delivery system* controls the delivery of the expression system to specific locations within the body. Plasmid-based gene expression systems contain a cDNA sequence coding for a therapeutic gene and several other genetic elements, including introns, polyadenylation sequences, and transcript stabilizers, to control transcription, translation, and protein stability. The *gene delivery system* distributes the plasmid to the desired target cells, after which the plasmid is internalized into the cells. Once inside the cytoplasm, the plasmid can then translocate to the nucleus, where gene expression begins, leading to the production of a therapeutic protein through the steps of transcription (synthesis of RNA from DNA into the nucleus) and translation (synthesis of protein from mRNA in the cytoplasm).

Index